NEW TRENDS IN OPTICAL SOLITON TRANSMISSION SYSTEMS

SOLID-STATE SCIENCE AND TECHNOLOGY LIBRARY

Volume 5

Aims and Scope of the Series

The aim of this series is to present monographs on semiconductor materials processing and device technology, discussing theory formation and experimental characterization of solid-state devices in relation to their application in electronic systems, their manufacturing, their reliability, and their limitations (fundamental or technology dependent). This area is highly interdisciplinary and embraces the cross-section of physics of condensed matter, materials science and electrical engineering.

Undisputedly during the second half of this century world society is rapidly changing owing to the revolutionary impact of new solid-state based concepts. Underlying this spectacular product development is a steady progress in solid-state electronics, an area of applied physics exploiting basic physical concepts established during the first half of this century. Since their invention, transistors of various types and their corresponding integrated circuits (ICs) have been widely exploited covering progress in such areas as microminiaturization, megabit complexity, gigabit speed, accurate data conversion and/or high power applications. In addition, a growing number of devices are being developed exploiting the interaction between electrons and radiation, heat, pressure, etc., preferably by merging with ICs.

Possible themes are (sub)micron structures and nanostructures (applying thin layers, multi-layers and multi-dimensional configurations); micro-optic and micro-(electro)mechanical devices; high-temperature superconducting devices; high-speed and high-frequency electronic devices; sensors and actuators; and integrated opto-electronic devices (glass-fibre communications, optical recording and storage, flat-panel displays).

The texts will be of a level suitable for graduate students, researchers in the above fields, practitioners, engineers, consultants, etc., with an emphasis on readability, clarity, relevance and applicability.

The titles published in this series are listed at the end of this volume.

New Trends in Optical Soliton Transmission Systems

Proceedings of the Symposium
held in Kyoto, Japan, 18–21 November 1997

Edited by

Akira Hasegawa
Research Professor,
Kochi University of Technology
and
Consultant,
NTT Science and Core Technology Laboratory Group,
Kyoto, Japan

SPRINGER SCIENCE+BUSINESS MEDIA, B.V.

A C.I.P. Catalogue record for this book is available from the Library of Congress.

ISBN 978-94-010-6161-2 ISBN 978-94-011-5141-2 (eBook)
DOI 10.1007/978-94-011-5141-2

Printed on acid-free paper

Originally published by Kluwer Academic Publishers in 1998
Softcover reprint of the hardcover 1st edition 1998

ORGANIZER AND PROGRAM COMMITTEE

Organizer

Research Group for Optical Soliton Communications (ROSC)
Chairman: Akira HASEGAWA, Osaka University

Program Committee

Kazuo AIDA, Nippon Telegraph and Telephone Corporation
Nick J. DORAN, Aston University
Masayuki MATSUMOTO, Osaka University
Tetsuya MIKI, The University of Electro-communications
Kazuo SAKAI, Kokusai Denshin Denwa Co., Ltd.

Secretariat

Hiroshi TAKEHARA and Minoru SHINOMIYA
Support Center for Advanced Telecommunications Technology Research, Foundation (SCAT)
Tel. +81-3-3351-0540, Fax. +81-3-3351-1624
E-mail : soliton@scat.or.jp

Preface

This book summarizes the proceedings of the invited talks presented at the International Symposium on New Trends in Optical Soliton Transmission Systems held in Kyoto during November 18 - 21, 1997. As a result of worldwide demand for ultra high bitrate transmissions and increased scientific interest from the soliton community, research on optical solitons in fibres has made remarkable progress in recent years. In view of these trends, the Research Group for Optical Soliton Communications (ROSC), chaired by Akira Hasegawa, was established in Japan in April 1995 to promote collaboration and information exchange among communication service companies, industries and academic circles in the theory and application of optical solitons. This symposium was organized as a part of the ROSC activities.

As with the 1st ROSC symposium, this symposium attracted enthusiastic response from worldwide researchers involved in the subject of soliton based communications and intensive discussions were held throughout the symposium. Particular emphases were made to dispersion managements of soliton transmission. I would like to note that in the 1st symposium the (adiabatic) dispersion managements just began to appear in reducing radiation at amplifiers and reducing collision effects in WDM system. These have become standard this time, but in addition new, non-adiabatic dispersion managements have been introduced independently by various scientists all over the world. Nick Doran wanted to enhance soliton power to increase the S/N ratio in the low average dispersion systems, Georges wanted to reduce soliton power in standard fibre, Jacob and Carter wanted to send soliton in NRZ fibre, and Kumar and Hasegawa wanted to produce stationary chirped soliton. Ironically all these attempts have come up with essentially the same answer; change dispersion periodically. The non-adiabatic dispersion managements have created new nonlinear stable pulse with frequency chirp which has yet to be named. But this new pulse remarkably has solved practically all the classic problems of ideal solitons, the Gordon-Haus effect, collision induced frequency shift in wavelength division multiplexing (WDM) and interaction of adjacent solitons. Richardson's naming of it as "power soliton" is quite appropriate; the power soliton can propagate in standard fibre, allows WDM, allows dense packing and increases bit-rate per channel and reduces the Gordon-Haus effect because of zero average dispersion without sacrificing the S/N ratio etc..

The final day of the symposium was devoted to the introduction of the consortia of soliton based communications, MIDAS, ESTHER, UPGRADE, AON and STAR which are producing very interesting and fruitful results all over the world.

In the preparation of the final draft of the proceedings, Messrs. H. Takehara and M. Shinomiya have again devoted a significant amount of their time to make the manuscript consistent and well coordinated. On behalf of ROSC, the editor would like to thank those efforts without which the publication of the book would have faced significant delay.

The editor would also like to express his appreciation for the support of the Support Center for Advanced Telecommunications Technology Research Foundation (SCAT) to which Messrs. Takehara and Shinomiya are affiliated, which has enabled us to hold this successful symposium.

Kyoto, February 1998

Akira HASEGAWA

TABLE OF CONTENTS

PART 2 : CONSORTIUM SESSION

PROPAGATION AND INTERACTION OF OPTICAL SOLITONS IN FIBERS WITH RANDOM PARAMETERS

F. KH. ABDULLAEV
Physical-Technical Institute, Uzbekistan Academy of Sciences
700084, Tashkent-84 G. Mavlyanov str. 2-b, Uzbekistan

J. G. CAPUTO
Laboratoire de Mathematique
INSA de ROUEN et UPRESA CNRS 60-85
B. P. 8, 76131 Mont-Saint-Aignan, Cedex, France

AND

M. P. SØRENSEN
Department of Mathematical Modelling
The Technical University of Denmark
Bldg. 321, DK-2800 Lyngby, Denmark

1. Introduction

The problem of propagation and interaction of optical solitons in fibers with variable parameters has attracted a lot of interest recently. Fibers with periodic amplification and/or dispersion are an important example [1, 2]. It was shown recently that fibers with periodic dispersion management can be used as long-haul optical transmission systems using solitons [3]. In a real fiber we always have together with the artificially produced inhomogeneities random modulations of the parameters. So it is interesting to investigate the influence of random modulations on optical solitons dynamics in fibers.

In this work we will consider the propagation and interaction of optical solitons in fibers. Two types of perturbations will be analyzed for the nonlinear Schrödinger equation (NLSE): random dispersive and nonlinear perturbations. In section 2 the process of modulational instability (MI) of electromagnetic waves in fibers with random nonlinearity and dispersion is investigated. In the investigation of pulse propagation in fiber with random parameters the variational approach is used. We will show that it provides

A. Hasegawa (ed.), New Trends in Optical Soliton Transmission Systems, 1–13.

together with the Inverse Scattering Theory (IST) when this one is applicable simple models which provide insight into the physics of the problem. Of course these simple models should be validated by direct comparison with the full numerical solution of the NLSE. So in section 3 the validity of the variational approach for the problems considered is analyzed. The propagation and interaction of solitons in randomly modulated fibers using the variational approach and inverse scattering theory are considered in the next sections.

2. Modulational Instabilities in the Fibers with Random Parameters

The modulational instability of electromagnetic waves is one of the factors which reduce the efficiency of long-haul optical transmission systems. In particular the MI can amplify the fluctuations of an initially partially coherent wave propagating in the region of negative group velocity dispersion [4]. This problem is described by NLSE with random initial condition. Random modulations of a fiber parameters lead to the problem of NLSE with parametrically acting noise perturbations. This perturbation can produce new regions of MI and modify the gain in known domains. For example random modulations of dispersion occur when going from one fiber piece to another in dispersion management schemes. As has been shown recently periodic modulations like the periodic amplification or periodic dispersion management can lead to the appearance of new sidebands in the gain curve for a nonlinear plane electromagnetic wave propagating in fiber. The case of periodic modulations of nonlinearity has been studied in [5]-[7], the periodic modulations of dispersion in [8]-[10].

For the periodic modulation case the sidebands correspond to the parametric resonance of the unstable mode with the modulation period. One of the consequences is the increase of the region of MI and the reduction of the gain in comparison with a fiber with constant parameters. In the case of randomly varying amplification or dispersion we have all frequencies in the spectrum of modulation, they have the same energy for white noise while for coloured noise the components are concentrated near one or few harmonics and will play an important role in the pulse dynamics.

The evolution of an electromagnetic wave in a fiber with inhomogeneous dispersion and amplification is described by the modified NLSE

$$iu_x + \frac{d(x)}{2}u_{tt} + g(x)|u|^2u = 0, \tag{1}$$

where $d(x)$ is the variable dispersion and $g(x)$ is the amplification. In the case of a random amplification $g = 1 + g_1$, where $g_1(x)$ is assumed to be a white Gaussian-distributed noise and constant dispersion $d(x) \equiv d_0$, we will search the field in the form

$$u = [A + \psi(x,\tau)] \exp(i\phi), \psi = 2A^2 \int_0^x g(x')dx', \tag{2}$$

where $\psi << A$. By expressing ψ as

$$\psi = C(x) \exp(i\Omega t) + B^*(x) \exp(-i\Omega t), \tag{3}$$

and introducing $c = C - B,\ b = C + B$, we find for the mean value

$$< b^2 >= \exp(\omega_0^2 \sigma^2 h^2 x)(b_1 + \frac{\sigma^2 b_2}{4} + \frac{b_3}{4\omega_0^2}) + \exp(-\omega_0^2 h^2 \sigma^2 x) \sin 2\omega_0 x + ... \tag{4}$$

where the parameters $b_1 = b_0^2,\ b_2 = 2b_0 b_{x0},\ b_3 = 2b_{xo}^2$ are defined by the initial conditions for b and b_x. The other parameters are the frequency $\omega_0^2 = \Omega^2(\Omega^2 - 4A^2\nu),\ h = 4A^2\nu/(\Omega^2 - 4A^2\nu),\ \nu = \pm 1$ for negative and positive dispersion correspondingly.

We perform numerical simulations of MI for the case when the initial amplitude of weak periodic modulation is $b_0 = 10^{-2}, A = 1.0, \Omega = 4.1888$, corresponding to the maximal gain in equation (4), $\sigma^2 = 0.1$. The average over 40 realizations for the stochastic NLSE has been performed. We observe an exponential growth of the mean square of the correction $b(x), c(x)$ with superimposed oscillations with frequency $2\omega_0$. This is the phenomenon of stochastic parametric resonance. The growth rate given by equation (4) $\gamma_n = \omega_0^2 \sigma^2 h^2 \cong 1.036$ is in very good agreement with the full numerical simulations. In order to investigate the stage beyond the framework of perturbation theory, we study the wave evolution up to an amplitude of order one. This case corresponds to large fluctuations of the nonlinearity parameter and large modulation frequency. The theoretical increment value is equal to $\gamma_n = 2.07$ and is in good agreement with the initial increment in our simulations.

The case of random dispersion, when $d_0(x) = 1 + d_1(x), g(x) \equiv 0$ can also be analyzed [9]. Studying the system of equations for the second moments the characteristic equation for the MI gain p can be obtained

$$p^3 + 2\Omega^4 \sigma^2 p^2 + \Omega^2(D + \Omega^6 \sigma^4)p - 4A^4 \Omega^4 \sigma^2 = 0\,. \tag{5}$$

Analysing this equation, we found the MI gain in the region of the normal dispersion ($\nu = +1$)

$$p_d = \frac{4\Omega^2 A^4 \sigma^2}{\Omega^2 + 4A^2 + \Omega^6 \sigma^4} \tag{6}$$

which coincides in the limit of small σ^2 with the one obtained from the random nonlinearity modulation case. In the region of anomalous dispersion a simple approximate solution like equation (6) cannot be found. The numerical analysis of the characteristic equation for the MI gain shows that the dispersion fluctuations lead to an increase of the MI domain and a decrease of the gain with respect to the uniform dispersion case.

Stochastic parametric resonance phenomena are responsible for growing sidebands. Because a random variation of dispersion contains all frequencies, the parametric resonance arises for any frequency of the modulation. As a consequence the MI gains equations (4), (6) occur.

3. Variational Approach to the Solitons Propagation in Fibers with Modulated Parameters

Here we will describe the propagation of chirped solitons by the variational approach developed in [11]. We assume the anzatz for the solution of equation (1)

$$u = A(x)\mathrm{sech}(\frac{t}{a(x)})\exp{(ib(x)t^2)}, \tag{7}$$

where A, a *and* b are the pulse amplitude, width and chirp correspondingly. Then writing the variational principle for the averaged Lagrangian, we obtain the equations for the pulse parameters

$$a_x = 2abd(x), \tag{8}$$

$$b_x = \frac{2d(x)}{\pi^2 a^4} - 2b^2 d(x) - \frac{2N^2 g(x)}{\pi^2 a^3}. \tag{9}$$

These equations describe the motion of a particle with a variable mass $1/d(x)$ in the field of the anharmonic nonstationary potential

$$U(a,x) = \frac{2d(x)}{\pi^2 a^2} - \frac{4N^2 g(x)}{\pi^2 a}. \tag{10}$$

Here N^2 is the energy of the pulse. We note that the unperturbed problem is the Kepler problem of a particle of unit mass moving in the field of the central potential equation (10) [12]. It is integrable and the action-angle variables J, θ can be found. Using these variables the equations of motion (8, 9) take the form

$$\frac{dJ}{dx} = -\epsilon(x)\frac{\partial V}{\partial \theta}, \tag{11}$$

$$\frac{d\theta}{dx} = \omega(J) + \epsilon(x)\frac{\partial V}{\partial J}. \tag{12}$$

We will use the Fourier component of the interaction part of the Hamiltonian $V(J,\theta)$. In the case of a modulated nonlinearity we have

$$V(J,\theta) = -\frac{4N^2}{\bar{b}} \sum_m \frac{2}{\bar{b}} J_m(e_0 m) \exp(-im\theta) + c.c., \tag{13}$$

where J_m is the Bessel function, $\bar{b} = 2N^2/\pi^2 E, e_0 = (1 - \pi^2|E|/2N^4)^{1/2}$ and E is the total energy of the particle.

The investigation of this effective particle model for the periodic modulations of dispersion $d = d_0 + f_0 \sin \Omega x$ has been performed in reference [13]. The low resonances $\Omega = (1/2, 1, 2)\omega_0$, where $\omega_0 = 2N^4/\pi^4$ have been investigated and decay of soliton has been predicted. A numerical investigation of the splitting vs. non splitting of solitons in the parameter plane modulation amplitude vs. energy of pulse has been performed for the partial differential equation (PDE) (1) in reference [14]. The numerical simulations of the high frequency limit has been considered in reference [15]. We compared the evolution given by equations (8, 9) with the numerical simulation of the PDE and give all the details in reference [16]. In Figure 1 we examine the case of the 1/2 resonance for a soliton of energy $N^2 = 1.2$ and $f_0 = 0.2$ and plot $a(x)$ and the phase plane (a, b) for the values estimated from the PDE solution and the ones obtained from the ordinary differential equations (ODE) (8, 9). The two systems show a fast decay of the soliton after 4 or 5 oscillations for the same values of the forcing amplitude f_0 indicating a good validity of the variational description. This result holds for a modulation frequency Ω below the soliton frequency ω_0.

When the modulation frequency Ω is increased and reaches ω_0 a resonance occurs for the soliton and radiation is emitted, this has been predicted using the inverse scattering theory [17, 18]. In this case the variational anzatz fails. The ODE's indicate a decay of the soliton after 5 oscillations for $f_0 = 0.2$ while the PDE solution shows a 10 % decay for the same time. As expected from reference [18] linear waves are emitted by the soliton and the decay rate of its amplitude is maximum for $\Omega \cong \omega_0$. Plotting the evolution of the pulse energy as a function of x shows a fast decay for the fundamental resonance. As Ω is increased this rate of decay is reduced. It is therefore expected that a simplified description will hold for very large frequencies for which the emitted radiation would be negligible. This could be related to the exponential decrease of the stochastic layer which is obtained for the ODE's [16].

To find the high frequency behavior of the pulse we perform the averaging of equations (7, 8) over fast oscillations of dispersion using the Kapitza approach [12] by considering the perturbation parameter $f_0^2/\Omega^2 \ll 1$. We obtain the following system for the averaged soliton width a and chirp b is obtained

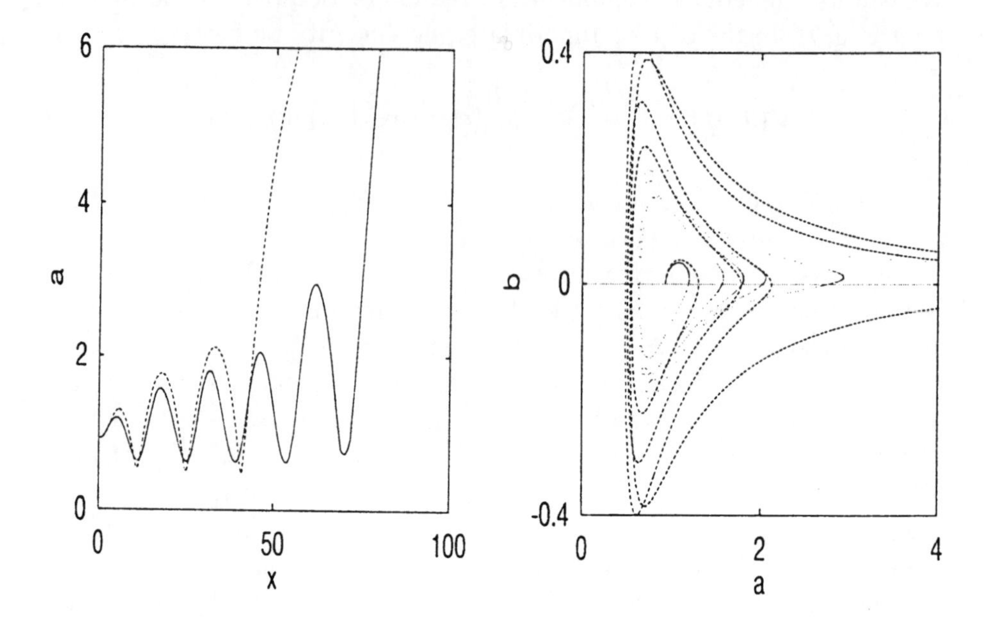

Figure 1. Oscillation of the width parameter for the soliton (left plate) and phase plane (a,b) (right plate). The estimates from the PDE are in full line (left) and dots (right) while the ones for the variational equations (8, 9) are in dashed lines.

$$a_x = 2ab - \frac{f_0^2}{\Omega^2} b\left(\frac{4}{\pi^2 a^3} + 2ab^2\right) + O\left(\frac{f_0^3}{\Omega_0^3}\right), \tag{14}$$

$$\begin{aligned} b_x = \frac{2}{\pi^2 a^4} - 2b^2 - \frac{2N^2}{\pi^2 a^3} + 4\frac{f_0^2}{\Omega^2}\left(\frac{14}{\pi^2 a^4} b^2 + \frac{3}{\pi^4 a^8}\right. \\ \left. - \frac{2}{\pi^2 a^4} b^2 - 2b^4 - \frac{6N^2 b^2}{\pi^2 a^3}\right) + O\left(\frac{f_0^3}{\Omega_0^3}\right). \end{aligned} \tag{15}$$

We investigate numerically the agreement with the solutions of full NLSE. There is a quantitative agreement for the width and the chirp for $x < 5$. Past this point the PDE solution gives a stationary width which is equal to the average of the width given by the ODE. The variational approach again provides a simple model from which one can analyze the PDE. It should be noted that many properties of the pulse propagation in fibers with strong dispersion management like the influence of a pre-chirp of the pulse and the fixed points for the amplitude and chirp may be described in the framework of such a simple model. For example we can obtain from this averaged system a formula for the enhanced power necessary for the stable propagation of a pulse in a two-step dispersion management scheme

[16]; this was found empirically from the numerical simulations in reference [19].

4. Propagation of Optical Soliton in Fiber with Random Parameters

The adiabatic dynamics of optical solitons under stochastic perturbations has been considered in references [20, 21]. In a fiber with modulated parameters we have a chirped soliton propagation and here the application of a variational approach seems reasonable. One of important examples when the randomness plays a role is the dispersion management for adiabatic soliton transmission. For such a system the proportionality of $d(x)$ to $g(x)$ can be difficult to achieve exactly and the deviation of the ratio can be random. Then introducing a new variable instead of x, assuming $d(x) \neq 0$, we have in the NLSE the random perturbation $R = (1 - g(x)/d(x))$. We will apply the variational approach to this problem. As we have seen in the previous section the evolution of the soliton width is analog to the motion of a particle of unit mass in the field of a stochastically perturbed central potential. Using the equations for this problem in the action-angle variables we rewrite the equation for the action J (11) in terms of the energy E [22],

$$\frac{dE}{dx} = -\epsilon(x)\omega(E)\frac{\partial V}{\partial \theta}. \tag{16}$$

Analytical treatment of this equation can be performed for values of E close to separatrix, $|E| \ll |E_c|$ and near the bottom of the potential well, $E \sim E_c = -2N^4/\pi^2$. For the analysis of the first case we will consider the values of e_0, such that $N_0 = (1 - e_0^2)^{-3/2} \gg 1$. N_0 defines the number of harmonics in the spectrum of the unperturbed motion. Averaging over the oscillations of θ we have for the mean energy of the effective particle

$$< E >= -|E_0|(1 - \frac{20\pi^4 J_1^2(1)}{N^4}|E_0|^3\sigma^2 x), \tag{17}$$

where J_1 is the Bessel function of order 1. In the second case $E \sim E_c$ we can assume $E = E_c + \delta$, where $\delta \ll E_c$. Solving the equation for δ and averaging over the period θ and all realizations of $\epsilon(x)$ we obtain for the energy growth

$$< \delta >= \delta_0 \exp\left(\frac{16N^8}{\pi^2}\sigma^2 x\right). \tag{18}$$

So we observe a diffusive growth of the energy of the particle in the potential well. This corresponds to the increase of the soliton width in the course of propagation along a randomly inhomogeneous fiber. Assuming that the effective particle escapes the well when its mean energy reaches the value $E = 0$ we can roughly estimate the escape length as follows

$$x_d \approx \frac{\pi^2}{16N^8\sigma^2}. \tag{19}$$

Let us compare this results with the one given by the IST approach. There is a point where the predictions of the variational approach can be compared with the exact methods. It is the stationary point in the bottom of potential well; then the radiative damping can be calculated [17, 18]. Taking into account that for equation (1) the number of particles is conserved, i.e.

$$2N^2 = \int_{-\infty}^{\infty} |u|^2 dt = 4\eta + \frac{1}{\pi}\int_{-\infty}^{\infty} \ln(1 - |b(x,\lambda)|^2) d\lambda, \tag{20}$$

where λ is the spectral parameter and b is the Jost coefficient we can find the radiative damping of soliton amplitude.

The calculations give the radiative damping of the amplitude of the soliton

$$\frac{d\eta}{dx} = -\pi\sigma^2\mu\eta^5, \quad \mu = \frac{128}{15}. \tag{21}$$

The solution is

$$\eta^2 = \frac{\eta_0^2}{\sqrt{1 + 4\pi\sigma^2\mu\eta_0^4 x}}. \tag{22}$$

It is seen that the amplitude of the soliton is decaying as $x^{-1/4}$. From this results we conclude that the decay length of the soliton is $x_d \approx 1/3\sigma^2 N^8$. This estimation is consistent with the one given by equation (19) obtained from the variational approach. However for a position near the separatrix which corresponds to large chirp values and/or significant deviations of the initial amplitude from the fixed point $a_c = 1/N^2$, the estimation (17) has only a qualitative character. This is connected with the superior weight of the continuous component. Here the anzatz (7) should be corrected by taking into account additional terms corresponding to the radiation field.

In the coloured noise case with the correlation function

$$B(x - y) = \frac{\sigma^2\gamma}{2}\exp(-\gamma|x - y|), \tag{23}$$

where $\gamma = 1/l_c$, l_c is the correlation length. When $4\eta^2 l_c >> 1$, the IST gives the result

$$\eta = \eta_0 \exp(-\frac{\pi\sigma^2}{16 l_c^2} x). \tag{24}$$

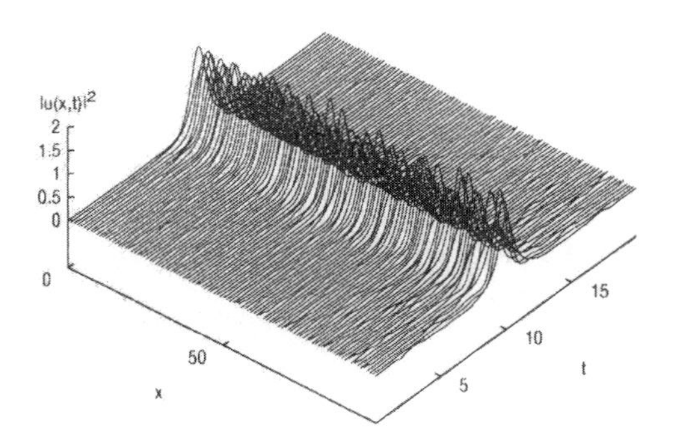

Figure 2. The stabilisation of a soliton due to coloured noise with σ^2=0.125 and γ=1.0. The initial condition is $2\eta_0 = 1.3$.

The equation shows that increasing values of the correlation length l_c leads to a smaller damping and a growth of the soliton decay length.

We have compared the analytical expressions for the soliton decay rate derived in the previous sections with numerical simulations [23]. The mean value $< max|u|^2 >$ was calculated from 150 realizations, as a function of the position x using white noise for a first order soliton. We have used the parameter values $2\eta_0 = 1, \sigma^2 = 0.01$, $\sigma^2 = 0.1$ and $\sigma^2 = 0.2$. The analytical expression we used is the one given in equation (22). Note that this expression for the soliton decay rate is only valid in the intial state of propagation $(x < \sigma^{-2})$. Apart from a sligth overestimation of the numerical results, the analytical expression is in a good agreement with the simulations.

In the coloured noise modulation case we observe the interesting fact that the soliton is stabilized. This phenomenon is illustrated in Figure 2 with the values of parameters $\sigma^2 = 0.125, \eta_0 = 0.65, \gamma = 1$.

The damping is strongly reduced as predicted by the variational approach. No splitting is observed contrary to the case of white noise. The same phenomena is observed when we use the parameters $\sigma_1^2 \equiv \gamma^2\sigma^2 = 0.25, \gamma = 0.2$ and $\eta_0 = 0.65$. It should be noted that the results are strongly dependent from the values of the correlation length. For small correlation lengths splitting is observed while for larger correlation lengths the soliton remains stable. In the case of a periodic modulation of the dispersion the stabilization effect has been numerically observed in reference [24] for a modulation period equal to the soliton period.

5. Interaction of Optical Solitons in Random Media

The interaction of optical solitons in fibers with periodic modulations has been considered in reference [25] and later in reference [24]. Resonant phenomena showing the existence of stable and unstable regions as a function of the modulation frequency and initial power have been observed. We focus here on the interaction of solitons in a medium with random nonlinearity. Our anzatz is a two-soliton configuration with a temporal separation of $2t_0$ and an arbitrary phase difference ψ

$$u = u_s(t+t_0) + \exp(i\psi)u_s(t-t_0). \tag{25}$$

For the parameters $p = \eta_1 - \eta_2, q = \xi_1 - \xi_2, \psi = \delta_1 - \delta_2, r = \zeta_1 - \zeta_2$ we obtain following the Karpman-Solov'ev approach [26] the reduced system of equations (16)

$$\frac{d^2 p_1}{dx^2} = -512\eta^4 \exp(-2\eta r_0)(2p_0 V(x) + p_1), \tag{26}$$

$$\frac{d^2 r_1}{dx^2} = 512\eta^4 \exp(-2\eta r_0) r_1. \tag{27}$$

Here we assume that the solution has the form $p = p_0 + p_1, r = r_0 + r_1 ...,$ where p_0, r_0 are the solutions of unperturbed system and p_1, r_1 are assumed to be small. The mean square difference of the soliton amplitude can be calculated from equation (25). We will assume that in the initial stage of propagation $r_0(x)$ changes slowly and can be replaced in first approximation by t_0. This is possible because for unequal soliton amplitudes, the variations of r_0 are small and the period of variation T is large. Then we get for the white noise modulation

$$< p_1^2 > \approx 2\omega_0^2 p_0^2 \sigma^2 [x - \frac{\sin(2\omega_0 x)}{2\omega_0}], \tag{28}$$

and for the coloured noise modulation

$$< p_1^2 > \approx 2\omega_0^2 p_0^2 \sigma^2 [\frac{x}{1 + l_c^2 \omega^2} - \frac{\sin(2\omega_0 x)}{2\omega_0}], \tag{29}$$

where $\omega_0^2 = 512\eta^4 \exp(-2\eta t_0)$.

The mean square of the amplitude difference grows diffusively during propagation. Therefore, an assymetry in the two soliton shape must be observed in a medium with random fluctuations. For larger propagation distances it is difficult to obtain analytical predictions from the Karpman-

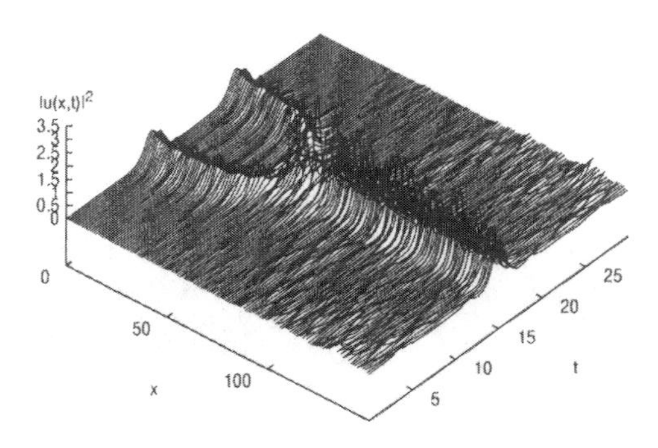

Figure 3. Two soliton interaction and the apperance of a bound state due to coloured noise, the parameters are given in the text.

Solov'ev approach. However, some estimations can be obtained by considering the effective potential for the soliton relative distance r. From equation (28) it follows that the potential is

$$U = 128\eta^2 \exp(-2\eta r) \cos\psi. \tag{30}$$

We assume that the amplitude difference changes very slowly, which is confirmed by our numerical simulations [23]. In a random media ψ is a Gaussian random value. Averaging over all realizations, we find

$$U = 128\eta^2 \exp(-2\eta r) \exp(- < \psi^2 > /2), \tag{31}$$

where $< \psi^2 >= 2\sigma_\psi^2 x$ and σ_ψ^2 is the standard deviation for the relative phase ψ. This expression shows that the averaged interaction potential decreases with the propagation distance. The behavior for longer distances is more complicated. Numerical simulations confirm the growth of the soliton amplitude difference as predicted by the theory up to distance $x \sim 5$ for $\sigma^2 = 1, \gamma = 1.0, 2\eta_1 = 1.0$ and $2\eta_2 = 0.8$. For larger distances we have a deviation from the behavior predicted by the Karpman-Solov'ev approach. We observe a decreasing of the amplitude difference. To investigate the effect of parameters beyond the validity of the perturbation theory we have performed numerical simulations on the dynamics of two soliton interactions. Depending on the parameter values and the type of noise chosen two phenomena occur. In the white noise case we have found break up of the oscillating two soliton state using $t_0 = 5$ and $\sigma^2 = 0.01$. However, in a medium with coloured noise soliton interactions can lead to an interesting fact which is the appearance of a bound state of solitons (fusion) as illustrated in Figure 3.

We demonstrate this process for $t_0 = 5$, $\gamma = 0.87$ and $\sigma^2 = 0.00435$, γ is the inverse of the correlation length l_c. We note that the interval of the parameter values where this phenomenon exists is narrow ($\gamma = 0.87 \pm 0.05$). A possible explanation is that here we have a narrow spectrum for the noise and the effective potential describing the interaction of solitons is close to the one for chirped solitons in a periodically modulated medium.

References

1. Hasegawa, A. and Kodama, Y.: *Solitons in Optical Communications*, Oxford Univ.Press, Oxford, (1995).
2. Hasegawa, A., Kodama, Y. and Maruta, A.: Recent progress in dispersion managed soliton transmission technologies, *Opt.Fiber Techn.*, (1997), in press.
3. Smith, N. J., Knox, F. M., Doran, N. J., Blow, K. J. and Benion, I.: Enhanced power solitons in optical fibers with periodic dispersion management, *Electron.Lett.*, **32**, (1996), pp.54-57.
4. Akhmanov, S. A. Dyakov, Y. and Chirkin, A.: *Introduction to Statistic Radiophysics and Acoustics*, Nauka Publishers, Moscow, (1986), in Russian.
5. Matera, F., Mecozzi, A., Romagnoli, M. and Settembre, M.: Sidebands instability induced in long distance fiber links, *Opt.Lett.*, **18**, (1993), pp.1499-1501.
6. Abdullaev, F. KH.: Modulational instability of electromagntic waves in fibers with modulated parameters, *Pisma JTP*, **20**, (1994), pp.25-28, in Russian.
7. Abdullaev, F. KH., Darmanyan, S. A., Bishoff, S. A. and Soerensen, M. P.: Modulational instability of electromagnetic waves in media with varying nonlinearity, *JOSA*, **B14**, (1997), pp.27-33.
8. Smith, N. and Doran, N. J.: Modulation instability in fibers with periodic dispersion management, *Opt.Lett.*, **21**, (1996), pp.570-574.
9. Abdullaev F. KH., Darmanyan S. A., Kobyakov, A. and Lederer F.: Modulational instability in optical fibers with variable dispersion, *Phys.Lett.*, **A220**, (1996), pp.213-218.
10. Bronski, J. C. and Kutz, J. N.: Modulational instability of plane waves in nonreturn-to zero communications systems with dispersion management,*Opt.Lett.*, **21**, (1996), pp.937-940.
11. Anderson, D., Lisak, M. and Reichel, T.: Asymptotic propagation of pulses in a soliton-based optical-fiber communication system, *JOSA*, **B5**, (1988), pp.207-210.
12. Landau, L. D. and Lifshitz, E. M.: *Mechanics*, Pergamon, London, (1992).
13. Malomed, B. A., Parker, D. F. and Smyth, N .F.: Resonant shape oscillations and decay of a soliton in a periodically inhomogeneous nonlinear optical fiber, *Phys.Rev.*, **E48**, (1993), pp.1418-1425.
14. Grimshaw, R., He, J. and Malomed, B. A.: Decay of a fundamental soliton in a periodically modulated nonlinear waveguide, *Phys.Scr.*, **53**, (1996), pp.385-393.
15. Gabitov, I., Shapiro, E. and Turitsyn, S.: Asymptotic breathing pulse in optical transmission system with dispersion compensation, *Phys.Rev.*, **E55**, (1997), pp.3624-3633.
16. Abdullaev, F. KH. and Caputo, J. G.: Validity of variational approach for chirped solitons in fibers with periodic dispersion, *Phys.Rev.E*, (1997), submitted.
17. Gordon, J. P.: Dispersive perturbations of solitons of the nonlinear Schrödinger equation, *JOSA*, **B9**, (1992), pp.91-97.
18. Abdullaev, F. KH., Caputo, J. G. and Flytzanis, N. : Envelope soliton propagation in media with temporally modulated dispersion, *Phys.Rev.*, **E52**, (1994), pp.1552-

1558.
19. Smith, N. J., Doran, N. J., Knox, F. M. and Forysiak, W.: Energy scaling characteristics of solitons in strongly dispersion managed fibers, *Opt.Lett.*, **21**, (1996), pp.1981-1983.
20. Abdullaev, F. KH.: Propagation of optical soliton in fiber with fluctuating parameters, *Sov.J.Tech.Phys.Lett.*, **9** , (1983), pp.49-54, in Russian.
21. Elgin, J. N.: Stochastic perturbations of optical solitons, *Opt.Lett.*, **18**, (1993), pp.10-14.
22. Abdullaev, F. KH., Abdumalikov, A. A. and Baizakov, B. B.: Propagation of chirped optical solitons in fibers with randomly varying parameters, *Opt.Commun.*, **138**, (1997), pp.305-310.
23. Abdullaev, F. KH., Bishoff, S., Hansen, A., Soerensen, M. P. and Smeltnik, J.: Propagation and interaction of optical solitons in random media, *JOSA B*, (1997), submitted.
24. Bauer, R. G. and Mel'nikov, L. A.: Multisoliton fission and quasi-periodicity in a fiber with periodically modulated core diameter, *Opt.Commun.*, **115**, (1995), pp.190-197.
25. Hasegawa, A. and Kodama, Y.: Guiding-center soliton, *Phys.Rev.Lett.*, **66** , (1991), pp.161-164.
26. Karpman, V. I. and Solov'ev, V. V.: A perturbation theory for soliton systems, *Physica*, **D3**, (1981), pp.142-164.

(Received November 17, 1997)

SOLITON JITTER AND COLLISIONS IN BIREFRINGENT FIBRES

J. ELGIN, J. SILMON-CLYDE, S. BAKER AND J. HARVEY
Department of Mathematics, Imperial College
180, Queen's Gate, London, SW7 2BZ, United Kingdom.

Abstract. We discuss two factors of importance in design considerations of soliton based communication systems, which are a consequence of the birefringent nature of the fibre. First, we discuss PMD-mediated jitter in a periodically amplified pulse train. Next we consider collisions between solitons in WDM systems and demonstrate that WDM is incompatible with polarization division multiplexing.

1. Introduction

We discuss two factors of importance in design considerations of soliton based communication systems, which are a consequence of the birefringent nature of the fibre. First, we analyse the effects of polarization mode dispersion on the soliton, and show this leads to a jitter in the pulse arrival times in a periodically amplified pulse train. Next we discuss the effects of pulse collision in wavelength division multiplexed (WDM) systems and demonstrate that polarization division multiplexing is incompatible with WDM, thereby supporting recent experimental observations in this area.

2. Soliton Jitter in Birefringent Fibres

Polarization mode dispersion (PMD) is an important factor in the design consideration of soliton-based communication systems. The random variation in the orientation of the birefringence axes along an optical fibre will affect the polarization state of the optical soliton. In consequence, the soliton's velocity is perturbed. Over an ensemble of fibres, this results in a jitter in the pulse arrival times (as characterised by the variance σ_F^2, below). For a single fibre, containing a periodic sequence of amplifiers, the amplifier

A. Hasegawa (ed.), New Trends in Optical Soliton Transmission Systems, 15–23.

noise will lead to a differing uncertainty in the polarization state of each pulse in the train, and therefore there will be a differing uncertainty in their velocities. It is this contribution to soliton jitter — i.e. that arising from polarization effects, and not simply Gordon Haus, that we discuss here.

It is well known that propagation of an ultrashort envelope pulse, $\mathbf{q}(z,t)$, through birefringent fibre can be described by the coupled nonlinear Schrödinger equations:

$$
\begin{aligned}
i\frac{\partial q_1}{\partial z} - \beta q_1 + i\beta'\frac{\partial q_1}{\partial t} + i\Gamma q_2 - \frac{\partial^2 q_1}{\partial t^2} - 2(|q_1|^2 + A|q_2|^2)q_1 - 2Bq_2^2q_1^* &= 0 \\
i\frac{\partial q_2}{\partial z} + \beta q_2 - i\beta'\frac{\partial q_2}{\partial t} - i\Gamma q_1 - \frac{\partial^2 q_2}{\partial t^2} - 2(A|q_1|^2 + |q_2|^2)q_2 - 2Bq_1^2q_2^* &= 0.
\end{aligned} \tag{1}
$$

$q_{1,2}(z,t)$ are the complex valued pulse envelopes corresponding to the two orthogonal polarization modes. z is propagation distance down the fibre and t is retarded time moving with the propagating pulse at the mean group velocity. All quantities are in normalised non-dimensional form. The parameter β is a measure of the difference in phase velocity between each mode, while β' corresponds to the group velocity difference, and is the derivative of β with respect to frequency ω. The term with coefficient $\Gamma(z)$ describes a random variation in the orientation of the axes of polarization along the fibre. Finally, the nonlinear coefficients A and B represent the strength of the nonlinear coupling, and in their nondimensional form satisfy $A + B = 1$.

If the parameters $\beta, \beta', \Gamma(z)$ and B are not too large, equations (1) support an approximate solution corresponding to a perturbed vector soliton pulse,

$$
\mathbf{q}(z,t) = \begin{pmatrix} q_1 \\ q_2 \end{pmatrix} = 2\eta \operatorname{sech} 2\eta(t-y)\exp(i\phi)\begin{pmatrix} \cos\theta\exp(-i\alpha) \\ \sin\theta\exp(+i\alpha) \end{pmatrix} \tag{2}
$$

Here $y(z)$ is the position of the peak of the soliton, and $\{2\theta(z), 2\alpha(z)\}$ describes the polarization state of the pulse, ($\phi(z,t)$ is a phase component).

The relationship between polarization angle θ, and the soliton velocity dy/dz, is of the form [1]

$$
\frac{dy}{dz} = \beta'\cos 2\theta \tag{3}
$$

In this article we analyse the statistical properties of solitons subjected to both PMD and amplifier noise. Consider a model where the fibre is a concatenated chain having N segments of equal length z_h, say, such that the polarization angle θ_n, in any section $(n-1)z_h \leq z \leq nz_h$ is constant in that segment but varies randomly from segment to segment. This model is consistent with real communication fibre where the orientation of the birefringence axes varies on a length scale, $z_h \sim 100$ m, a much shorter length scale than dispersive and nonlinear effects [3]. We choose one member from an ensemble of all such fibres and between all segments of the chosen fibre an amplifier is inserted.

As discussed by Evangelides *et al.* [3], the amplifier noise results in a small cumulative uncertainty in the polarization angle of the pulse in the nth segment by an amount, $\Delta\theta_n = \sum_{i=1}^{n} \delta\theta_i$, say. Suppose that the fluctuations $\delta\theta_n$, from the nth-amplifier have a root mean square deviation which takes the common value θ_W, where $\theta_W \ll 1$, for all amplifiers. Now consider an optical pulse propagating down such a fibre. The statistical properties which describe its arrival time at the exit from the fibre follow from an ensemble average over $\delta\theta_n$ (not θ_n which are fixed for different segments). The statistical properties of the ensemble of the fibres are such that the PMD places an upper limit on the variance of the soliton jitter, as we now show.

Assuming Gaussian statistics for the amplifier noise [2], a straightforward calculation for the variance of pulse arrival times, $\sigma^2 = y^2\rangle_a - y\rangle_a^2$, then gives

$$\sigma^2 = \sigma_F^2 \frac{1}{N} \sum_{j,k=1}^{N} \frac{(h-g)}{h} \left[(h-g)\cos 2\theta_j \cos 2\theta_k + (h+g)\sin 2\theta_j \sin 2\theta_k\right] , \tag{4}$$

where $g = \exp\left(-(j+k)\frac{\theta_W^2}{2}\right)$, $h = \exp\left(-|j-k|\frac{\theta_W^2}{2}\right)$ and $\sigma_F^2 = \frac{\beta'^2 z_h}{2} z$ where $z = Nz_h$. It is found that σ_F^2 is the variance that would describe the uncertainty in pulse position over an ensemble of fibres in the absence of amplifiers [1]. This expression for σ_F^2 is the same as that in work by Mollenauer *et al.* [6] — there it describes the variance in the soliton's width over an ensemble of fibres as it broadens due to random birefringence.

In the limit, $\sqrt{N}\theta_W \gg 1$, equation (4) simplifies to $\sigma \sim \sigma_F$, the uncertainty in the polarization state saturates (i.e. the polarization state of the emerging pulses are uniformly distributed over the Poincaré sphere). The effects of the amplifier noise are such that the pulses effectively explore the ensemble of fibres.

In the opposite limit, $\sqrt{N}\theta_W \ll 1$, equation (4) becomes

$$\sigma = \sigma_F \, \Delta\theta\chi , \tag{5}$$

where $\Delta\theta = \sqrt{N}\theta_W$ and

$$\chi = \left\{ \frac{2}{N^2} \sum_{i=1}^{N} \sum_{j,k=i}^{N} \sin 2\theta_j \sin 2\theta_k \right\}^{1/2} . \tag{6}$$

The parameter $\Delta\theta$ is the angular spread in the polarization states caused by amplifier noise [3]. The parameter χ contains all the relevant information about the fibre selected from the ensemble and will of course change value from member to member of the ensemble. We can define an average value for χ, say $\bar{\chi} = \left(\langle\chi^2\rangle_F\right)^{1/2}$ where the suffix F dentoes an average over the ensemble of fibres, and it can be shown that $\bar{\chi} \sim O(1)$. With an additional average over the ensemble of fibres, equation (5) becomes $\sigma^2 \to \sigma^2\rangle_F = \sigma_F^2 \Delta\theta^2 \propto z^2$, as found using the more phenomenological approach of Mueller matrices in work by Mollenauer *et al.* [7].

Equation (5) above shows that a statistical distribution in polarization states caused by amplifier noise necessarily leads to a jitter in the pulse arrival times. For example, following Evangelides *et al.* [3], using dimensional parameter values, $\beta' \sqrt{z_h} = 0.2$ ps·km$^{-1/2}$, $Z = 9{,}000$ km, we find $\sigma_F \sim 19$ ps, $\Delta\theta \sim 7^\circ = 0.12$ rad, $\bar{\chi} \sim 1$ and so $\sigma = 2.3$ ps.

3. Incompatibility of Polarization-Division Multiplexing with Wavelength-Division Multiplexing

Mollenauer *et al.* [8] have demonstrated experimentally that polarization-division multiplexed solitons depolarize making polarization-division multiplexing incompatible with wavelength-division multiplexing (WDM). They argue plausibly that a series of two-soliton collisions can lead to this depolarization. Here we construct a simple theoretical model which confirms their assertion.

All communication fibres are subject to a rapidly and randomly varying birefringence. The effect of this on pulse velocity was discussed in the previous section. When the effects of polarization mode dispersion are negligible (i.e. the birefringence is low enough and varies sufficiently rapidly) soliton propagation in the fibre is correctly described by the Manakov equation [4, 11, 12, 5]. As this equation is integrable [4] the fiber system in this limit is actually simpler to analyse than the coupled nonlinear Schrödinger equations which applies to fibre sections with fixed birefringence. A nice property of integrable systems is that an N-soliton collision can be constructed from a superposition of 2-soliton collisions. After deriving an expression for the effect of a 2-soliton collision on the Stokes vectors of the pulses concerned we will use this property to examine the case when two soliton trains interact, which, when coupled with weak birefringence, results in a depolarization of the solitons in both trains.

We begin with the Manakov equation [4] which is obtained from equations (1) when the parameters β, β', Γ and B are set to zero. We now write the single soliton solution in the form

$$\mathbf{q}(\mathbf{z},\mathbf{t}) = \frac{2\eta\mathbf{c}\exp[-i(2\xi\mathbf{t} + 4(\eta^2-\xi^2)\mathbf{z}]}{\cosh 2\eta(\mathbf{t}-4\xi\mathbf{z})}\,, \tag{7}$$

where

$$\mathbf{c} = \begin{bmatrix} \cos\theta\exp(i\alpha_1) \\ \sin\theta\exp(i\alpha_2) \end{bmatrix}. \tag{8}$$

The eigenparameter $\zeta = \xi + i\eta$ determines the velocity and amplitude of the pulse. The vector $\mathbf{c}$ contains all the information about the polarization state of the pulse, where θ is the polarization angle and α_1 and α_2 are phases. Of course the polarization and phase angles are relative to a set of axes located at the input to the fibre; the unitary transformation used to convert from the coupled nonlinear Schrödinger equations to the Manakov equations causes a rapid and random variation of the axes along the fibre [11, 12].

We introduce the Stokes vector $\mathbf{S}$, whose components $\mathrm{S_i}$ are given by

$$\mathrm{S_i} = \mathbf{c}^\dagger\sigma_\mathbf{i}\mathbf{c}\,, \tag{9}$$

where σ_i are the Pauli matrices (using the notation of Mollenauer *et al.* [8] rather than the standard notation). Manakov used Inverse Scattering Theory to determine the effect of collisions on the polarization states (and positions) of solitons. In terms of the Stokes vectors, Manakov's result reads [9]

$$\begin{aligned} \mathbf{S}_1^+ &= a\mathbf{S}_1^- + b\mathbf{S}_2^- + c\mathbf{S}_1^- \times \mathbf{S}_2^- \\ \mathbf{S}_2^+ &= a\mathbf{S}_2^- + b\mathbf{S}_1^- + c\mathbf{S}_2^- \times \mathbf{S}_1^-\,, \end{aligned} \tag{10}$$

where the parameters a, b and c are

$$\begin{aligned} a &= 1/(|\phi|^2\cos^2\gamma + \sin^2\gamma) \\ b &= 1-a \\ c &= a\sqrt{|\phi|^2-1} \end{aligned} \tag{11}$$

and

$$\phi = (\zeta_2 - \zeta_1^*)/(\zeta_2 - \zeta_1)\,. \tag{12}$$

Here 2γ is the angle between the Stokes vectors $\mathbf{S}_1^-$ and $\mathbf{S}_2^-$ so that $\cos 2\gamma = \mathbf{S}_1^- \cdot \mathbf{S}_2^-$.

We will only examine the case of where the colliding pulses have equal amplitudes, so that $\eta_1 = \eta_2 = \eta$. Then, $\phi = 1 + i\mu$ where $\mu = 2\eta/(\xi_2 - \xi_1)$,

$a = 1/(\mu^2 \cos^2 \gamma + 1)$ and $c = a\mu$. For this equal amplitude case , it is easily shown that $\mathbf{S_1^+}$ and $\mathbf{S_2^+}$ are obtained from $\mathbf{S_1^-}$ and $\mathbf{S_2^-}$ by a simple precession about the vector $\mathbf{S_0} = (\mathbf{S_1^-} + \mathbf{S_2^-})/2$ through an angle β where

$$\cos \beta = 2a - 1 . \tag{13}$$

In the limit $\mu \ll 1$ and $\gamma \ll 1$ we recover the approximate result reported by Mollenauer *et al.* [8].

We now extend equations (10) so that they apply to the general case of solitons in different pulse trains, as appropriate for WDM systems. We choose a reference frame such that the solitons in one train are observed to move to the left, while those in the other train move to the the right. Denote by $\mathbf{L_j^i}$ the Stokes vector of the ith such soliton in the left-moving train after it has experienced j collisions with solitons in the right-moving train; similarly for $\mathbf{R_j^i}$. Then for the collision between $\mathbf{L_{j-1}^i}$ and $\mathbf{R_{i-1}^j}$ equations (10) are generalised to read:

$$\mathbf{L_j^i} = a_{ij}\mathbf{L_{j-1}^i} + b_{ij}\mathbf{R_{i-1}^j} + c_{ij}\mathbf{L_{j-1}^i} \times \mathbf{R_{i-1}^j} \tag{14}$$

$$\mathbf{R_i^j} = a_{ij}\mathbf{R_{i-1}^j} + b_{ij}\mathbf{L_{j-1}^i} + c_{ij}\mathbf{R_{i-1}^j} \times \mathbf{L_{j-1}^i} , \tag{15}$$

where, as before, a_{ij}, b_{ij} and c_{ij} are defined in terms of an angle γ_{ij} which is given by:

$$\cos 2\gamma_{ij} = \mathbf{L_{j-1}^i}.\mathbf{R_{i-1}^j} . \tag{16}$$

For the colliding pair of pulses previously discussed, $\mathbf{L_0^1}$, $\mathbf{R_0^1}$ are respectively $\mathbf{S_1^-}$ and $\mathbf{S_2^-}$, while $\mathbf{L_1^1}$ and $\mathbf{R_1^1}$ are $\mathbf{S_1^+}$ and $\mathbf{S_2^+}$.

Consider now the general case where two trains of soliton pulses interact. Suppose these are initially in the polarisations $\mathbf{L^1}$ and $\mathbf{R^1}$. Using equations (14) and (15) we can evaluate numerically the degree of polarization d_{pol}, in the WDM channel containing the N pulses of the $\mathbf{R}$-train as a function of the number of collisions M where d_{pol} is defined as

$$d_{\text{pol}} = |\sum_{i=1}^{N} \mathbf{R_M^i}|/\mathbf{N} .$$

Figure 1 shows d_{pol} plotted against possible number of collisions for three different values of γ_{11}, where $\cos 2\gamma_{11} = \mathbf{L^1} \cdot \mathbf{R^1}$, indicated in the caption. For these simulations all pulses were present in the $\mathbf{R}$-train while the $\mathbf{L}$-train consisted of a random bit pattern of pulses and $\mu = 0.5$, N = 40. The number of possible collisions is the number which each $\mathbf{R}$-pulse would have experienced had all the $\mathbf{L}$-pulses been present; it is effectively a measure

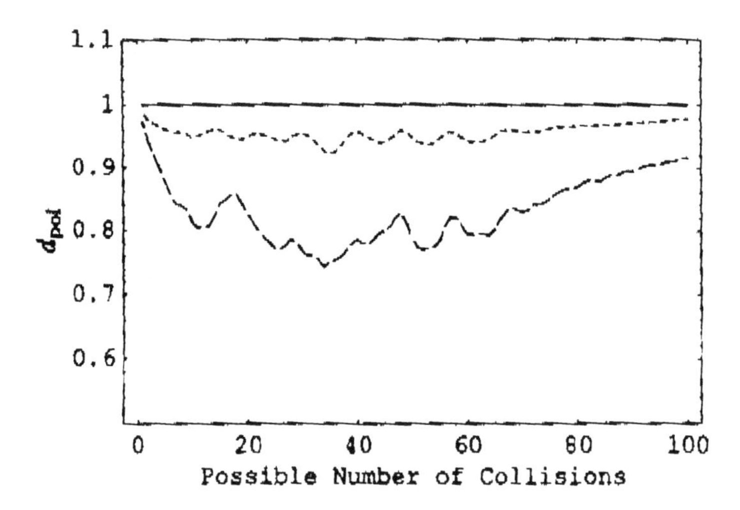

Figure 1. Numerically generated plot of d_{pol} for **R**-train as a function of possible number of collisions. The **L**-train has a random bit pattern. (i)Solid line: $\gamma_{11} = 0$. (ii)Dotted line: $\gamma_{11} = \pi/8$. (iii)Dashed line: $\gamma_{11} = \pi/4$. For all plots $\mu = 0.5$, N = 40.

of the distance travelled in the fibre. As expected, for $\gamma_{11} = 0$ the trains remain copolarized. For trains which are not initially copolarized d_{pol} drops quickly before increasing toward 1 as the number of collisions increases; this behaviour may be explained by consideration of the single pulse interaction with a train discussed earlier.

Every real fibre will exhibit weak birefringence, causing an initially linearly polarized pulse to evolve through a sequence of elliptic polarization states before returning to its initial state, over a length scale of tens of metres. For two pulses separated in frequency by an amount $\Delta\omega$ (as in a WDM system), there will be a differential rotation rate, $\Delta\kappa$ say. If we assume that a unitary tranformation has been carried out to remove the effects of weak birefringence from the **R**-train, the **L**-train will see a relative evolution of its initial polarization state over a length scale $l \approx 2\pi/\Delta\kappa$. We incorporate this additional effect into our model by letting $\mathbf{L}^{\mathbf{i}} \to \mathbf{U}\mathbf{L}^{\mathbf{i}}$ between collisions where the unitary operator $\mathbf{U}$ is

$$\mathbf{U} = \begin{bmatrix} 1 & 0 & 0 \\ 0 & \cos\Delta\alpha & \sin\Delta\alpha \\ 0 & -\sin\Delta\alpha & \cos\Delta\alpha \end{bmatrix}.$$

Here, $\Delta\alpha = \Delta\kappa Z_c$ where Z_c is the distance between soliton collisions. We now determine the final polarization states of the **R**-train by numerically

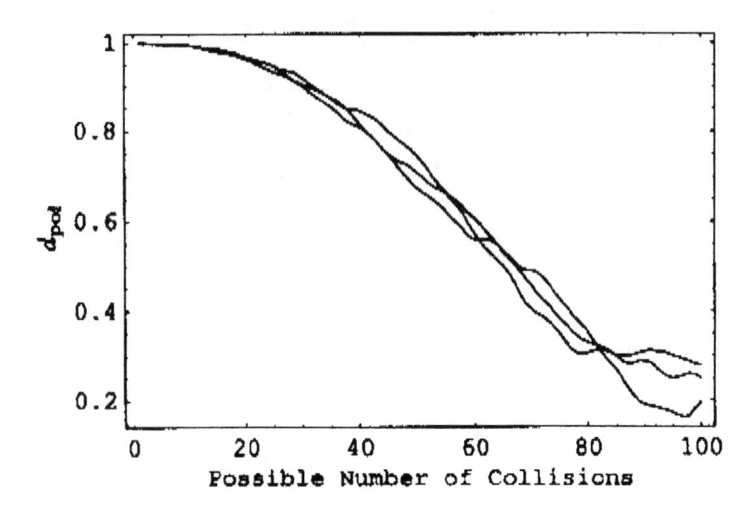

Figure 2. Numerically generated plot of d_{pol} for **R**-train as a function of possible number of collisions. The effects of weak birefringence are incorporated and the **L**-train has a random bit pattern; results for three different random bit patterns are plotted. For all plots $\mu = 0.5$, N $= 40$, $\Delta\alpha = 0.08$ rad.

evaluating equations (14) and (15) letting $\mathbf{L^i} \to \mathbf{UL^i}$ between collisions.

To estimate $\Delta\kappa$, we note that $\Delta\kappa \approx \kappa'\Delta\omega$, and that the polarization mode dispersion coefficient for a dispersion shifted fibre is $\kappa'/h^{1/2} \approx$ 0.2ps/km$^{1/2}$ [3], where h is the length scale for polarization decorrelation which we assume to be $\approx$ 100 m [10]. Assuming the pulses in each train have a duration $\approx$ 10 ps and are separated by $\approx$ 50 ps, this gives $\Delta\kappa \sim 0.05$ rad/km. Following Mollenauer *et al.* [8], we assume the distance between collisions to be $Z_c \approx 100$ km, so that $\Delta\alpha$ is of the order 2π. Given the uncertainty in the numerical values for the parameters discussed, and having shown that it can be at least 2π we assume that $\Delta\alpha$ may take any value. We have investigated the case when $\Delta\alpha = 0.08$ rad. Results are shown in Figure 2 for the case when N $= 40$ and $\mu = 0.5$. Initially all pulses have the same polarization state and as before all pulses were present in the **R**-train while the **L**-train consisted of a random bit pattern of pulses (results for three different bit patterns are shown). d_{pol} now markedly decreases with increasing number of collisions, and the qualitative behaviour of the decrease is entirely consistent with the experimental results discussed in work by Mollenauer *et al.* [8].

In conclusion, we have used the integrability of the Manakov system to devise a simple analytic model which confirms an assertion based on the results of experimental investigation by Mollenauer *et al.* [8], that WDM is incompatible with polarization-division multiplexing.

References

1. Baker S.M. and Elgin J.N., (1998), Solitons in Randomly Varying Birefringent Fibre, *Journal of European Optical Society B*, Special issue: Polarization Effects in Lasers and Spectroscopy.
2. Elgin J.N., (1995), Inverse Scattering Theory with Stochastic Initial Potentials, *Physics Letters*, **Vol.110A, No.9**, pp.441-443.
3. Evangelides S.G., Mollenauer L.F., Gordon J.P. and Bergano N.S., (1992), Polarization Multiplexing with Solitons, *IEEE Jounal of Lightwave Technology*, **Vol.10, No.1**, pp.28-35.
4. Manakov S. V., (1973), On the theory of two-dimensional stationary self focusing of electromagnetic waves. Zh. Eksp. Teor. Fiz. **65**, 505 [Sov. Phys. JETP **Vol.38, No.2**, pp.248-253 (1974)]
5. Marcuse D., Menyuk C. R., Wai P. K. A., (1997), Application of the Manakov-PMD Equation to Studies of Signal Propagation in Optical Fibers with Randomly Varying Birefringence, *Journal of Lightwave Technology*, **Vol.15, No.9**, pp.1735-1746.
6. Mollenauer L.F., Smith K, Gordon J.P., Menyuk C.R., (1989), Resistance of solitons to the effects of polarization dispersion in optical fibres, *Optics Letters*, **Vol.14, No.21**, pp.1219-1222.
7. Mollenauer L.F. and Gordon J.P., (1992), Birefringence-mediated timing jitter in soliton transmission, *Optics Letters*, **Vol.19, No.6**, pp.375-377.
8. Mollenauer L. F., Gordon J. P., Heismann F.,(1995), Polarization Scattering by soliton-soliton collisions, *Optics Letters*, **Vol.20 No.20**, pp.2060-2062.
9. Silmon-Clyde J. P., Elgin J. N., (1997), Incompatibility Of Polarization Division Multiplexing with Wavelength-Division Multiplexing in Soliton Transmission Systems, To appear in *Optics Letters* November/December 1997
10. Ueda T., Kath W. L., (1994), Stochastic simulation of pulses in non-linear optical fibres with random birefringence, *Journal of Optical Society of America B* **Vol.11 No.5**, pp.818-825.
11. Wai P. K., Menyuk C. R., Chen H. H., (1991), Stability of solitons in randomly varying birefringent fibers, *Optics Letters*, **Vol.16 No.16**, pp.1231-1234.
12. Wai P. K., Menyuk C. R., (1996), Polarization Mode Dispersion, Decorrelation, and Diffusion in Optical Fibers with Randomly Varying Birefringence, *IEEE Journal of Lightwave Technology*, **Vol.14 No.2**, pp.148-157 .

(Received November 17, 1997)

SOLITON STABILITY
IN PMD PERTURBED TRANSMISSION SYSTEMS

M. KARLSSON, X. ZHANG AND P. A. ANDREKSON
Department of Optoelectronics and Electrical Measurements
Chalmers University of Technology
SE-412 96 Gothenburg, Sweden

AND

K. BERTILSSON
Ericsson Telecom AB, Transport Network Application Lab.
SE-126 25 Stockholm, Sweden

Abstract. This paper presents a detailed investigation of how solitons broaden in the presence of polarization-mode dispersion (PMD). For a given degree of PMD and pulse width there exists an optimum choice of group-velocity dispersion, for which solitons are most resistant to the effects of PMD. We also optimize the soliton power enhancement factor and find that solitons can propagate roughly twice the distance compared to linear pulses in the absence of group-velocity dispersion. Finally we investigate the PMD robustness of dispersion managed solitons, and in the case simulated we find no quantitative difference as compared to conventional solitons.

1. Introduction

Recent experiments have shown that it is possible to upgrade the installed fiber-optic telecommunication networks by high bit-rate soliton transmission, which can provide a longer transmission distances than conventional non-return to zero transmission [1]. However, polarization-mode dispersion (PMD) may be large in installed fibers; typical PMD-coefficients of 0.1–1 ps/km$^{1/2}$ have been measured [2, 3]. Dispersion-shifted fibers (which are of particular interest for soliton transmission) have been found to have particularly high values ($\sim$ 1 ps/km$^{1/2}$) [3]. Neglecting group velocity dispersion

A. Hasegawa (ed.), New Trends in Optical Soliton Transmission Systems, 25–37.

(GVD), pure PMD causes pulses to broaden at a relative broadening rate [4]

$$\tau = \sqrt{\tau_0^2 + \frac{\mathrm{PMD}^2 \mathrm{z}}{4}}, \tag{1}$$

where $\tau(z)$ is the RMS-width of the pulse, τ_0 is the initial RMS-width and z is the propagation distance along the fiber.

It is generally believed that such high values of the PMD will ultimately limit the transmission bandwidth of linear transmission systems. Solitons, however, have for quite a while been known to be resilient against PMD [5, 6, 7, 8, 9], the physical reason being the nonlinear attraction due to cross-phase modulation (XPM) between the two orthogonal polarization components (so called self-trapping). This will also enable the solitons to maintain the degree of polarization during propagation to a larger extent than linear pulses [7, 9]. In reference [5] it was argued that solitons can avoid excessive broadening, provided the PMD is below a certain threshold value, $\mathrm{PMD} < 0.27\sqrt{|\beta_2|}$. In reference [8] this expression was generalized to take into account also the dependence on fiber length. However, solitons have also been found to shed a significant amount of dispersive radiation when perturbed by PMD [5, 6], since they are periodically perturbed at a very rapid rate by the random birefringence. Obviously, this will reduce the soliton content (as defined by the solution to the Zakharov-Shabat, or Manakov scattering problems [10, 11]) of the pulse during propagation, with a subsequent broadening (or equivalently, energy loss) of the soliton as a result. This broadening was considered in a recent paper [9], where an analytical theory predicted PMD-perturbed solitons to broaden according to equation (1), but with a lower value of the PMD cefficient; it was found that $\mathrm{PMD_{sol}} = \pi/(3\sqrt{3})\mathrm{PMD_{lin}} \approx 0.605\ \mathrm{PMD_{lin}}$. From these findings, the critical system designer might wonder what is gained by using solitons, since solitons and linear pulses will broaden at almost the same rate. However, it is well known that solitons losing energy to dispersive waves can be stabilised by suitable filtering mechanisms; e.g. synchronous modulation in amplitude [12] or phase [13], sliding filters [12, 14] or saturable absorbtion [15] (for an exhaustive list see reference [13]). In particular, it has been demonstrated that synchronous amplitude modulation or sliding filters can be quite useful for the suppression of PMD-induced soliton broadening [9].

We therefore believe that solitons perturbed by PMD deserves a more detailed investigation. We do not consider filtering issues in this paper, since we believe that the pure soliton dynamics is not yet properly understood. For instance, how does the relative (with respect to PMD) amount of GVD (or equivalently, nonlinearity) affect the soliton broadening? Is it possible to improve performance by launching pulses with a slightly larger amplitude than that corresponding to the N = 1 soliton? Can performance be improved with dispersion managed solitons? These questions will be adressed in this paper. In particular, we find that a sole threshold value or a sole broadening rate does not give a complete picture of the soliton stability in the presence of PMD [16].

2. Confidence Intervals and Necessary Number of Averages

Another obstacle when analyzing PMD lies in its statistical nature. This implies that all simulations should be carried out a large number of times with different stochastic seeds so that an accurate average can be computed. In previous works this has only been effectuated in references [8] and [9] where, respectively, 50 and 10 runs where averaged. We have found in our simulations that this is not enough to get reliable results; hence in all simulations we use an average of 500 different fibers with the same PMD coefficient. The reason for this is that the differential group delay (DGD) induced by PMD is a statistical quantity with a rather broad (Maxwellian) distribution function, so that many samples are required to get a reliable average. More precisely, for a Maxwellian variable, the averaging over M runs enables us, with 90 % (99 %) probability, to get an average with a relative confidence interval of $2.70/\sqrt{M}$ ($4.24/\sqrt{M}$), see reference [17]. For $M = 500$ this implies confidence intervals of ± 6 % with 90 % probability, or ± 10 % with 99 % probability. Even though we cannot expect the expectation values of the random variables we compute in this paper to be Maxwellian, we have no reason to believe that their probability distribution should be significantly more peaked, and we have to settle with the above estimates for the accuracy.

3. Solitons Perturbed by PMD

The theoretical framework for the study of solitons affected by PMD is based on two (nonlinearly) coupled nonlinear Schrödinger equations for the linear polarization states according to reference [6]:

$$i\frac{\partial u}{\partial x} + i\delta\frac{\partial u}{\partial t} + \frac{1}{2}\frac{\partial^2 u}{\partial t^2} + (|u|^2 + \frac{2}{3}|v|^2)u = 0 \tag{2}$$

$$i\frac{\partial v}{\partial x} - i\delta\frac{\partial v}{\partial t} + \frac{1}{2}\frac{\partial^2 v}{\partial t^2} + (\frac{2}{3}|u|^2 + |v|^2)v = 0\,. \tag{3}$$

Here it is assumed that the linear birefringence is so large that the the two polarization states u and v are uncorrelated in phase. The difference in walk-off between the two polarization states is accounted for with the normalized group velocities $\pm\delta = \pm\Delta\beta_0' t_0/(2\beta_0'')$ in the respective polarization component (see e.g. reference [9]), where $\Delta\beta_0'$ is the group-velocity difference of the polarization components, β_0'' is the GVD, and t_0 is the time scale of the normalized time coordinate. Note that the x-coordinate is normalized to indicate the number of dispersive lengths, i.e. $x = z/L_d$, where $L_d = t_0^2/\beta_0''$. The numerical solution procedure follows the standard technique [5, 6, 7, 8, 9] of randomly rotating the birefringence axes at increments L_c in the simulation, together with the addition of a random phase angle between the polarization components. This is the so-called correlation length, or mode coupling length, over which one polarization state can be completely coupled to its orthogonal polarization state. This method of solution is valid if the birefringence beat length of the fiber is much shorter than all other length scales of the problem (of the order of meters). In our simulations we generally took $L_c = 0.5$ km which equaled the simulation step size, but we made sure by reducing the step size (but maintaining L_c) that the accuracy was sufficient, in particular for the cases with short dispersive lengths. In all simulations we used an initial soliton $u(0,t) = \sqrt{9/8}\mathrm{sech}(t/t_0)$, where the factor of 9/8 is used to excite the first order Manakov soliton of the averaged coupled equations [6, 9]. Since the time and space coordinates of the NLS equations are normalized with respect to the soliton parameters, the only free parameters of the problem is δ and L_c. The coefficient δ determines the relative influence of PMD and nonlinearity. In the limit of low dispersion and nonlinearity (for which δ is large) we verified a relative pulse broadening (in the RMS-sense) according to equation (1), where PMD $= \Delta\beta_0'\sqrt{L_c} = \delta 2\beta_0''/t_0\sqrt{L_c}$, see reference [18], and τ_0 is the initial (RMS-)width, and z is the propagation distance (km) . We will refer to this case of pulse broadening as the linear case. Now, by using nonzero values of the GVD (corresponding to values of δ less than infinity) we are able to monitor how the width depends on the propagation distance z for solitons, and the result can be viewed in Figure 1. It is clearly seen that solitons broaden in qualitatively the same way as linear PMD pulses (as pointed out in reference [9]). But we also see that the initial rate of broadening is less than that of the linear pulses and that it decreases with increasing dispersion (or nonlinearity). That means the PMD influence can be reduced by increasing the GVD, which was found in the early works [5, 6]. Note that this slightly differs from the analytical theory of reference [9], which predict no dispersion dependence in the broadening rate. The reason for this discrepancy is that the expansion theory of reference [9] is valid only when the soliton is weakly perturbed [6], so that the analytic theory can be interpreted as giving a lower bound on the broadening rate. Moreover, from a physical standpoint there must also be a dispersion dependence in the broadening rate, since for weak dispersions (large values

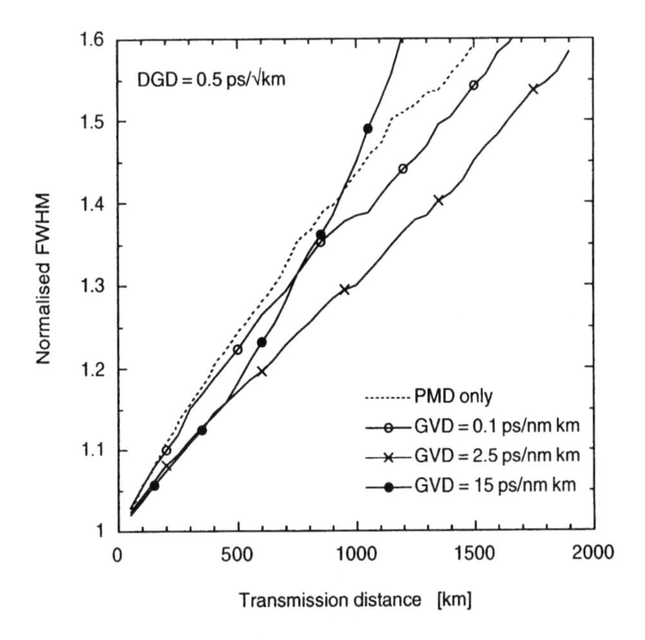

Figure 1. PMD-induced pulse broadening (FWHM normalized to its initial value) as a function of distance for linear pulses and solitons of various GVD. The initial pulse width (FWHM) is 20 ps, and the PMD is 0.5 ps/km$^{1/2}$.

of δ) the PMD influence is so strong that the solitons should broaden similarly to linear pulses. Indeed this is what we see in Figure 1. The novel feature of Figure 1 can be seen in the case of large dispersion (i.e. small values of δ). We observe that the broadening rate eventually increases, even to the extent of exceeding the linear pulse broadening. The reason for this behavior is that for a given distance L, the number of dispersive lengths propagated, L/L_D, increases with increasing GVD, and concomitantly the amount of dispersive wave generation. An important consequence of this is that the Manakov approximation will be more and more violated. Gradually the solitons will lose their (Manakov-) soliton character and GVD will dominate the broadening at a rate far above the PMD-broadening rate. We can thus draw the important conclusion, to our knowledge previously unnoticed, that *for a system with given length and PMD chararcteristics, an optimum GVD exists which minimizes the soliton pulse broadening.* As a consequence of this, it is not correct to state that solitons with a large GVD, in general, are more resistant to the effect of PMD than solitons with a small GVD.

The question then arises as to what this optimum value of the dispersion is. Obviously the optimum dispersion will depend on how we define it. From

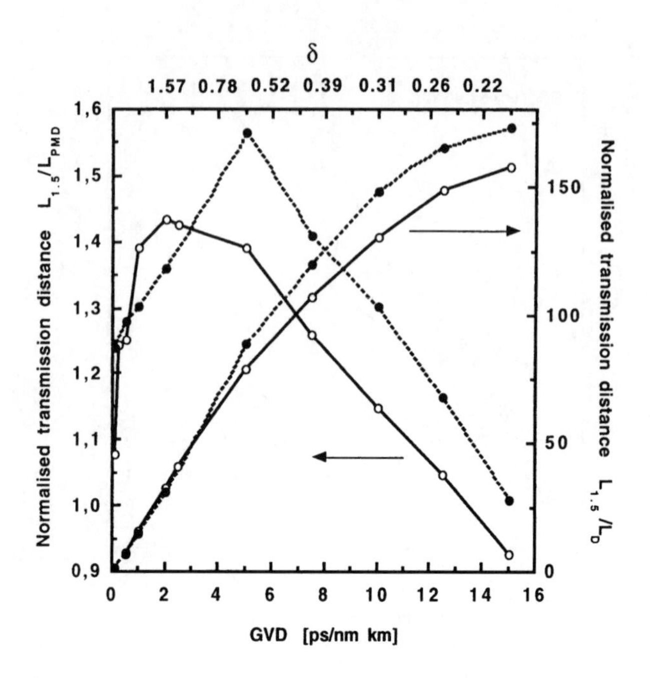

Figure 2. Maximum transmission distance $L_{1.5}$ for a 50 % FWHM increase. The scale to the left plots $L_{1.5}$ relative to the corresponding length L_{PMD} for a system limited by linear PMD. The scale to the right plots $L_{1.5}$ in number of dispersive lengths, for the case of PMD $= 0.5$ ps/km$^{1/2}$ and initial pulse width equal to 20 ps. The solid (dotted) line has mode coupling length $L_c = 0.5$ (0.25) km.

a communication system point of view it is relevant to consider a 1 dB eye-opening penalty, which, roughly speaking, corresponds to a FWHM-width of 1.5 the initial value. We therefore calculate the distance $L_{1.5}$ at which the FWHM-width have increased by 50 %, and use this as a measure of how much solitons can improve the transmission. For a linear system, we define a corresponding length L_{PMD}. Assuming a functional form of the broadening as in equation (1) also for the FWHM-width, we find from Figure 1 that $L_{PMD} \approx 0.69T_0^2/\mathrm{PMD}^2$, where T_0 is the initial FWHM-width. Figure 2 shows $L_{1.5}$ normalized to L_{PMD} as a function of the dispersion D (lower scale) and δ (upper scale). The particular case simulated used a 20 ps wide (FWHM) soliton pulse (corresponding to a bit rate of 10–15 Gbit/s), with a PMD of 0.5 ps/km$^{1/2}$ and an $L_c = 0.5$ km. We emphasize that losses were neglected in this simulation (as in Figure 1), so all degradations are due to PMD alone. The linear broadening length L_{PMD} were found (from Figure 1) to be 1,100 km. As stressed above, the only free parameters of the problem are δ and L_c, so there will be an optimum value of δ for each

L_c. In the two cases simulated, the enhancement in propagation distance by using solitons is around 1.5 times, and occurring at $\delta \approx 1.6$ for $L_c = 0.5$ km, and at $\delta \approx 0.6$ for $L_c = 0.25$ km. By using the definition of δ, we can express the optimum dispersion in terms of PMD, pulsewidth and mode coupling length as

$$\beta''_{0,opt} = \frac{t_0 \text{PMD}}{2\delta_{opt}(L_c)\sqrt{L_c}} \,. \tag{4}$$

The function $\delta_{opt}(L_c)$ is unknown, and our simulations has revealed a few values only. However, the dependence of the mode coupling length in this expression is evident. The measurements of this important fiber parameter that is available to date have large uncertainties, but values of L_c of the order 100 m –1 km for cabled fibers and 1 m –10 m for spooled fibers have been obtained [2]. Obviously solitons will perform better with shorter correlation lengths, since then the polarization state rotates more rapidly, and the Manakov system is a better approximation to the propagation equations. The right-hand scale of Figure 2 shows the propagation distance in dispersive lengths, and it is clearly seen that there is a very small dependence on dispersion when PMD is dominating, whereas for large values of the dispersion, the propagation distance appears to be limited to a specific number of dispersive lengths, which will depend on the coupling length L_c. This underlines our interpretation that in this regime it is the shedding of dispersive waves that limit the performance rather than the random fluctuation of the group velocity. Fiber loss have been neglected in these simulations, and we argue that we have amplifiers so densely spaced that we can consider average guiding-center solitons. Fortunately, at realistic values of the PMD, the optimum value of the GVD is rather low, being of the order of 1–5 ps/nm·km. Finally, we shall point out that if the dispersive length is of the order of the correlation length, the polarization states in a dispersive length (or a soliton period) may not be uniformly distributed over the Poincaré sphere, and nonlinear PMD will be important [19, 20]. The problem with nonlinear PMD is that it is determined by length scales (such as the mode coupling length and the birefringence beat length) that are difficult to measure (and even estimate) experimentally, and that in addition vary along the fiber. Future transmission experiments up to 100 Gbit/s, may shed more light on the polarization state evaluation, but for the moment it seems like beat lengths of the order of 1–10 m and correlation lengths of the order of 0.1–1.0 km are relevant figures.

Because the Kerr nonlinearity counteracts the PMD influence, slightly increasing the soliton power may enhance the soliton robustness. Indeed this is the case, as is shown in Figure 3. We see that for $\delta = 1.57$ it is possible to enhance the propagation distance $L_{1.5}$ another 20 %, which is optimum for soliton order $N = 1.1$. Note that N is defined from the

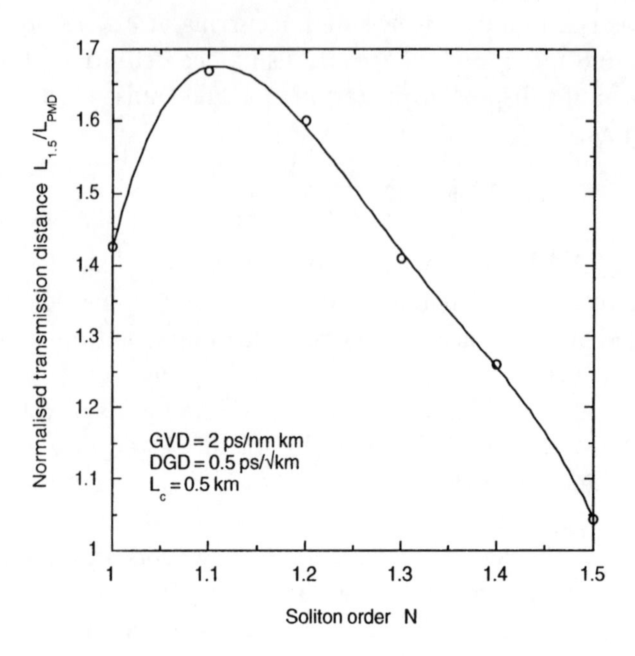

Figure 3. The distance $L_{1.5}$, normalized to the corresponding distance for linear pulses as a function of soliton order N. We consider the optimized value of δ at $L_c = 0.5$ km, i.e. $\delta = 1.57$.

initial condition as $u(0,t) = N\sqrt{9/8}\text{sech}(t/t_0)$. This optimum value can be expected to vary slightly with δ but to what extent is unknown.

In Figure 4 we consider inclusion of fiber loss (0.25 dB/km) and periodic amplification. We see in Figure 4 that the broadening rate of the PMD-perturbed solitons is increases with the amplifier spacing. It is well known that periodic amplification of solitons will lead to the generation of dispersive waves at a frequency phase-matched by the perturbation wavenumber [21]. For a fixed amplifier spacing the emission of dispersive waves will decrease with the dispersive length [21] and hence the pulses will broaden more rapidly for higher dispersions. Since the higher values of the dispersion will be degraded by this effect, the optimum GVD discussed above will be shifted to lower values. However, it seems that for an amplifier spacing of one half a dispersive length the amount of dispersive waves generated is negligible in comparison to the amount generated by the random birefringence.

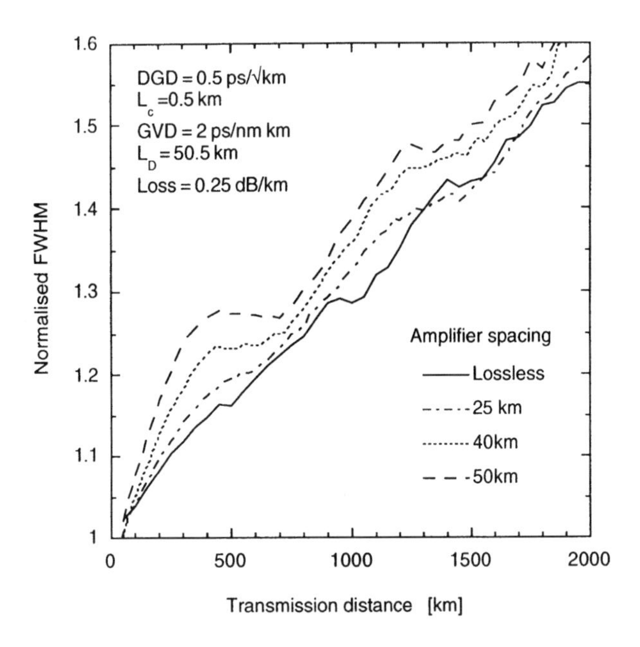

Figure 4. The normalized FWHM as a function of propagated distance of solitons having GVD = 2 ps/nm·km (i.e. at the optimum value of δ) for various amplifier spacings.

4. Dispersion-managed Solitons Affected by PMD

The next point to consider is whether dispersion-managed solitons are more robust to PMD than the conventional solitons considered in the previous section. It is well-known that dispersion managed solitons have many advantages over conventional solitons, e.g. increased power margin [22], less interaction, less dispersive wave generation, less jitter, etc. [23, 24]. Moreover, since dispersion-managed solitons have a power enhancement factor that increases the nonlinearity locally, one would expect those solitons to allow improved PMD suppression. There are some factors complicating a direct comparison between the two kinds of soliton propagation, however. The pulse shape differs; it is well known that Gaussian pulses perform better in dispersion-managed systems. Therefore we use $u(0,t) = N\sqrt{9/8}\mathrm{sech}(t/t_{0,s})$ for the conventional solitons and $u(0,t) = N\sqrt{9/8}\exp(-t^2/(2t_{0,g}^2))$ for the dispersion managed solitons. The pulse widths are chosen so that the widths (FWHM) T_0 are equal and 20 ps; $t_{0,g} = T_0/1.665$ and $t_{0,s} = T_0/1.763$. The factor N^2 is known as the power enhancement factor, and needs to be optimized for each dispersion map. In order to evaluate performance we now consider the recieved eye diagram of a 10 GBit/s sequence of 8 bits (11101000). Since the interaction between the solitons and the dispersive waves may be considerable in dispersion managed systems we consider the eye closure of the recieved signal. We use two reciever filters; an optical Fabry-Perot filter with a bandwidth of 40 GHz, and an electric, 8 GHz fourth-order Bessel filter. The eye closure is defined as the eye opening relative to the back-to-back eye opening.

In absence of PMD, we have used the eye closure as a measure to optimize the power enhancement factor [25]. We have found that a power enhancement factor $N^2 \approx 2$ is optimal for all investigated cases were there is a compensating fiber in each amplifier span. Higher optimal power enhancement factors were obtained for dispersion maps with the compensating fiber covering several amplifier spans. In those cases the optimal value of N^2 was of the order of 4–10, and dependent of the dispersion perturbation, see reference [25] for details. Since those latter cases are the most nonlinear, we expect them to be most suitable for suppressing PMD. In the simulated case we used a dispersion compensating fiber every 3 amplifier span, with a dispersion map according to 40+38+2+40 km, where the long fibers have $D = 1$ ps/nm·km and the 2 km compensating fiber has $D = -53$ ps/nm·km, and amplifiers are inserted every 40 km. The average dispersion using this configuration is 0.1 ps/nm·km. The PMD was crudely assumed to be 0.5 ps/km$^{1/2}$ for all fibers, but it is straightforward to modify the simulation to account for changing PMD during propagation. The mode coupling length as set to $L_c = 0.5$ km for all fibers. Figure 5 shows the eye closure with and without PMD. In absence of PMD we establish an optimum power enhancement factor of 2.5 (dotted curve). With PMD present this case is slightly better than the conventional soliton case, but the difference is really marginal and within the statistical inaccuracy. As can be seen from the simulation, the dispersion managed solitons show no obvious advantage over the conventional soliton, but obviously there are many more cases to consider. It will certainly be possible to optimize the performance of a system with respect to the depth of the dispersion map, the average dispersion, the power enhancement factor and other free design parameters, and more work on these issues need definitely to be carried out. From these initial simulations, however, we cannot conclude a stronger PMD robustness of dispersion managed solitons. We can however not rule out that dispersion managed solitons, in a properly designed system might show improved performance over conventional solitons.

5. Conclusions

In conclusions, we have numerically reassessed soliton stability for optical fibers with PMD. By varying the dispersion for given fiber parameters such as PMD, coupling length and pulse width it is possible to optimize the

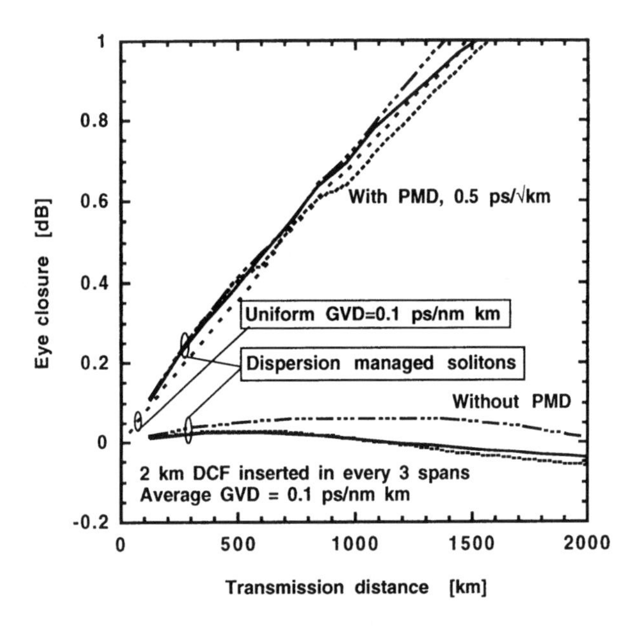

Figure 5. The eye closure of a dispersion managed system with compensting fibers every third amplifier span, with and without PMD. The solid, dotted and dashed-dotted curves correspond to power enhancement factors $N^2 = 2$, 2.5 and 3, respectively. The long-dotted curve in the PMD-case is a conventional soliton having uniform GVD and optimised $N^2 = 1.21$. The mode coupling length was 0.5 km.

dispersion and the soliton order so that a propagation length of roughly twice that of a linear PMD-broadened pulse induces the same broadening. A means of increasing this value further is to use some form of soliton transmission control, e.g. sliding filters, that can remove the dispersive waves and effectively eliminate the pulse broadening. This was recently exemplified in reference [9], but we have shown that by optimizing the dispersion and soliton amplitude the performance can be improved. We wish to emphasize the importance of these issues, since *controlled solitons offer at present the only known way of combatting large amounts of PMD.* When considering dispersion managed solitons we have found no major differences as compared to conventional solitons, but further research is necessary to make definite conclusions.

Acknowledgements

This work was in part financially supported by the Swedish Research Council for Engineering Sciences (TFR), the Swedish Board for Industrial and Technical Development (NUTEK) and the ACTS project Multi-gigabit Interconnection using Dispersion compensation and Advanced Soliton techniques (MIDAS).

References

1. E. Brun-Maunnand, J. Hamaide and E. Desurvire: Recent progress on soliton systems, and references therein, *22nd European Conference on Optical Communication*, Oslo, Sept.15-19, (1996), vol.3, pp.93-100.
2. M. C. de Lignie, H. G. J. Nagel, and M. O. van Deventer: Large polarization mode dispersion in fiber optic cables, *IEEE J. Lightwave Technol.*, **12**, (1994), pp.1325-1329.
3. A. Galtarossa, G. Gianello, C. G Someda and M. Schiano: In-field comparison among polarization-mode-dispersion measurement techniques, *IEEE J. Lightwave Technol.*, **14**, (1996), pp.42-48.
4. N. Gisin, R. Passy and J. P. Von der Weid: Definition and measurement of polarization mode dispersion: interferometric versus fixed analyzer methods, *IEEE Phot. Technol. Lett.*, **6**, (1994), pp.730-732.
5. L. F. Mollenauer, K. Smith, J. P. Gordon, and C. R. Menyuk: Resistance of solitons to the effects of polarization mode dispersion in optical fibers, *Optics Letters*, **14**, (1989), pp.1219-1221.
6. P. K. A Wai, C. R. Menyuk, and H. H. Chen: Stability of solitons in randomly varying birefringent fibers, *Optics Letters*, **16**, (1991), pp.1231-1233.
7. S. G. Evangelides Jr., L. F. Mollenauer, J. P. Gordon, and N. S. Bergano: Polarization multiplexing with solitons, *IEEE J. Lightwave Technol.*, **10**, (1992), pp.28-35.
8. F. Matera, and M. Settembre: Compensation of polarization mode dispersion by means of the Kerr effect for nonreturn-to-zero signals, *Optics Letters*, **20**, (1995), pp.28-30.
9. M. Matsumoto, Y. Akagi, and A. Hasegawa: Propagation of solitons in fibers with randomly varying birefringence; Effects of soliton transmission control, *IEEE J. Lightwave Technol.*, **15**, (1997), pp.584-589.
10. V.E. Zakharov and A.B. Shabat: Exact theory of two-dimensional self-focusing and one-dimensional self-modulation of waves in noninear media, *Sov. Phys. JETP*, **34**, (1972), pp.62-69.
11. S.V. Manakov: On the theory of two-dimensional stationary self-focusing of electromagnetic waves, *Sov. Phys. JETP*, **38**, (1974), pp. 248-253.
12. H. Kubota and M. Nakasawa: Soliton transmission control in time and frequeny domains, *IEEE J. Quantum Electron.*, **29**, (1993), pp.2189-2196.
13. N. J. Smith and N. J. Doran: Evaluating the capacity of phase-modulator controlled long-haul soliton transmission, *Optical Fiber Technology*, **1**, (1995), pp.218-235.
14. L. F. Mollenauer, J. P. Gordon, and S. G. Evangelides: The sliding-frequency guiding filter: an improved form of soliton jitter control, *Opt. Lett.*, **17**, (1992), pp.1575-1577.
15. Y. Kodama, M. Romagnoli and S. Wabnitz: Soliton stability and interactions in fibre lasers, *Eletron. Lett.*, **28**, (1992), pp.1981-1983.
16. X. Zhang, M. Karlsson, P. Andrekson, and K. Bertilsson: Soliton stability in optical fibers with polarization mode dispersion, to appear in *IEEE Phot. Technol. Lett.*
17. A. Papoulis, ch. 9-2 in ***Probability, random variables and stochastic processes***, 3rd ed., McGraw-Hill, New York (1991).
18. Actually, when using the model of randomly rotated birefringent fiber pieces every L_c, then $< \mathrm{PMD}^2 >= \Delta\beta_0'^2 \mathrm{L_c}$, but since the DGD is Maxwellian, $< \mathrm{PMD} >= \sqrt{8/(3\pi)}\Delta\beta_0'\sqrt{\mathrm{L_c}}$. For details, see C. D. Poole and D. Favin: Polarization mode dispersion measurements based on transmission spectra through a polarizer, *IEEE J. Lightwave Technol.*, **12**, (1994), pp.917-929.
19. P. K. A. Wai, and C. Menyuk: Polarization mode dispersion, decorrelation, and diffusion in optical fibers with randomly varying birefringence, *IEEE J. of Lightwave Technol.*, **14**, (1996), pp.148-157.
20. P. Wai, W. Kath, C. Menyuk and J. Zhang: Analysis of nonlinear polarization mode dispersion in optical fibers with randomly varying birefringence, pp.257-258, *Conference on Optical Fiber Communication*, Feb.16-21 (1997), Dallas, USA.

21. J. P. Gordon: Dispersive perturbations of solitons of the nonlinear Schrödinger equation, *J. Opt. Soc. Am. B*, **9**, (1992), pp.91-97.
22. M. Nakasawa, H. Kubota, A. Sahara, and K. Tamura: Marked increse in the power margin through the use of dispersion managed solitons, *IEEE Phot. Technol. Lett.*, **8**, (1996), pp.1088-1090.
23. N. Smith, F. Knox, N. Doran, K. Blow and I. Bennion: Enhanced power solitons in optical fibers with periodic dispersion management, *Electronics Lett.*, **32**, (1996), pp.54-55.
24. A. Hasegawa, Y. Kodama, and A. Maruta: Recent progress in dispersion managed soliton transmission technologies, *Optical Fiber Technology*, **3**, (1997), pp.197-213.
25. X. Zhang and P. Andrekson: Characteristics of power enhancement in dispersion managed soliton systems, *Optical Fiber Technology* (in press).

(Received December 25, 1997)

ROLE OF THE DISPERSIVE WAVE IN SOLITON DYNAMICS AND INTERACTIONS

M. ROMAGNOLI, L. SOCCI
Fondazione Ugo Bordoni,
via B. Castiglione, 59, 00142 Roma, Italy

I. CRISTIANI
Dipartimento di Elettronica, Universita di Pavia,
via Ferrata 1, 27100 Pavia, Italy

AND

P. FRANCO
Pirelli Cavi s.p.a., viale Sarca 222, 20146, Milano, Italy

Abstract. We describe the interplay between solitons and their continuum. The long-range soliton interactions caused by the dispersed continuum arises from a cross-phase modulation mechanism and could severely hamper soliton transmission. In this work we highlight frequency chirping as major responsible of continuum generation and interaction with the soliton stream in the transmission channel.

1. Introduction

Average soliton propagation has been exploited to overcome the periodical energy variation occurring in transmission systems and in fiber lasers [1]. The concept of average soliton is valid whenever the typical length scale over which a soliton evolves is considerably longer than the period of the energy variation. This approach is very useful because permits to keep small the coherent continuum generation, even though this imposes a limit to the minimum pulsewidth supported by the system [2]. In fact for a given value of group-velocity dispersion (GVD) shorter is the pulse larger is the amount of radiation emitted upon propagation, in particular whenever the pulse dispersion length is comparable or shorter than the periodicity of the energy variation, the self-generated continuum causes a strong effect that

A. Hasegawa (ed.), New Trends in Optical Soliton Transmission Systems, 39–51.

affects soliton stability and interactions between neighbouring and/or distant solitons [3]. In practical terms these effects deteriorate the performance of the system. For instance in mode-locked fiber lasers the shortest achievable pulsewidth is determined by the amount of soliton energy converted into radiation, i.e. greater is the conversion more is the loss of the cavity [4]. This is tolerable to a certain extent, after that the active medium can't sustain anymore the energy requirement for the shortening of the circulating solitons. In transmission systems the amount of radiation generated through the resonant interaction with the driven-dumped system, on one hand worsens the signal to noise ratio, and on the other hand, being coherent radiation, that induces a nonlinear interaction leading to novel type of soliton-soliton interactions. In both these cases the possible solution without resorting to in-line control elements (filters, modulators, saturable absorbers, etc.) is twofold: reduction of the amplified span length or GVD compensation, that is a straightforward mean to restore the average-soliton condition. For instance in a dispersion managed systems a suitable design of the compensation map could minimize the emission of linear radiation and therefore the soliton parameter deviations [5].

The scope of this work is to go through the properties of the self- generated continuum with the aim to quantify the interplay between continuum of radiation and soliton pairs. We find that dispersed continuum generated by one soliton [6], being wide spectrum and linear, is sufficiently spread in time that can interfere with all sorrounding solitons [7]. The blue-shifted wing of the continuum spectrum travels faster than the generating soliton and interferes with previous solitons and viceversa for the red-shifted wing [8]. The effect resulting from the interference between the red/blue shifted wing with other solitons is analogous to WDM soliton collisions [9], with the consequence that, due to dispersive waves, the temporal position of solitons is not preserved upon propagation. Finally we show that this interaction is particularly effective if caused by the linear radiation emitted from an initially chirped pulse, in fact even in a few dispersion lengths long-range and direct soliton-soliton interactions are enhanced by the presence of the continuum released by the chirped soliton. The continuum induced interaction is large enough to corrupt a transmission channel.

2. Theory

Pulse propagation in an anomalously dispersive periodically amplified system is described by the nonlinear Schrödinger (NLS) equation with the inclusion of a perturbation that takes into account the map of gain and loss. The perturbed NLS equation reads

$$i\frac{\partial Q}{\partial z}+\frac{1}{2}\frac{\partial^2 Q}{\partial t^2}+|Q|^2Q=-i\Gamma Q+i(G-1)\sum_{n=1}^{N_{amp}}\delta(z-nZ_a)Q\,, \quad (1)$$

where Q is the propagating field. equation (1) is conveniently expressed in soliton units, the unit length is $\mathrm{L}_d = \mathrm{T}_0^2/|\beta_2|$ that refers to dispersion length, the group-velocity dispersion (GVD) parameter is β_2 (ps^2/km) and the arbitrary time is T_0 (ps). The fiber loss coefficient is $\Gamma = \alpha \mathrm{L}_d$, the net gain at each amplification stage is $G = Q(Z_a^+, T)/Q(Z_a^-, T)$, the amplifier spacing is Z_a (km), and finally the total number of amplification stages is $\mathrm{N}_{amp.}$ equation (1) can be reduced to a dimensionless Hamiltonian form

$$i\frac{\partial u}{\partial z}+\frac{1}{2}\frac{\partial^2 u}{\partial t^2}+|u|^2u=-A(z)|u|^2u\,, \quad (2)$$

where $Q = (A(z)+1)u$ [1] and $A(z) = a_0^2 exp[-2\Gamma(\mathrm{z}-\mathrm{nz_a})]-1$ is a real function with a_0 corresponding to the initial amplitude of the field.

If $A(z) \ll 1$ or if the periodicity of the system is shorter than the characteristic length L_d, the small perturbation leads to a small phase modulation on the propagating field $u(z)$.

Considering the interaction between two adjacent solitons, we know that in the limit of small soliton spacing, $\Delta_0 < 10$, the superposition of the soliton tails causes the interaction within a reasonable propagation distance. In the case of large separations, $\Delta_0 \gg 10$, in combination with the average soliton condition, $A(z) \sim 0$, there is no interaction [10, 11]. Nevertheless, we show, that whenever linear radiation participates to the interaction the soliton experiences a very rich dynamics.

In order to study the continuum induced soliton-soliton interaction we express the field as

$$u(z,t) = u_s + u_p = (u_{s1}+u_{s2})e^{i\theta(z)} + u_{p1} + u_{p2}\,, \quad (3)$$

where the initial values at $z = 0$ are respectively

$$\begin{aligned} u_s(0,t) &= u_{s1}(0,t)+u_{s2}(0,t) = \nu\{\mathrm{sech}[\nu(t-\xi/2)]+\mathrm{sech}[\nu(t+\xi/2)]\} \\ u_p(0,t) &= u_{p1}(0,t)+u_{p2}(0,t) = 0\,, \end{aligned} \quad (4)$$

for the soliton and for the perturbing field respectively.

The use of relations (3) and (4) inserted into equation (2) leads to a set of perturbed NLS equations that reads

$$
\begin{aligned}
&i\frac{\partial u_{si}}{\partial z}+\frac{1}{2}\frac{\partial^2 u_{si}}{\partial t^2}+|u_{si}|^2 u_{si}= \\
&A(z)(1-|u_{si}|^2)u_{si}-(1+A(z))(2|u_{si}|^2 u_p+u_{si}^2 u_p^*), \quad i=1,2\,.
\end{aligned} \tag{5}
$$

In equation (5) we have neglected terms in u_p of O(2), in fact $|u_p(z,t)|^2 \ll |u_s(z,t)|^2$.

In order to show the role of the continuum in the soliton-soliton interaction we studied the case where the pair of solitons are largely spaced ($\Delta_0 \gg 10$). Within this scheme the superposition of the tails of the soliton pair is small, of the order of $\exp(-\Delta_0)$, and therefore the direct interaction can be neglected without affecting the accuracy of the solution. On the contrary the orthogonality condition between soliton and its continuum, that strictly holds in the limit of single soliton propagation [12], in the presence of a second soliton is no more valid. This is because the linear superposition of the dispersive waves produced by the pair is a field that with reference to the central position of each soliton is temporally asymmetric. We are going to show below that the lack of orthogonality renders the interaction of the soliton with its continuum very effective and therefore whereas we can neglect the effect of one of the two solitons we no more can do the same for its continuum.

Soliton evolution in the presence of linear radiation resulting from the linear superposition of the continuum generated by itself and that from another soliton spaced by Δ_0 is studied by means of perturbation theory. This approach yields the evaluation of the evolution of the whole set of soliton parameters along the propagation coordinate. The expression for the perturbing fields has been found by Gordon [12] by linearizing the one soliton solution of the NLS equation. This result is conveniently expressed by resorting to an associate field $f(z,t)$

$$
u_p=-\frac{d^2 f}{dt^2}+2\gamma\frac{df}{dt}-\gamma^2 f+u_s^2 f^*\,, \tag{6}
$$

where $\gamma=-\left(\frac{1}{u_s}\frac{\partial u_s}{\partial t}\right)^*$ and the associate field itself satisfies

$$
i\frac{\partial f}{\partial z}+\frac{1}{2}\frac{\partial^2 f}{\partial t^2}=A(z)u_s\,. \tag{7}
$$

Following the Gordon's procedure we find that in a periodically amplified system the far field solution of equation (7) is the superposition of a dispersive u_{pd}, and a resonant u_{pr}, part. The resonant part is responsible for the coupling of the soliton energy to the continuum. With the initial condition $f(0,t)=u_p(0,t)=0$, solution of equation (7) reads

$$u_{pr} = i\frac{\pi}{2}\sum_k k\frac{A_k^*}{\omega_d}\operatorname{sech}\left(\frac{\pi}{2}\omega_d\right) e^{i\left[\omega_d|t| - \frac{\omega_d^2}{2} + 2\tan^{-1}\left(\frac{1}{\omega_d}\right)\right]}, \qquad (8)$$

where the disperive wave frequency is $\omega_d = \sqrt{4\pi N/z_a - 1}$ and N is integer.

Once the analytic expression for u_p is known we are able to compute the soliton parameter evolution by means of the perturbation theory [11]. The problem to be solved is that of equation (5) where the perturbation is that reported on the right-hand side. Each soliton is expressed as $u_{si}(z,t) = \nu i\operatorname{sech}(\zeta_i)e^{i\varphi i}$ and the four soliton parameters are $\zeta_i = \nu_i(t - \xi i), \varphi_i = \mu_i\xi_i/\nu_i, \xi_i = \mu_i z + \Delta_{0i}/2, \delta_i = (\mu_i^2 + \nu_i^2) + \delta_{0i}, i = 1, 2$. Nevertheless since we intend to describe the interaction of a pair solitons it is convenient to combine the parameters of the pair, in order to do this we take into account the mirror symmetry of the problem, i.e. $[\nu_1(z), \mu_1(z), \xi_1(z), \varphi_1(z)] = [\nu_2(z), -\mu_2(z), -\xi_2(z), \varphi_2(z)]$. The relevant parameters for the description of the continuum induced soliton-soliton interaction are therefore the average amplitude $\eta(z) = \{\nu_1(z) + \nu_2(z)\}/2 = \nu_1(z)$, the frequency difference $q(z) = \mu_1(z) - \mu_2(z) = 2\mu_1(z)$, and the separation of the pair $\Delta(z) = \xi_1(z) - \xi_2(z) = 2\xi_1(z)$. The evolution o.d.e. for the soliton pair parameters read

$$\begin{aligned}
\frac{d\eta}{dz} &= -\eta^3\int_{-\infty}^{\infty} dt\,\operatorname{sech}(\zeta)^3|u_p|\sin\{\lambda(z,t)\} \\
\frac{dq}{dz} &= 6\eta^3\int_{-\infty}^{\infty} dt\,\operatorname{sech}(\zeta)^3\tanh(\zeta)|u_p|\cos\{\lambda(z,t)\} \\
\frac{d\Delta}{dz} &= 2q - 2\eta\int_{-\infty}^{\infty} dt\,\zeta\operatorname{sech}(\zeta)^3|u_p|\sin\{\lambda(z,t)\},
\end{aligned} \qquad (9)$$

with $\lambda = \varphi_s - \varphi_p$ is the phase difference between soliton and perturbing wave.

3. Numerical Solutions and Discussion

The set of equations (9) has been numerically solved togheter with equations (6) and (7) for the perturbation. Equations (9) were solved by resorting to a fourth-order Runge-Kutta routine. Numerical solution of equation (7) was obtained by means of a split-step spectral routine. For comparison with the perturbation theory we numerically solved equation (1).

The first problem that we dealed with is the propagation of a pair of solitons spaced apart by Δ_0 , in Figure 1 we report the collision distance as a function of the initial separation for the case of periodically amplified system with a fiber loss of 0.2 dB/km, amplifier spacing $Z_a = 50$ km and a

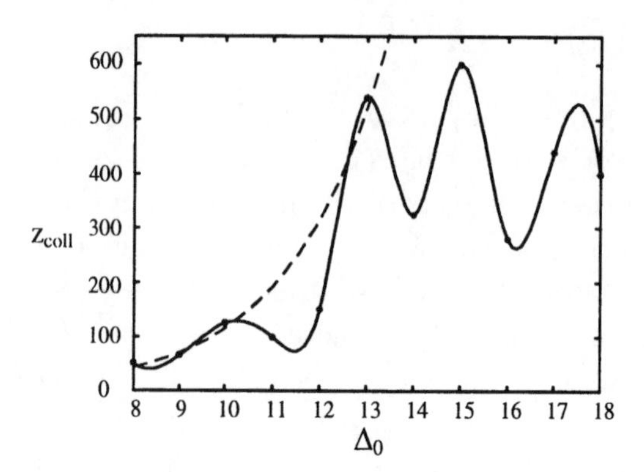

Figure 1. Collision distance of a soliton pair as a function of initial separation Δ_0. Solid line correponds to a quasi resonant condition, $z_a = 1$ ($Z_a = 50$ km, $\beta_2 = -20$ ps^2/km, $T_0 = 32$ ps) and attenuation $\Gamma = 10$ dB ($\alpha = 0.2$ dB/km), whereas the dashed line refers to soliton-soliton interaction in the lossless case (no continum) [10]

pulsewidth such that $z_a = 1$ ($L_d = 50$ km). The solid curve with circles is obtained from the numerical solution of equation (1) whereas for comparison the dashed curve refers to soliton-soliton interaction in the lossless case [10, 11]. The effect of the fiber loss is that of inducing the generation of a continuum of radiation [3, 12]. The continuum is linearly dispersed all along the fiber and collides with the neighboring solitons. We show for istance in Figure 2 the soliton parameter evolution, computed from solution of the set of equations (11), that shows an overall sinusoidal behaviour due to the interaction with the linear radiation released from an in-phase soliton spaced apart by $\Delta_0 = 18$.

To obtain the results of Figure 2 we solved a problem with amplifier spacing $Z_a = 10$ km, and an overall attenuation for amplified span of 2 dB, this was done to keep the perturbation small. In Figure 2 we show the mean amplitude $\eta(z)$, the frequency difference $q(z)$ and the soliton separation $\Delta(z)$. To verify the accuracy of the perturbative approach we show in Figure 2 a), c) the evolution of position and amplitude computed from equation (1) confirming a good agreement. The result in Figure 2 describes an attractive interaction as it is expected for in-phase solitons. As shown in Figure 2 b), c), at the beginning of the propagation, say for $z < 5$, the frequency difference and soliton separation are constant even though the dispersive wave has been already generated, see the interference of the soliton with its own dispersive wave in Figure 2 a). Conversely for $z > 5$ the dispersive waves generated by the pair begin to cross each other and the solitons as well. In this condition the orthogonality of the soliton with

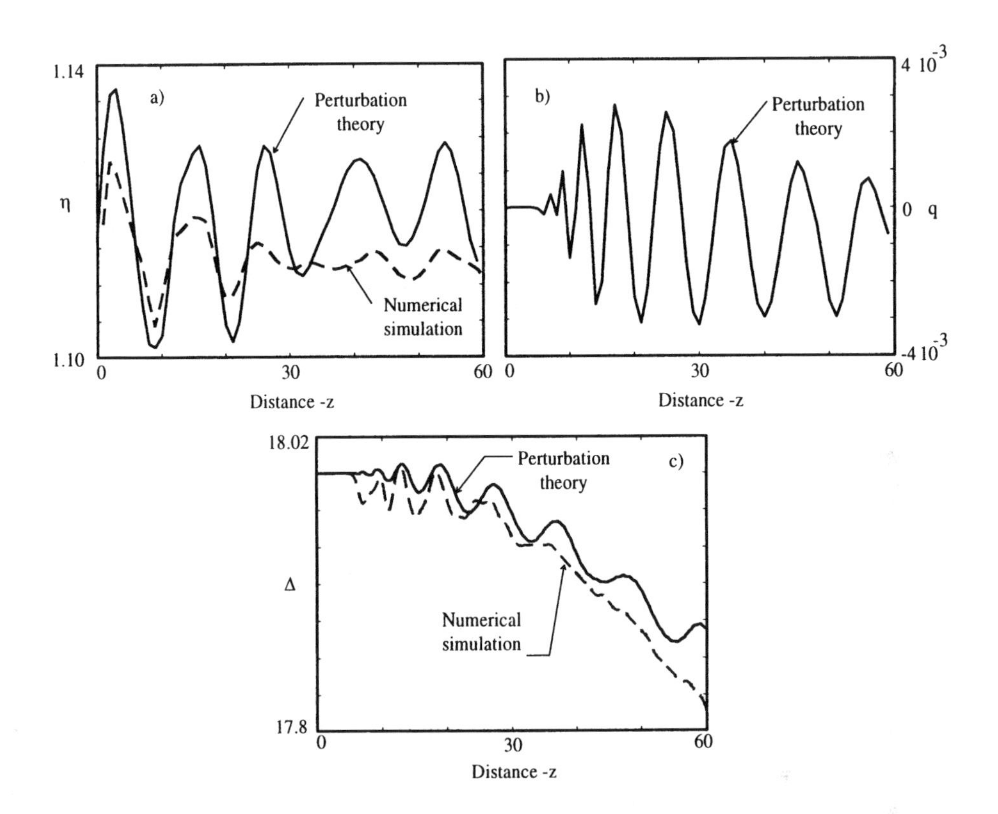

Figure 2. Evolution of the parameters of an attracting soliton pair initially separated by $\Delta_0 = 18$ (576 ps): a) mean amplitude η, b) frequency difference q, c) separation Δ. Perturbative approach (solid lines), direct solution of equation (1) (dashed line). The amplifier spacing was $z_a = 1$ (Z_a =10 km, $\beta_2 = -20$ ps^2/km, $T_0 = 32$ ps), and the fiber attenuation $\Gamma = 2$ dB ($\alpha = 0.2$ dB/km).

its own dispersive wave breaks up and both frequency and position begin to evolve. The sinusoidal evolution of these parameters is related to the temporal profile of the dispersive field u_p.

In Figure 3 we report a second interesting example of dispersive wave mediated soliton-soliton interaction, the initial conditions are the same as those of Figure 2, the solitons are in-phase with initial separation $\Delta_0 = 25$. Whereas for direct soliton-soliton interaction we should observe an attraction that ends up into a collision, here the solitons experience a repulsive interaction. This is simply because at $\Delta_0 = 25$ the interaction is only due to the dispersive fields. In Figure 3 b) we show that the soliton interaction with only the dispersive field generated from the adjacent soliton is incorrect. In fact the evolution of the soliton separation, $\Delta = \Delta(z)$, computed by

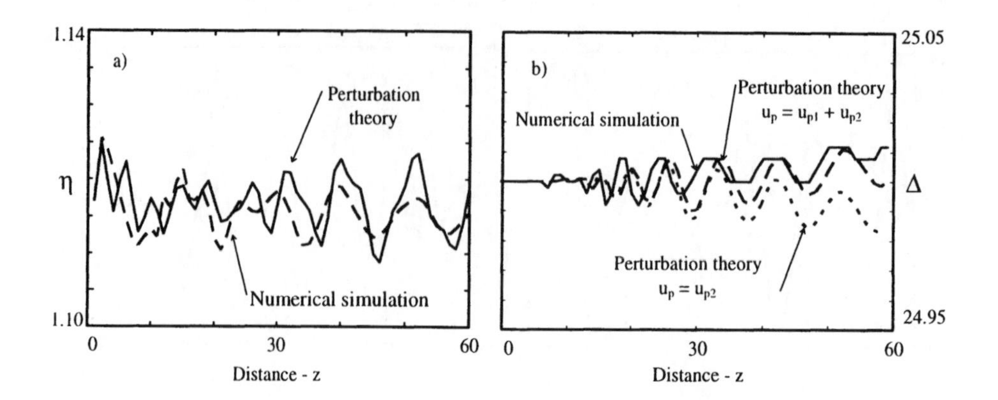

Figure 3. Same as in the previous figure but $\Delta_0 = 25$ (800 ps). The amplifier spacing was $z_a = 1$ (Z_a =10 km, $\beta_2 = -20$ ps^2/km, $T_0 = 32$ ps), and the fiber attenuation $\Gamma = 2$ dB ($\alpha = 0.2$ dB/km). Here the initially in-phase solitons repel each other. The perturbative approach fits the solution of equation (1) for $u_p = u_{p1} + u_{p2}$. This demonstrates the failure of the orthogonality between soliton and its own continuum whenever a second soliton is involved in the interaction.

direct solution of equation (1) is fitted by the perturbation theory whenever $u_p = u_{p1} + u_{p2}$. On the contrary the simple interaction of each soliton with only the dispersive field associated to the adjacent soliton, $u_{s1}(u_{s2})$ with $u_{p2}(u_{p1})$, cannot even fit the sign of the interaction, thus confirming the break-up of the soliton-continuum orthogonality that occurs when dealing with more than a single soliton.

The physical mechanism that produces the oscillations of the soliton parameters upon propagation may be understood by an inspection of Figure 4. In this figure we report the evolution of both amplitude and separation of a soliton pair spaced by $\Delta_0 = 16$ propagating in a line with amplification period $z_a = 4$ and an overall attenuation for amplified span of 10 dB. The reason of the strong perturbation that we used is to show more apparently the effect of the interaction of the soliton with the continuum. In this example we solved numerically equation (1). The solid lines indicate the evolution of the soliton parameters whereas the circles the location of the lumped amplifiers. Figure 4 a) shows the evolution of the amplitude.

Figure 4 b) shows an overall oscillation around the initial separation Δ_0 = 16. From $z = 0$ to 50 the average slope of Δ (z) is constant. From $z = 50$ to 100 the slope becomes negative, here the interaction is attractive, then at about $z = 100$ we observe a second net change of slope corresponding to an increase of attraction. The observed changes of slope originate from cross-phase modulation (XPM) between continuum and soliton.

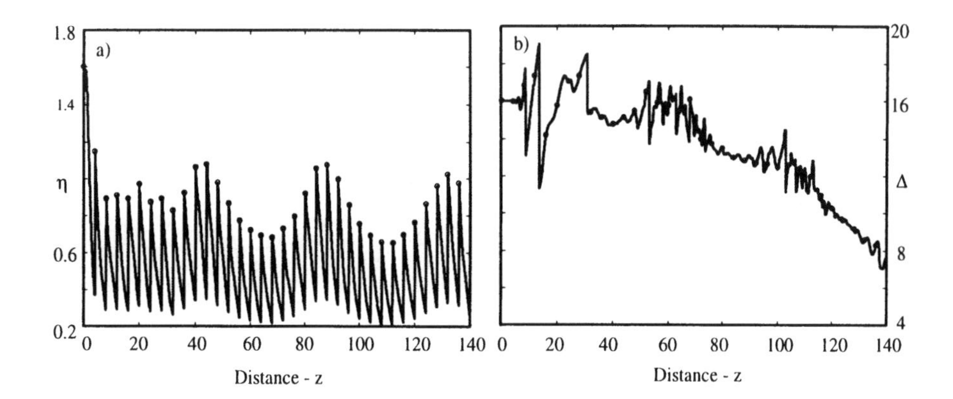

Figure 4. Mean amplitude η and separation Δ, for a propagating soliton pair spaced by $\Delta_0 = 16$ (256 ps) in a system with amplifier spacing $z_a = 4$ ($Z_a = 50$ km, $\beta_2 = -20$ ps^2/km, $T_0 = 16$ ps) and attenuation $\Gamma = 10$ dB ($\alpha = 0.2$ dB/km). The coherent superposition of a soliton with its own continuum and that released from an adjacent soliton may lead to soliton net frequency shift whenever the interaction occurs in correspondence of an amplification stage as for WDM soliton collisions [9]

The linear radiation emitted by the solitons splits in two spectral wings, the blue-shifted wing travels faster than the soliton and viceversa the red-shifted wing. The interaction of the adjacent soliton occurs therefore with only one of the two wings, analogously to the collision of a pair of solitons with different center-wavelength [9]. This type of collision can be either elastic or unelastic as for WDM soliton collisions. In a lossless transmission line the XPM experienced by the leading edge of the soliton leads to a net frequency shift of the carrier such to increase the frequency difference between the soliton and the perturbing field, and viceversa during the interaction with the trailing edge of the soliton. Once the collision is over the soliton center frequency is restored to the initial value while the position of the soliton is shifted by a finite value. Conversely in those cases where the collision occurs when the soliton goes through an amplifier some energy is lost, and the soliton emerges from the collision with a net frequency shift that produces a continuous temporal drift.

The unelastic nature of the collision over the amplifier is easily explained in terms of XPM efficiency, in fact the soliton power before and after the amplifier changes by a factor corresponding to the net gain of the amplifier.

This power difference determines a post-amplification frequency shift that overcompensates the shift experienced by the soliton just before amplification.

In Figure 4 b) we observe that the oscillation of the soliton separation

Δ occurs whenever the dispersive wave crosses the soliton far from the amplification stages whereas we have the change of slope every time the collision occurs in coincidence with an amplifier. This explanation improves the comprehension of the collision distance shown in Figure (1), in fact the direct interaction between two solitons [10] disappears as soon as the soliton separation grows larger than $\Delta \sim 10$, on the contrary the presence of dispersive radiation is such to induce the interaction even for widely spaced solitons.

The strength of the interaction depends on the power of the dispersive wave, in normal system designs the pulsewidth is such to fulfill the average soliton condition and the continuum can be neglected. Nevertheless the amount of linear radiation can even be generated by a chirped soliton. Frequency chirp is a typical problem arising from the use of non transform-limited soliton sources, for instance gain switched diode lasers or electro-absorption modulators [13]. In this case the amount of radiation emitted by the soliton is such to produce an important effect even in the framework of the average soliton condition.

To study the effect of the energy released by an initially chirped soliton we solved equation (1) with the initial condition $u_s(0,t) = \nu\{\text{sech}[\nu(t - \Delta_0/2)] + \text{sech}[\nu(t+\Delta_0/2)]\}\exp(-iCt^2)$, note the inclusion of a linear chirp factor whose parameter is C = ± 0.3. We will show that with this chirp factor the time-bandwidth product slightly increases by 4.4 % that is sufficient to contribute to a non-negligible transmission degradation. The transmission line in the example was made of 50 km long amplified spans. The pulsewidth we used was such to have a dimensionless amplifier spacing z_a = 1. The overall span loss was 12.5 dB but we neglected it in one case to gain insight on the pure effect due to the emission of radiation from the chirped soliton.

The case with inclusion of the attenuation was considered to understand the situations including the radiation generated by the lumped amplification. The initial separation of the soliton pair was $\Delta_0 = 20$. The result of the simulation for C = 0.3 are reported in Figure 5 where we have the evolution along z of either the soliton intensity $|Q|^2$, and separation Δ. Both in the case with (dashed curve) or without loss (solid curve) the amplitude variation, reported in Figure 5 a), shows an initial pulse compression (amplitude increase) due to the positive chirp [14] that is followed by pulse broadening (amplitude decrease) corresponding to the dispersive energy release and then by other expected intensity oscillations [14]. The separation of the pair, reported in Figure 5 b), starts an oscillation at z = 4 and increases with distance. The maximum excursion of oscillation is of the order of ±7.5 % of Δ_0 that is sufficiently large especially if referred to a transmission system where the bit separation reasonably could be $\Delta_{bit} = 10$ or

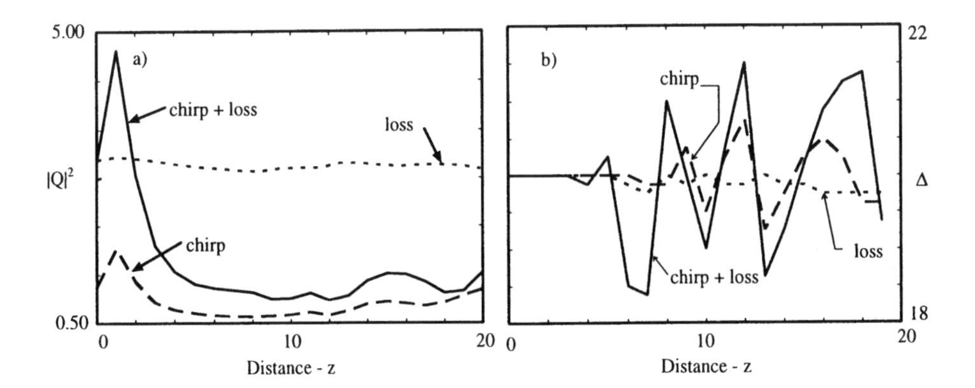

Figure 5. Intensity $|Q|^2$, and separation Δ, evolution for a chirped soliton pair spaced by $\Delta_0 = 20$ (640 ps) in a system with amplifier spacing $z_a = 1$ ($Z_a = 50$ km, $\beta_2 = -20$ ps^2/km, $T_0 = 32$ ps) and attenuation $\Gamma = 12.5$ dB ($\alpha = 0.25$ dB/km). The chirp parameter was C = 0.3.

less. In this case the separation excursion is as large as ± 15 % of Δ_{bit} and this could be a relevant change. Concerning negative chirp values, C = $-$ 0.3 for comparison, we observed a smaller effect either for the amplitude and for the pair separation.

The effect of the dispersive wave that we observed for long-range soliton-soliton interaction is effective for direct soliton-soliton interactions too. In Figure 6 we report the results of simulations carried out in the case of $\Delta_0 =$ 10, span attenuation $\Gamma = 12.5$ dB and frequency chirp C = 0.2. Concerning the intensity evolution we can observe in Figure 6 a) that the positive chirp yields an initial compression followed by broadening as in the case described in Figure 5.

Concerning the soliton separation we verified that without loss and chirp the solitons do not interact within the distance we are considering. Nevertheless the inclusion of the attenuation causes emission of continuum and a beginning of repulsion is observed (dotted curve), with the inclusion of the sole frequency chirp the interaction becomes fairly large and the separation Δ oscillates (dashed curve), finally, in the more realistic condition where both chirp and loss are considered, the soliton-soliton interaction is the largest (solid curve). Notwithstanding the relatively small value of C = 0.2, the maximum excursion of separation is $\Delta_{max} = 14$ % of Δ_{bit} that is as large as that of Figure 5 assuming a time slot coincident with the initial separation $\Delta_0 = 10$.

In conclusion we investigated the role of the continuum generated by solitons upon propagation. The interplay of the soliton with its own radia-

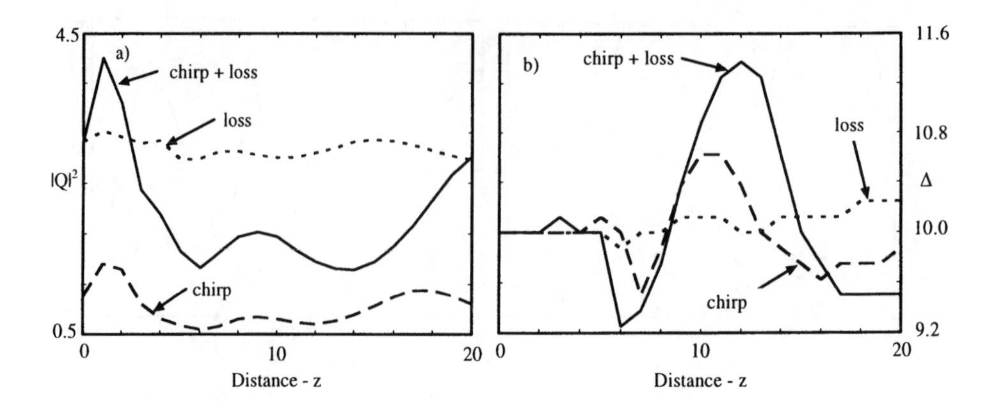

Figure 6. Intensity $|Q|^2$ and separation Δ, evolution for a chirped soliton pair spaced by $\Delta_0 = 10$ (320 ps) in a system with amplifier spacing $z_a = 1$ ($Z_a = 50$ km, $\beta_2 = -20$ ps^2/km, $T_0 = 32$ ps) and attenuation $\Gamma = 12.5$ dB ($\alpha = 0.25$ dB/km). The chirp parameter was C = 0.2.

tion is null as long as a single soliton is considered, this interplay becomes relevant when more than one soliton propagates. The interaction mechanism is analogous to soliton collisions in WDM systems and its efficiency is roughly independent of the relative separation of solitons. In practical conditions, we found that the energy released by weakly chirped solitons is sufficient to contribute to transmission degradation.

Acknowledgements

M. Romagnoli and L. Socci have carried out this work in the framework of the agreement between Fondazione 'Ugo Bordoni'and the Italian Post and Telecommunication Administration. This work was also supported by the European Community program ACTS/ESTHER.

References

1. Hasegawa, A. and Kodama, Y.: Guiding-center soliton in optical fibres, *Opt. Lett.*, **15**, (1990), pp.1443-1445.
2. Charbonnier, B. and Georges, T.: Numerical investigation of the path-averaged soliton validity domain without soliton control, *Electron. Lett.*, **32**, (1996), pp.126-127.
3. Midrio, M., Romagnoli, M., Wabnitz, S. and Franco, P.: Relaxation of guiding center solitons in optical fibers, *Opt. Lett.*, **21**, (1996), pp.1351-1353.
4. Kärtner, F. X., Kopf, D. and Keller, U.: Solitary-pulse stabilization and shortening in actively mode-locked lasers, *J. Opt. Soc. Am.*, **B 12**, (1995), pp.486-496.
5. Georges, T.: Extended path-averaged soliton regime in highly dispersive fibers, *Opt. Lett.*, **22**, (1997), pp.679-681.

6. Kuznetsov, E. A., Mikhailov, A. V. and Shimokhin, I. A.: Nonlinear interaction of solitons and radiation, *Phisica*, **D 87**, (1995), pp.201-215.
7. Loh, W. H., Grudinin, A. B., Afanasjev, V. V. and Payne, D. N.: Soliton interaction in the presence of a weak nonsoliton component, *Opt. Lett.*, **19**, (1994), pp.698-701.
8. Haus, H. A., Khatri, F. I., Wong, W. S., Ippen, E. P. and Tamura, K. R.: Interaction of soliton with sinusoidal wavepacket, *IEEE J. Quant. Electron.*, **32**, (1996), pp.917-924.
9. Mollenauer, L. F., Evangelides, S. G. and Gordon, J. P.: Wavelength division multiplexing with solitons in ultra-long distance transmission using lumped amplifiers, *IEEE J.Quant. Electron.*, **9**, (1991), pp.362-367.
10. Gordon, J. P.: Interaction forces among solitons in optical fibers, *Opt. Lett.*, **8**, (1983), pp.596-598,
Kodama, Y. and Nozaki, K.: Soliton interaction in optical fibres. *Opt. Lett.*, **12**, (1987), pp.1038-1040.
11. Karpman, V. I., Solov'ev, V. V.: A perturbational approach to the two soliton systems, *Phisica*, **3D**, (1991), pp.487-502.
12. Gordon, J. P.: Dispersive perturbation of solitons of the nonlinear Schrodinger equation, *J. Opt. Soc. Am.*, **B 9**, (1991), pp.91-97.
13. Lazaridis, P. and Debarge, G.: Time-bandwidth product of chirped sech^2 pulses: application to phase-amplitude coupling factor measurement, *Opt.Lett.*, **20**, (1995), pp.1160-1162.
14. Agrawal, G. P.: *Nonlinear Fiber Optics*, Academic Press, New York.

(Received November 17, 1997)

GENERATION OF HIGH-REPETITION-RATE DARK SOLITON TRAINS AND FREQUENCY CONVERSION IN OPTICAL FIBERS

G. MILLOT, S. PITOIS, E. SEVE, P. TCHOFO-DINDA,
AND S. WABNITZ
Laboratoire de Physique, Université de Bourgogne
B.P. 400, 21011 Dijon, France

S. TRILLO
Fondazione Ugo Bordoni
via B. Castiglione 59, 00142 Rome, Italy

M. HAELTERMAN
Université Libre de Bruxelles
50, Av. Roosevelt, CP 194/5, 1050 Bruxelles, Belgium

AND

J. M. SOTO CRESPO
Instituto de Optica
Consejo Superior de Investigaciones Cientificas
Serrano 121, 28006 Madrid, Spain

Abstract. Induced modurational polarization instability in birefringent fibers leads to trains of dark soliton-like pulses. Optimal large-signal cw and soliton frequency conversion is also analysed.

1. Introduction

A well-known mechanism of pattern formation in nonlinear physics is the modulational instability (MI) [1] of wave trains against low-frequency perturbations. MI leads to self-filamentation [2] or modulation [3] of light beams in optical dielectrics. MI may be induced into an optical fiber by mixing two lasers: in the anomalous group-velocity dispersion (GVD) regime, trains of solitons with THz repetition rates were generated [4]. In the normal GVD regime, MI may also be induced by cross-phase modulation

A. Hasegawa (ed.), New Trends in Optical Soliton Transmission Systems, 53–67.

(CPM) between two copropagating pumps [5, 6, 7]. In low-birefringence fibers, instability to low-frequency polarization modulations, contrary to the Benjamin-Feir instability, only occurs for beam powers above a certain threshold value [8, 9]. In this work we present the experimental observation of striking light-activated polarisation symmetry-breaking by inducing the MI with a weak signal beam [10].

As it occurs in the anomalous GVD regime, the CPM-induced MI in the normal GVD regime may lead to time and space periodic sequences of pulses [11]. For polarization MI, the link between the instability and the stationary vector solitary wave structures of the birefringent fiber [12] was also investigated [13]. We demonstrate experimentally here that polarization MI in the normal GVD regime may lead to high-repetition-rate dark soliton-like pulse trains.

Stimulated Raman scattering (SRS) in fibers may compete and interact with the parametric scattering. Raman gain may be exploited for periodic signal amplification in long-distance fiber links. On the other hand, SRS sets a strict limitation to the power budget in unrepeatered fiber transmissions. One may achieve suppression of SRS in optical fibers by favouring parametric four-wave mixing (FWM): in high-birefringence fibers, this occurs through a mixed-mode pump configuration [14, 15, 16]. We have studied the suppression of the Raman Stokes radiation in a selective and controlled manner in any of the axes of a birefringent fiber: as we shall see, this is achieved by changing the sign and the magnitude of the group-velocity mismatch (GVM) between the two pumps, by simply tuning the wavelength separation of two orthogonal pumps [17].

FWM in optical fibres is an important nonlinear limiting factor in lightwave telecommunications, and it may be exploited for the compensation of link distortions through mid-span spectral inversion. For a degenerate interaction (i.e, signal at ω_s mixing with two pump photons at ω_p and an idler photon at $\omega_i = 2\omega_p - \omega_s$), it is commonly believed that efficient interactions require the phase-matching condition $k_i + k_s = 2k_p$. We shall demonstrate here that, in a birefringent fiber, the signal power level has a key role in determining the four-wave mixing efficiency. As a matter of fact, a large increase of the frequency conversion efficiency occurs for signal powers above a given threshold value (typically a few percents of the pump power). This effect only takes place for signal detunings outside the small-signal parametric gain bandwidth. The measures are well reproduced by the solutions of the nonlinear three-mode equations [18].

Optical phase conjugation undoes optical signal distortions upon propagation in dispersive and nonlinear medium [19]. By generating the conjugate replica of a signal at the middle of fiber-optic transmission link, dispersive and nonlinear broadenings may be compensated for reference [20].

When applied to soliton systems, mid-span spectral inversion also removes other important transmission capacity-limiting effects such as pulse interactions [21]. Four-wave mixing in fibers may also be exploited for wavelength-shifting and time-demultiplexing of high-bit-rate signals. For efficient and broadband frequency conversion, a pump near the zero-dispersion wavelength of a fiber is required. As a result, the fiber GVD has opposite signs at the signal and conjugate wavelengths. In the conjugation of ultrashort solitons, the GVD sets a fundamental limit to the conversion efficiency. We shall consider here the parametric coupling of a cw pump with short signal and conjugate pulses [22], in order to search for stable two-component solitary waves that permit the propagation of a soliton-like pulse at the conjugate frequency, thanks to the parametric coupling with a pulse at the signal frequency inside a laser cavity. The dissipative and parametrically coupled waves may be generated for any signal GVD sign, and correspond to the optimal frequency conversion of ultrashort pulses.

2. Modulational Instability in Normal Dispersion

In the presence of a nonlinear third-order polarization, Maxwell's equations lead to two coupled nonlinear Schrödinger (NLS) equations for the circular polarization components u and v of the field [7]

$$
\begin{aligned}
i\frac{\partial u}{\partial \xi} &= \beta\frac{\partial^2 u}{\partial \tau^2} + \frac{\Delta}{2}v - (|u|^2 + 2|v|^2)u\,, \\
i\frac{\partial v}{\partial \xi} &= \beta\frac{\partial^2 v}{\partial \tau^2} + \frac{\Delta}{2}u - (|v|^2 + 2|u|^2)v\,.
\end{aligned}
\tag{1}
$$

Here $\beta = \pm 1/2$ for normal or anomalous GVD, and Δ is the linear birefringence. The stability analysis of equations (1) with a linearly polarized pump along a fiber axis shows that orthogonal modulations grow for $0 < \mu(\Delta, \Omega)/P_p < 2/3$, where $\mu \equiv \beta\Omega^2 + \Delta$ and $P_p \equiv |u|^2 + |v|^2$ is the pump power. The signature of this instability is the exponential growth (with a spatial rate, or intensity gain, G) of optical intensity into the sidemodes. In the normal GVD regime, a pump beam may be modulationally unstable with either $\Delta < 0$ (slow mode) or $\Delta > 0$ (fast mode). In the last case, the pump power P_p must be larger than a threshold value $P_t \equiv 3\Delta/(2R)$, where $R = 2\pi n_2/(\lambda_p A_{eff})$, λ_p is the pump wavelength, n_2 is the nonlinear refractive index and A_{eff} is the effective area of the fiber mode. Figure 1 illustrates the frequency dependence of G for $P_p = 112$ W: solid and dashed curves are obtained for a pump aligned with the slow or the fast axis of the

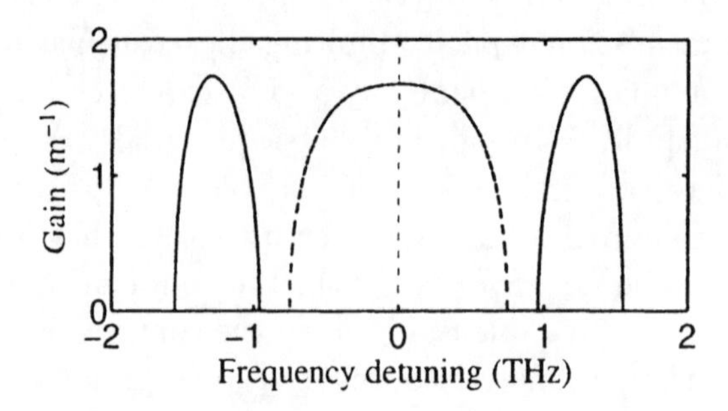

Figure 1. MI gain vs. signal detuning: solid (dashed) curve for slow (fast) pump.

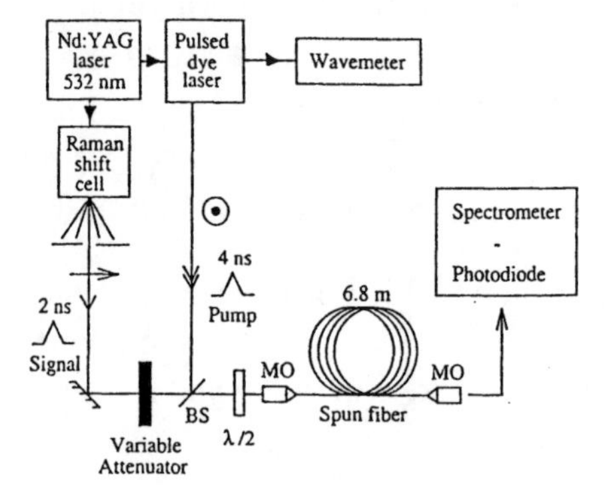

Figure 2. Schematic of experimental apparatus.

birefringent fiber, respectively. As can be seen, the polarization symmetry-breaking leads to non-overlapping, strongly asymmetric gain profiles.

In the experiment (see Figure 2), quasi-cw operation was achieved with nanosecond pulses from two distinct lasers [10]. The pump was cw tunable ring dye laser, subsequently amplified by a three-stage dye cell. The signal was obtained by frequency-shifting (through self-stimulated Raman scattering) an intense beam (1 MW peak power) from a frequency-doubled, injection-seeded and Q-switched Nd:YAG laser (at the wavelength $\lambda =$ 532.26 nm), operating at a repetition rate of 25 Hz. Orthogonally polarized and synchronized pump and signal pulses were launched in L= 6.8 m of weakly birefringent fiber (with beat length $L_b \simeq L$) whose cw polarization instability threshold power was $P_t = 70$ W.

Figure 3 shows measured output light spectra with a signal detuning of 1.2 THz, and a pump (of power $P_p = 112$ W) parallel to either the fast (Figure 3(a)) or the slow (Figure 3(b) axis. As can be seen, in agreement with the predictions of Figure 1, the stability of the fast fiber mode prevents a transfer of photons from the pump into the signal and the idler. Whereas

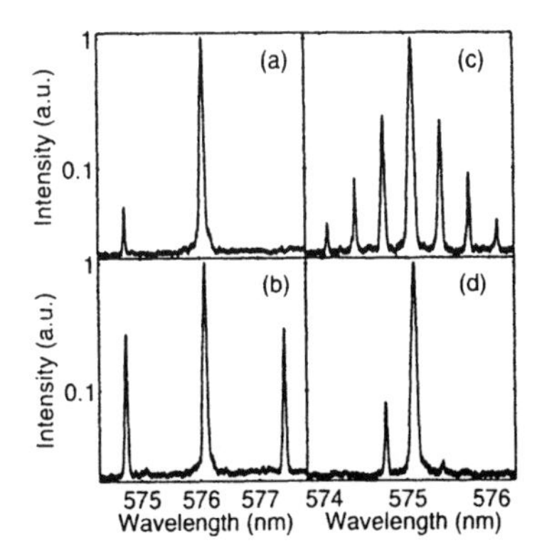

Figure 3. Output spectra with pump along fast (a) or slow (b) axis, a 112 W pump and $f = 1.2$ THz; (c) and (d): as in (a)-(b), with a $f = 0.3$ THz.

for slow mode pumping a deep polarization modulation of the pump wave is observed.

Figures 3(c)-(d) show that the portrait of the instability is drastically changed if the signal frequency detuning from the pump is reduced down to 0.3 THz. In this case (see Figure 1), the slow mode is stable and the fast mode is unstable. This leads to the strong polarization modulations in Figure 3(c). Since two harmonics of the input modulation fall in the MI bandwidth, a cascade of higher-order sidebands is observed. Figure 3(d) confirms that the slow mode is stable with respect to the injected signal.

3. Dark Pulse Train Generation

In this section, we will show experimentally, in good agreement with the theoretical predictions, that CPM-induced MI in birefringent and in two-mode fibers leads to the formation of ultra-high repetition rate soliton-like pulse trains in the normal GVD regime.

3.1. BIREFRINGENT FIBER

MI-induced break-up of an intense beam into a train of ultrashort solitary wave pulses is well displayed by Figure 4(a), where the pump power on the fast axis was 268 W, and the signal power was 1.9 W. From the nearly triangular envelope of the comb associated with the spectral intensity in Figure 4 one may infer that the pulse train has wavefronts with 400 fs duration. Figure 4(b) also shows that a rotation of the pump polarization by 90 degrees leads to spontaneous MI.

Figure 5 displays the corresponding intensity profiles and spectra versus the angular frequency Ω, in the fast and slow polarization components at

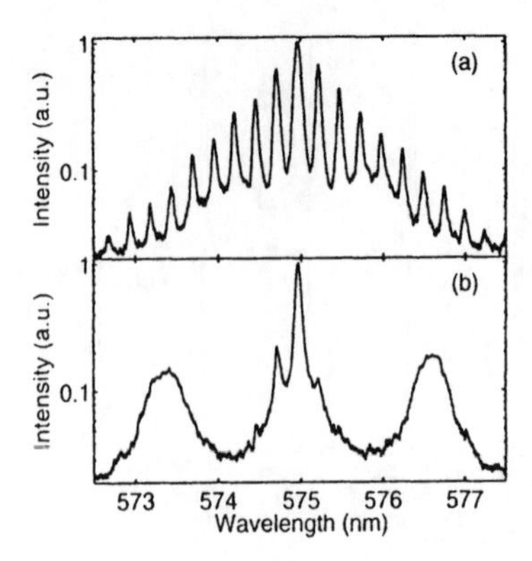

Figure 4. As in Figure 3, for $P_p = 268$ W and $f = 0.23$ THz.

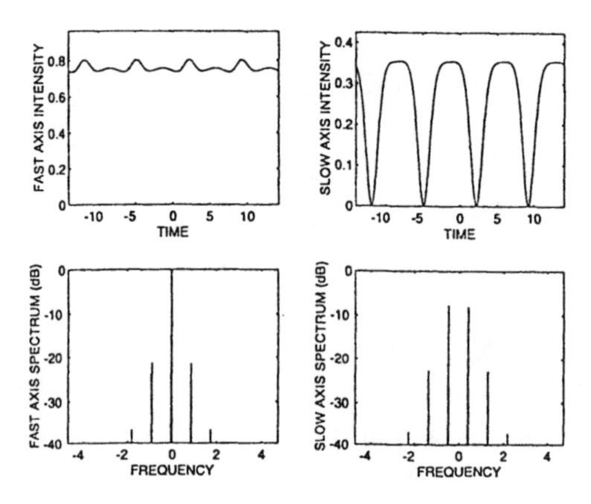

Figure 5. Output intensities vs. time τ and frequency Ω in fast and slow polarization components at $Z = 3.2$ m.

$\xi = 3.5$. As can be seen, the sidebands on the fast (slow) mode are the even (odd) harmonics of the initial modulation. In other words, the electric field components on the two orthogonal axes have opposite time-inversion parities. The pulses on the slow axis closely resemble a periodic train of dark solitons. This is clearly shown by the associated phase profiles in Figure 5: a complete phase rotation of π occurs between two consecutive maxima of the pulse train.

Figure 6 displays experimental spectra of light emerging from the slow (a) and the fast (b) fiber axis. In the left side, pump and signal were aligned with the fast and slow fiber axis, respectively. The signal detuning was $f = 0.3$ THz, while the pump and signal powers were tuned to $P_p = 96$ W (or $P_p = 1.37P_t$) and $P_s = 50$ mW, respectively. In the right side of Figure 6, the pump and signal were rotated by 90 degrees, with $P_p = 96$ W, $P_s = 100$ mW and $f = 1$ THz. In both cases, the general behavior of the

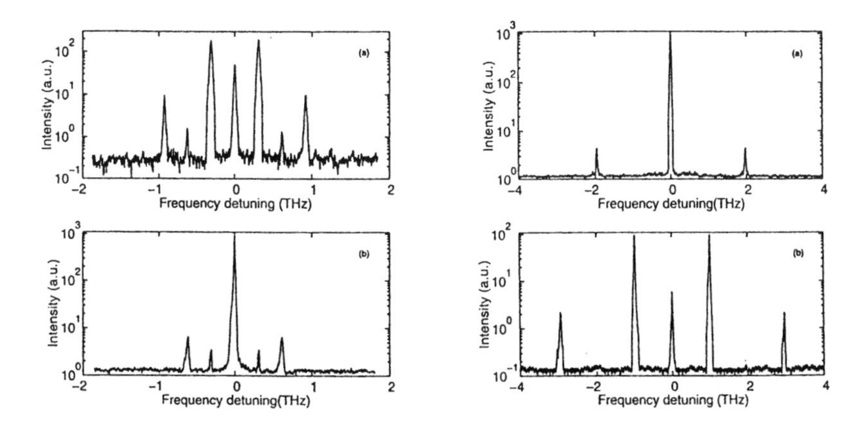

Figure 6. Experimental spectra of light from (a) slow and (b) fast axis, with a pump on the fast (left side) or slow (right side) axis.

observed spectra is in good agreement with theoretical predictions. Weak additional sidebands in the experimental spectra are due to cw-polarization MI-induced pump depolarization.

3.2. TWO-MODE FIBER

In a two-mode fiber, the coupled NLS equations for the amplitudes u and v of the LP_{01} and LP_{11} modes read as

$$\begin{aligned} i\frac{\partial u}{\partial \xi} + i\delta\frac{\partial u}{\partial \tau} &= \beta_1\frac{\partial^2 u}{\partial \tau^2} - (|u|^2 + 2c|v|^2)u\,, \\ i\frac{\partial v}{\partial \xi} - i\delta\frac{\partial v}{\partial \tau} &= \beta_2\frac{\partial^2 v}{\partial \tau^2} - (b|v|^2 + 2c|u|^2)v\,, \end{aligned} \tag{2}$$

where b and c are determined from the mode profiles. We shall consider here the wavelength range where the two modes have equal group velocities (i.e. $\delta \simeq 0$) and $2c > 1$, $2c/b > 1$. In this regime, pulse propagation as described by equations (2) with two equally excited pump modes is roughly equivalent to that of a birefringent fiber (see equations (1)) with a pump on the fast axis and $P_p \gg P_t$.

In Figure 7 we show an example of experimental break-up of the pump beam into a train of ultrashort pulses by induced MI in a two-mode fiber. In this case, the pump wavelength was set to 624.8 nm, and about 100 W were coupled into the fundamental LP_{11} and the first-order LP_{11} modes; whereas the signal power was equal to 19 W. The signal was obtained from the Raman cell at 624.57 nm (i.e. 210 GHz from the pump) by selecting the second-order Stokes wave of the Nd:YAG laser.

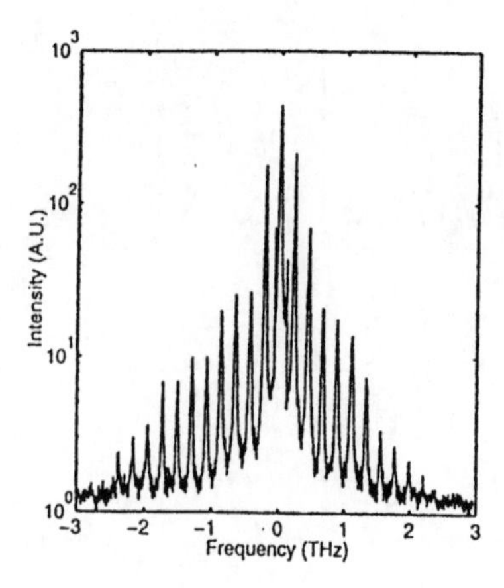

Figure 7. Spectrum of induced MI for $f = 210$ GHz, pump and signals of 201 W and 19 W.

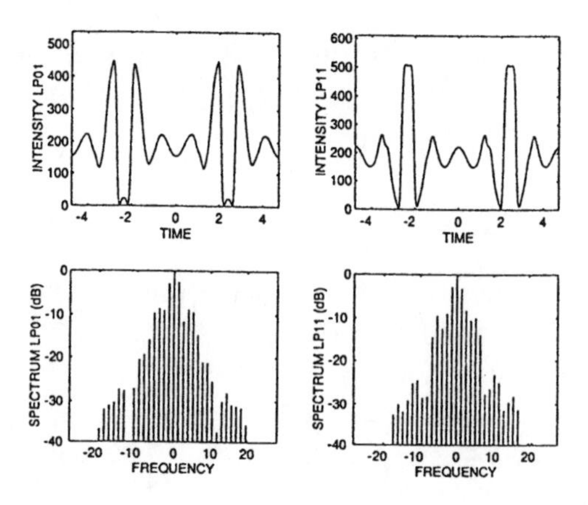

Figure 8. Output intensities and spectra from 8 m of two-mode fiber.

Figure 8 provides a numerical simulation of equations (2), showing the intensities of the field envelopes in the two modes vs. time (ps) and frequency ($2\pi THz$), respectively. Here the input power in each mode was equal to 200 W and 1.5 W in the pump and signal, respectively, with a frequency spacing of 210 GHz. As can be seen, the large number of harmonics of the initial modulation corresponds in the time domain to a periodic train of square pulses. These pulses closely resemble to the modal kink solitons or domain walls, which are the stable vector solitary-wave solutions in the normal GVD regime for a two-mode fiber.

4. Stimulated Raman Scattering Suppression

Parametric suppression of SRS may be obtained on each axis for sufficiently large GVM between the pumps. Whereas for small GVM, suppression and

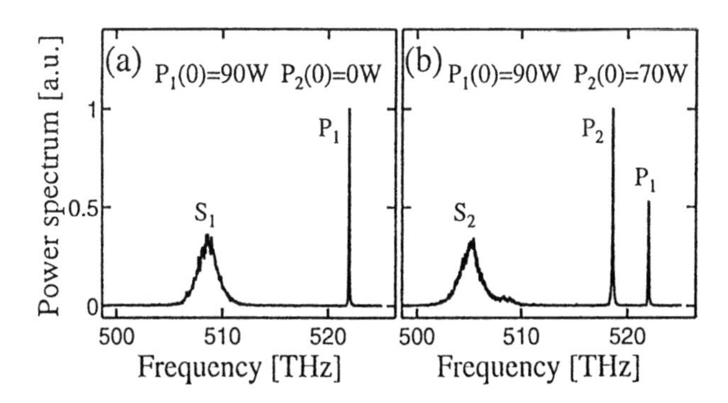

Figure 9. Measured spectra with (a) single and (b) dual-frequency pump and no GVM; $P_{1(2)}$ and $S_{1(2)}$ are pump and Stokes waves along the fast (slow) axis.

polarization switching of SRS is obtained by exploiting the energy transfer between the pumps through the orthogonal component of Raman gain. Experimental spectra clearly demonstrate the effectiveness of such method, which permits to suppress SRS in a given polarization without generating unwanted parametric sidebands.

In the experiments, we set at first the low-frequency pump power $P_2(0) = 0$; $P_1(0)$ was then raised well above the SRS threshold value (see Figure 9(a)). Next, the power $P_2(0)$ of the low-frequency pump was increased until full suppression of SRS for the pump P_1 was observed (see Figure 9(b)). Note that the linear birefringence of the present fiber is such that the frequency interval between the pumps for which the GVM vanishes is close to the frequency detuning for peak orthogonal Raman gain. Although for nonzero values of the GVM, both FWM and Raman transfer between the pumps concur to polarization switching of SRS, we found that orthogonal Raman gain remains the dominant mechanism for SRS suppression at relatively small GVMs.

5. Strong-signal Phase Conjugation

In this section, we discuss the novel peculiar effects observed in the regime of strong frequency conversion of quasi-cw pulses in a bifrefringent optical fiber. The frequency-dependence of the conversion efficiency in the large-signal regime is shown by Figure 10. Here $P_p = 160$ W, and the input signal fractional power $\alpha = P_s/P_p = 5$ %. As can be seen, initially the largest idler growth rate is obtained for frequency detunings around 1.4 THz, which is the parametric phase-matching frequency. As the idler energy fraction reaches a few percents of the pump power, however, pump depletion effects lead to a rapid shift of the peak conversion frequency, say, f_{opt}, towards

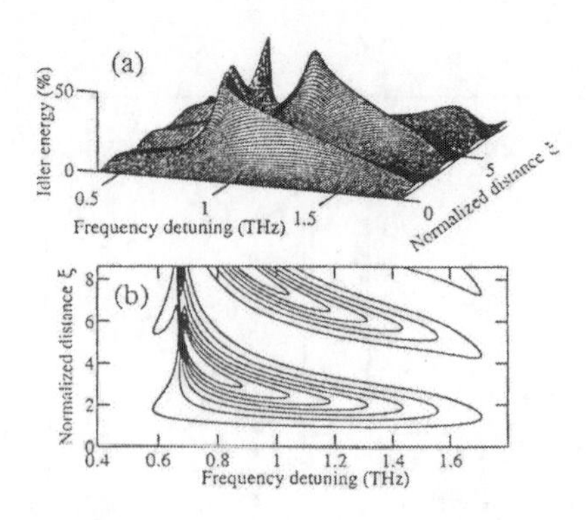

Figure 10. (a) Idler conversion vs. detuning f and distance; (b) Contour plot.

progressively lower values. After about three meters of fiber, f_{opt} stabilizes to a value slightly lower than 0.7 THz, that is well below the parametric gain bandwidth between 0.96 and 1.7 THz.

At a given detuning f, the large-signal regime of MPI is characterized by a threshold value of the signal fraction α. The dramatic effect that crossing this threshold signal power level has on the frequency conversion efficiency is clearly displayed by the phase-space portrait (Figure 11). Here f = 0.83 THz and P_p = 160 W. The solid separatrix curve in Figure 11 (corresponding to the threshold value of α = 1.46 %) divides the phase-plane in different domains of low and high energy conversion. In fact, the dashed curve in Figure 11 corresponds to signal power levels below the threshold value (here α = 1 %): as can be seen, the degree of frequency conversion is negligible at any distance. In contrast, the dot-dashed curve in Figure 11, obtained for a signal power larger than the threshold (e.g. α = 2 %), almost complete conversion is achieved.

A global display of the output idler energy fraction measured against detuning f is given in Figure 12, where α = 5 %, and the pump peak power is 160 W. The experimental data (stars) are compared with the predictions of the three-wave model (solid line). Highest idler energies ($\simeq$ 18 %, corresponding to 40 % of pump depletion) are measured around 0.9 THz. As can be seen in Figure 12, peak frequency conversion is achieved outside the small-signal parametric gain curve (dashed curve).

One of the most remarkable signatures of the large-signal MPI is the sudden increase of the frequency conversion from the pump into the sidebands, as a result of a small increase of the signal power. This effect is clearly displayed in the experimental results of Figure 13, where the output idler energy is measured as a function of α, for a given P_p = 160 W, and various pump-signal detunings. Namely, f = 0.83 THz (Figure 13(a)),

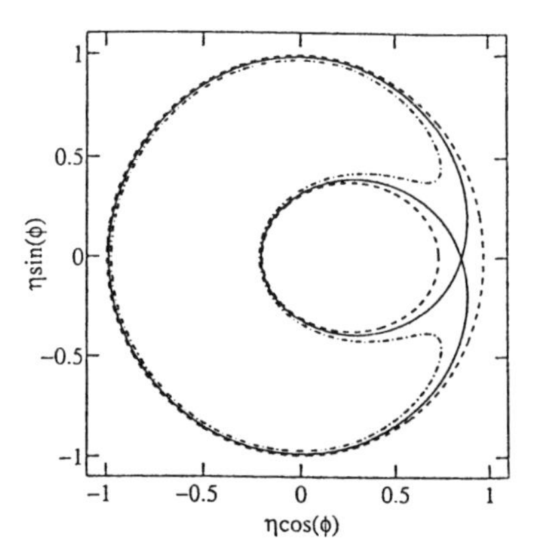

Figure 11. Dashed curve: $\alpha = 1$ % (low conversion); solid curve: $\alpha = 1.5$ % (separatrix); dot-dashed curve: $\alpha = 2$ % (high conversion).

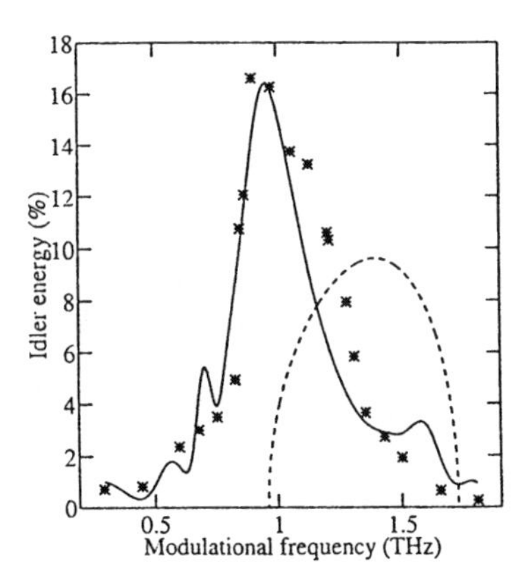

Figure 12. Measured (stars) and theoretical (solid curve) energy conversion vs. detuning f, for $P_p = 160$ W and $\alpha = 5$ %. Dashed curve: small-signal MPI gain bandwidth.

$f = 1.055$ THz (Figure 13(b)) and $f = 1.432$ THz (Figure 13(b)): the left and right columns show the experimental results and the (pulse averaged) theoretical predictions, respectively. A strong increase of conversion occurs whenever the input signal power level crosses the threshold value (i.e. $\alpha \simeq 5$ %) is only observed for f outside the parametric gain bandwidth (i.e. the fiber is operated in the large-signal regime of MPI), in good agreement with the theoretical predictions.

6. Soliton Frequency Conversion

Let us consider here the frequency conversion of a soliton pulse in the presence of a strong cw pump, i.e. we neglect pump depletion (e.g. soliton frequency conversion in a fiber loop laser). We also take λ_p close to the

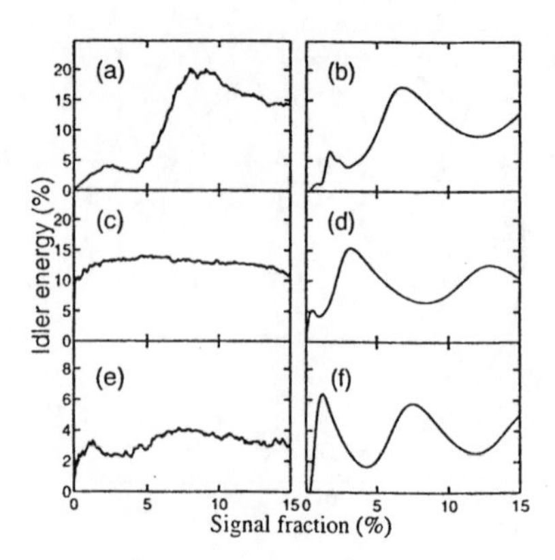

Figure 13. Experimental (left) and theoretical (right) output idler energy vs. input signal fraction α for (a) $f = 0.83$ THz; (b) $f = 1.055$ THz; (c) $f = 1.35$ THz.

fiber zero dispersion wavelength, so that the GVM between signal and idler vanishes, and the two fields have opposite GVDs. One may then reduce the signal and idler propagation equations to the two coupled NLS equations

$$\begin{aligned} i\frac{\partial u}{\partial Z} + \frac{1}{2}\frac{\partial^2 u}{\partial T^2} + i\alpha u + i\gamma v - \rho u - \sigma\left(|u|^2 + 2|v|^2\right)u &= 0\,, \\ i\frac{\partial v}{\partial Z} + \frac{1}{2}\frac{\partial^2 v}{\partial T^2} + i\alpha v + i\gamma u + \rho v + \sigma\left(|v|^2 + 2|u|^2\right)v &= 0\,, \end{aligned} \tag{3}$$

where u and v are in the anomalous and normal GVD regime, respectively. In equations (3) $\sigma = -1$, the loss coefficient $\alpha \equiv \Gamma z_0$, the parametric gain coefficient is $\gamma \equiv RP_p z_0$, $\rho \equiv -(\delta k/2 + \eta RP)z_0$, z_0 is the dispersion distance, and δk is the linear wave vector mismatch.

Numerical integrations of equations (3) with an initial condition given by a stationary wave profile of the type $u(Z,T) = U(T)\exp\{iqZ\}$, $v(Z,T) = V(T)\exp\{iqZ\}$ plus a small perturbation led to the formation of stable coupled pulse-like solutions (see Figure 14). Note that the V component of the vector soliton has a dip in its center and is temporally broader than its U counterpart. This stable vector soliton represents a pair of signal and conjugate pulses of comparable amplitude and time width. Therefore, under such conditions the signal pulse propagates along the fiber unchanged, along with a similar phase-conjugate replica.

From the applicative viewpoint, it is important to generate conjugate solitary waves by a single pulse injected at the signal wavelength (along with the cw pump). We performed extensive numerical solutions of equations (3) with the initial signal $u(Z = 0, T) = A\text{sech}(\omega T)$ and $v(Z = 0, T) = 0$.

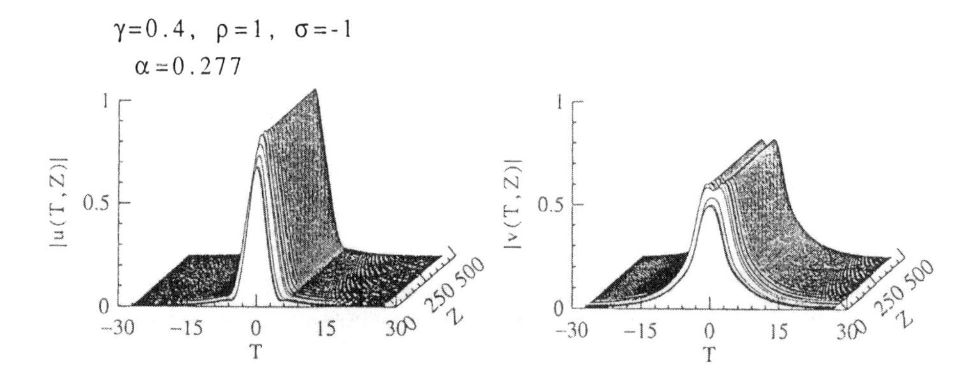

Figure 14. Stable coupled signal and conjugate pulses for $\alpha = 0.277$, $\gamma = 0.4$ and $\rho = -\sigma = 1$.

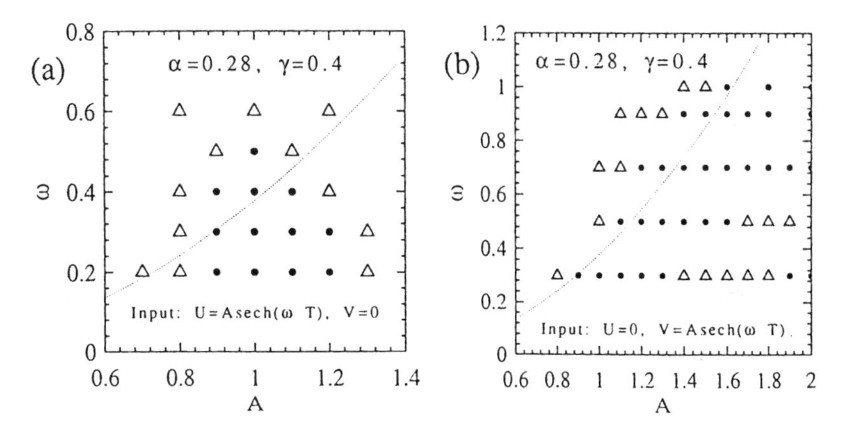

Figure 15. Basin of attraction of conjugate solitons vs. initial amplitude A and inverse witdth ω, and signal in (a) anomalous or (b) normal GVD regime.

The filled circles in Figure 15(a) illustrate the basin of attraction from the above initial conditions into stable coupled solutions, as a function of the initial amplitude A and the inverse witdth ω. On the other hand, the open triangles represent the initial conditions that led to decay of both pulses. The dotted curve in Figure 15(a) illustrates the relation $2A^2/\omega = E_{ss}$, where E_{ss} is the energy corresponding to the stable stationary solution. Quite remarkably, Figure 15(b) reveals that the basin of attraction into stable conjugate solitary waves is broader for a signal in the normal GVD regime.

7. Conclusions

In short summary, in this work we presented several new results on the frequency conversion of pulses and solitons in optical fibers. We demonstrated that induced-MI in a birefringent or two-mode fiber in the normal GVD regime is subject to strong polarization or modal instabilities. This instability leads to the first observation of femtosecond pulse trains at repetition rates of 200–300 GHz in the normal GVD regime. Simulations reveal that these pulse trains have the structure of dark solitons or domain walls. The interplay between Raman and parametric scatterings in the fiber leads to novel possibilities for the selective suppression and polarization switching of SRS in a given polarization by use of a dual-frequency pumping scheme. In the case of strong frequency conversion, the experiments reveal a marked dependence of the conversion efficiency on the signal power level. Quite contrary to intuition, largest conversions are achieved for relatively large signals with a detuning outside the MI small-signal gain curve. Finally, we revealed the existence of stable coupled conjugate pulses in fiber four-wave mixing. These waves may be generated by injecting a single pulse at the signal wavelength, and permit to maintain time localization in the field of the ultrashort conjugate pulse.

Acknowledgement

The Conseil Régional de Bourgogne, the Centre National de la Recherche Scientifique, and the Ministère de la Recherche et de la Technologie are gratefully acknowledged for their financial support of this work. The work of S. Trillo was carried out under the agreement with the italian post and telecommunications administration. The work of M. Haelterman was supported by the Belgian National Fund for Scientific Research (FNRS) and the Interuniversity Attraction Pole Program of the Belgian Government (under grant P4-07). The work of J. M.Soto Crespo was supported by the Comunidad de Madrid under contract 06T/039/96 and by the CICyT under contract TIC95-0563-03.

References

1. Benjamin, T. J. and Feir, J. E.: *J. Fluid Mech.*, **27**, (1967), p.417.
2. Bespalov,V. I. and Talanov, V.I.: *Pis'ma Zh. Eksp. Teor. Fiz.*, **3**, (1966), p.471 [*JETP Lett.*, **3**, (1966), p.307].
3. Tai, K., Hasegawa, A. and Tomita, A.: *Phys. Rev. Lett.*, **56**, (1986), p.135.
4. Tai, K., Tomita, A., Jewell, J. L. and Hasegawa, A.: *Appl. Phys. Lett.*, **49**, (1986), p.236.
5. Agrawal, G. P.: *Phys. Rev. Lett.*, **59**, (1987), p.880.
6. Berkhoer A. L. and Zakharov, V. E.: *Zh. Eksp. Teor. Fiz.*, **58**, (1970), p.903 [*Sov. Phys. JETP*, **31**, (1970), p.486].

7. Wabnitz, S.: *Phys. Rev. A*, **38**, (1988), p.2018.
8. Daino, B., Gregori, G. and Wabnitz, S.: *J. Appl. Phys.*, **58**, (1985), p.4512; Winful, H. G.:*Opt. Lett.*, **11**, (1986), p.33; Trillo, S., Wabnitz, S., Stolen, R. H., Assanto, G., Seaton, C.T. and Stegeman, G. I.: *Appl. Phys. Lett.*, **49**, (1986), p.1224.
9. Feldman, S. F., Weinberger, D.A. and Winful, H. G.: *Opt. Lett.*, **15**, (1990), p.311; Murdoch, S. G., R. Leonhardt, R. and Harvey, J. D.: *Opt. Lett.*, **20**, (1995), p.866.
10. Millot, G., Seve, E. and Wabnitz, S.: *Phys. Rev. Lett.*, **79**, (1997), p.661.
11. Akhmediev, N. N., Korneev, V. I. and Mitskevich, N. V.: *Izvestiya Vysshikh Uchebnykh Zavedenii, Radiofizika*, **34**, (1991), p.84.
12. Christodoulides, D. N.: *Phys. Lett. A*, **132**, (1988), p.451.
13. *Optics Commun.*, **111**, (1994), p.86.
14. Golovchenko, E. and Pilipetskii, A. N.: *Sov. Lightwave Commun.*, **1**, (1991), p.271.
15. Mamyshev, P. V. and Vertikov, A. P.: *Technical Digest Series (O.S.A, Washington, D.C., 1992)*, **13**, (1992), p.130.
16. Trillo, S. and Wabnitz, S.: *J. Opt. Soc. Am. B*, **9**, (1992), p.1061.
17. Tchofo Dinda, P., Millot, G. and Wabnitz, S.: *Opt. Lett.*, **22**, (1997).
18. G. Cappellini and S. Trillo, *Opt. Lett.*, **16**, (1991), p.895.
19. Yariv, A., Fekete, D. and Pepper, D.: *Opt. Lett.*, **4**, (1979), p.52.
20. Watanabe, S., Chikama, T., Ishikawa, G., Terahara, T. and Kuwahara, H.: *IEEE Phot. Techol. Lett.*, **4**, (1993), p.1241.
21. Forysiak, W. and Doran, N. J. : *Electron. Lett.*, **30**m (1994), p.154.
22. Franco, P., Fontana, F., Cristiani, I., Zenobi, M., Midrio, M. and Romagnoli, M.: *Opt. Lett.*, **21**, (1996), p.788.

(Received November 17, 1997)

STUDIES ON SIGNAL PROPAGATION IN OPTICAL FIBERS WITH RANDOMLY VARYING BIREFRINGENCE

P. K. A. WAI
Department of Electronic Engineering
The Hong Kong Polytechnic University
Hung Hom, Kowloon, Hong Kong, China

D. MARCUSE AND C. R. MENYUK
Department of Computer Science and Electrical Engineering
University of Maryland Baltimore County
Baltimore, Maryland 22128-5398, USA.

AND

W. L. KATH
Department of Engineering Sciences and Applied Mathematics
McCormick School of Engineering and Applied Science
Northwestern University
2145 Sheridan Rd., Evanston, Illinois 60208-3125, USA.

Abstract. Randomly varying birefringence leads to nonlinear polarization-mode dispersion (PMD) in addition to the well-known linear PMD. Here, we calculated the variance of the field fluctuations produced by this nonlinear PMD. The effects of fiber nonlinearity, linear PMD and nonlinear PMD are studied numerically using the Manakov-PMD equation which models signal propagation in optical fibers with randomly varying birefringence. Nonlinear PMD is shown to be unimportant in present day communication systems.

1. Introduction

Real single mode optical fibers are birefringent [1]. In the situation where the birefringence axes and magnitude are fixed as a function of distance along the fiber, the evolution of the optical pulse envelope has been shown to be governed by the coupled nonlinear Schrödinger (CNLS) equation [2]-

A. Hasegawa (ed.), New Trends in Optical Soliton Transmission Systems, 69–81.

[4]. In standard communication fibers, however, both the orientation of the principle birefringence axes and the magnitude of the birefringence vary randomly with distance along the fiber. In this case the mean pulse evolution is governed by a special case of the CNLS equation known as the Manakov equation [5, 6]. The deviations in the polarization state from the mean behavior are random. Because the two principle polarizations have slightly different group velocities, these polarization fluctuations produce linear or intensity-independent signal pulse broadening and distortion known as polarization mode dispersion (PMD). Such linear PMD has been studied in great detail [7]-[11]. In addition to the linear PMD, however, the polarization fluctuations also produce intensity-dependent signal pulse distortions known as nonlinear PMD [12]-[14]. In this paper we determine the statistics of the pulse distortions produced when nonlinear PMD acts upon a propagating signal. We will show that nonlinear PMD effects are negligible for present-day, practical communication systems.

Several length scales are important for the discussion of PMD effects. The first is the fiber decorrelation length [12], h_{fiber}, which is the autocorrelation length scale associated with variations in the principal axes of birefringence and which is an intrinsic property of the fiber. The second is the birefringence beat length, L_B, which is a measure of the birefringence strength and is the length scale over which a 2π phase shift would be produced between the two principal polarizations if the birefringence axes were fixed. The calculation for the statistics of the nonlinear PMD fluctuations simplifies considerably when either $h_{\text{fiber}} \ll L_B$ or $h_{\text{fiber}} \gg L_B$. Finally, there is the polarization decorrelation length, h_E, which is the autocorrelation length associated with the electromagnetic field fluctuations induced by the random birefringence variations [12].

The formulation of the paper is as follows. In Section 2, we review the Manakov-PMD equation governing the evolution of nonlinear optical signals in the presence of random birefringence. In Section 3, we discuss the nonlinear PMD. In Section 4, we apply the Manakov-PMD equation to an example of NRZ communication signals. Section 5 gives our conclusions.

2. Manakov-PMD Equations

The dimensionless set of equations describing the evolution of the electromagnetic field envelope in a nonlinear optical fiber with random birefringence are the Manakov-PMD equation [12]-[14],

$$i\frac{\partial\overline{\Psi}}{\partial z} \pm \frac{1}{2}\frac{\partial^2\overline{\Psi}}{\partial t^2} + \frac{8}{9}|\overline{\Psi}|^2\overline{\Psi} = -ib'\overline{\Sigma}\frac{\partial\overline{\Psi}}{\partial t} - \frac{1}{3}(\widehat{\mathbf{N}} - \langle\widehat{\mathbf{N}}\rangle), \tag{1}$$

where $\overline{\Psi} = (\bar{U}, \bar{V})^{\mathrm{T}}$, the $\pm$ sign is for either anomalous or normal second order dispersion, $b' = \partial b / \partial \omega$ is the group delay difference per unit length, and $2b$ is the birefringence strength. An alternate measure of the amount of birefringence is the beat length, $L_B = \pi / b$. Also,

$$\overline{\Sigma} = \begin{pmatrix} x_1 & x_2 - ix_3 \\ x_2 + ix_3 & -x_1 \end{pmatrix}, \tag{2}$$

$$\begin{aligned} \widehat{N}_1 &= z_1^2(2|\bar{V}|^2 - |\bar{U}|^2)\bar{U} - z_1(z_2 - iz_3)(2|\bar{U}|^2 - |\bar{V}|^2)\bar{V} \\ &\quad - z_1(z_2 + iz_3)\bar{U}^2\bar{V}^* - (z_2 - iz_3)^2\bar{V}^2\bar{U}^*, \qquad (3a) \\ \widehat{N}_2 &= z_1^2(2|\bar{U}|^2 - |\bar{V}|^2)\bar{V} + z_1(z_2 + iz_3)(2|\bar{V}|^2 - |\bar{U}|^2)\bar{U} \\ &\quad + z_1(z_2 - iz_3)\bar{V}^2\bar{U}^* - (z_2 + iz_3)^2\bar{U}^2\bar{V}^*, \qquad (3b) \end{aligned}$$

and

$$\begin{aligned} \langle \widehat{N}_1 \rangle &= \tfrac{1}{3}(2|\bar{V}|^2 - |\bar{U}|^2)\bar{U} \qquad (4a) \\ \langle \widehat{N}_2 \rangle &= \tfrac{1}{3}(2|\bar{U}|^2 - |\bar{V}|^2)\bar{V}. \qquad (4b) \end{aligned}$$

The random variables x_j, z_j are the first and third components of the solution to the vector equation

$$\frac{d}{dz}\begin{pmatrix} x_j \\ y_j \\ z_j \end{pmatrix} = \begin{pmatrix} 0 & g_\theta & 0 \\ -g_\theta & 0 & 2b \\ 0 & -2b & 0 \end{pmatrix}\begin{pmatrix} x_j \\ y_j \\ z_j \end{pmatrix}, \tag{5}$$

for $j = 1, 2, 3$. At $z = 0$ the initial conditions $x_j = \delta_{j1}$, $y_j = \delta_{j2}$, $z_j = \delta_{j3}$ are satisfied, where δ_{jk} is the Kronecker delta, i.e. $\delta_{jk} = 1$ if $j = k$ and $\delta_{jk} = 0$ otherwise. The vectors $\mathbf{R}_j = (x_j, y_j, z_j)$ are vectors on the Poincaré sphere [12, 13] and $\mathbf{R}_1$ is initially pointed in the x-direction, $\mathbf{R}_2$ is initially pointed in the y-direction and $\mathbf{R}_3$ is initially pointed in the z-direction. Here $\overline{\Psi}$ is the electric field envelope measured relative to axes which follow the linear polarization state in the fiber. In the simplest birefringence model, it is assumed that the birefringence strength $2b$ remains constant and the orientation angle θ is driven by the white noise process,

$$\frac{d\theta}{dz} = g_\theta(z), \tag{6}$$

where

$$\langle g_\theta(z) \rangle = 0, \qquad \langle g_\theta(z) g_\theta(z') \rangle = \sigma_\theta^2\, \delta(z - z'), \tag{7}$$

and σ_θ^2 is a constant. This model has been shown to capture all of the essentials of the more realistic case where both the orientation angle and the birefringence strength vary randomly [11, 15]. Technically speaking, the averaging here should be done over a random ensemble of fibers, or rather over a random ensemble of the functions $g_\theta(z)$ which determine the birefringence angle $\theta(z)$; physical systems such as this one that can be described by Markov processes are generally ergodic [16, 17]; however, in which case ensemble averages can be replaced by spatial averages.

3. PMD statistics

The statistics of the coefficients x_j and z_j can be determined either from their probability distributions or from the equations for their statistical moments. The former method gives complete information about the coefficients' statistics, but finding the probability distribution involves solving a partial differential equation known as the Fokker-Planck equation [17]. Using averages and statistical moments of the coefficients is more direct, and yields useful information about the behavior of the linear and nonlinear PMD.

The equations for the means and second moments of the various coefficients have already been determined; they are [12]

$$\frac{d}{dz}\langle x_j\rangle = -\tfrac{1}{2}\sigma_\theta^2\langle x_j\rangle, \tag{8a}$$

$$\frac{d}{dz}\langle y_j\rangle = -\tfrac{1}{2}\sigma_\theta^2\langle y_j\rangle - 2b\langle z_j\rangle, \tag{8b}$$

$$\frac{d}{dz}\langle z_j\rangle = 2b\langle y_j\rangle, \tag{8c}$$

where $j = 1, 2, 3$, and

$$\frac{d}{dz}\langle x_j x_k\rangle = -\sigma_\theta^2[\langle x_j x_k\rangle - \langle y_j y_k\rangle], \tag{9a}$$

$$\frac{d}{dz}\langle y_j y_k\rangle = \sigma_\theta^2[\langle x_j x_k\rangle - \langle y_j y_k\rangle] - 4b\langle \tfrac{1}{2}(y_j z_k + y_k z_j)\rangle, \tag{9b}$$

$$\frac{d}{dz}\langle z_j z_k\rangle = 4b\langle \tfrac{1}{2}(y_j z_k + y_k z_j)\rangle, \tag{9c}$$

$$\frac{d}{dz}\langle \tfrac{1}{2}(y_j z_k + y_k z_j)\rangle = 2b[\langle y_j y_k\rangle - \langle z_j z_k\rangle] - \tfrac{1}{2}\sigma_\theta^2\langle \tfrac{1}{2}(y_j z_k + y_k z_j)\rangle. \tag{9d}$$

where $j, k = 1, 2, 3$. These equations are the same ones that the means and second moments of the local Stokes parameters $\tilde{S}_j$, $j = 1, 2, 3$ satisfy [12].

In particular,

$$\begin{aligned}
&\langle x_j\rangle \leftrightarrow \langle\tilde{S}_1\rangle, \quad \langle y_j\rangle \leftrightarrow \langle\tilde{S}_2\rangle, \quad \langle z_j\rangle \leftrightarrow \langle\tilde{S}_2\rangle,\\
&\langle x_j x_k\rangle \leftrightarrow \langle\tilde{S}_1^2\rangle, \quad \langle y_j y_k\rangle \leftrightarrow \langle\tilde{S}_2^2\rangle, \quad \langle z_j z_k\rangle \leftrightarrow \langle\tilde{S}_3^2\rangle,\\
&\langle \tfrac{1}{2}(y_j z_k + y_k z_j)\rangle \leftrightarrow \langle\tilde{S}_2\tilde{S}_3\rangle.
\end{aligned} \tag{10}$$

Not all of the first and second moments of x_j, y_j and z_j appear in equations (2) and (3); the only terms which are needed are x_1, x_2, x_3, $z_1^2 - \frac{1}{3}$, $z_1 z_2$, $z_1 z_3$, $z_2^2 - z_3^2$ and $z_2 z_3$. When one talks about the variations induced in the field envelope $\overline{\Psi}$, however, it is really the statistics of the *integrals* of these quantities that are needed. The variations in $\overline{\Psi}$ depend upon the integrals of the coefficients because the length scale over which the fluctuations in x_j, y_j and z_j occur (the polarization or field decorrelation length [12], h_E) is much shorter than the natural length scale over which the pulse envelope evolves, namely the dispersion length (Note in equation (1) all spatial variables have been scaled using this length).

If the birefringence fluctuations produce uniform mixing on the Poincaré sphere, then the field decorrelation length is proportional to the fiber decorrelation length, $h_E \propto h_{\text{fiber}}$ [12]. This is the situation when the fiber decorrelation length is much longer than the beat length, $h_{\text{fiber}} \gg L_B$. When $h_{\text{fiber}} \ll L_B$, however, azimuthal mixing on the Poincaré sphere is poor and $L_B^2/(12\pi^2 h_{\text{fiber}})$ becomes the relevant lengthscale for fluctuations in the coefficients z_j [12]. In this limit, then, nonlinear PMD becomes much more significant than linear PMD. Nevertheless, even for values which lie at the edge of the realistic parameter range, e.g. $h_{\text{fiber}} \approx 0.3$ m and $L_B \approx 100$ m, the nonlinear PMD lengthscale is smaller than dispersion lengths typical in standard communication systems, and the disparate lengthscales can be exploited to determine the statistics of the induced field fluctuations.

Because the field decorrelation length is short compared with the dispersion length, the field envelope $\overline{\Psi}$ is not able to respond to any individual changes in these coefficients, but rather only to their cumulative effects. Alternatively, since $d\overline{\Psi}/dz$ is $O(1)$, the envelope is able to build up appreciable changes only after distances of the order of the dispersion length, and it experiences only small variations over shorter distances. Thus, over a short distance interval from z_0 to $z_0 + l$, in the terms on the right-hand side of equation (1), $\overline{\Psi}$ can be replaced by its initial value, $\overline{\Psi}(z_0, t)$, and only the random coefficients themselves are integrated.

The means, variances and covariances of the integrals of these various nonlinear PMD coefficients can be calculated by extending the system of moments, equations (8) and (9), to include these new terms (alternatively, the Fokker-Planck equation can be extended). For example, the variance of the linear PMD coefficient, $\mathrm{Var}\Big(\int_{z_0}^{z_0+l} x_j(\zeta)\,d\zeta\Big)$, can be determined from

$$\frac{d}{dz}\Big\langle\Big(\int_{z_0}^{z_0+l} x_j(\zeta)\,d\zeta\Big)^2\Big\rangle = 2\Big\langle x_j\Big(\int_{z_0}^{z_0+l} x_j(\zeta)\,d\zeta\Big)\Big\rangle \tag{11a}$$

$$\frac{d}{dz}\Big\langle x_j\Big(\int_{z_0}^{z_0+l} x_j(\zeta)\,d\zeta\Big)\Big\rangle = -\tfrac{1}{2}\sigma_\theta^2\,\Big\langle x_j\Big(\int_{z_0}^{z_0+l} x_j(\zeta)\,d\zeta\Big)\Big\rangle + \langle x_j^2\rangle \tag{11b}$$

and equations (9). The moments equations describe the evolution of the variance for all values of z_0, even within a few field decorrelation lengths of the initial condition at $z_0 = 0$ where the process is not stationary. The variance for the nonlinear PMD coefficients can be determined in a similar way but the number of equations grows very quickly because additional unneeded moments must be included to make the set of equations complete. If we are primarily concerned with distance intervals l which are long compared with the mixing length scale of the various coefficients, a simpler method can be employed [14]. The results in the limit $h_{\text{fiber}} \ll L_B$ and $h_{\text{fiber}} \gg L_B$ are summarized as follows.

In the limit $h_{\text{fiber}} \ll L_B$, we have $\sigma_\theta^2 \gg b$. Using the mixing length $h_E = \sigma_\theta^2/24b^2 = L_B^2/(12\pi^2 h_{\text{fiber}})$, we obtain [14]

$$\mathrm{Var}\Big(\int_{z_0}^{z_0+l} (z_1^2 - \tfrac{1}{3})\,d\zeta\Big) \sim \frac{8}{45} h_E(l - h_E)\,, \tag{12}$$

$$\mathrm{Var}\Big(\int_{z_0}^{z_0+l} z_j z_k\,d\zeta\Big) \sim \frac{2}{15} h_E(l - h_E)\,, \tag{13}$$

for $(j,k) = (1,2)$, $(j,k) = (1,3)$ or $(j,k) = (2,3)$ and

$$\mathrm{Var}\Big(\int_{z_0}^{z_0+l} (z_2^2 - z_3^2)\,d\zeta\Big) \sim \frac{8}{15} h_E(l - h_E)\,. \tag{14}$$

In the limit $h_{\text{fiber}} \gg L_B$, or equivalently $\sigma_\theta^2 \ll b$, we define the mixing length $h_E = 1/(6\sigma_\theta^2) = h_{\text{fiber}}/12$. We obtain [14]

$$\mathrm{Var}\Big(\int_{z_0}^{z_0+l} (z_1^2 - \tfrac{1}{3})\,d\zeta\Big) \sim \frac{8}{45} h_E(l - 4h_E)\,, \tag{15}$$

$$\mathrm{Var}\Big(\int_{z_0}^{z_0+l} z_j z_k\,d\zeta\Big) \sim \frac{2}{15} h_E(l - 4h_E)\,, \tag{16}$$

for $(j,k) = (1,2)$, $(j,k) = (1,3)$ or $(j,k) = (2,3)$ and

$$\mathrm{Var}\Big(\int_{z_0}^{z_0+l} (z_2^2 - z_3^2)\,d\zeta\Big) \sim \frac{8}{15} h_E(l - 4h_E)\,. \tag{17}$$

All asymptotic covariances of the integrated nonlinear PMD terms are zero.

A significant contribution from nonlinear PMD can only occur when mixing on the Poincaré sphere is poor, i.e. this mixing occurs on the same length scale or even a longer length scale than is associated with the Kerr effect, chromatic dispersion, and polarization mode dispersion. There are two limits in which this situation can occur [12, 15]. The first and most obvious is when h_{fiber}, the fiber correlation length, becomes comparable to the nonlinear and dispersive scale lengths. Since the mixing length on the Poincaré sphere must be longer than h_{fiber}, it is apparent that mixing will be poor. This limit, which applies to sub-picosecond pulses, does not occur in present-day communication systems.

In the second limit, $h_{\text{fiber}} \ll L_B$. The mixing length h_E is given by $L_B^2/(12\pi^2 h_{\text{fiber}})$. When the mixing length approaches the nonlinear and dispersive scale lengths, mixing will be poor [12]-[15]. This limit corresponds to decreasing the magnitude of the birefringence, the birefringence becomes sufficiently small that the polarization state is nearly unaffected by the randomly varying birefringence and, in a fixed frame, the electric field remains correlated over long lengths. This increased mixing length only affects the nonlinear PMD; the strength of the linear PMD is proportional to h_{fiber}. It has been proposed to significantly decrease the linear PMD by rapidly shifting the orientation of the birefringence axes in a fiber as it is drawn, which amounts to reducing h_{fiber} while keeping L_B fixed [12]. This approach raises the specter that nonlinear PMD might become important in these fibers, but that is not the case because the factor of $1/(12\pi^2)$ is so small that it is difficult to achieve a significant mixing length even under extreme circumstances. We will demonstrate this in the next section.

4. Numerical simulations

The parameters that we selected in our simulations are: the wavelength, $\lambda = 1.55$ μm, the effective mode area, $A_{\text{eff}} = 52\mu\text{m}^2$, the Kerr coefficient, $n_2 = 2.6 \times 10^{-20}$ m^2/W, the linear PMD coefficient, $D_{\text{PMD}} = 3$ ps/km$^{1/2}$, an FFT with $2^{10} = 1024$ grid points, and a total fiber length, $L = 500$km. We use here a dispersion map consisting of a periodic sequence of normal dispersion anomalous dispersion fiber sections. The period length is 100 km, consisting of 80 km sections with normal dispersion coefficient, $D = -1$ ps/nm·km, and of 20 km compensation sections with $D = 4$ ps/nm·km.

As the input pulse pattern we used a 16 bit pseudorandom word in NRZ (non-return to zero) format with a 0.2 ns pulse width corresponding to a signaling rate of 5 Gbit/s. These pulses are generated by passing rectangularly-shaped pulses through a low-pass Bessel filter of $B = 8.75$GHz width, which gives the pulses slightly rounded edges. These optical input pulses are shown in Figure 1(a) in units of mW. The peak pulse power is $P_{\text{peak}} = 20$ mW. This power is somewhat high for communications purposes but was chosen here to produce a sizable nonlinear effect in 500 km of fiber. The received pulses is filtered with a $B_{\text{elect}} = 5$ GHz "electrical" Bessel filter. Figure 1(b) shows the square-law-detected, filtered input pulse. All simulations reported here were run without fiber losses, therefore there are also no optical amplifiers in the system and, consequently, no amplifier (ASE) noise. Higher order dispersions were also neglected.

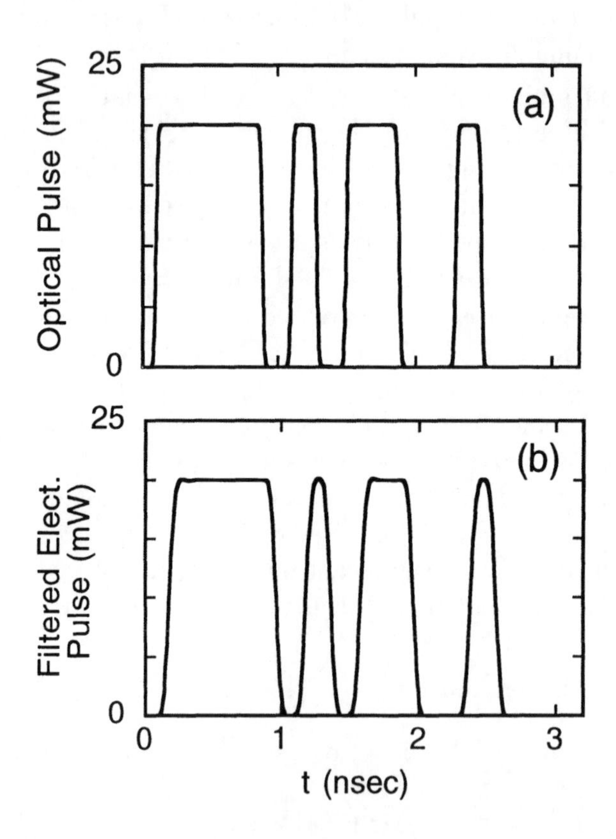

Figure 1. The train of pseudorandom input pulses consisting of 16 bits. (a) Power of the optical pulses in mW; (b) Detected and filtered current in mW units.

The behavior of the system is computed by solving the Manakov-PMD equation. The CNLS equation is also solved for comparison. Computation of the CNLS equation is extremely slow because it is necessary to solve the equation on the short length of the random birefringence. It requires 1.5 days for the job at hand on a SUN SPARCstation 10. In contrast, it only takes two minutes to solve the Manakov-PMD equation [13]. Figure 2(a) shows the output pulses for $h_{\text{fiber}} = 100$ m and $L_B = 10$ m of the

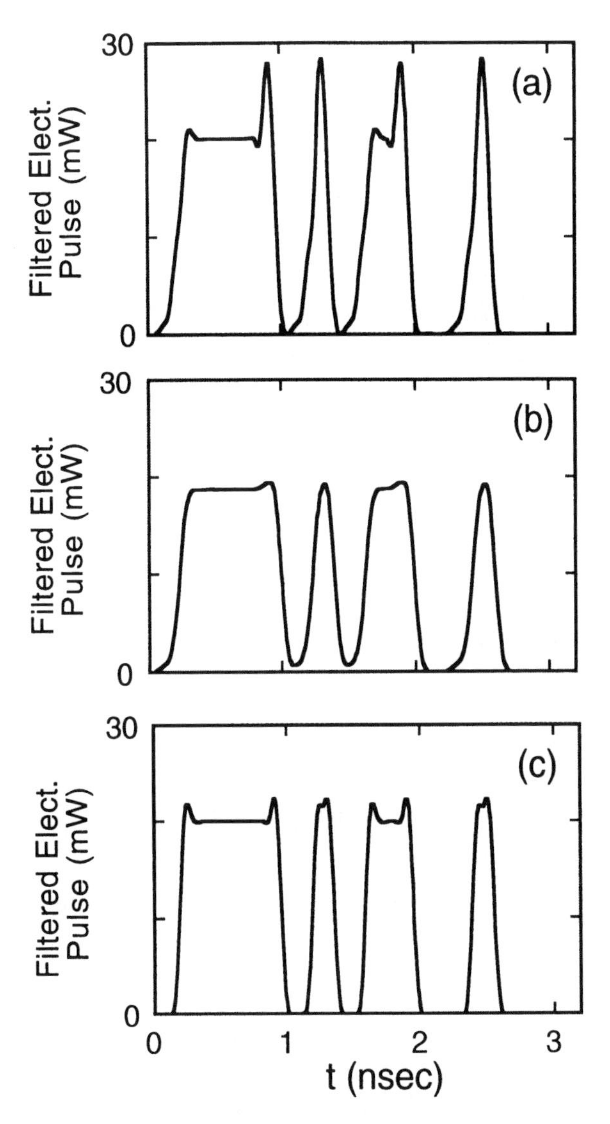

Figure 2. Detected and filtered current of the output pulses for $h_{\text{fiber}} = 100$ m and $L_B = 50$ m. (a) This figure shows the solution of the CNLS equation and the Manakov-PMD equation both with and without the nonlinear PMD term. All figures look alike; (b) Same as (a) but for a linear fiber; (c) Same as (a) but with PMD $= 0$.

CNLS equation and the Manakov-PMD equation both with and without the nonlinear PMD terms. All figures look alike. The solutions of the CNLS and the Manakov-PMD equation are identical. From Figure 2(a), the nonlinear PMD does not make any discernible difference. The shape of the pulses is strongly dependent on the random number sequence used for computing the random variations of the fiber orientation angle. The overshoot seen at the trailing edges of the pulses in Figure 2(a) can be somewhat larger or smaller. They can even switch sides and appear at the leading edges of the pulses. These differences correspond physically to the observation that different fibers with different random variations of the birefringence will have somewhat different behaviors.

To identify the source of the pulse distortion, we "turn off" the nonlinearity, obtaining Figure 2(b). Since in a linear system the compensated dispersion map prevents dispersion from distorting the pulses, the remaining slight pulse distortion seen in Figure 2(b) is caused entirely by linear PMD which is not affected by the dispersion map. However, not all the distortion in Figure 2(a) is due to the interaction of PMD and the fiber nonlinearity, as can be seen in Figure 2(c), which was computed with nonlinearity but with $D_{\mathrm{PMD}} = 0$. This figure shows that the nonlinearity conspires with dispersion to distort the pulses even though the overall cumulative linear dispersion of the system is zero. The slight asymmetry of the pulses in Figure 2(c) is attributable to the "electrical" filter which shows this tendency already in Figure 1(b), but which is enhanced by the substantial broadening of the spectrum at the end of the fiber. A comparison of Figure 2(a) through 2(c) makes it clear that the bulk of the pulse distortion is caused by the interaction of the linear PMD with the fiber nonlinearity. However, this interaction is almost entirely due to the interaction of the linear PMD with the Manakov nonlinearity, i.e. the Kerr nonlinearity averaged over the Poincaré sphere, since inclusion of nonlinear PMD leads to no "observable" effect.

To study the effects of nonlinear PMD, we carried out simulations with the same parameter set except that we changed h_{fiber} from 100 m to 10 m and L_B from 50 m to 10 km. Obviously, a beat length of 10 km is far larger than can be presently achieved in practice. Nonetheless, the mixing length is only 100 km in this case, which while approaching the nonlinear and dispersive scale lengths is not larger than them. Even under these fairly extreme conditions, we found no significant effect from the nonlinear PMD terms. To observe the effect of the nonlinear PMD terms, we had to further lower h_{fiber} from 10 m to 1 m. In this case, the mixing length equals 1,000 km which is larger than the nonlinear and dispersive scale lengths.

The pulse shape computed for these conditions with the CNLS as well as with the Manakov-PMD equation, including the nonlinear correction term, is shown in Figure 3(a). The figures computed with the two equations are indistinguishable. Figure 3(b) shows what happens when we neglect the nonlinear PMD. The pulse shape in Figure 3(b) is distinctly different from that of Figure 3(a). This example clearly shows that the nonlinear correction term in equation (1) does make a contribution if the inequality $h_{\mathrm{fiber}} \ll L_B$ is sufficiently-well satisfied. However, this example represents

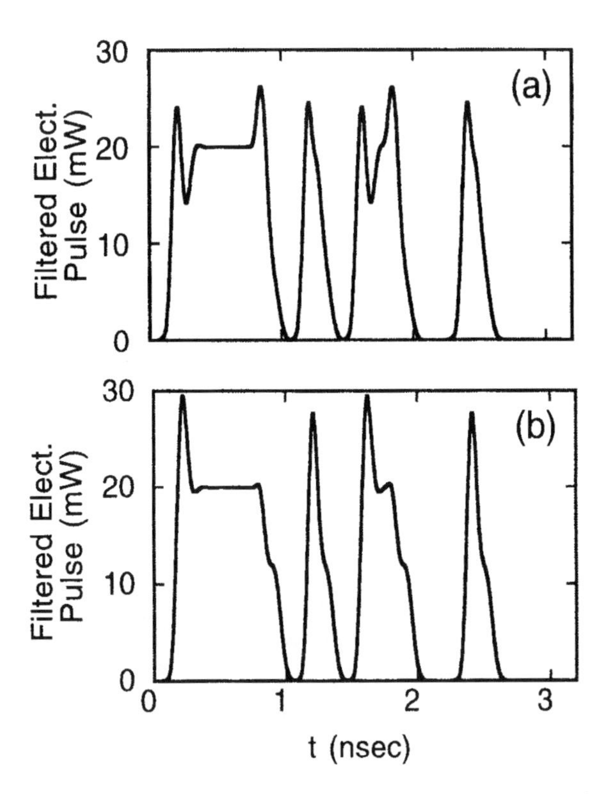

Figure 3. Detected and filtered current of the output pulses for $h_{\text{fiber}} = 10$ m and $L_B = 10,000$ m. (a) The solution of the CNLS equation and the Manakov-PMD equation with nonlinear PMD; (b) The solution of the Manakov-PMD equation without nonlinear PMD.

a highly extreme case that is unlikely to be found in practice so that the nonlinear correction term can safely be neglected for all communications fibers in use today.

5. Conclusions

In this paper we have calculated the variances of the nonlinear PMD fluctuations produced by random birefringence in an optical fiber. It was shown that for distances that are long compared with the polarization decorrelation length h_E, the nonlinear PMD corrections have zero mean and a variance given by $2Rh_E l$, where R is an $O(1)$ constant ranging from $1/15$ to $4/45$ depending on the particular correction term being considered. The mixing length h_E was found to be $L_B^2/(12\pi^2 h_{\text{fiber}})$ when $h_{\text{fiber}} \ll L_B$ and $h_{\text{fiber}}/12$ when $h_{\text{fiber}} \gg L_B$.

Situations in which the nonlinear PMD is not negligible occur when the mixing length h_E is not small in comparison with the dispersive or nonlinear length scales, so that incomplete mixing on the Poincaré sphere results. Since fiber decorrelation lengths for standard communication fiber are typically in the range 0.3–300 m, and beat lengths range from 10 to 100 km, such a situation is not likely to occur in the limit $h_{\mathrm{fiber}} \gg L_B$ where $h_E = h_{\mathrm{fiber}}/12$, since the resulting mixing lengths are quite short. Even when $h_{\mathrm{fiber}} \ll L_B$ and extreme values for h_{fiber} and L_B are considered, the resulting amount of nonlinear PMD is still negligible under conditions typical for practical, present-day communication systems. For example, for $h_{\mathrm{fiber}} = 0.3\,\mathrm{m}$ and $L_B = 100\,\mathrm{m}$, one obtains $h_E = 280\,\mathrm{m}$. This is much shorter than present-day dispersive and nonlinear scale lengths which are typically hundreds of kilometers.

References

1. Agrawal, G. P.: *Nonlinear Fiber Optics*, 2nd ed., Academic Press, San Diego, (1995).
2. Menyuk, C. R.: Nonlinear pulse propagation in birefringent optical fibers, *IEEE Journal of Quantum Electronics*, **23**, (1987), pp.174-176.
3. Menyuk, C. R.: Stability of solitons in birefringent optical fibers. I: Equal propagation amplitudes, *Opt. Lett.*, **12**, (1987), pp.614-616.
4. Menyuk, C. R.: Stability of solitons in birefringent optical fibers. II: Abitrary amplitudes, *J. Opt. Soc. Amer. B*, **5**, (1988), pp.392-402.
5. Wai, P. K. A., Menyuk, C. R. and Chen, H. H.: Stability of solitons in randomly varying birefringent fibers, *Opt. Lett.*, **16**, (1991), pp.1231-1233.
6. Evangelides Jr., S. G., Mollenauer, L. F., Gordon, J. P. and Bergano, N. S.: Polarization multiplexing with solitons, *J. Lightwave Technology*, **10**, (1992), pp.28-35.
7. Poole, C. D.: Statistical treatment of polarization dispersion in single-mode fiber, *Opt. Lett.*, **13**, (1988), pp.687-689.
8. Poole, C. D.: Measurement of polarization-mode dispersion in single-mode fibers with random mode coupling, *Opt. Lett.*, **14**, (1989), pp.523-525.
9. Foschini, G. J. and Poole, C. D.: Statistical theory of polarization dispersion in single mode fibers, *J. Lightwave Technology*, **9**, (1991), pp.1439-1456.
10. Menyuk, C. R. and Wai, P. K. A.: Polarization evolution and dispersion in fibers with spatially varying birefringence, *J. Opt. Soc. Amer. B*, **11**, (1994), pp.1288-1296.
11. Wai, P. K. A. and Menyuk, C. R.: Anisotropic diffusion of the state of polarization in optical fibers with randomly varying birefringence, *Opt. Lett.*, **20**, (1995), pp.2493-2495.
12. Wai, P. K. A. and Menyuk, C. R.: Polarization mode dispersion, decorrelation and diffusion in optical fibers with randomly varying birefringence, *J. Lightwave Technol.*, **14**, (1996), pp.148-157.
13. Marcuse, D., Menyuk, C. R. and Wai, P. K. A.: Application of the Manakov-PMD equation to the studies of signal propagation in optical fibers with randomly varying birefringence, *J. of Lightwave Technol.*, **15**, (1997), pp.1735-1746.
14. Wai, P. K. A., Kath, W. L., Menyuk, C. R. and Zhang, J. W.: Nonlinear polarization-mode dispersion in optical fibers with randomly varying birefringence, to appear in *Journal of Lightwave Technology.*
15. Wai, P. K. A. and Menyuk, C. R.: Polarization decorrelation in optical fibers with randomly varying birefringence, *Opt. Lett.*, **19**, (1994), pp.1517-1519.
16. Lichtenberg, A. J. and Lieberman, M. A.: *Regular and Stochastic Motion*, Springer-Verlag, New York, (1983).
17. Van Kampen, N. G.: *Stochastic Processes in Physics and Chemistry*, North Holland,

Amsterdam, (1992).
18. Papanicolaou, G. C.: Introduction to the asymptotic analysis of stochastic equations, in *Modern Modeling of Continuum Phenomena* (Lectures in Applied Mathematics), R. C. DiPrima, (Ed.), American Mathematical Society, 1977, **Vol. 16**, pp.109-147.

(Received November 17, 1997)

ULTRA-LONG DISTANCE SOLITON TRANSMISSION LINE DESIGN BY TWO-DIMENSIONAL PULSE WIDTH EVALUATION METHOD

K. SHIMOURA, Y. KANAOKA AND S. SEIKAI

Technical Research Center
The Kansai Electric Power Co., Inc.
3-11-20 Nakoji, Amagasaki, Hyogo 661, Japan

Abstract. We propose "Two-dimensional pulse width evaluation method" for the design of soliton communication systems. We applied this method to 20 Gbit/s transoceanic distance soliton lines. The amplifier output power becomes almost constant for all optimally designed soliton lines. The optimal averaged dispersion is reduced about 0.1 ps/nm·km and two-soliton interaction is suppressed in periodically dispersion compensated case. Stable nonlinear transmission can be achieved even if the averaged dispersion of 0.0 ps/nm·km. The dispersion compensation period should be longer than the amplifier spacing and 3 to 5 amplification spans are optimal.

1. Introduction

Many analytical, numerical and experimental investigations have been done to clarify the optimal condition of transoceanic distance soliton communication systems [1]-[3]. In such systems, we must optimize more than one parameter simultaneously. Two-dimensional pulse width evaluation method is a contour mapping of root mean square (rms) pulse width on the fiber dispersion–amplifier output power plane, and an effective method to know the transmission condition of soliton systems.

For the near future global communication, one-channel soliton systems have many practical advantages. Therefore we applied this method to the design of 20 Gbit/s transoceanic distance one-channel systems.

A. Hasegawa (ed.), New Trends in Optical Soliton Transmission Systems, 83–92.

2. Two-dimensional Pulse Width Evaluation Method

To clarify the optimal transmission condition, Q-factor contour mapping is proposed [4]. But for soliton systems, signal interactions between separated pulses are weak and pulse width analysis is effective to the first order estimation of system performance [5].

The pulse width evaluation method considers only one pulse, therefore the result is only applicable to the well pulse separated case. The condition for satisfactory transmission is evaluated whether the rms pulse width $Trms$ is less than 30 % of the bit slot or not. This condition is sufficient to attain a required S/N for various pulse shapes. For example, Gaussian function, this condition means 90.5 % of pulse energy is in a bit slot, therefore the S/N = 9.53. Q-factor represents the S/N at the receiver decision circuit and requires Q > 7 for the BER of 10^{-12} [11].

The signal component is evaluated by the pulse energy in the correct bit slot and the noise is evaluated by the energy out of the bit slot. This definition of Q can be applied to ultra-high speed RZ systems since the time constant of receiver electrical circuit becomes comparable to the bit rate and pulse energy is integrated at the decision circuit.

The pulse width analysis requires very small calculation time. We have executed below simulations with "Mathematica Ver.3.01"on a personal computer (Pentium-II, 266 MHz, WinNT 4.0). The calculation time for most cases are about 4 hours, where the distance step size is 5,000 m, time step is 1 ps and 441 (21×21) calculation points are used for contour maps drawing.

3. Uniform Dispersion Case

Figure 1 to Figure 3 show the transmissible condition for the 20 Gbit/s soliton line with uniform chromatic dispersion but different amplifier spacing La = 30 km, 50 km and 80 km. The contour lines indicate the areas where $Trms$ is less than 30 % of the bit slot. Transmission distances are (a) 2 Mm and (b) 10 Mm. The initial pulse shape is a sech^2-type and 10 ps FWHM is assumed. The Gaussian-shaped filter with 3 nm width is used in each amplifier. The nonlinear Schrödinger equation with a third order dispersion term (0.07 ps/nm^2·km) is solved by the split step Fourier method. Amplifier noise figure (6 dB) is also included. 4-bit slots (200 ps) are considered for the calculation.

From Figures 1(a), 2(a) and 3(a), $Trms$ becomes minimal at the dispersion value D_{opt} ps/nm·km of

$$D_{opt} = 0.20 \frac{T^2}{La}, \tag{1}$$

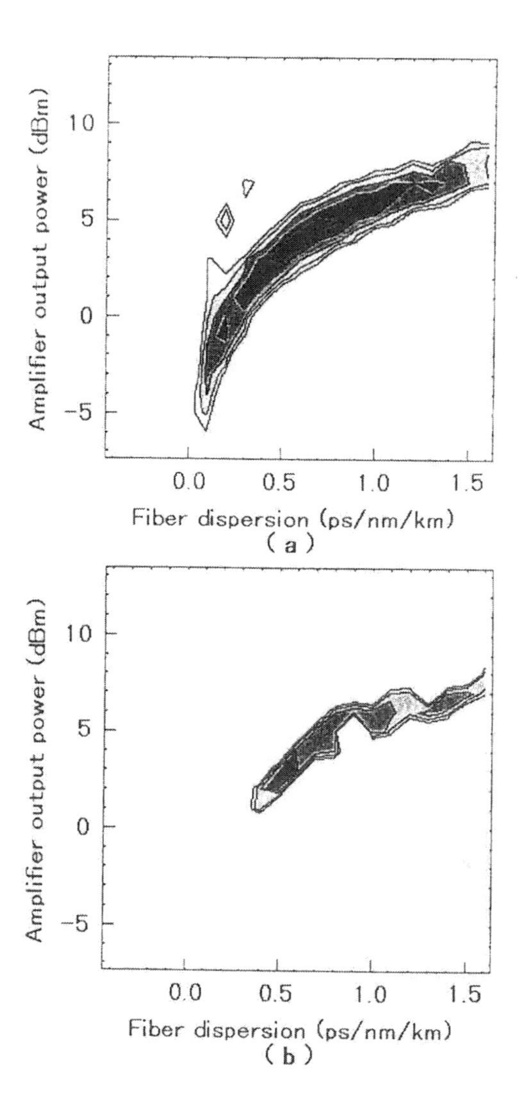

Figure 1. Transmissible areas of 20 Gbit/s uniform dispersion line for the distances of (a) 2 Mm and (b) 10 Mm. Amplifier spacing is 30 km. Contour lines indicate $Trms$ are (from outside) 30 %, 25 %, 20 %, 15 % and 12 % of the bit slot.

where T ps is a initial pulse width and La km is a amplifier spacing. This relation means that the optimal transmission can be achieved when the ratio of La to dispersion distance Ld is approximately 0.8. From guiding center soliton theory, this value should be much less than 1, but S/N and dispersion margins affect this value [6].

The optimal amplifier output power is about 3.6 dBm and the power margin is about 3.6 dB in every case. This fact can be derived from the

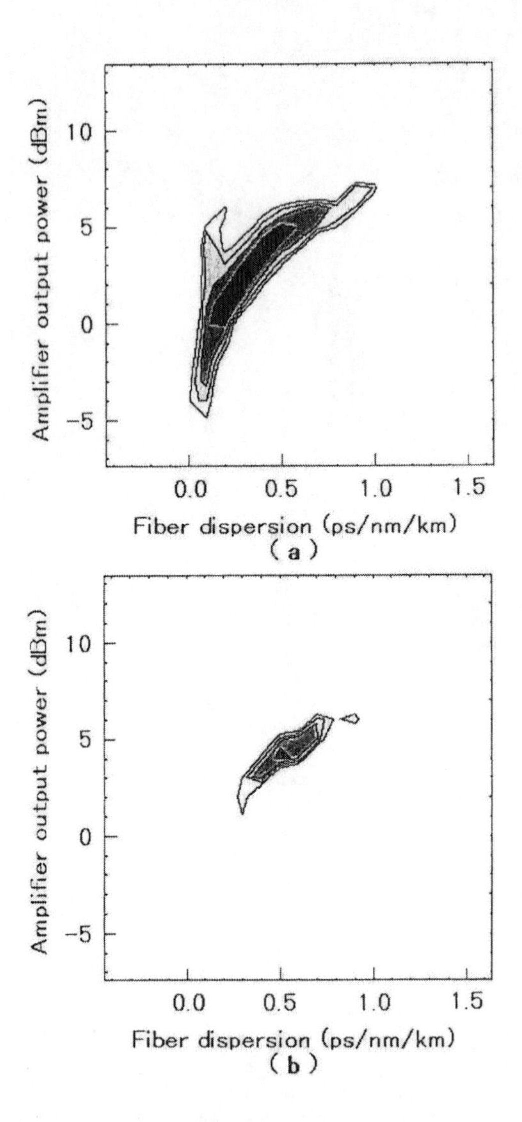

Figure 2. Transmissible areas of 20 Gbit/s uniform dispersion line for the distances of (a) 2 Mm and (b) 10 Mm. Amplifier spacing is 50 km. Contour lines indicate $Trms$ are (from outside) 30 %, 25 %, 20 %, 15 % and 12 % of the bit slot.

soliton peak power theory. The peak power of dynamic soliton Po mW for ordinary dispersion shifted fiber is approximated by

$$Po = 2.19 \times 10^3 \frac{D}{T^2} \ln Go \, , \tag{2}$$

where Go is the amplifier gain related with La.

$$\ln Go = \alpha La \, . \tag{3}$$

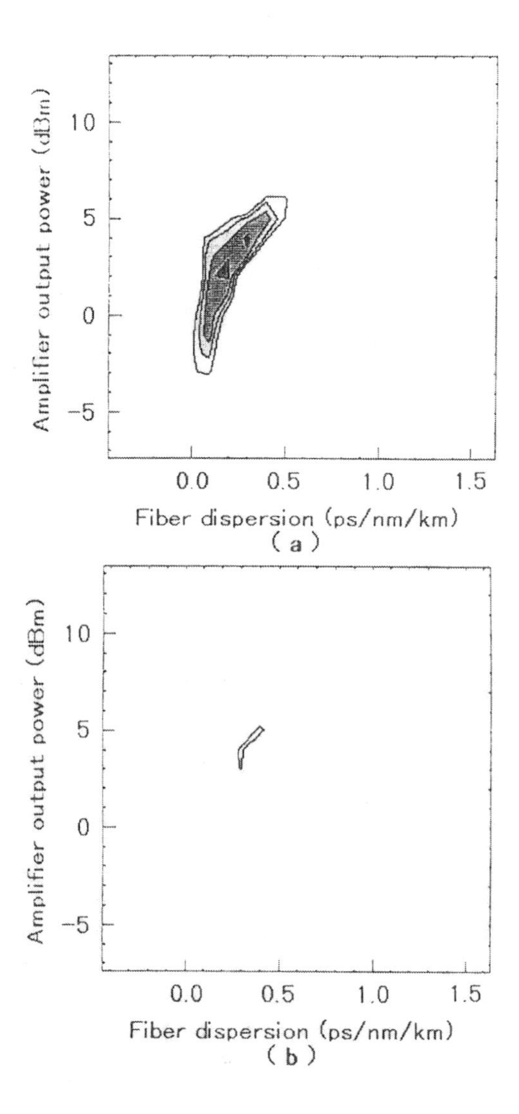

Figure 3. Transmissible areas of 20 Gbit/s uniform dispersion line for the distances of (a) 2 Mm and (b) 10 Mm. Amplifier spacing is 80 km. Contour lines indicate $Trms$ are (from outside) 30 %, 25 %, 20 % and 15 % of the bit slot.

Then Po is expressed by

$$Po = 1.28 \times 10^2 \alpha_0 \frac{La}{Ld}, \tag{4}$$

where α_0 is a loss factor in dB/km, related α by $\alpha = 0.2303\alpha_0$. This equation indicates that if we design the fiber dispersion as

$$\frac{La}{Ld} = const. \,, \tag{5}$$

then the soliton peak power doesn't depend on the amplifier spacing or the pulse width [12]. This relation is important for the design and operation of soliton systems because we can easily optimize all soliton systems by using the same output power amplifier. For example, non-equal amplifier spacing line can be designed easily and bit rate can be upgraded without changing the amplifier output power [7].

If we design the dispersion as $La/Ld = 0.8$, then the optimal peak power calculated from equation (4) is 20 mW (13 dBm) and the averaged amplifier output power becomes 2.3 mW (3.6 dBm), where α_0 of 0.2 dB/km and the mark ratio of 50 % are assumed. This value is adequate for the communication thanks to the Kerr coefficient $n_2 = 2.24 \times 10^{-20}$ m^2/W is an appropriate value [8].

4. Periodically Dispersion Compensated Case

We consider the transmission line model that dispersion compensation fibers (DCF) are periodically installed in uniform dispersion shifted fiber (DSF) line. In this model the compensation effect can be expected when the dispersion compensation Dc ps/nm greater than

$$Dc > 0.25T^2 \,, \tag{6}$$

because below this value the compensation period Lc becomes shorter than the dispersion length of the DSF at zero-averaged dispersion condition and the dispersion management effects are absorbed. For $T = 10$ ps pulse, $Dc >$ 25 ps/nm is required.

Figure 4 shows the transmissible areas for dispersion compensated case. DCF of −40 ps/nm is installed in every 3 amplifiers. $La = 50$ km and Lc = 150 km. In figure 4(a), the transmissible area is clearly appeared over the dispersion of 0.27 ps/nm·km, where the averaged dispersion becomes anomalous. The dynamic range is expanded to more than 10 dB, but the optimal power is almost same as the uniform dispersion case.

Figure 5 shows the 4-bit '1101'waveform development for the uniform dispersion case (a) and dispersion compensated case (b). The DSF dispersions are 0.25 ps/nm·km for (a) and 0.32 ps/nm·km for (b). The averaged dispersion of (b) is 0.05 ps/nm·km. The amplifier output powers are 1 dBm in both cases. In these cases rather small dispersions are used to suppress interaction and jitter, but two-soliton interaction is observed in (a). By the

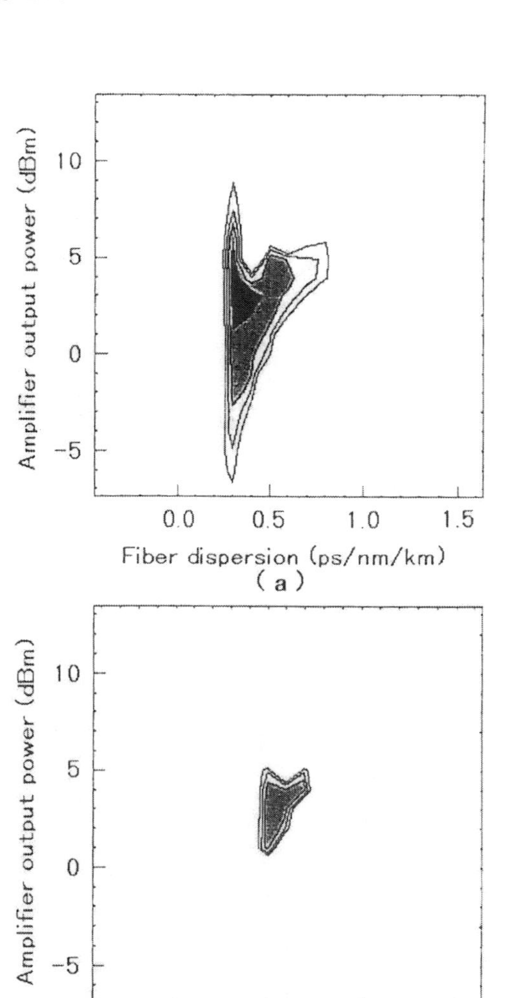

Figure 4. Transmissible areas of 20 Gbit/s dispersion compensated line for the distances of (a) 2 Mm and (b) 10 Mm. Amplifier spacing is 50 km and dispersion compensation fiber of −40 ps/nm is installed in every 3 amplification periods. Contour lines indicate *Trms* are (from outside) 30 %, 25 %, 20 % and 15 % of the bit slot.

periodic dispersion compensation, the optimal averaged dispersion can be reduced and interaction is suppressed without changing the signal power.

The limitation of equation (1) is determined by the two-soliton interaction theory and the collision distance is given by reference [12]

$$Lt < La \exp\left(\frac{0.882}{r}\right), \tag{7}$$

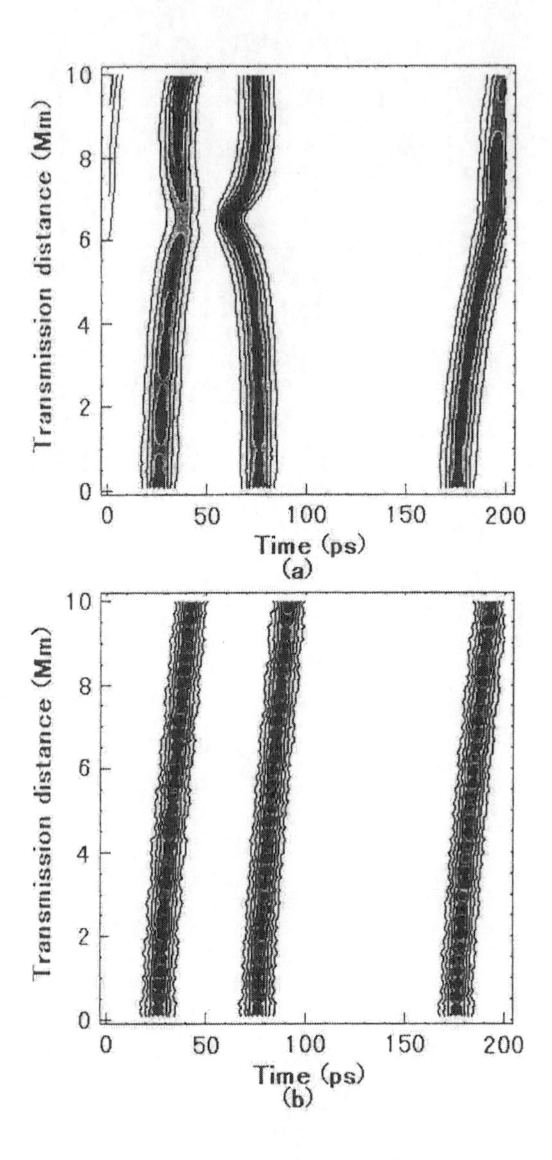

Figure 5. Waveform development of (a) uniform dispersion case ($D = 0.25$ ps/nm·km), and (b) dispersion compensated case ($D = 0.32$ ps/nm·km, $Dav = 0.05$ ps/nm·km). Amplifier output powers are 1 dBm in both cases.

where r is a pulse duty ratio defined by $r = T_{FWHM}/T_{SLOT}$. Beyond this distance, dispersion management becomes effective.

5. Nonlinear Transmission at Zero-averaged Dispersion

Figure 6 shows the zero-averaged dispersion transmission at different compensation period. The same dispersion compensation fiber of -40 ps/nm

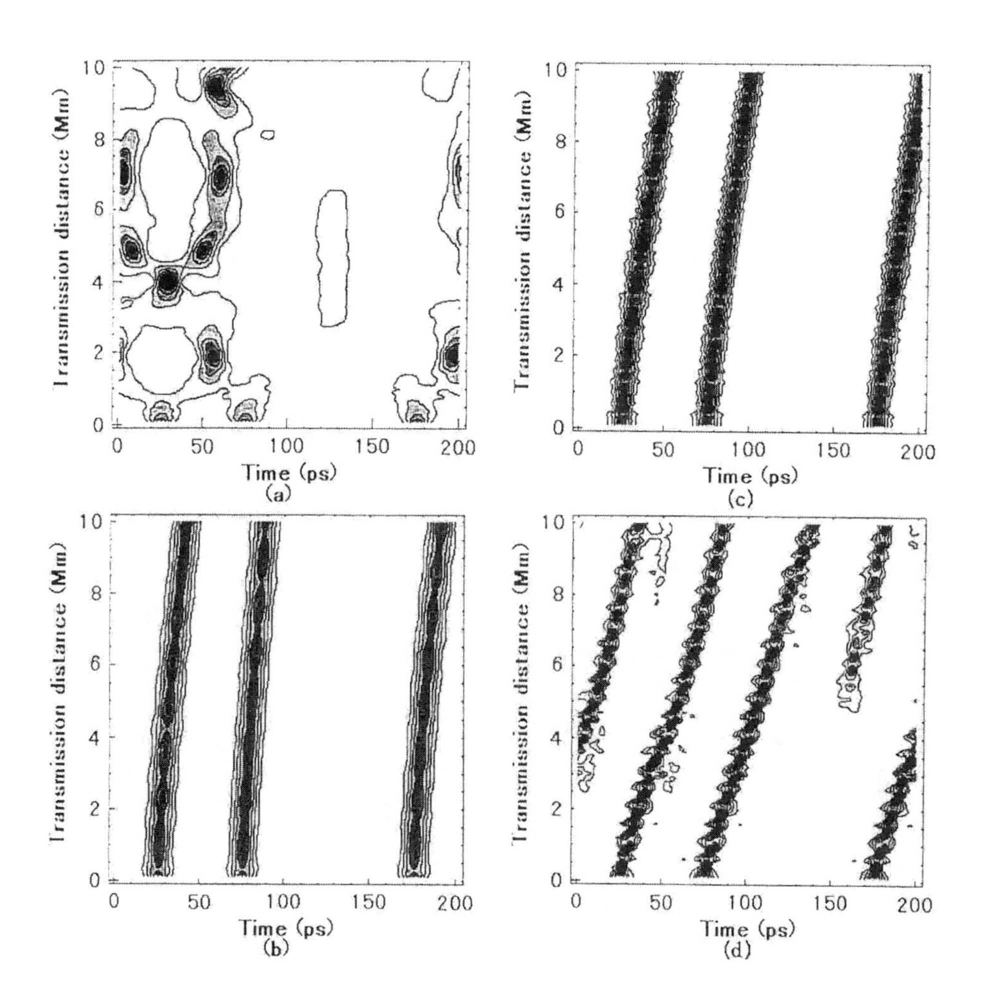

Figure 6. Nonlinear dispersion compensated transmission at zero-averaged dispersion. Dispersion compensation fiber of −40 ps/nm are installed in every (a) 1, (b) 2, (c) 3 and (d) 10 amplification spans. Amplifier output powers are 1 dBm and 50 km spacings.

is used in each case, but Lc are (a) 50 km, (b) 100 km, (c) 150 km and (d) 500km. The amplifier spacings are 50 km and the amplifier output powers are 1 dBm in all cases.

Nonlinear pulse stabilization effects are observed and stable transmissions are achieved in Figure (b) and (c). The transmission characteristics become very sensitive to the fiber dispersion in case (b), therefore 3 to 5 amplification spans seems to appropriate for the compensation period. This scheme is important for the transoceanic communication systems [9, 10].

6. Conclusion

Two-dimensional pulse width evaluation method is proposed to the design of one-channel soliton systems. We find the amplifier output power becomes almost constant for all optimally designed soliton lines and confirmed the remarkable advantage of the periodic dispersion compensation scheme for ultra-long distance transmission.

This simulation method should be combined with multi-bit waveform calculation, but gives significant information about soliton systems. A sample code can be downloaded from next URL.

http://www.asahi-net.or.jp/ ix6k-smur/soliton.html.

References

1. Suzuki, M., Edagawa, N., Morita, I., Yamamoto, S., Taga, H. and Akiba, S.: *Multi-ten Gbit/s soliton transmission over transoceanic distances*, in A. Hasegawa (Ed.), Physics and Applications of Optical Solitons in Fibres'95, Kluwer Academic Publishers, (1995), pp.375-391.
2. Farre, J. and Ono, T.: (1995) *Implementable soliton system design*, in A. Hasegawa (Ed.), Physics and Applications of Optical Solitons in Fibres'95, Kluwer Academic Publishers, (1995), pp.351-363.
3. Mollenauer, L. F., Evangelides Jr., S. G. and Haus, H. A.: Long-distance soliton propagation using lumped amplifiers and dispersion shifted fiber, *J. Lightwave Technol.*, **9**, (1991), pp.194-197.
4. Sahara, A., Kubota, H. and Nakazawa, M.: Q-factor contour mapping for evaluation of optical transmission systems: soliton against NRZ against RZ pulse at zero group velocity dispersion, *Electron. Lett.*, **32**, (1996), pp.915-916.
5. Agrawal, G. P.: *Nonlinear Fiber Optics*, Academic Press, (1995), Chap.3.
6. Hasegawa, A. and Kodama, Y.: *Solitons in Optical Communications*, Oxford University Press, (1995).
7. Nakazawa, M., Kimura, Y., Suzuki, K, Kubota, H., Komukai, T., Yamada, E., Sugawa, T., Yoshida, E., Yamamoto, T., Imai, T., Sahara, A., Yamauchi, O. and Umezawa, M.: Soliton transmission at 20 Gbit/s over 2000 km in Tokyo metropolitan optical network, *Electron. Lett.*, **31**, (1995), pp.1478-1479.
8. Namihira, Y.: Effect of optical pulse broadening on nonlinear refractive index measurements by self-phase modulation at 1.55 μm, *Optoelectronics and Communications Conference Technical Digest*, **17C2-3**, (1996), pp.148-149.
9. Suzuki, M., Morita, I., Edagawa, N., Yamamoto, S., Taga, H. and Akiba, S.: Reduction of Gordon-Haus timing jitter by periodic dispersion compensation in soliton transmission, *Electron. Lett.*, **31**, (1995), pp.2027-2029.
10. Nakazawa, M., Yamada, E., Kubota, H., Yamamoto, T. and Sahara, A.: Numerical and experimental comparison of soliton, RZ pulse and NRZ pulses under two-step dispersion allocation, *Electron. Lett.*, **33**, (1997), pp.1480-1482.
11. Bergano, N. S., Kerfoot, F. W. and Davidson, C. R.: Margin Measurements in Optical Amplifier Systems, *IEEE Photon. Technol. Lett.*, **5**, (1993), pp.304-306.
12. Shimoura, K. and Seikai, S.: Fiber dispersion and amplifier output power design for soliton transmission systems, *IEICE Trans.*, **E81-C, No.2**, (1998), to be published.

(Received December 25, 1997)

PRACTICAL DESIGN OF DISPERSION MANAGED SOLITON SYSTEM

G. C. GUPTA, Y. YANO, T. ONO AND K. EMURA
C&C Media Research Laboratories, NEC Corporation
4-1-1, Miyazaki, Miyamae-ku, Kawasaki 216, Japan

Abstract. Here we present a model and simulation results for a dispersion-managed soliton system. In this system, both the dispersion map and launched power are optimized to maintain the soliton initial phase at each span length to reduce soliton interaction and soliton break-up. A practical design of the system using a dispersion compensation fiber is investigated for 10 ps-pulsewidth, 20 Gbit/s soliton signals transmitted over 18,000 km with 60 km amplifier spacing. To demonstrate the suppression of soliton interaction by this technique, we kept the pulsewidth at 10 ps and increase the bit rate. We found that for a higher bit rate, a dispersion compensation fiber should be placed in each amplifier span to suppress soliton interaction. Simulation results suggest that by optimizing technical parameters, 40 Gbit/s soliton signals can be stably transmitted over 15,000 km and 50 Gbit/s soliton signals over the trans-oceanic distance of 9,000 km.

1. Introduction

In the last couple of years great interest has been focused on investigating soliton-based dispersion management (DM) techniques both by simulations and experiments [1]-[9]. The most apparent advantage of using DM technique is that it is not necessary for every fiber segment to have anomalous dispersion. As long as the average group velocity dispersion (GVD) is zero or within an anomalous region for each span length, a soliton can propagate through a fiber with normal GVD. Recently, a quasi-soliton technique had been proposed, in which the soliton pulse is chirped at the transmitter end using a phase modulator [10]. This technique increases the complexity of the system and requires an exponentially dispersion-decreasing fiber. Here we present a practical design for a DM soliton system. We use the

A. Hasegawa (ed.), New Trends in Optical Soliton Transmission Systems, 93–102.

practical optical pulse shape (10-ps wide) through fiber with an optimized dispersion-arrangement. The way to optimize it to maintain the initial soliton phase condition at each span in order to reduce the soliton interaction. The simulation results presented here suggest potential advantages.

This paper is organized as follows: in the next section we outline the system design issues and then describe the design of the DM soliton system. In Section 3 we present simulation results. First, we compared the system design of 20 Gbit/s soliton system for different dispersion management span. In this comparison we kept the span length, pulsewidth and bit rate constant for all cases while other technical parameters were optimized. Then to demonstrate the suppression of soliton interaction by this technique, we kept the pulse width at 10 ps while increase the bit rates from 20 Gbit/s up to 50 Gbit/s (twice the pulse separation) for the same span length (60 km).

2. System design issues

The main issue in designing the soliton transmission system is to find the optimum system parameters that give the largest margins in terms of power, chromatic dispersion and pulse-width that allow stable transmission over the longest distance. Even though a soliton is quite a robust pulse, the management of these effects is the key to designing soliton communication systems. The maximum transmission length of the system is mainly set by three limits related to the signal-to-noise ratio, pulse jitter and soliton interaction. Especially, for the higher bit rates, the effect of soliton interaction will be critical to system performance degradation. Thus, finding a method of reducing the soliton interaction has become very importance, which is the issue we focus on here.

In a soliton transmission system, a soliton is confined to the time slot $T_B = 1/B$, determined by the bit rate B, and it accompanies other solitons in the neighboring time slots. When two solitons separated in time by a few soliton widths propagate together inside an optical fiber, the overlap and interference of their tails become a source of interaction. To achieve higher-bit-rate transmission, it is thus necessary to determine how close two solitons can come without interacting with each other. Interaction between them depends not only on the pulse distance but also on their relative phase and amplitude. Using short pulses reduces the dispersion length (Zc), and shortening the Zc increases the generation of the dispersive wave emission. On the other hand, reducing the dispersion increases the instability of the system to chromatic dispersion fluctuations, which chirp the solitons and cause pulses to tend to separate into soliton and non-soliton components.

Reference [3] describes the basic idea for reducing soliton interaction.

It has been shown that introducing a phase difference between neighboring solitons, unequal amplitudes in adjacent pulses, and active soliton transmission control techniques can reduce soliton interaction. In reference [10], the authors required an exponentially decreasing fiber and chirped the soliton pulse at the transmitter end using a phase modulator. We, on the other hand, used a non-chirp soliton pulse at the transmitter end. Using a chirped soliton pulse at the transmitter end increases the complexity of the system, requires additional active components, and increases the cost of the system. In this paper, we present a simple system design without any active control of the DM soliton system. This offers promising results for suppressing soliton interaction and maintaining soliton stability to transmit high-bit-rate signals over a long distance without soliton break-up.

3. Dispersion Managed Soliton System

Here we outline a DM optimization technique for maintaining the soliton phase at its initial state in each dispersion management period. Soliton extra ordinary property is that the optical phase remains uniform across the entire pulse (intensity independent) because of the interplay between non-linear and dispersive effects. The phase of the non-soliton components is intensity-dependent because of the self-phase modulation (SPM) induced phase shift, since the phase of the pulse is modulated by its own intensity. As the soliton propagates, its shape does not change, but its phase changes linearly with peak power and propagation distance [1, 2]. Since non-soliton components are partially coherent, when they propagate inside the fiber they becomes progressively less coherent. Such a coherent degradation can be understood by noting that SPM converts the intensity fluctuation into additional phase fluctuations, making light less coherent [1, 2].

Thus, to reduce the effect of non-soliton components as well as soliton interaction, one powerful technique is dispersion management to control the optical phase of solitons. Dispersive elements, such as a dispersion compensation fiber (DCF), are used in each span length to maintain the soliton phase in its initial state. The dispersion-managed technique reduces the soliton interaction. It also reduces the coupling between soliton and non-soliton component, so soliton remain stable, over a long distance without breaking up into soliton and non-soliton components.

Simulation results showed that there exist criteria in order to provide dispersive elements at each span length to maintained soliton phase in its initial state. Optimizing both the dispersion map and launch power can reduce soliton interaction; this was due to providing chirp at every span and this attribute to pulse rapid phase rotation. Thus, higher-bit-rates signals can be stably transmitted over a long distance without soliton break-up. Using dispersive element at each span length to control the soliton initial phase, we also partially achieve dispersion compensation at each span length.

4. Simulation Results

4.1. SYSTEM DESIGN

4.1.1. *Simulation method*

First, to examine the advantages of the DM system, we compare its transmission characteristic with different span of the dispersion management. There are many system parameters in the design of soliton systems, including fiber dispersion, transmission power, average GVD, fiber loss and span length. In such a case, optical non-linear simulation analysis is a very powerful tool for obtaining the optimum transmission distance for a high-speed soliton transmission system. The propagation behavior of the ultra-short pulse in a single-mode fiber is described by the Non-linear Schrödinger Equation. We have developed an optical non-linear simulator using the split-step Fourier Transform method to solve the Non-linear Schrödinger Equation. First, we consider the case for the 10 ps, 20 Gbit/s soliton signal transmission. We used a span length of 60 km in the following calculation. Used parameters are, the pulsewidth 10 ps, 20 Gbit/s bit rate, fiber area, 50 μm^2, fiber attenuation 0.21 dB/km, slope of dispersion 0.07 ps/nm^2·km, Kerr coefficient 2.62×10^{-20} m^2/W and pseudorandom bit sequence (PRBS) 2^7. For each of these cases, dispersion map and launch power was optimized. The initial phase at the transmitter end was non-chirp for all four cases. The average GVD of the four configurations was an anomalous region in the center wavelength of 1.552 μm, so it's satisfies the soliton condition. At the receiver end, the signal was received by a photo-diode and filtered by a second-order Butterworth-type low-pass filter with a bandwith of 0.8 B (bit rates). In these simulation results, to simplify the problem we chose to neglect ASE noise of EDFA, polarization mode dispersion and polarization hole burning.

4.1.2. *System configuration*

Many system parameters are involved in determining the optimum condition for the dispersion management of soliton transmission using a DCF. In the presence of soliton interactions, the generation of dispersive waves may lead to strong signal degradation. Therefore we decided to put the DCF in front of EDFA in this simulation [11, 12]. We compared the different spans for the dispersion management as shown in Figure 1. In Case I, which is single-span DM case, dispersion is in positive (anomalous fiber) region and negative dispersion (dispersive, DCF) in each span to control to maintain

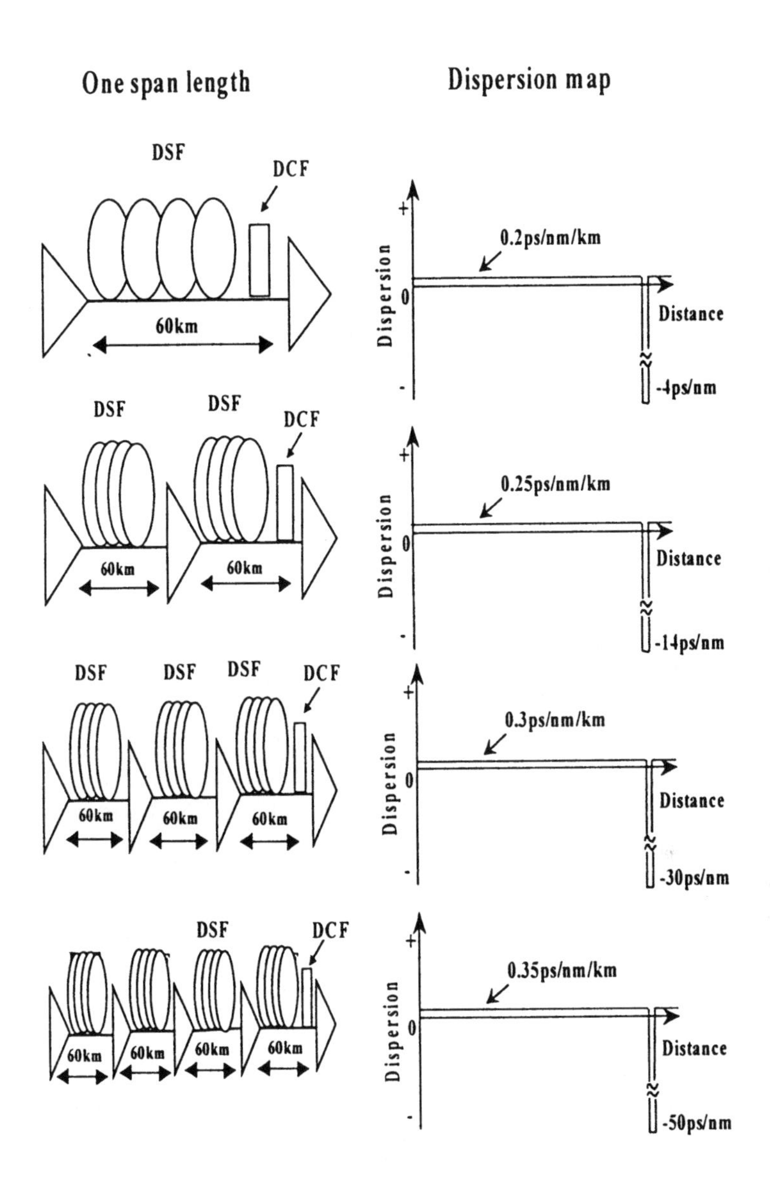

Figure 1. One span length configuration and corresponding dispersion map for four different span DM techniques (four cases, 10 ps, 20 Gbit/s).

the soliton initial phase. In Cases II, III and IV we used a DCF after two, three and four spans respectively. In each case, the optimization of the dispersion map as well as launch power is very important parameters. The optimized dispersion values and fiber lengths for 20 Gbit/s transmissions are shown in Figure 1.

4.1.3. *Results*

Figure 2 shows the eye-opening penalty (dB) versus the distance (km) for all four cases. From these results, we can estimate that the 10 ps pulsewidth, 20 Gbit/s with pulse spacing five times, dispersion managed soliton signals can be stably transmitted over the trans-oceanic distance of 9,000 km with a penalty of less than 1 dB in every case. For all four cases, results are close. We kept the pulsewidth at 10 ps and increase the bit rates from 20 to 40 Gbit/s (2.5 time pulse separations), and launch power was increased by 3 dB, (Cases I, II, III and IV in Figure 3). With a large dispersion step in the compensation period, (Cases II, III and IV), we found that it was not possible to transmit at 40 Gbit/s bit rates with 2.5 times pulse spacing over trans-oceanic distance. Because the large dispersion step in the compensation period introduced oscillation tails in the pulse, and the interaction between neighboring soliton pulses seriously degraded the system performance. In Case I (dispersion compensation in each span), we found that the required optimized dispersion value of the DCF is twice as compared with 20 Gbit/s system. In Case I, the interaction between neighboring pulses was reduced compared with other cases, because of the pulse attributes to their rapid phase rotation in each span. This indicates that the frequency of dispersion management is very important for higher-bit-rate soliton systems in order to suppress soliton interaction. This result is clearly indicates that a properly optimized DM technique improves soliton stability and reduces soliton interaction. This was due to optimization of the system parameters and pulse attributed to rapid phase rotation, which resulted in less coupling between soliton and non-soliton components.

4.2. TRANSMISSION CHARACTERISTICS

4.2.1. *Effect of span length*

In order to maintain the soliton stability, the amplifier spacing has to be much shorter than the dispersion length [1, 2]. If amplifier spacing is large, soliton are strongly perturbed and cannot survive over a long distance. A perturbed soliton generates dispersive wave radiation of low power that is not part of the soliton. Therefore, it's spread out as any linear wave during traveling through in dispersive medium. For soliton communication, one must try to minimize the energy loss to dispersive wave because their interaction with the remaining solitons and as a result will lead to soliton instability. To check the soliton stability, we kept the 10 ps pulsewidth and increased the span from 60 to 75 km for the 20 Gbit/s system. For each different span length, the dispersion maps of fibers were arranged optimally, and launch power was also optimized. Figure 3 shows eye-opening penalty (dB) versus distance (km) for different amplifier spacing (65, 70 and 75

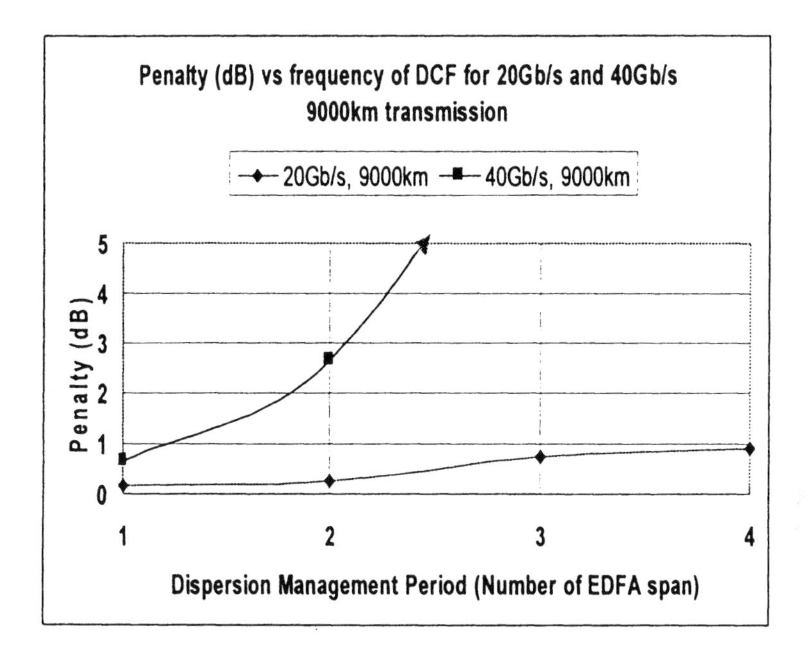

Figure 2. Penalty (dB) vs. dispersion management period (number of EDFA span) for all four cases, 20 Gbit/s and 40 Gbit/s for 9,000km (10 ps pulsewidth).

km). From these results, we can estimated that 10 ps pulsewidth, 20 Gbit/s dispersion-managed soliton signals can be stably transmitted over 10,000 km with a 75 km amplifier spacing with a penalty of less than 1 dB.

4.2.2. *Results for 40 Gbit/s and 50 Gbit/s*

A relatively large spacing between neighboring soliton pulses is necessary to avoid soliton interaction, which limits the bit rate of soliton communication systems. We considered the 40 and 50 Gbit/s (twice the pulse separation) for Case I and optical power was optimized. The average GVD were 0.06 and 0.041 ps/nm·km for 40 Gbit/s and 50 Gbit/s system in the center wavelength 1.552 μm. The other technical parameters were optimized for the same span length (60 km), this simulation result is presented here. Figure 4 shows the eye opening penalty (dB) versus the distance (km) for 10 ps, 20 Gbit/s, 40 Gbit/s and 50 Gbit/s DM soliton systems. The result indicate that 10 ps pulsewidth, 40 Gbit/s soliton signals can be stably transmitted over 15,000 km with 0.6 dB penalty and 50 Gbit/s ones over trans-ocean distance 9,000 km with 1 dB penalty. Amplifier spacing was 60 km for both above cases. In the 20 Gbit/s case (Case I), a further simulation checked even longer distance and indicated that transmission was possible

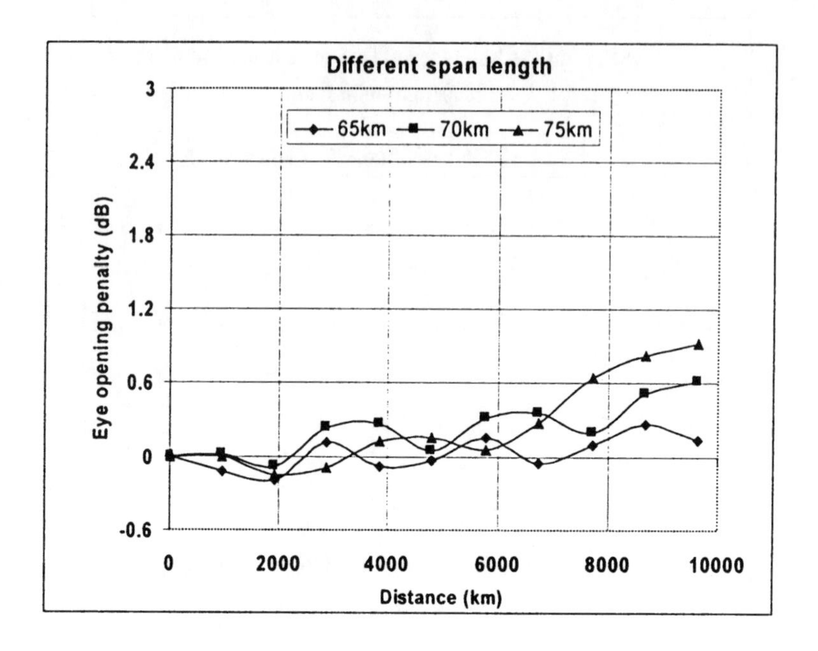

Figure 3. The eye opening penalty (dB) vs. distance (km) for 65, 70 and 75 km span length for 10,000 km distance using Dispersion Managed Soliton System (10 ps, 20 Gbit/s).

up to 18,000 km with a penalty less than 1 dB (Figure 4).

5. Discussion

In Figure 4, we showed that sech 10 ps pulsewidth soliton signals can be stably transmitted over trans-oceanic distance with a bit rates of 50 Gbit/s (twice the pulse-separation). From this analysis, we found that for a higher bit rate a DCF should be placed in each amplifier span, before EDFA's in order to suppress soliton interaction. These results indicate that a properly optimized DM technique reduces soliton interaction and enables soliton to maintain their stability over trans-oceanic distances. Since this technique uses only passive in-line components, the proposed system is easy to implements and maintain, provides a large amplifier spacing of 60 km, which reduces the cost of the transmission system, and can be upgrade in the future. For high-bit-rates trans-oceanic distance transmission application, DM technique could be one of the best candidates. Although ASE noise was not considered in this analysis, its effects should be considered in real system.

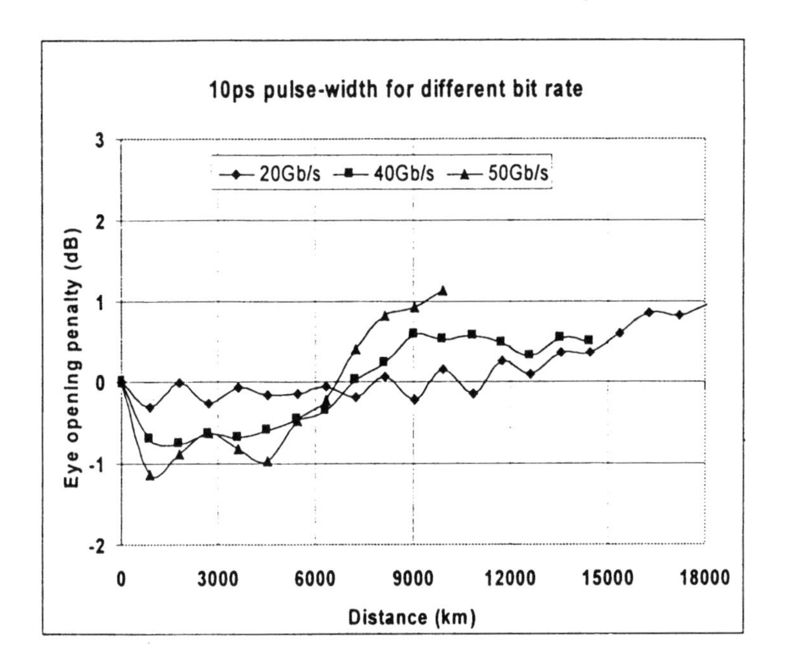

Figure 4. Eye opening penalty (dB) vs. distance (km) for 20 Gbit/s, 40 Gbit/s and 50 Gbit/s Dispersion Managed Soliton System (10 ps pulse-width).

6. Conclusion

The dispersion management technique can overcome the fundamental limitation imposed by earlier proposed soliton techniques. By reducing the soliton interaction it enables the system to stably transmit at higher bit rates over a long distance without soliton break-up. A system designed using this technique offers good transmission characteristics in terms of high bit rate up to 50 Gbit/s and over trans-oceanic distances. It can overcome the transmission limit imposed by the non-linear nature of the solitons. Furthermore, this technique uses only passive in-line components, so the system should be easy to implement in practice.

Acknowledgements

The authors would like to express their sincere gratitude to Dr. S. Goto, Dr. J. Namiki and Dr. M. Shikada for their encouragement and Mr. T. Suzaki and Mr. K. Fukuchi for their many useful suggestions. This work was done for the "Soliton based all-optical communication network research"project supported by Telecommunications Advancement Organization of Japan (TAO).

References

1. Hasegawa, A. and Kodama, Y.: *Soliton and optical communication*, Oxford press, (1995).
2. Agrawal, G. P.: *Nonlinear fiber optics*, Academic press, (1995).
3. Taylor, J. R.: *Optical soliton-theory and experiment*, chapter-5, soliton-soliton interaction, (1992), Cambridge University press.
4. Suzuki, M., et al.: *Electron. Letter*, **31**, (1995), p.2027.
5. Haus, H. A., et al.: *J. Quantum Electronic*, **31**, (1995), p.591.
6. Gabitov, I., et al.: *Opt. Letter*, **21**, (1996), p.327.
7. Smith, N. J., et al.: *Electron. Letter*, **32**, (1996), p.54.
8. Nakazawa, M., et al.: *Jpn. J. Appl. Phys.*, **34**, (1995), L681.
9. Nakazawa, M., et al.: *Electron. Letter*, **31(5)**, (1995), p.216.
10. Kumar, S. and Hasegawa, A.: *Optics Letter*, **22(6)**, (1997), p.372.
11. Wabnitz, S., et al.: *Photonics Tech. Letters*, **8(8)**, (1996), p.1091.
12. Wald, M., et al.: *Photonics Tech. Letters*, **9(12)**, (1997), p.1670.

(Received January 12, 1998)

PHOTON-NUMBER NOISE OF SPECTRALLY FILTERED OPTICAL SOLITONS

A. SIZMANN, S. SPÄLTER, M. BURK, U. STRÖßNER,
M. BÖHM AND G. LEUCHS
Lehrstuhl für Optik
Physikalisches Institut der Universität Erlangen-Nürnberg
Staudtstr. 7/B2, D-90158 Erlangen, Germany

1. Introduction

Fundamental optical solitons in fibers are stationary objects if described within the framework of classical electrodynamics. Quantum optical solitons, however, have uncertainties associated with their position, momentum, phase and amplitude. These uncertainties become mutually correlated, broaden or contract as the quantum soliton propagates. Thus, the fundamental quantum soliton is not a stationary object during propagation [1]-[4].

Studying the quantum dynamics of optical fiber solitons is interesting and useful in many ways. First, fundamental bounds of signal-to-noise ratios in pulse propagation, optical functions and devices can be established through quantum-noise-limited measurements. Knowing the physical limitations of a fiber-optic communication system serves as a reference to which the actual performance can be compared and beyond which it cannot be improved. Second, the physics of a nonlinear fiber-optic system can be used for passive, in-line reduction of classical and quantum noise instead of using active stabilization loops that are engineered around the system. The latter can reach at best the shot-noise level within a limited bandwidth. Another advantage of the passive low-noise techniques used in measurements below the shot-noise limit is their enormous bandwidth. Based on the Kerr and Raman effect in fibers, noise reduction is expected to be as broadband as the corresponding nonlinearity. Finally, quantum measurements on solitons is stimulating research on new devices, such as an all-optical limiter that uses spectral filtering.

A. Hasegawa (ed.), New Trends in Optical Soliton Transmission Systems, 103–115.

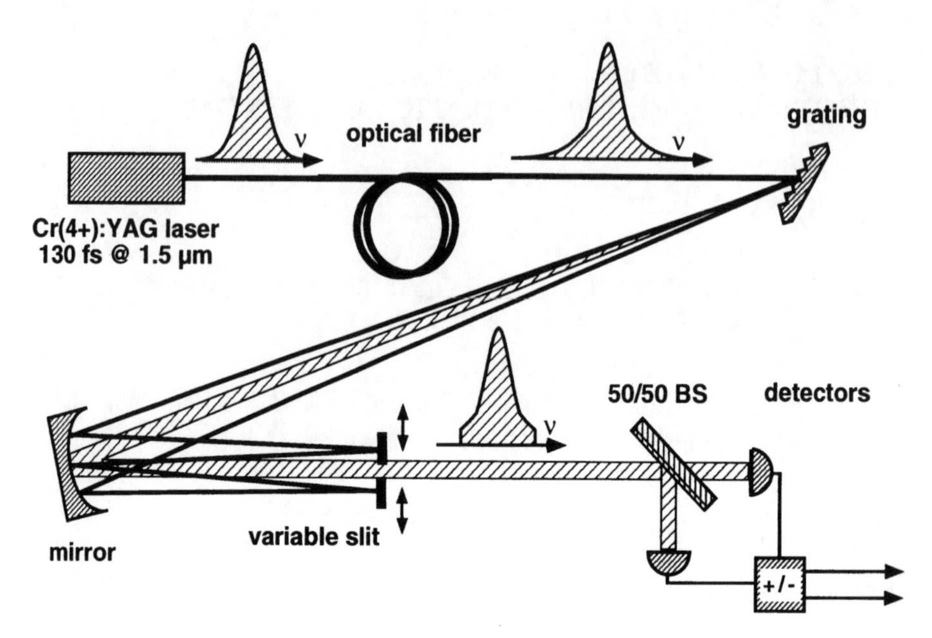

Figure 1. Experimental setup for photon-number noise reduction.

Spectral filtering of solitons may constitute an effective ultrabroadband noise reduction device. If not designed properly, however, a spectral filter in a soliton line can raise vacuum fluctuations to more than 10 dB above the shot-noise level. This paper reviews the state of the art of quantum measurements using spectrally filtered solitons. First, the basic experimental setup and results are briefly presented. Next, models that describe the observed noise reduction are discussed. Then, spectral filtering experiments using picosecond and sub-picosecond pulses are reviewed in Sections 4 and 5. The final section summarizes the paper.

2. Experimental Setup and Results: An Overview

Recently, an unexpected intensity noise reduction mechanism was discovered by NTT researchers S. Friberg and coworkers [5, 6]. The experimental setup used by Friberg et al. and later by Spälter et al. [7, 9], is shown schematically in Figure 1.

An optical soliton is spectrally filtered after propagating through a fiber over several soliton periods. Depending on pulse width, energy and fiber length, the pulse spectrum will broaden or contract. With a bandpass filter centered around the output pulse spectrum, more or less energy is transmitted through the filter. As a result, a nonlinear input-output transfer function for the pulse energy is obtained (Figure 2).

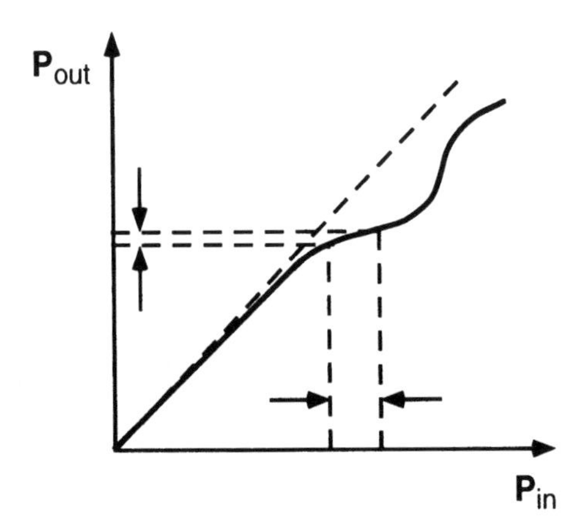

Figure 2. Schematic diagram of intensity noise reduction from a nonlinear input-output energy-transfer function.

At certain input energies, input fluctuations should be deamplified due to the reduced slope of the energy transfer function. It was found that even the quantum uncertainty of the intensity, i.e. the photon-number uncertainty, is reduced at certain input energies, leading to directly detectable sub-shot-noise photon-number fluctuations at the output. Table 1 summarizes experimental demonstrations and theoretical predictions for noise reduction in such a system. Up to 3.8±0.2 dB (59 %) of noise reduction below the shot-noise was measured recently [9], using sub-picosecond solitons. Extrapolating from an overall detection efficiency of 76 % in this measurement, loss-free detection would yield a quantum noise reduction of 6.4±0.8 dB (77 %) in this system.

An important difference between experiments with picosecond (Section 4) and sub-picosecond pulses (Section 5) is the influence of the Raman effect. Sub-picosecond pulses show a Raman self-frequency shift [10, 11] already in short pieces fiber. For picosecond pulses the Raman effect eventually becomes significant for the long distances used in fiber-optic communication systems. In Section 5, the influence of the Raman effect on the noise measurements is discussed. For instance, high-pass filtering is an interesting application, because it may be used to limit the Raman-self-frequency shift in fiber channels. Recent noise measurements with self-frequency shifted solitons show that high-pass filtering is also an effective method for noise reduction far below the standard quantum limit.

TABLE 1. Experimental and theoretical results for noise reduction in photon-number or intensity-1/f noise of spectrally filtered solitons. The pulse widths are FWHM values. C is the chirp parameter as defined in G. P. Agrawal: "Nonlinear Fiber Optics" (Academic Press 1995). The photon-number noise reduction values in parentheses are: * extrapolated for 100 % detection efficiency (experiment), ** includes Raman scattering at 300 K (theory).

reference	noise reduction	pump light
experiment		
Friberg et al., 1996 [6]	2.3 (3.7)* dB below shot-noise	2.7-ps soliton, 1.46 μm
Spälter et al., 1997 [7]	2.2 (3.5)* dB below shot-noise	133-fs soliton, 1.50 μm
Spälter et al., 1997 [8]	3.2 (4.5)* dB below shot-noise	160-fs soliton, 1.51 μm
Friberg and Machida, 1997 [20]	23 dB reduction of 1/f-noise	1.5-ps soliton, 1.54 μm
Spälter et al., 1997 [9]	3.8 (6.4)* dB below shot-noise	135-fs soliton, 1.51 μm
theory		
Werner, 1996 [13]	6.5 dB below shot-noise	N=1.0 soliton
Werner, 1996 [14]	5.2 (4.3)** dB below shot-noise	N=1.0, 1.8-ps soliton
Werner, 1996 [14]	6.8 (4.7)** dB below shot-noise	N=1.1, 1.8-ps soliton C=-0.5
Friberg *et al.*, 1996 [6]	5.9±0.2 dB below shot-noise	N=1.1, 2.7-ps soliton
Mecozzi and Kumar, 1997 [17]	6.5 dB below shot-noise	N=1.0 soliton
Werner and Friberg, 1997 [15]	4.8** dB below shot-noise	N=1.0, 1.8-ps soliton
Werner and Friberg, 1997 [15]	7.1±0.2 dB below shot-noise	N=1.2 soliton
Werner, 1997 [16]	8.1±0.3 dB below shot-noise	N=1.3 soliton

3. Amplification and Deamplification of Input Noise

Noise reduction at certain input energies can be understood in terms of nonlinear pulse evolution in the fiber and subsequent filtering. The classical pulse dynamics in the fiber tend to stabilize the soliton. If the input pulse energy does not exactly match the fundamental soliton energy, oscillations in the temporal and spectral width will occur in the initial pulse evolution because of the self-stabilization dynamics [12]. In the long run, a certain fraction of the input energy ends up in a soliton with increased or reduced spectral width. If a spectral band-pass filter is located at a propagation distance where excess energy leads to spectral broadening, the excess energy can be removed through the filter (Figure 3). In the other case of lower pulse energies, spectral narrowing will lead to increased transmission through the filter. Therefore, an optical limiting effect is expected to occur which deamplifies small input fluctuations (Figure 2) – even fluctuations at the quantum level. A series of such filters may be an effective optical limiter with high transmission at low energies and constant energy throughput

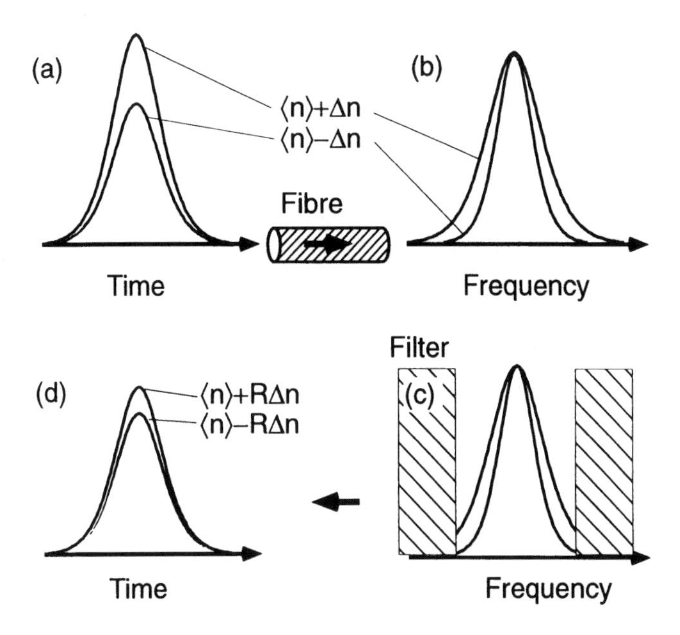

Figure 3. (a) The photon-number uncertainty Δn of the coherent input soliton in the time domain appears in the out-lying parts of the spectrum (b) after propagation through the fiber. When those sidebands are removed through spectral filtering (c), the photon-number uncertainty is reduced by a factor $R < 1$ below the shot-noise limit in the transmitted pulse (d). Fundamental soliton noise reduction of 6.5 dB, i.e. R=0.22 was predicted for a fiber length of 3 soliton periods and 18 % losses through the bandpass filter [13,17].

above a certain threshold energy.

In contrary to a reduced slope, an enhanced slope in the nonlinear input-output curve should amplify input fluctuations. As has been observed in experiments, more than 10 dB excess noise through amplified vacuum fluctuations can be generated in the filtering process [6].

The classical output noise properties can be calculated from input energy fluctuations if the nonlinear transfer relationship is known. The maximum noise reduction occurs around the point of zero slope of the input-output function.

Spectral filtering of quantum solitons was modeled using the generalized positive P representation [6], [13]-[16] or a linearized approach [17]. Numerical simulations of fundamental solitons of 1.8 ps input pulse width (FWHM) predicted up to 4.8 dB noise reduction below the shot-noise limit, if Raman noise at room temperature is included [15], and up to 8.1 dB noise reduction without Raman noise [16]. The analytical results of the linearized model of spectrally filtered solitons reveal that the interference of the soliton with the quantum-noise continuum produces oscillations in photon-number noise reduction along the fiber and limit the amount of noise reduction that can be achieved. In a classical analysis similar oscillations with decaying amplitude were predicted for the interference of the soliton with a dispersive wave [18]. For fundamental solitons of the nonlinear Schrodinger equantion (without Raman noise), a maximum noise reduction of 6.5 dB below the shot-noise limit was predicted by numerical evaluation using the generalized positive P representation [13] and the linearized model [17]. In these predictions a bandpass filter was positioned at a distance of 3 soliton periods with an optimum loss of 18 %.

4. Spectral Filtering of Picosecond Pulses

In the pioneering experiment, optical 2.7 ps pulses from a mode-locked NaCl color center laser were propagated down a 1.5 km fiber with energies slightly above the fundamental soliton (Figure 1) [6]. After spectral filtering through a grating and a slit, the transmitted light was detected by a balanced two-port detector. The sum and difference of the photocurrent fluctuations, measured with a delay line for frequency-dependent phase-shifting of the photocurrents, were recorded between 5 and 30 MHz. The sum of the photocurrent fluctuations measure the photon-number fluctuations whereas their difference represents the shot-noise reference level. Filtering away outlying components of the broadened spectra produced 2.3 dB (41 %) of noise reduction below the photocurrent shot-noise limit in direct detection. Accounting for measurement and detection losses, this corresponds to a photon-number noise reduction of 3.7 dB (57 %).

Alternating amplification and deamplification of vacuum fluctuations was clearly observed as the input energy was varied from slightly below to far above the fundamental soliton energy. Also, as the filter bandwidth was narrowed down at fixed input energy, a transition from noise reduction to enhancement was observed. Up to 2.7 dB (46 %) of noise reduction below the shot-noise limit and up to 10.9 dB (1130 %) of excess noise above the shot-noise limit were measured at a filter bandwidth of 1.4 nm and 0.4 nm, respectively. The input pulses were shot-noise limited in both cases. This shows that strong filtering in the presence of nonlinear pulse evolution can rapidly deteriorate the signal-to-noise ratio. All these phenomena were predicted by the quantum soliton model [6, 14].

5. Spectral Filtering of Sub-picosecond Pulses

When operating with sub-picosecond pulses, stimulated Raman scattering becomes a dominating effect. Strong noise reduction was observed in this case, too. Solitons from a saturable-absorber modelocked all-solid-state Chromium-YAG laser [19] were launched into 10 m of polarization-maintaining single-mode optical fiber at a rate of 194 MHz.

The pulses were centered at 1506 nm with a spectral width of 14.5 nm (FWHM) and a pulse duration of 160 fs (FWHM). Spectral filtering was achieved with the beam dispersed by a diffraction grating and then focused onto a variable slit consisting of two knife edges (Figure 1). In this way, variable band-, high- and low-pass filters were realized. The sum and difference of the photocurrent fluctuations from the balanced two-port detector were recorded at 20 MHz by two spectrum analysers synchronously. The sum photocurrent measures the photon-number fluctuations whereas the difference photocurrent is used as the shot-noise reference. In several series of measurement, the power, fiber length and filter losses were varied in order to find optimum noise reduction parameters.

5.1. NOISE REDUCTION AND ENHANCEMENT AS A FUNCTION OF PULSE ENERGY AND FILTER CUT-OFF WAVELENGTH

In a first experiment, a band-pass filter was used. Before the noise measurement, the unfiltered spectra were recorded at various launch powers (Figure 4). Here, spectral broadening and the Raman self-frequency shift are clearly visible for power levels above 25 mW in the 10 m fiber. At a launched power of 20.4 mW, which was 80 % of the soliton energy, up to 3.2 dB (52 %) noise reduction in the photocurrent fluctuations were obtained [8].

Next, a low-pass filter was applied to the short-wavelength side where SPM-induced sidebands appear in the spectra shown in Figure 4. Clearly, the recurring noise reduction as a function of the launch power is related to the slope of the nonlinear energy-transfer function (Figure 5).

The impact of the Raman effect on the nonlinear filter mechanism was investigated with a variable edge filter at fixed power. Figure 6 shows the noise levels relative to shot noise as a function of the filter cut-off wavelength, at a fixed pump power of 20 mW. A comparison with the unfiltered power spectrum clearly demonstrates the asymmetry with respect to the center wavelength of the pulse. The interpretation is simple in the nonlinear input-output picture: a self-frequency shift at the quantum level supports an optical limiting effect when the edge filter is placed on the long-wavelength side, improving the noise reduction. Even if the Raman self-frequency shift and asymmetry is not yet clearly visible in the power spectrum of the out-

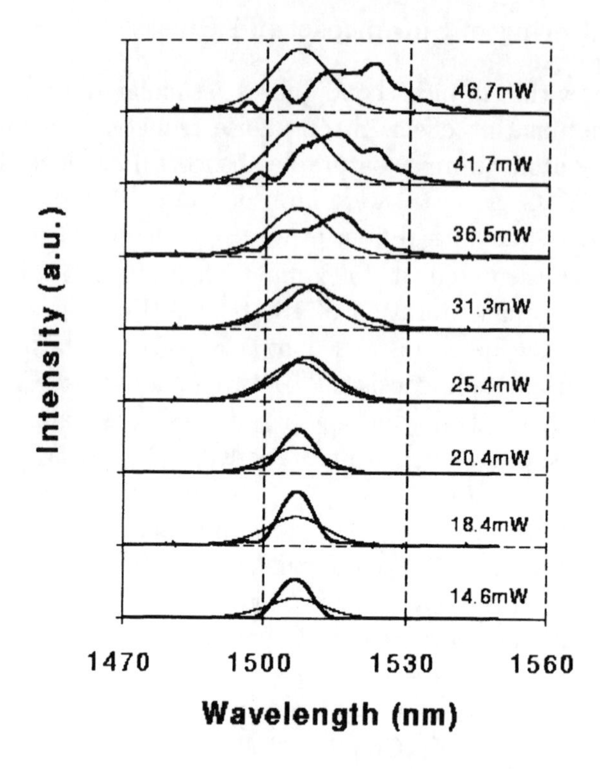

Figure 4. Power spectra measured at the fiber output (solid line) for various input powers. For comparison, input spectra (thin line) are shown. The input spectrum was measurend once and is scaled up in energy in proportion to the pump power [8].

put pulse (Figure 4), the quantum noise inside the pulse spectrum seems to be significantly redistributed. Therefore, Raman-self-pumping seems to contribute to amplification or deamplification of intensity noise even at power levels below the onset of a macroscopic self-frequency shift.

5.2. NOISE REDUCTION AS A FUNCTION OF FIBER LENGTH

The above measurements have been performed with 160 fs pulses in 10 m of fiber. The optimum noise reduction also depends on the fiber length. Theoretically, an optimum fiber length of approximately 3 soliton periods is predicted for maximum noise reduction of fundamental solitons [13]-[17].

In another series of measurements [9], a fiber was gradually shortened. The laser output pulses were centered at 1504 nm with spectral width and pulse duration kept constant at 18.5 nm (FWHM) and 130 fs (FWHM) respectively. The energy of the pulses launched into the fiber was 52±1 pJ, i.e. approximately the energy required for a fundamental soliton (54±2 pJ).

Similar to the measurement with 160 fs pulses shown in Figure 6(a),

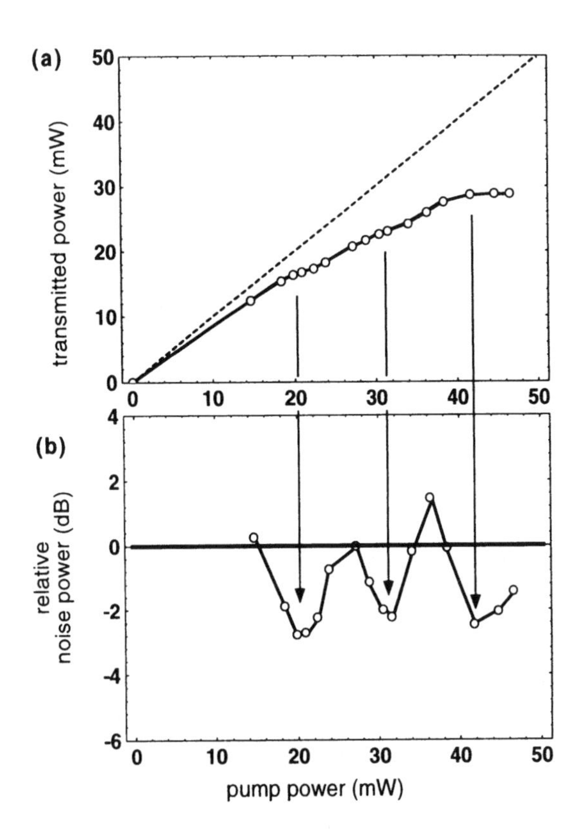

Figure 5. Photon-number noise reduction: (a) Nonlinear input-output behavior of the transmission through a low-pass filter (cut-off at 1498 nm) after 10 m of fiber. (b) Photon-number noise reduction occurs where the slope of the input-output transfer is reduced. Up to 2.4 dB noise reduction below the shot-noise is observed [8].

a frequency high-pass and a low-pass filter was applied, at either side of the spectrum. The photocurrent fluctuations at 20 MHz were recorded for different cut-off wavelengths, and only the value for the noise minimum for each filter was then plotted against the particular fiber length, while the fiber was shortened from 90 m down to 36 cm (Figure 7).

At a fiber length of 2.4 m and 2.9 m we found noise minima, corresponding to 1.3 dB and 2.3 dB noise reduction for the low-pass and the high-pass filtering respectively. The minima might correspond to the first half-period of the oscillations of noise reduction as a function of fiber length, as predicted by theory [14, 17]. The noise minima were observed at a fiber length of 3.6±0.4 and 3.0±0.3 soliton periods for red- and blue-filtered pulses (frequency high- and low-pass filtering) respectively, values coming close to the theoretical prediction. Furthermore, the noise levels approach the shot noise level as expected in the limit of vanishing propagation distance for

(a)

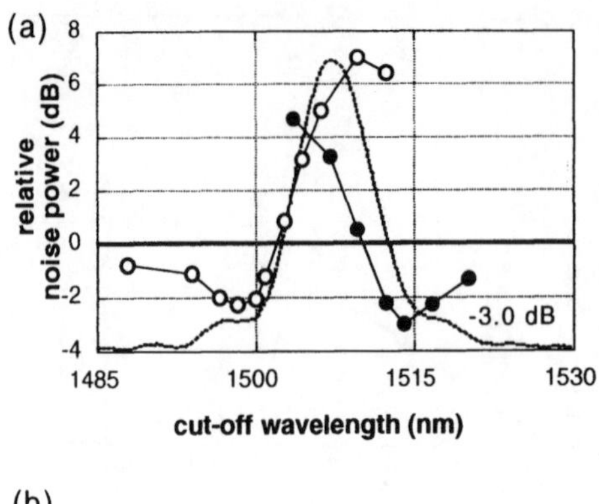

(b)

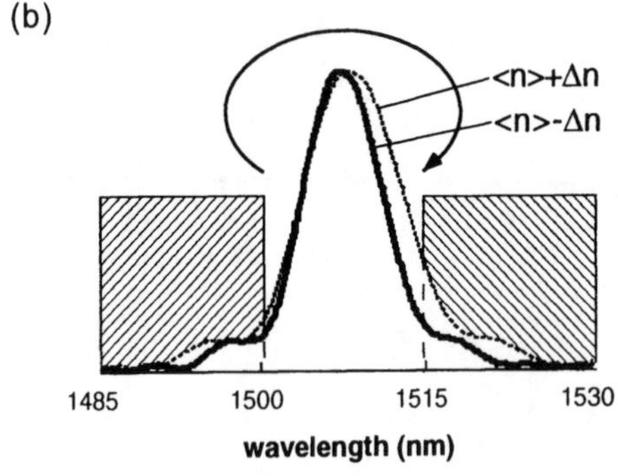

Figure 6. (a) Noise levels relative to shot-noise achieved with low- and high-pass filtering (circles and dots, repectively): Compared to the unfiltered power spectrum (dotted line), the measured noise levels are spectral asymmetric. (b) The schematic figure of the output power spectrum before filtering illustrates why high-pass filtering is more efficient than low-pass filtering in noise reduction: Raman-self-pumping (arrow) leads to asymmetric broadening on the quantum level, carrying more photon-number noise into the low-frequency side of the spectrum. This produces the observed asymmetry in noise reduction when high-/low-pass filters are applied [8].

both high- and low-pass filtering, thus providing yet another check of the shot noise level.

For longer fibers, there is no clear evidence for oscillations by the data points. In the limit of fibers much longer than the soliton period, substantially stronger noise reduction (3.2 dB, corresponding to 4.5 dB if corrected for linear losses) is observed than for short fibers if the red wavelength components are filtered. In contrast, the blue-filtered solitons approach the shot-noise limit with increasing fiber length. The Raman effect is believed to be responsible for the difference both in the amount of noise reduction as well as in the position of maximum noise reduction for high- and low-pass filtering.

If theoretical predictions for fundamental solitons over short propaga-

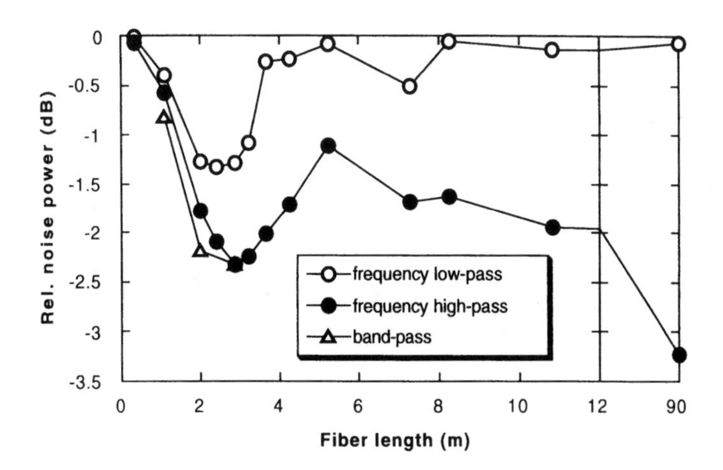

Figure 7. Photon-number noise reduction as a function of fiber length, for high-pass (dots) and for low-pass filtering (circles). Using an optimized band-pass filter instead of single edge filters improved the noise reduction only for fiber lengths smaller than 3 m (triangles) [9].

tion distances [13]-[15], [17] are extrapolated to long propagation distances, the 3.2 dB noise reduction in a 90 m fiber (more than 100 soliton periods) represents an unexpectedly large value. For higher-order solitons more noise reduction is predicted: up to 8.1±0.3 dB for a N=1.3 soliton [16]. Indeed, stronger noise reduction than for the fundamental soliton was observed at a pulse energy close to a N=1.3 soliton (135 fs FWHM, 85 pJ pulse energy), again in 90 m of fiber [9]. The photocurrent fluctuations at 20 MHz were reduced to 3.8 dB (59 %) below shot-noise, which is to our knowledge the best noise reduction observed with filters. Taking an overall detection efficiency of 76 % into account, the inferred quantum noise reduction is 6.4 dB (77 %).

6. Conclusion

What can quantum optics do for soliton-based optical communication? The nonstationary dynamics of fundamental quantum solitons have been intensively studied by measurements below the shot-noise limit. Several noise reduction techniques have been found that might work passively within a nonlinear fiber-optic system and may become useful as low-noise techniques in fiber-optic communication systems.

Quantum noise in a measurement is a consequence of the Heisenberg uncertainty relation. Therefore, quantum noise can only be deamplified in one measured variable if noise is enhanced in the conjugate variable. For optical solitons, these conjugate pairs are intensity and phase, or position and momentum. The mechanism that allows for noise reduction below the shot-noise limit also leads to amplification of vacuum fluctuations, either in the variable conjugate to the measurement variable or, if not designed properly, in the measured variable itself.

This brief review focuses on a recently demonstrated low-noise technique which uses the combined effect of nonlinear pulse propagation and spectral filters. Up to 3.8 dB noise reduction below the shot-noise limit and more than 10 dB amplification of vacuum fluctuations above the shot-noise level were observed. Based on the Kerr effect and on Raman scattering, both amplification and deamplification of noise are expected to be broadband to up to approx. 100 THz.

As far as applications in fiber-optic communication systems are concerned, the spectral filtering method may be useful in many ways: First, when spectral filters are used, nonlinearities may significantly affect the signal-to-noise ratio. If the system is designed right, the nonlinearities lead to photon-number or intensity rms-noise reduction in the pulses transmitted through the filter. Second, enhanced signal-to-noise ratios may be realized with low-noise soliton sources. In a low-noise application, the nonlinear filter technique has produced 23 dB reduction of the 1/f classical noise of a mode-locked erbium-doped fiber laser [20]. Furthermore, the noise reduction mechnism of spectral filtering may be built into the fiber using Bragg gratings. Such an environmentally protected all-in-fiber system would introduce fewer losses and would be particularly robust. Finally, the nonlinear input-output behavior may stimulate new research on optical limiting in soliton transmission systems or other nonlinear systems, where precise intensity control is important.

Acknowledgements

The authors would like to thank M.J. Werner for enlightening discussions. An EU grant in the framework of ESPRIT, Project No. 20029 (ACQUIRE) and a travel grant by the Deutsche Forschungsgemeinschaft are gratefully acknowledged.

References

1. Drummond, P. D., Shelby, R. M., Friberg, S. R. and Yamamoto, Y.: Quantum solitons in optical fibres, *Nature*, **365**, (1993), p.307.
2. Friberg, S. R.: Quantum Nondemolition Detection via Sucessive Back-Action-Evasion Measurements: A Step Towards the Experimental Demonstration of Quantum State Reduction, in Special Section on nonlinear Theory and its Applications, *IEICE Trans. Electron.*, (1996).

3. Spälter, S., van Loock, P., Sizmann, A. and Leuchs, G.: Quantum non-demolition measurements with optical solitons, *Appl. Phys.*, **B 64**, (1997), p.213.
4. Sizmann, A.: Quantum solitons: experimental progress and perspectives, *Appl. Phys.*, **B 65**, (1997), p.745.
5. Friberg, S. R., Machida, S. and Levanon, A.: Invited Paper **TuF2**, *Conference on Lasers and Electro-Optics, CLEO/Pacific Rim '95*, Chiba, Japan, (July 10-14, 1995).
6. Friberg, S. R., Machida, S., Werner, M. J., Levanon, A. and Mukai, T.: Observation of Optical Soliton Photon-Number Squeezing, *Phys. Rev. Lett.*, **77**, (1996), p.3775.
7. Spälter, S., Sizmann, A., Strosner, U., Burk, M. and Leuchs, G.: Photon number squeezing of spectrally filtered solitons, *Quantum Electronics and Laser Science Conference (QELS'97)*, Baltimore, Maryland, (May 18-23 1997), **QThB4**.
8. Spälter, S., Burk, M., Strosner, U., Bohm, M., Sizmann, A. and Leuchs, G.: Photon number squeezing of spectrally filtered sub-picosecond optical solitons, *Europhys. Lett.*, **38**, (1997), p.335.
9. Spalter, S., Burk, M., Strosner, U., Sizmann, A. and Leuchs, G.: Propagation of broadband sub-picosecond soliton squeezing in a fiber, submitted.
10. Mitschke F. M. and Mollenauer, L. F.: Discovery of the soliton self-frequency shift, *Opt. Lett.*, **11**, (1986), p.659.
11. Gordon, J. P.: Theory of the soliton self-frequancy shift, *Opt. Lett.*, **11**, (1986), p.662.
12. Hasegawa A. and Tappert, F.: Transmission of stationary nonlinear optical pulses in dispersive dielectric fibers. I. Anomalous dispersion, *Appl. Phys. Lett.*, **23**, (1973), p.142.
13. Werner, M. J.: *The Physical Society of Japan Spring Meeting*, **1a-YK-10** , (1996).
14. Werner, M. J.: Quantum statistics of fundamental and higher-order coherent quantum solitons in Raman-active waveguides, *Phys. Rev.*, **A 54**, (1996), R2567.
15. Werner M. J. and Friberg, S. R.: Phase Transitions and the Internal Noise Structure of Nonlinear Schrödinger Equation Solitons, *Phys. Rev. Lett.*, **79**, (1997), p.4143.
16. Werner, M. J.: private communication.
17. Mecozzi A. and Kumar, P.: Linearized quantum-fluctuation theory of spectrally-filtered optical solitons, *Opt. Lett. 22*, (1997), p.1232.
18. Gordon, J. P.: Dispersive perturbations of solitons of the nonlinear Schrödinger equatuion, *J. Opt. Soc. Am.*, **B 9**, (1992), p.91.
19. Spalter, S., Bohm, M., Burk, M., Mikulla, B., Fluck, R., Jung, I. D., Zhang, G., Keller, U., Sizmann, A. and Leuchs, G.: Self-starting soliton-modelocked femtosecond Cr(4+):YAG laser using an antiresonant Fabry-Perot saturable absorber, *Appl. Phys.*, **B 65**, (1997), p.335.
20. Friberg S. R.and Machida, S.: Soliton spectral filtering for suppression of intensity noise: demonstration of $>$ 23 dB reduction of 1/f noise, preprint.

(Received January 5, 1998)

VARIATIONAL APPROACH TO TRANSMISSION IN DM LONG OPTICAL LINKS

K. HIZANIDIS AND N. EFREMIDIS
Department of Electrical and Computer Engineering
National Technical University of Athens, 157 73 Athens, Greece

B. A. MALOMED
Department of Interdisciplinary StudieS, Faculty of Engineering
Tel Aviv University, Tel Aviv 69978, Israel

AND

H. E. NISTAZAKIS AND D. J. FRANTZESKAKIS
Department of Physics, Applied Physics Section
University of Athens, Panepistimiopolis, 157 84 Athens, Greece

Abstract. A simple variational model is introduced and the dynamical system of the evolving pulse characteristics is integrated numerically in a case of a line with periodically modulated second order dispersion and third order dispersion. The results obtained are compared with ones of the full numerical integration of the corresponding generalized nonlinear Schrödinger equation. It is demonstrated that the variational model is in a fairly good agreement with the numerical exact one, although the influence of the third order dispersion seems rather weak in the variational model.

1. Introduction

The use of the periodically modulated dispersion (dispersion management, DM) has recently attracted a lot of attention as a means to improve operation of soliton-based long-haul optical communication lines. A number of promising results, both experimental [1]-[4] and theoretical [5]-[10] have been published. However, the detailed investigation of the effect of the parameters which characterize the initial (at the launching node) pulse (such as power, chirp, full-width-half-maximum, t_{FWHM}) in conjunction with the characteristics of the DM line [i.e. values of the anomalous and normal

A. Hasegawa (ed.), New Trends in Optical Soliton Transmission Systems, 117–129.

group velocity dispersion (GVD), periodicity, etc.] remains a tedious task. Due to the complexity of the problem of the long DM transmission, numerical simulation is the most handy line of approach, although considerations of the asymptotic behavior of the optical pulse have also started gaining popularity [11, 12]recently. The standard approach to the analytical investigation is the variational one [13, 14]which frequently renders a dynamical (in the sense of ordinary differential equations with the propagation distance as the dynamical variable) system to be investigated instead.

The objective of this work is to obtain some general conclusions about the validity of the variational approach, that is, its capability of providing useful information about the proper choice of the parameters which characterize the initial pulse in a long DM communication line. Along this line, a simple variational model (ansatz) is introduced and the results obtained are compared with the ones produced by directly simulating the corresponding generalized nonlinear Schrödinger equation. Since the role of the third order dispersion (TOD) in a fairly long (about 12,000 km) DM communication line may become important, especially in a nearly compensated one, TOD is also incorporated in both the numerical simulations and the variational model.

This paper is organized as follows: In Section 2 the variational model is formulated. In Section 3 the results of the full numerical solution of the problem are discussed and compared with results from the numerical integration of the dynamical equations provided by the variational approximation. Finally in Section 4 the main conclusions are recapitulated.

2. The Variational Formulation

A nearly compensated transmission line consists of alternating fiber segments with anomalous and normal dispersions, so that the mean dispersion is anomalous, but its absolute value is much smaller than the actual dispersion in either segment. The model describing the nonlinear pulse transmission in the presence of DM and TOD is,

$$iq_z + ik'q_t - \frac{k''}{2}q_{tt} - i\frac{k'''}{6}q_{ttt} + \kappa q|q|^2 = 0, \qquad (1)$$

where $q(z,t)$ is a slowly varying function of time t and z, the coordinate along the transmission line, κ is the nonlinear Kerr coefficient, $k' = \partial k/\partial\omega$, $k'' = \partial^2 k/\partial\omega^2$ and $k''' = \partial^3 k/\partial\omega^3$ are, respectively, the inverse group velocity, the usual dispersion (group velocity dispersion, GVD) assumed to take different values in different segments of the line and third order dispersion (TOD) coefficient, which may also take different values. All coefficients are evaluated at the carrier frequency. This equation can readily be normalized as follows,

$$iu_\zeta + i\beta_1(\zeta)u_\tau + \frac{\beta_2(\zeta)}{2}u_{\tau\tau} - \frac{i}{6}\beta_3(\zeta)u_{\tau\tau\tau} + \gamma(\zeta)u|u|^2 = 0\,, \qquad (2)$$

where $\zeta \equiv |k_0''|z/t_0^2$, $\tau \equiv (t - k_0'z)/t_0$, $t_0 \equiv t_{\rm FWHM}/2\ln(\sqrt{2}+1)$, $u \equiv qt_0|\kappa_0/k_0''|^{1/2}$, $\beta_1(\zeta) \equiv (k'(z) - k_0')t_0/|k_0''|$, $\beta_2(\zeta) \equiv k''(z)/|k_0''|$, $\gamma(\zeta) \equiv \kappa(z)/|\kappa_0|$, $\beta_3(\zeta) \equiv k'''(z)/t_0|k_0''|$ where the subscript "0" refers to the values at $z = 0$. Then the initial ($\zeta = 0$) values of the normalized coefficients are, $\beta_1(\zeta = 0) = 0$, $\beta_2(\zeta = 0) = 1$, $\gamma(\zeta = 0) = 1$, while $\beta_3(\zeta = 0)$ is a given constant. This equation can be derived from a Lagrangian function $L = L(u, u_\tau, u_\zeta, u_{\tau\tau}; u^*, u_\tau^*, u_\zeta^*, u_{\tau\tau}^*; \zeta)$, given by the following expression,

$$\begin{aligned} L \;=\; & i\frac{\beta_3(\zeta)}{12}\left(u_\tau^* u_{\tau\tau} - u_\tau u_{\tau\tau}^*\right) + \frac{\beta_2(\zeta)}{2}|u_\tau|^2 + \\ & i\frac{\beta_1(\zeta)}{2}\left(u^* u_\tau - u u_\tau^*\right) + \frac{1}{2}\left(u^* u_\zeta - u u_\zeta^*\right) + \frac{\gamma(\zeta)}{2}|u|^4. \end{aligned} \qquad (3)$$

The ansatz

$$u(\zeta) = A(\zeta)\text{sech}\left(\frac{\tau'}{T(\zeta)}\right)e^{i\left[\Phi(\zeta) - \Omega(\zeta)\tau' + B(\tau')\tau'^2\right]} \qquad (4)$$

is now introduced with $\tau' = \tau - U(\zeta)$. The pulse is characterized predominantly by the real parameter functions $A(\zeta)$, $T(\zeta)$ and $\Phi(\zeta)$ which are respectively the amplitude, temporal width and phase of the propagating pulse. The remaining three, also real, parameter functions are responsible for temporal distortions of the pulse, such as shifts of the central frequency and maximum amplitude [$\Omega(\zeta)$ and $U(\zeta)$] and chirp [$B(\zeta)$]. Upon integrating the Lagrangian over time, the resulting variational equations render the following dynamical equations which govern the evolution of the parameter functions,

$$\frac{dT}{d\zeta} = -\frac{2b}{T}(\beta_2 + \beta_3\Omega_0)\,, \qquad (5)$$

$$\frac{dA}{d\zeta} = -\frac{A}{2T}\frac{dT}{d\zeta}\,, \qquad (6)$$

$$\frac{db}{d\zeta} = -\frac{2}{T^2\pi^2}(1 + b^2\pi^2)(\beta_2 + \beta_3\Omega) - \frac{2A^2\gamma}{\pi^2}\,, \qquad (7)$$

$$\frac{d\Phi}{d\zeta} = \frac{b_2}{3T^2}\left(1 - \frac{3}{2}\Omega^2T^2\right) + \frac{\beta_3\Omega}{6T^2}\left(1 - b^2\pi^2 - 2\Omega^2T^2\right) + \frac{5}{6}A^2\gamma\,, \qquad (8)$$

$$\frac{dU}{d\zeta} = \beta_1 + \beta_2\Omega + \beta_3\frac{1}{6T^2}\left(1 + b^2\pi^2 + 3\Omega^2T^2\right) , \tag{9}$$

$$\Omega = \Omega_0 , \tag{10}$$

where Ω_0 is a constant: In other words, the parameter $\Omega(\zeta)$ remains constant along the propagation distance. The parameter function $b(\zeta)$, used in the dynamical equations instead of the chirp function $B(\zeta)$, is B/T^2.

3. Discussion of the Numerical Results

In both numerical integrations, of the exact equations and the integration of the dynamical equations of evolution of the pulse shape, equations (6)-(9), the numerical values of both GVD and TOD and the initial t_{FWHM} and pulse power are the same and taken so that the DM mode is fixed: That is, 50 km anomalous −100 km normal −50 km anomalous, while the propagation distance is approximately 12,000 km in all simulations. A large number of runs have been performed with different maps, concluding that a particular choice is not crucial as long as the dispersion is anomalous in the first segment. The anomalous dispersion is fixed to -3 ps^2/km (which corresponds to a dispersion-shifted fiber) and the value of the normal dispersion is used as a parameter to control the mean value. The nonlinear coefficient (γ, in normalized form) is taken to be constant (at the value of 2.65 rad/W·km) over the entire transmission line while the TOD coefficient (β_3, in normalized form) is either zero or alternates between 0.088 ps^3/km and 0.074 ps^3/km which are quite realistic values to assume. For simplicity, the group velocity is also considered constant over the whole transmission line and the results are presented in the moving (with that velocity) frame of reference.

We first display the results of the full numerical simulations. In all runs performed, both Gaussian and soliton initial pulses have been used. The choice between these two shapes is not a crucial factor since there are only minor (negligible, in some instances) differences as far as the macroscopic and asymptotic characteristics of the DM transmission are concerned. The initial peak power, p_0, is taken to be either 0.65 mW or 1.2 mW. At these values of the peak power, the corresponding nonlinearity lengths are much larger than length characteristic of the applied DM (the period of 200 km) which indeed allows one to consider the nonlinearity as a relatively weak perturbation on the background of a strong DM. The initial t_{FWHM} is used as a control parameter. Below, systematic results will be presented for two different and practically relevant widths, 20 ps and 10 ps. Simulations

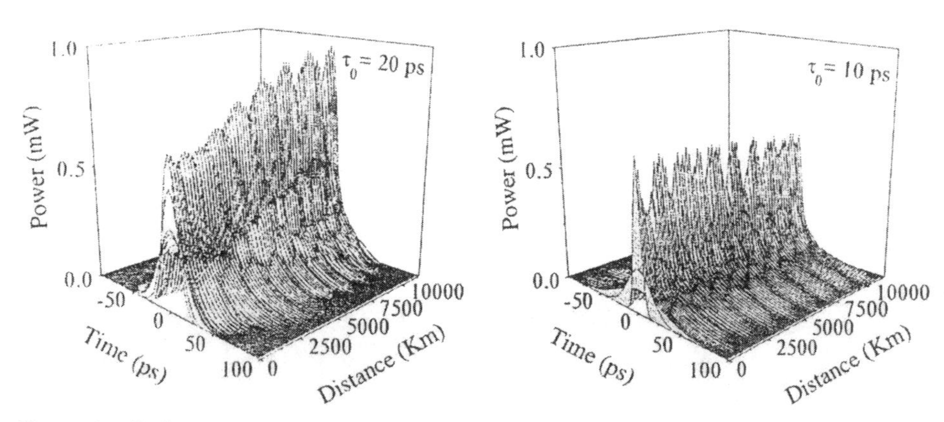

Figure 1. Pulse propagation in the case of exactly zero mean dispersion and with $\beta_3 = 0.083$ ps^3/km and for $t_{\mathrm{FWHM}} = 20$ ps, 10 ps.

for other values of t_{FWHM}, e.g. 5 ps, have produced very similar results. The simulations start with the case of zero TOD (in every segment of the transmission line), which has already be considered in a number of works (see, e.g. reference [11]). In accordance with the previously published numerical and analytical results [10], a steady transmission of the pulse is only possible in the presence of a nonzero anomalous (negative) mean dispersion. However, in the case where the mean dispersion is zero ($\langle\beta_2\rangle = 0$) the pulse does not disintegrates, decaying into radiation: Instead, it conserves practically all its initial energy, but its propagation becomes very irregular (especially for the narrower pulse), demonstrating quasi-periodic or chaotic shape oscillations (Figure 1).

For a slightly normal mean dispersion, the pulse eventually decays into radiation, but not in a straightforward way: Up to a distance 9,000 km it exhibits chaotic behavior similar to (for $t_{\mathrm{FWHM}} = 20$ ps) or more chaotic [for $t_{\mathrm{FWHM}} = 10$ ps (first figure in Figure 2)] than what can be seen in Figure 1. Then, at the final stage of the evolution (between 9,000 km and 12,000 km) it finally suffers a decay. Inclusion of TOD with the previously mentioned values ($\langle\beta_3\rangle = 0.083$ ps^3/km) always produces either neutral or definitely positive effect: The unstable propagation regime either remains approximately equally unstable, or exhibits essential stabilization. For the broader pulse, ($t_{\mathrm{FWHM}} = 20$ ps), the effect of TOD is conspicuous only in the case $\langle\beta_2\rangle = 0$. In this case, TOD suppresses the oscillations seen in the first figure in Figure 1, though not completely. For the narrower pulse, $t_{\mathrm{FWHM}} = 10$ ps, the effect is more impressive.

Figure 2 demonstrates the fact that TOD strongly stabilizes pulse propagation in the case when the usual mean dispersion is normal (+0.02 ps^2/km). In such a case, TOD prevents eventual decay of the pulse (at

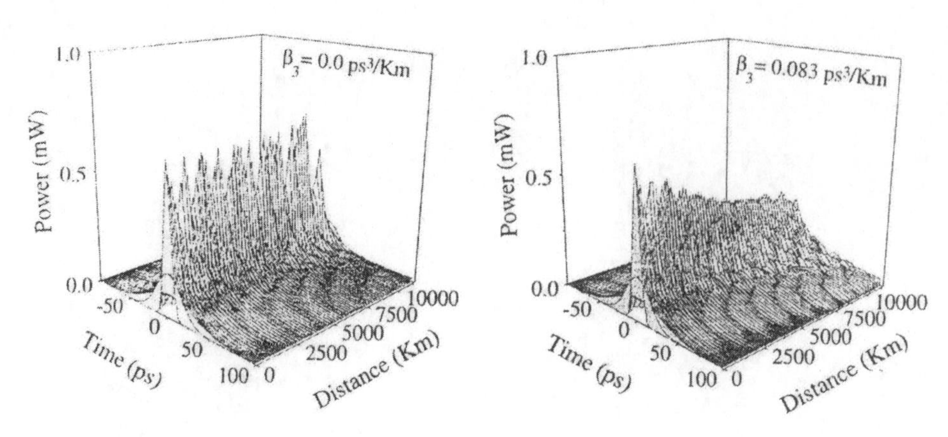

Figure 2. Propagation with $t_{\text{FWHM}} = 10$ ps, $\langle\beta_2\rangle = +0.02$ ps^2/km and $\langle\beta_3\rangle = 0.0$ ps^3/km, 0.083 ps^3/km.

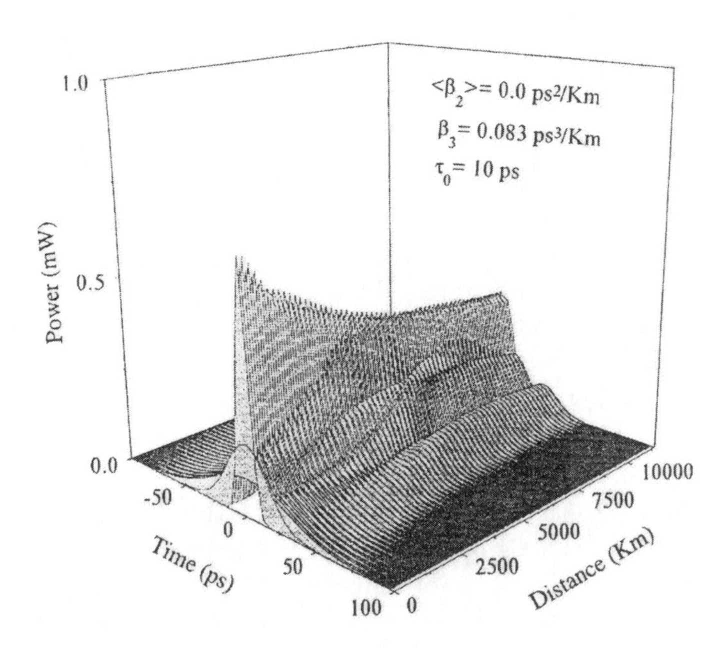

Figure 3. Propagation with $t_{\text{FWHM}} =10$ ps, $\langle\beta_2\rangle = 0.0$ ps^2/km and $\langle\beta_3\rangle = 0.083$ ps^3/km.

least, over the propagation distance 12,000 km). The best result is obtained for the case of exactly dispersion-compensated fiber: Comparing the corresponding propagation picture displayed in Figure 3, and the picture for the same case without TOD, second figure in Figure 1, one clearly sees a strong stabilizing effect of TOD.

Next, we investigate the effect of varying the mean dispersion over a range of values, that is $\langle\beta_2\rangle = -0.2$ ps^2/km $\div + 0.1$ ps^2/km at fixed in-

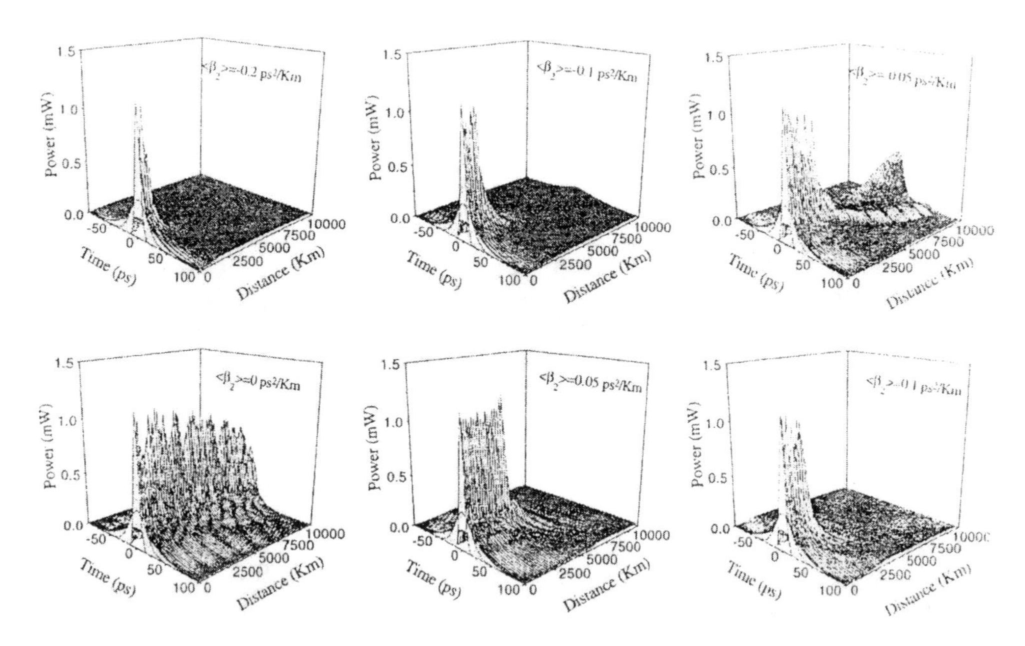

Figure 4. Propagation with $t_{\rm FWHM}$ =10 ps, $\langle\beta_2\rangle = -0.2 \div +0.1$ ps^2/km at p_0 =1.2 mW and $\langle\beta_3\rangle = 0.083$ ps^3/km.

creased power and narrow pulse (small $t_{\rm FWHM}$), 1.2 mW and 10 ps respectively. The results are illustrated in Figure 4. It is clear that a transition, from complete decay of the pulse at long distances (about 2,500 km $\div$ 5,000 km) to a robust propagation at a finite power (though, lower than the input power in some instances) and back, occurs between the values $\langle\beta_2\rangle = -0.1$ ps^2/km and $\langle\beta_2\rangle = 0.05$ ps^2/km. It is also clear that the most robust pulse propagation is achieved for a fully compensated line.

On the other hand, the effect of varying the initial peak power in the range of $1.0 \div 1.5$ mW at fixed (anomalous) mean dispersion ($\langle\beta_2\rangle = -0.1$ ps^2/km) and fixed $t_{\rm FWHM}$ =10 ps is illustrated in Figure 5. Again, depending upon the value of $\langle\beta_2\rangle$, there is a critical value of input peak power below of which the pulse decays at long distances. For instance, as it is clear from Figure 5, this value is about $p_0 = 1.0$ mW for the chosen value of $\langle\beta_2\rangle$, while there exists robust, though at a much lower than the input peak power level, propagation for $p_0 = 1.2$ mW. Above a second critical value (around $p_0 = 1.3$ mW) the propagation is unstable in the sense that, after a period of low peak power propagation, an increase occurs at long distances, though at a level lower than the input one.

We now assess the results of the variational method. In Figure 6 the effect of the input chirp on the asymptotic behavior of the propagating pulse is illustrated. The rest of the parameters chosen are typical of the

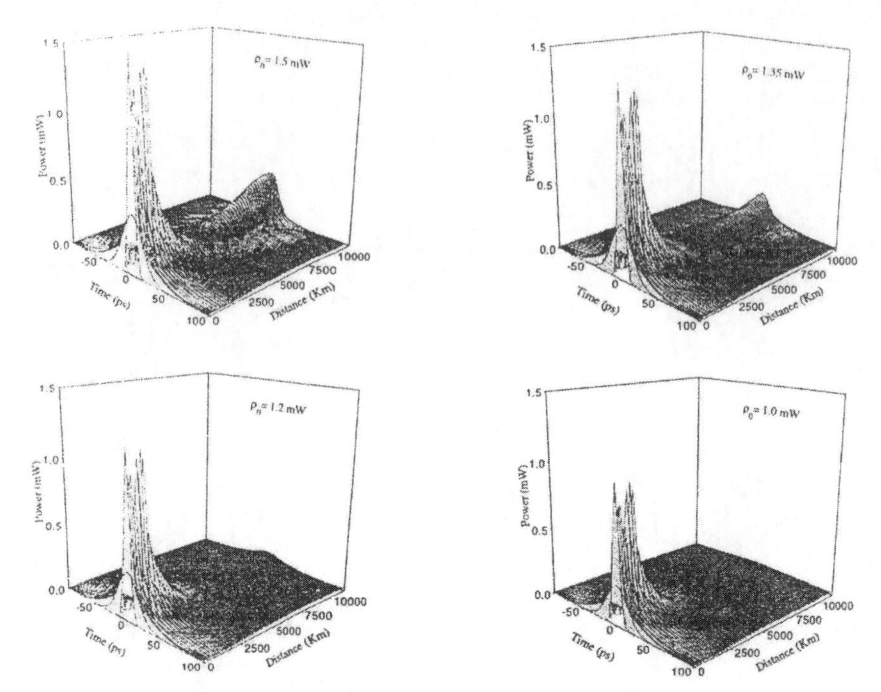

Figure 5. Propagation with $t_{\mathrm{FWHM}} = 10$ ps, $\langle\beta_2\rangle = -0.1$ ps^2/km with $p_0 = 1.5 \div 1.0$ mW (from bottom right to top left) and $\langle\beta_3\rangle = 0.083$ ps^3/km.

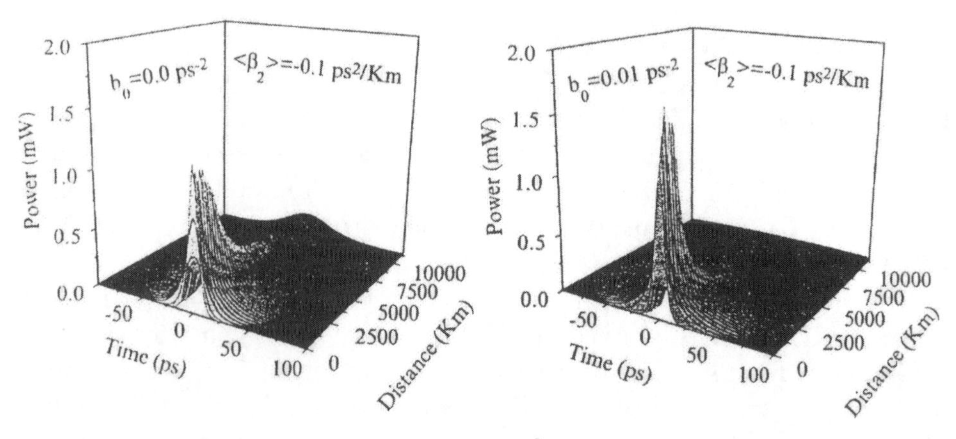

Figure 6. Pulse propagation with $\langle\beta_2\rangle = -0.1$ ps^2/km, $t_{\mathrm{FWHM}} = 10$ ps and $\langle\beta_3\rangle = 0.083$ ps^3/km with initial chirp $b_0 = 0.0$ ps^{-2}, 0.01 ps^{-2}, according to the variational model.

full numerical simulations presented earlier: That is, $p_0 = 0.65$ mW, $\langle\beta_2\rangle = -0.1$ ps^2/km, $t_{\mathrm{FWHM}} = 10$ ps and $\langle\beta_3\rangle = 0.083$ ps^3/km. It is clear that the presence of chirp ($b_0 = 0.01$ ps^{-2}) is detrimental for the pulse robustness. On the other hand, in the absence of chirp at the launching node, the result is quite close to that obtained by full integration (see in Figure 4).

In Figure 7 the evolution of the parameters of the ansatz are plotted

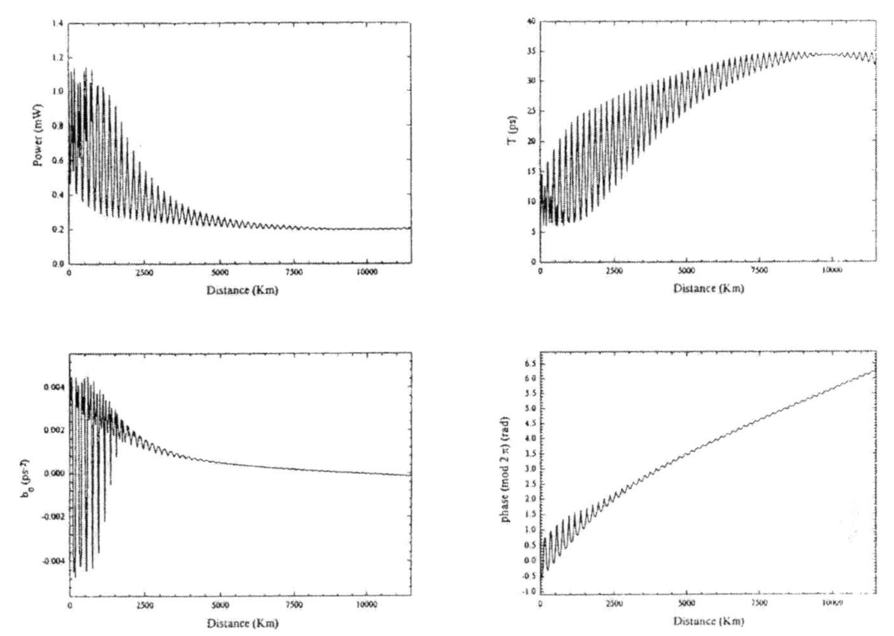

Figure 7. Evolution of peak power, temporal width, chirp and phase for a pulse propagating with initial $t_{\mathrm{FWHM}} = 10$ ps, initial chirp $b_0 = 0.0$ ps^{-2} along a line with $\langle\beta_2\rangle = -0.1$ ps^2/km and $\langle\beta_3\rangle = 0.083$ ps^3/km according to the variational model.

versus the propagation distance for the case $b_0 = 0.0$ ps^{-2} of Figure 6 (the robust one). It is observed that, although chirp develops during propagation (first figure in the bottom row), its asymptotic value is zero. This fact, along with the stabilization of the peak power at low level and at large distances (first figure in the top row), is quite suggestive of the asymptotic robustness of the pulse propagation at the chosen parameter values. The temporal width of the pulse, on the other hand (second figure in the top row), seems to exhibit a long-periodic pulsation on top of the short periodic one (of the order of the DM length scale): It certainly saturates at a distance of about 10,000 km. However the pulsation reappears at longer propagation distances. We do not pursue this issue here since this macro-periodic effect has a length scale beyond practical significance. The fourth figure in Figure 7 depicts the evolution of the phase: After a distance of about 5,000 km it becomes a linear function of the propagation distance.

In Figure 8 the evolution of the parameters of the ansatz are plotted versus the propagation distance for a case where $b_0 = 0.0$ ps^{-2} while $\langle\beta_2\rangle = 0.05$ ps^2/km. This case corresponds to a propagation during which the pulse practically decays after a distance of about 5,000 km whose evolution according to the exact integration (also close to the variational one) appears in Figure 4 (second figure in the bottom row). We observe that, besides

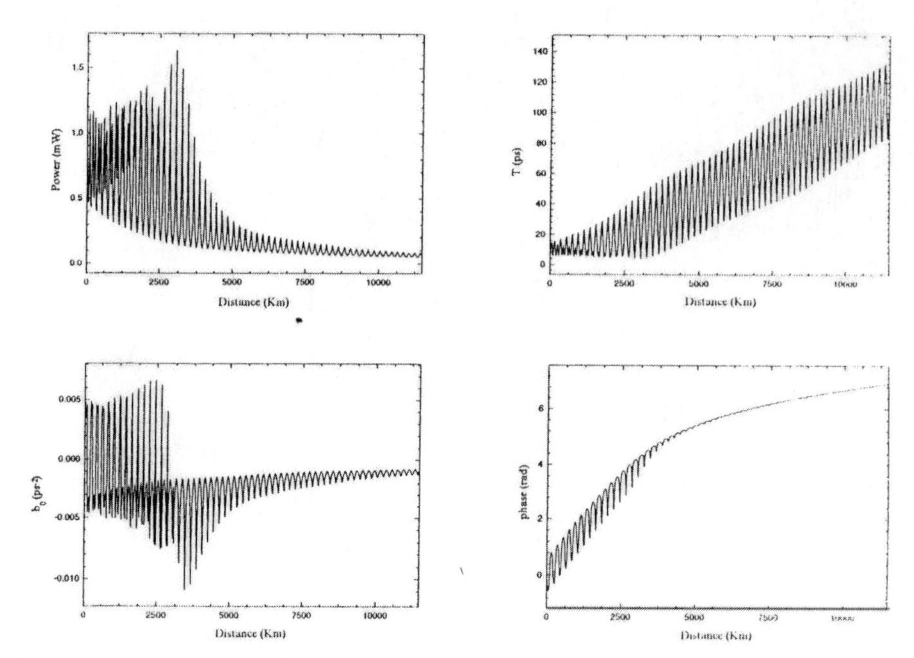

Figure 8. Evolution of peak power, temporal width, chirp and phase for a pulse propagating with initial $t_{\mathrm{FWHM}} = 10$ ps, initial chirp $b_0 = 0.0$ ps^{-2} along a line with $\langle\beta_2\rangle = +0.05$ ps^2/km and $\langle\beta_3\rangle = 0.083$ ps^3/km according to the variational model.

the much lower asymptotic value for the peak power (about one third) compared with the previous robust case, the pulsation of the temporal width persists and its mean value increases constantly with distance with no evidence of saturation (second figure in the top row). The rate of change of the phase, on the other hand, saturates at a much lower value than the previous case, rendering, as before, the phase function a linear function of the propagation distance.

The assessment of the credibility of the variational method is illustrated in Figure 9. It depicts the propagation of a pulse with $t_{\mathrm{FWHM}} = 10$ ps along a DM line with $\langle\beta_2\rangle = -0.05 \div +0.05$ ps^2/km at $\rho_0 = 1.2$ mW and $\langle\beta_3\rangle = 0.083$ ps^3/km. By comparing these figures with the respective figures in Figure 4 (third figure on the first row and first two in the bottom row) we conclude that the agreement is fairly good, although the variational model seems to slightly overemphasize the peak power (asymptotic and intermediate). However, the exclusion of TOD in this case does not affect, at least macroscopically, the result. Therefore, the variational model (in its present form) appears, besides the aforementioned minor overestimation of the peak power, be insensitive to the TOD, in contrast to the full numerical model.

Finally, in Figure 10, the parameter responsible for the shift in time

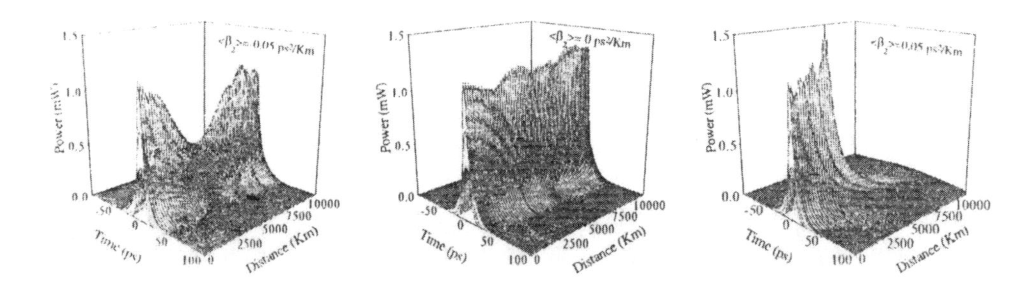

Figure 9. Propagation with $t_{\mathrm{FWHM}} = 10$ ps, $\langle\beta_2\rangle = -0.05 \div +0.05$ ps^2/km at $\rho_0 = 1.2$ mW and $\langle\beta_3\rangle = 0.083$ ps^3/km according to the variational model.

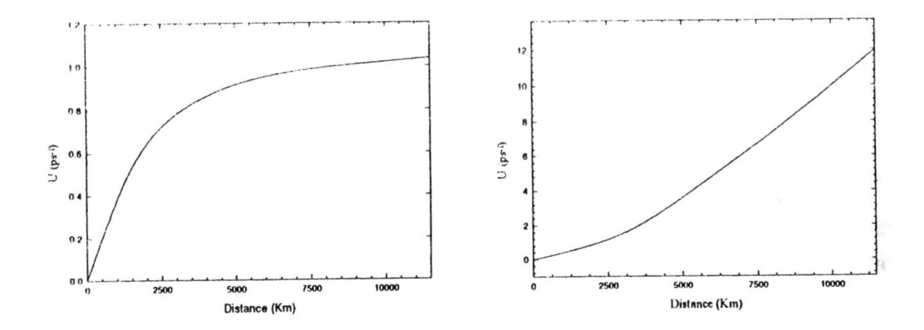

Figure 10. The parameter U as a function of the propagation distance for a pulse with $t_{\mathrm{FWHM}} = 10$ ps, initial power $p_0 = 1.2$ mW and zero initial chirp propagating in a line with $\langle\beta_2\rangle = -0.1, +0.05$ ps^2/km and $\langle\beta_3\rangle = 0.083$ ps^3/km according to the variational model.

of the peak amplitude (remember that the results are presented in the reference frame moving at the assumed constant group velocity), namely $U(\zeta)$, is plotted versus the propagation distance for a pulse with $t_{\mathrm{FWHM}} =$ 10 ps, initial power $p_0 = 1.2$ mW and zero initial chirp propagating in a DM line with $\langle\beta_2\rangle = -0.1$ (robust case) and $\langle\beta_2\rangle = +0.05$ ps^2/km. TOD is again $\langle\beta_3\rangle = 0.083$ ps^3/km. It is observed that this function, as well, seems to saturate in an asymptotic sense, while tends to increase unboundedly for the case of propagation characterized by asymptotic decay.

4. Conclusions

The main result of the full numerical method, based on systematic numerical simulations, is that TOD can be quite helpful to further improve operation of a line: It was demonstrated that it has a clear trend to stabilize pulse transmission in the cases when, otherwise, the transmission would be unstable. This pertains to the cases when the mean value of the usual dispersion is exactly equal to zero or slightly positive (normal). That is, TOD stabilizes the pulse transmission regime, making stable propagation possible even when the mean value of the usual dispersion is exactly zero. This result may be quite useful: It clearly shows that there is no need to maintain a prescribed mean value of the anomalous dispersion in the long line with a high accuracy in order to prevent destabilization and decay of the pulse since TOD will play that role.

The variational model, on the other hand, as compared with full numerical solution, is very reliable, although it seems to slightly overemphasize the peak power (asymptotic and intermediate) and it is rather insensitive to the TOD, in contrast to the full numerical model. The main, quite useful result of the variational model, however, is a tendency the characteristic parameters of the ansatz exhibit: To saturate only in the cases of the numerical asymptotically robust propagation; a result which reflects this robustness. This fact, along with the incomparable computational speed it possesses, makes the variational model an adequate tool for optimizing the pulse characteristics at the launching port for the purpose of robust propagation along a DM line.

References

1. Nakazawa, M. and Kubota, H.: *Electron. Lett.*, **31**,(1995), p.216.
2. Suzuki, M., Morita, I., Yamamoto, S., Edagawa, N., Toda, H. and Akiba, S.: *Optical Fiber Communications (OFC'95)*, OSA, Washington, D.C., (1995).
3. Edagawa, N., Morita, I., Suzuki, M., Yamamoto, S., Taga, H. and Akiba, S.: *Proceedings of the 21st European Conference on Optical Communications (ECOC'95)*, p.983, Brussels, (1995).
4. Carter, G. M., Jacob, J. M., Menyuk, C.R., Golovchenko, E.A. and Pilipetskii, A.N.: *Opt. Lett.*, **22**, (1997), p.513.
5. Smith, N. J., Forysiak,W., and Doran, N. J.: *Electron. Lett.*, **32**, (1996), p.2085.
6. Gabitov, I., and Turitsyn, S. K.: *Opt. Lett.*, **21**, (1996), p.327.
7. Moores, J. D.: *Opt. Lett.*, **21**, (1996), p.555.
8. Kumar, S. and Hasegawa, A.: *Opt. Lett.*, **22**, (1997), pp.372-374.
9. Yu, T., Golovchenko, E. A., Pilipetskii, A. N., and Menyuk, C. R., *Opt. Lett.*,**22**,(1997), p.793.
10. Malomed, B. A.: *Opt. Comm.*, **136**,(1997), p.313.

11. Gabitov, I., Shapiro, E. G. and Turitsyn, S. K.: *Opt. Comm.*, **134**, (1997), pp.317-329.
12. Gabitov, I., Shapiro, E. G. and Turitsyn, S. K.: *Phy. Rev. E*, **55**, (1997), pp.3624-3633.
13. Anderson, D.: *Phys. Rev. A*, **27**,(1983), pp.3135-3145.
14. Malomed, B. A.: *Physica Scripta*, **47**, (1993), pp.797-799.

(Received January 5, 1998)

NONLINEAR CHIRPED RZ AND NRZ PULSES IN OPTICAL TRANSMISSION LINES

Research on Optical Soliton Communications at Osaka University

Y. KODAMA
Graduate School of Engineering
Osaka University
2-1 Yamadaoka, Suita, Osaka 565, Japan

1. Introduction

In this paper, we summarize some of the recent results obtained by the soliton group at Osaka University. The main objective is to study nonlinear pulse propagation in an optical transmission system with dispersion management. This paper consists of two parts dealing with the properties of soliton-like RZ (return-to-zero) and conventional NRZ (non-return-to-zero) pulses. The main difference in the systems using those pulses is the sign of the group velocity dispersion. However recent studies show that both systems require similar dispersion managements for improving system performance.

In the case of the RZ pulse, the main purpose of the dispersion management is to reduce several effects such as radiations from the pulse due to lumped amplifiers compensating the fiber loss [5, 15], modulational instability [3], jitters caused by the collisions between signals in different channels of the wavelength-division-multiplexing (WDM) [11, 21, 24], and the Gordon-Haus effect resulting from the interaction with noise [6, 23], and to set a desired average value of the dispersion [19, 3]. It was pointed out in numerical studies [22, 8] that in such a line the pulse is deformed from the ideal soliton, and it has a chirp and requires an enhanced power when compared with the soliton case with a uniform dispersion. We then start in Section 2 to explain the effect of the dispersion managemant for the solitn-like RZ pulse. We first describe in the Subsection 2.1 the effect of the management for a linear system, which will give a basic concept of the management. In the Subection 2.2, we show that the average dynamics of the pulse can be described by the NLS equation having a nontrapping quadratic potential due to the chirp (frequency modulation of the pulse).

A. Hasegawa (ed.), New Trends in Optical Soliton Transmission Systems, 131–153.

Because of this potential, a pulse propagating in such a system eventually decays into dispersive waves in a way similar to tunneling effect of quantum mechanics. We here introduce the effective dispersion and the effective Kerr coefficient to describe the the pulse dynamics in the dispersion compensated system. Then in the Subsection 2.3 we take a simple example of the dispersion compensated system, and show explicitly that there exists a soliton-like pulse even in the case of normal averaged dispersion [20]. This result has been also announced independently in this ROSC conference by N. Doran, T. George, N. Kutz and myself [see this proceeding]. Some of the results discussed in this Section can be also found in references [12, 16].

In the case of NRZ pulse, the system should be operated at a small dispersion regime. Then, based on a weak dispersion limit of the nonlinear Schrödinger (NLS) equation, we have recently proposed in [13] a hydrodynamic model to describe nonlinear effects in the NRZ pulse propagation. With this model, we have also discussed a method to control the pulse broadening by means of initial phase modulation (pre-chirping) [14]. In Section 3, we present a mathematical formulation of the model in the framework of the Whitham averaging method [25], which describes the slow modulation of the amplitude and phase of quasi-periodic solution of the NLS equation. The resulting average equation is called the NLS-Whitham equation, and it determines the dynamics of the Riemann suface associated with the quasi-periodic solution. We then study the effect of pre-chirping for reducing the broadening of NRZ pulse due to the nonlinearity. As a result we find that the Riemann surface becomes singular (generation of a periodic solution observed as a formation of a *shock*), and its genus changes from zero to either one or two depending on the strength of the chirp. The detail of the present result on this part can be found in reference [17].

2. RZ Pulse Propagation in a Dispersion Compensated System

As a model of dispersion compensated system, we consider the NLS equation with variable coefficients in normalized units (see for example [10]),

$$i\frac{\partial q}{\partial Z} + \frac{d(Z)}{2}\frac{\partial^2 q}{\partial T^2} + a^2(Z)|q|^2 q = 0 \,, \tag{1}$$

where q represents the complex envelope of electric field in a fiber, and T and Z are normalized time and distance. The functions $a(Z)$ and $d(Z)$ represent for example the periodic amplification and dispersion management. We here assume that the periods Z_a and Z_d for $a(Z)$ and $d(Z)$ are the same. We denote the average dispersion $\langle d \rangle = d_0$, where $\langle \cdot \rangle$ represents the path-average over the period. The main objective here is to study a pulse propagation through the system described by equation (1).

2.1. DISPERSION COMPENSATION FOR LINEAR SYSTEM

To understand the effect of the dispersion compensation, we first consider the case of a linear system, that is, equation (1) with $a(Z) = 0$. In particular, we solve the initial value problem,

$$i\frac{\partial q}{\partial Z} + \frac{d(Z)}{2}\frac{\partial^2 q}{\partial T^2} = 0 \,, \tag{2}$$

with the input chirped Gaussian pulse,

$$q(T,0) = Q_0 \exp\left\{-\frac{T^2}{2T_0^2}(1 - iC_0)\right\} \,. \tag{3}$$

Then the problem can be solved explicitely and we have the solution,

$$q(T,Z) = Q_0\sqrt{\frac{T_0}{T_1(Z)}} \exp\left\{-\frac{T^2}{2T_1^2(Z)}(1 - iC(Z)) + i\theta(Z)\right\} \,, \tag{4}$$

where $T_1(Z), C(Z)$ and $\theta(Z)$ are given by

$$\left\{\begin{array}{ll} T_1(Z) & = T_0\left[\{1 + C_0F(Z)\}^2 + F^2(Z)\right]^{1/2} \,, \\ C(Z) & = C_0 + (1 + C_0^2)F(Z) \,, \\ \theta(Z) & = -\frac{1}{2}\tan^{-1}\left(\frac{F(Z)}{1 + C_0F(Z)}\right) \,. \end{array}\right. \tag{5}$$

Here the function $F(Z)$ represents the accumulated dispersion along the line,

$$F(Z) \equiv \frac{1}{T_0^2}\int_0^Z d(Z')dZ' \,. \tag{6}$$

For convenience, we introduce the variable $p(Z) \equiv 1/T_1(Z)$ proportional to the inverse of pulse width, and also $\tau \equiv p(Z)T$ to represent the change of the pulse width. Then the solution (4) can be written in the form,

$$q(T,Z) = \alpha_0\sqrt{p(Z)} \exp\left[-\frac{\tau^2}{2} + i\theta(Z)\right] e^{\frac{i}{2}C(Z)\tau^2} \,, \tag{7}$$

where $\alpha_0 = Q_0\sqrt{T_0}$. Also we note that $p(Z)$ and $C(Z)$ satisfy

$$\frac{dp}{dZ} = -Cp^3 d \,, \tag{8}$$

$$\frac{dC}{dZ} = C^2p^2d + p^2d \,. \tag{9}$$

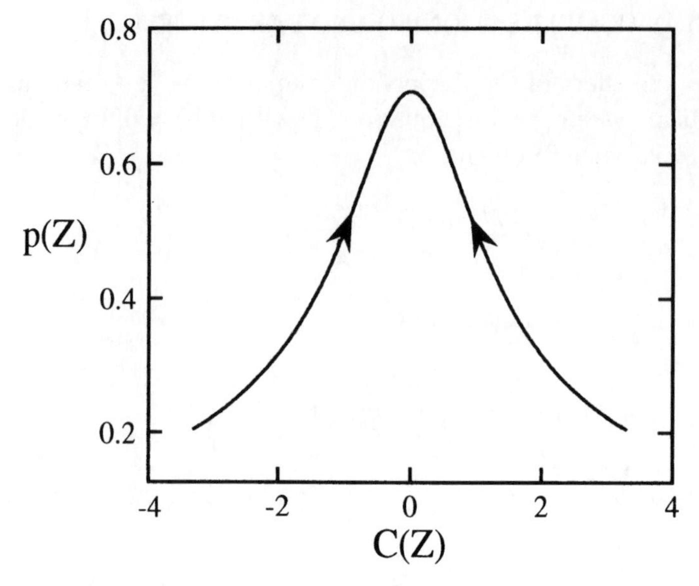

Figure 1. A periodic orbit for linear system.

The solution of these equations can be described by

$$(C^2 + 1)p^2 = H_0 = p^2_{max} \; , \tag{10}$$

where H_0 is a constant giving the maximum value of p^2 when $C = 0$. Figure 1 shows the solution curve (10). Here we took a case where the orbit is symmetric in the p-axis, but the general periodic orbit for a linear system can be represented by a finite and connected piece of the curve given by equation (10). Note that the functions $p(Z)$ and $C(Z)$ are periodic if and only if the dispersion compensation is complete, that is, $d_0 = \langle d \rangle = 0$ and $F(Z)$ is periodic.

2.2. NONLINEAR RZ PULSE IN A DISPERSION COMPENSATED SYSTEM

We now consider a nonlinear pulse propagation in the dispersion compensated system with a large dispersion variation. Because of the large variation, the guiding center soliton [9] described as perturbation from ideal soliton can not exist. The main feature of the pulse in such a system is a large deformation from sech-shape of soliton and appearence of strong chirp due to the large values of local dispersions. Because of the strong chirp, we first assume the solution q of equation (1) to take a form similar to the linear case (7),

$$q(T, Z) = \sqrt{p(Z)}v(\tau, Z)e^{\frac{i}{2}C(Z)\tau^2} \; , \tag{11}$$

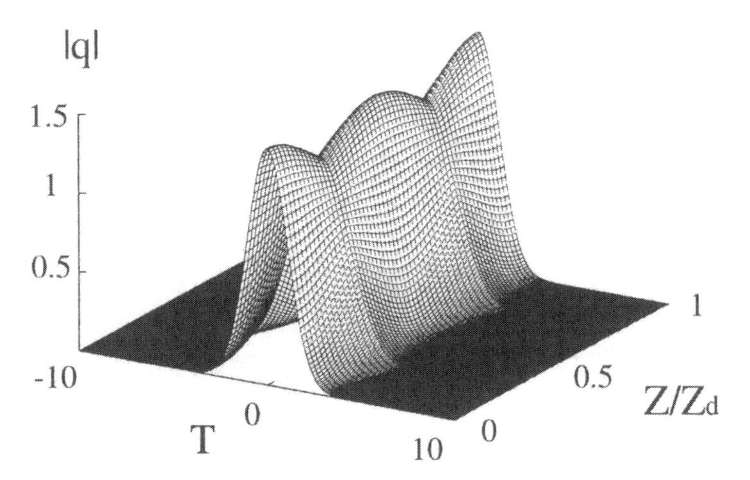

Figure 2. Pulse evolution in $|q|$ with $\Delta D = 58$ and $Z_d = 0.156$.

with $\tau = p(Z)T$. Then equation (1) becomes

$$i\frac{\partial v}{\partial Z} + \frac{p^2 d}{2}\frac{\partial^2 v}{\partial \tau^2} + pa^2|v|^2 v - \frac{\kappa}{2}\tau^2 v = 0 \ , \tag{12}$$

with the equations for $p(Z)$ and $C(Z)$ similar to equations (8) and (9),

$$\frac{dp}{dZ} = -Cp^3 d \ , \tag{13}$$

$$\kappa(Z) \equiv \frac{dC}{dZ} - C^2 p^2 d \ . \tag{14}$$

To demonstrate the significance of the transformation (11), we plot in Figures 2 and 3 the pulse shape in $|q|$ and in $|v|$ respectively for a 2 step periodic dispersion compensation with the dispersion map $d(Z) = d_0 + \Delta D/2$, for $0 < Z < Z_d/4$ and $3Z_d/4 < Z < Z_d$, and $d(Z) = d_0 - \Delta D/2$, for $Z_d/4 < Z < 3Z_d/4$, where $d_0 = 1, \Delta D = 58$ and $Z_d = 0.156$. As we can see from these figures, the new variable v is almost stationary, though the coefficients in equation (12) rapidly oscillate in large amounts. This indicates that the v can be well approximated by the solution of its averaged equation,

$$i\frac{\partial V}{\partial Z} + \frac{D_0}{2}\frac{\partial^2 V}{\partial \tau^2} + A_0|V|^2 V - \frac{K_0}{2}\tau^2 V = 0 \ , \tag{15}$$

where $D_0 = \langle p^2 d\rangle$, $A_0 = \langle pa^2\rangle$ and $K_0 = \langle\kappa\rangle = -\langle C^2 p^2 d\rangle$. Namely V gives the core of the pulse solution for a dispersion compensated system, and these coefficients determine the pulse dynamics as in the case of a soliton

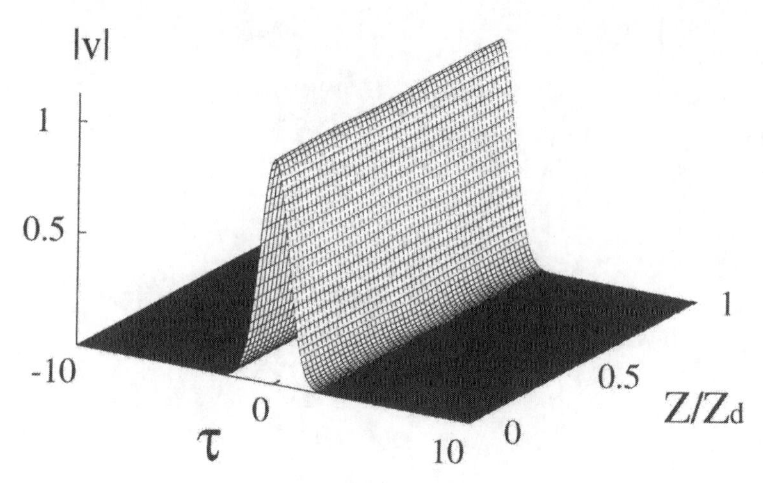

Figure 3. Pulse evolution in $|v|$.

system. We here call them as the effective dispersion D_0, the effective Kerr coefficient A_0, and the effective chirp K_0. Note here that D_0 is different from the simple path-average dispersion $d_0 = \langle d \rangle$, and it may take a positive value even in the case $\langle d \rangle \leq 0$. This implies that a (quasi-) stationary soliton-like pulse exists in the normal average dispersion regime. We will discuss this point further in subsection 2.3. Equation (15) makes sense only if equations (13) and (14) admit a periodic solution with the period Z_d, and the average is taken over this period. Equation (15) indicates that the chirp introduces a quadratic potential in addition to the usual self trapping term. The sign of K_0 is then essential to determine the tail character of the pulse. Namely $K_0 > 0$ gives a trapping potential, and the tail decays like a Gaussian pulse, while the case $K_0 < 0$ gives a nontrapping one, and the tail has an oscillation which decays only algebraically (see also Figure 10 below). The latter phenomena is known as the tunneling effect in the quantum mechanics. In the present case, we can show that K_0 is always negative for periodic solution of equations (13) and (14). In order to do this, we first determine the formula of $\kappa(Z)$ for the periodic solution. This may be obtained by using the variational principle. The Lagrangian for equation (1) is given by

$$L = \int \left\{ \frac{i}{2}(q_Z q^* - q q_Z^*) - \frac{d(Z)}{2}|q_T|^2 + \frac{a^2(Z)}{2}|q|^4 \right\} dT \,. \tag{16}$$

With the transformation (11), we rewrite L in the form,

$$L = \int \left\{ \frac{i}{2}(v_Z v^* - v v_Z^*) - \frac{p^2 d}{2}|v_\tau|^2 + \frac{pa^2}{2}|v|^4 \right. \tag{17}$$

$$-\left(\frac{\dot{C}}{2}+\frac{C\dot{p}}{p}+\frac{C^2p^2d}{2}\right)\tau^2|v|^2+\frac{i}{2p}(\dot{p}+Cp^3d)\tau(v_\tau v^*-vv_\tau^*)\bigg\}\,d\tau\ .$$

Then assuming a stationary solution for v in the form $v(\tau,Z)=f(\tau)e^{i\lambda(Z)}$, we obtain

$$L=-\frac{p^2d}{2}I_D+\frac{pa^2}{2}I_N-\left(\frac{C\dot{p}}{p}+\frac{C^2p^2d}{2}\right)I_C\ , \tag{18}$$

where we have dropped the terms of total derivative in Z, and the dots in p and C indicate the Z-derivatives. The constants I_k for $k=D,N,C$ are positive and defined by

$$I_D\equiv\int(f_\tau)^2\,d\tau,\quad I_N\equiv\int f^4\,d\tau,\quad I_C\equiv\int(\tau f)^2\,d\tau. \tag{19}$$

Then taking the variations with respect to C and p, we obtain equations (13) and (14) with

$$\kappa(Z)=\alpha_1p^2d-\alpha_2pa^2\ , \tag{20}$$

where α_1 and α_2 are positive constants defined by $\alpha_1=I_D/I_C$ and $\alpha_2=I_N/(2I_C)$. In the case of linear system, we have $\kappa(Z)=p^2d$ as in equation (9), and the average $\langle\kappa\rangle=0$. Then we can state:

Proposition 1 *The average $K_0\equiv\langle\kappa\rangle$ is negative for the periodic solution of equations (13) and (14) with (20), and we have*

$$K_0=\langle\alpha_1p^2d\rangle-\langle\alpha_2pa^2\rangle=-\Big\langle\frac{\alpha_2C^2pa^2}{\alpha_1+C^2}\Big\rangle\ <\ 0 \tag{21}$$

Proof. From equation (14) with (20), we have

$$\frac{1}{1+C^2/\alpha_1}\frac{dC}{dZ}=\alpha_1p^2d-\frac{\alpha_2}{1+C^2/\alpha_1}pa^2\ . \tag{22}$$

Since the left hand side of this equation is a total derivative of Z, its average over the period of the solution vanishes, i.e.

$$\langle\alpha_1p^2d\rangle=\Big\langle\frac{\alpha_2}{1+C^2/\alpha_1}pa^2\Big\rangle\ <\ \langle\alpha_2pa^2\rangle\ . \tag{23}$$

From equation (23), we easily obtain the result. □

A physical meaning of this proposition is that the chirp generated by the nonlinear term can not be compensated only by the (linear) dispersion compensation. The best performance is obtained for the complete linear (ideal) case. In this sense, a short compensation fiber with large dispersion like a grating [18] gives a better performance, since the second term representing the nonlinear effect in equation (20) may be ignored in the part of compensation.

2.3. EXAMPLE OF DISPERSION COMPENSATED SYSTEM

We here consider the following simple example of dispersion compensated system. The system is assumed to be lossless and have a two-step dispersion map,

$$d(Z) = \begin{cases} d_1 , & \text{for} \quad 0 < Z - nZ_d < Z_d/4 , \\ d_2 , & \text{for} \quad Z_d/4 < Z - nZ_d < 3Z_d/4 , \\ d_1 , & \text{for} \quad 3Z_d/4 < Z - nZ_d < Z_d . \end{cases} \tag{24}$$

We first study the equations for $p(Z)$ and $C(Z)$ which are now given by

$$\frac{dp}{dZ} = -Cp^3 d , \tag{25}$$

$$\frac{dC}{dZ} = (C^2+1)p^2 d - \frac{E_0}{\sqrt{2\pi}} p , \tag{26}$$

where we have assumed the stationary solution $f(\tau)$ to be Gaussian, $f(\tau) = a_0 \exp(-\tau^2/2)$, which gives $\alpha_1 = 1, \alpha_2 = E_0/\sqrt{2\pi}$ in equation (20), and E_0 represents its total energy related to the (enhanced) peak power as $E_0 = \sqrt{\pi} a_0^2$. As in the linear case (10), the solution can be expressed as

$$(C^2+1)p^2 = 2\beta(Z)p + H(Z) , \tag{27}$$

where $\beta(Z) = E_0/(\sqrt{2\pi} d(Z))$, and $H(Z)$ is also a 2-step perodic function with H_1 in the region with $d = d_1$ and H_2 in $d = d_2$. Note that $\beta = 0$ implies the linear case and $H(Z)$ becomes a constant H_0 as in equation (10). For the case $d_1 > 0$, one should choose $C(0) < 0$ for a pulse compression, and the minimum pulse width is obtained at the distance of zero chirp ($Z = 0, Z_d/2$ i.e. the mid point of each fiber in the present (lossless) case). We then obtain the following results for equations (25) and (26):

Proposition 2 *Let $d_1 > 0$ and $d_2 < 0$. Then a) $H_2 - H_1 = 2(\beta_1 - \beta_2)p_* > 0$, b) $p_1 := p(0) > p_2 := p(Z_d/2)$, where β_1, β_2 are the values of β in the anomalous and normal dispersion regions respectively, and p_* is the minimum value of p which realizes at the edges of each fiber, i.e. $Z = Z_d/4, 3Z_d/4$.*

Proof. From equation (27) with $Z = Z_d/4 - 0$ and $Z = Z_d/4 + 0$, we have $2\beta_1 p_* + H_1 = 2\beta_2 p_* + H_2$ which leads to a). To prove b), we first define the function $B_i(p) := C^2p^2 = -p^2 + 2\beta_i p + H_i$. Because of $\beta_1 > 0$

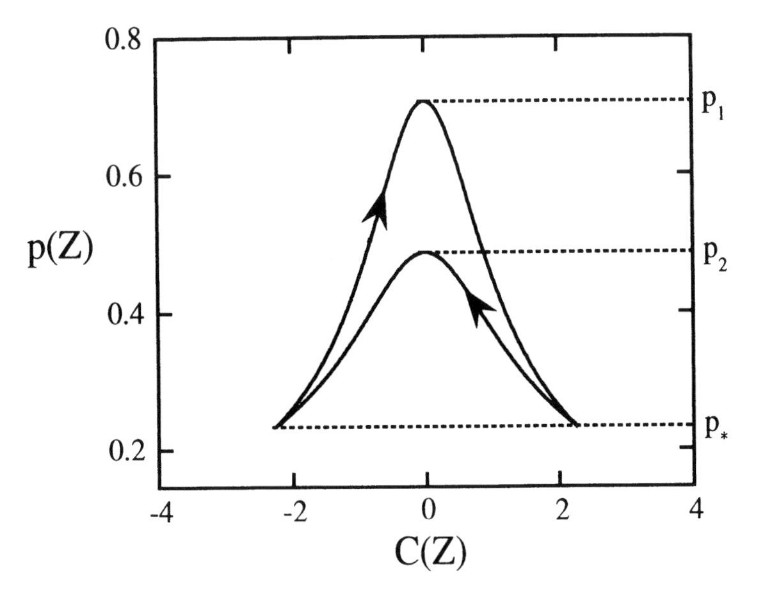

Figure 4. A periodic orbit in p-C plane.

and $\beta_2 < 0$, the maximum values of $B_1(p)$ and $B_2(p)$ are in the positive and negative p, respectively. It is then easy to see the relation b). Note that $B_1(p_*) = B_2(p_*)$ which gives a), and $B_1(p_1) = B_2(p_2) = 0$ since $C = 0$. □

Figure 4 shows a periodic orbit in the p-C plane with $d_1 = 27.5, d_2 = -25.5, Z_d = 1$. In Figure 5, we plot the corresponding solutions $p(Z)$ and $C(Z)$, and the dotted curves show the periodic solution for the linear system with $d_1 = 26.5, d_2 = -26.5$ as in Figure 1. We can see from these figures that the nonlinearity gives the asymetric deformation in the solutions $p(Z)$ and $C(Z)$, and this leads to $K_0 < 0$. The asymmetry also leads that the effective dispersion $D_0 = \langle p^2 d \rangle$ satisfies

$$\langle p^2 d \rangle > \langle p^2 \rangle \langle d \rangle \ . \tag{28}$$

Physically, this implies that the average pulse width is shorter in the anomalous dispersion regime ($d_1 > 0$) than that in the normal one ($d_2 < 0$). Because of the larger effective dispersion than the simple average dispersion, i.e. $D_0 > d_0$ for $\langle p^2 \rangle = 1$, we have a power enhancement to compensate the effective dispersion. One also notes that the total energy E_0 satisfies the following equation obtained by the average of equation (20),

$$E_0 = \frac{\sqrt{2\pi}}{\langle p \rangle}(D_0 - K_0) \ . \tag{29}$$

Since $K_0 < 0$ (Proposition 1), the enhanced power also results from the background nontrapping potential. In Figure 6, we plot the pulse energy

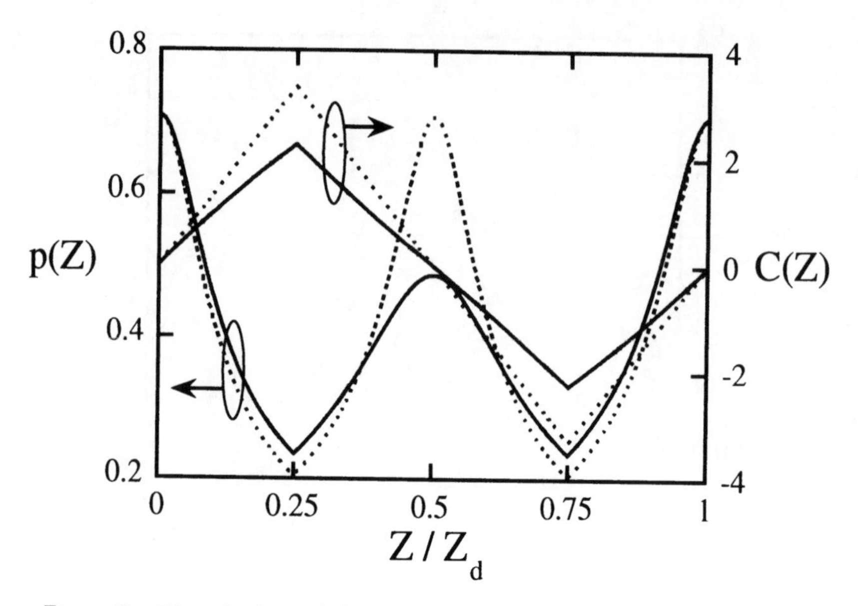

Figure 5. The solutions $p(Z)$ and $C(Z)$ in linear and nonlinear systems.

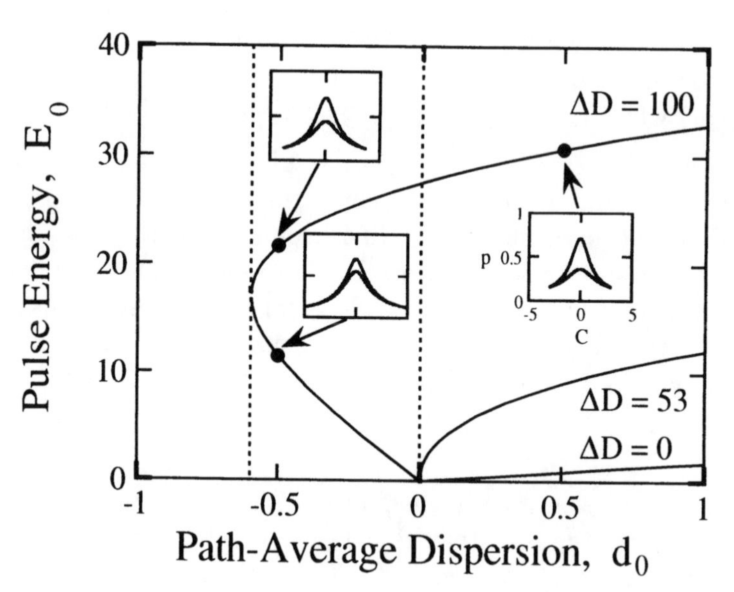

Figure 6. Periodic solutions of Eqs(25) and (26).

E_0 versus the average dispersion $d_0 = \langle d \rangle$ for periodic solutions of equations (25) and (26) with $Z_d = 1$. Note here that there exist two periodic solutions in the normal dispersion regime for $\Delta D := d_1 - d_2 > \Delta D_c \approx 53$. Here we have assumed $p(0) = 1/\sqrt{2}, C(0) = 0$. The existence of a soliton-like pulse in an normal average dispersion regime has been observed in reference [20], and the critical strength ΔD_c of the compensation has been analytically

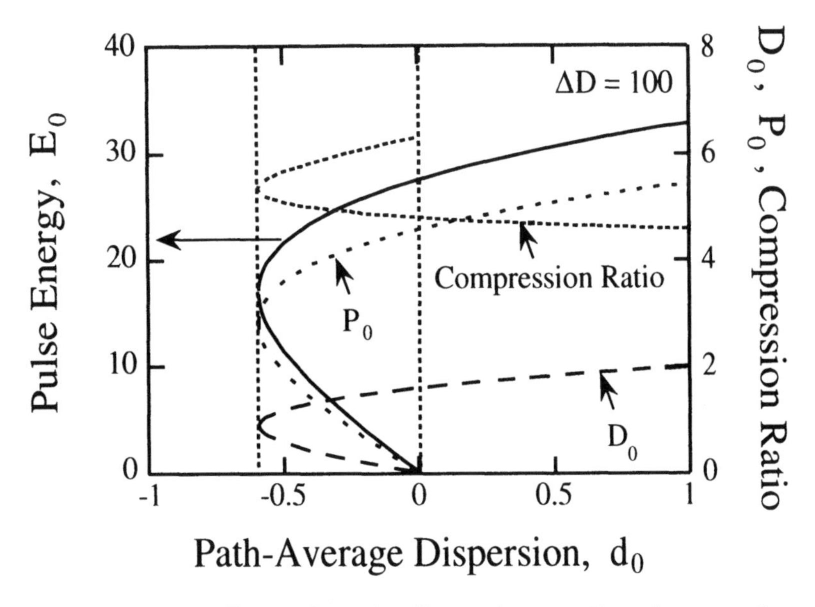

Figure 7. Energy E_0, effective dispersion D_0, peak power P_0 and compression ratio for the periodic solution with $\Delta D = 100$.

obtained in reference [2]. In this figure we also show the closed orbit in the p-C plane for the periodic solution corresponding to the point on the curve with $\Delta D = 100$. From equation (23) we see that D_0 is positive on the E_0-curve, and it is almost proportional to the peak power for each value of average dispersion d_0. Then the limit $D_0 = 0$ is obtained at the linear point where $p_1 = p_2$ and $d_0 = 0$. In Figure 7, we plot the pulse energy E_0, the effective dispersion D_0, the peak power $P_0 := A_0 a_0^2 = \langle p \rangle a_0^2$ and the compression ratio for this case. One can show that we need to have the ratio to be more than 3 for a periodic solution in the normal average dispersion. Note with the peak power P_0 that equation (29) is expressed by

$$P_0 = \sqrt{2}(D_0 - K_0) \ . \tag{30}$$

In the table, we list the values of the pulse energy, the peak power, the effective dispersion and the coefficient of the untrapped potential for several values of ΔD. The ratio $\sqrt{2}D_0/P_0$ indicates the percentage of the (effective) dispersion in the balance with the nonlinearity, and the rest of the percentage is contributed by the untrapped potential due to the chirp as shown in equation (30).

We thus have two solutions for $d_0 < 0$. However the general theory of dynamical system may predict as in the case of the saddle-node bifurcation that the solutions in the lower branch of the E_0-curve in Figures 6 and 7 for $d_0 < 0$ are unstable. This may be seen in Figure 8 where we plot the Poincaré maps of the solutions of equations (25) and (26) for $d_0 = -0.5$

TABLE 1. Energy E_0, effective dispersion D_0, peak power P_0, the effective chirp K_0, and the percentage of the dipersion contribution in the balance with the nonlinearity.

ΔD	d_0	E_0	P_0	D_0	K_0	$\sqrt{2}D_0/P_0[\%]$
0	0.5	0.886	0.353	0.25	0	100
53	0.5	9.071	2.067	0.823	-0.639	56.3
100	0.5	30.665	5.081	1.830	-1.763	50.9
	-0.5	21.634	3.603	1.189	-1.359	46.7
	-0.5	11.483	1.895	0.570	-0.770	42.5

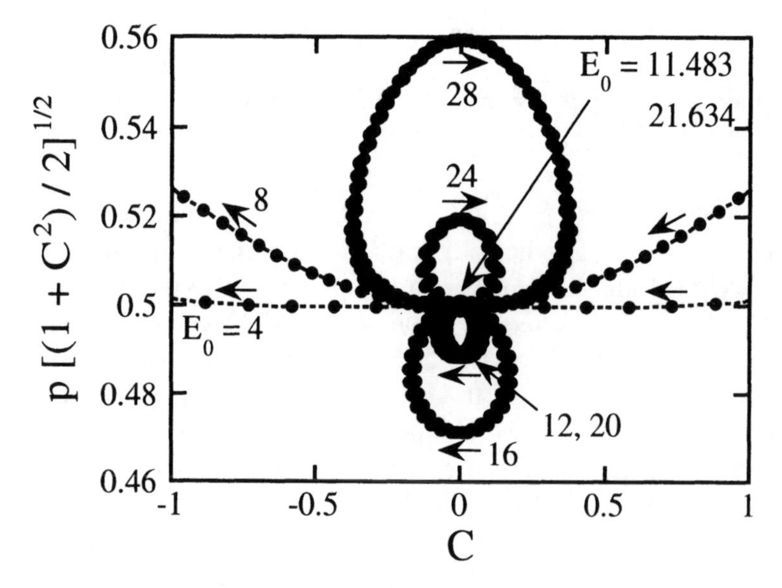

Figure 8. Poincaré maps for the solutions of equations (25) and (26) with the initial data, $p(0) = 1/\sqrt{2}, C(0) = 0$ and the various values of E_0.

with the initial data $C(0) = 0, p(0) = 1/\sqrt{2}$ and several different values of E_0. Corresponding to the periodic solutions on the E_0-curve in Figure 6, we have two fixed points $E_0 = 11.483$ and 21.634, and the solutions with $E_0 < 11.483$ seem to be unstable. However the unstable character is weak, and one can find the corresponding solution in the numerical solution of the NLS equation (1) (see Figure 9).

With the periodic solution for equations (25) and (26), we now discuss the solution of the averaged equation (15). A stationary solution of equation (15) may be obtained in the form $V(\tau, Z) = f(\tau)e^{i\lambda_0 Z}$ with a real function

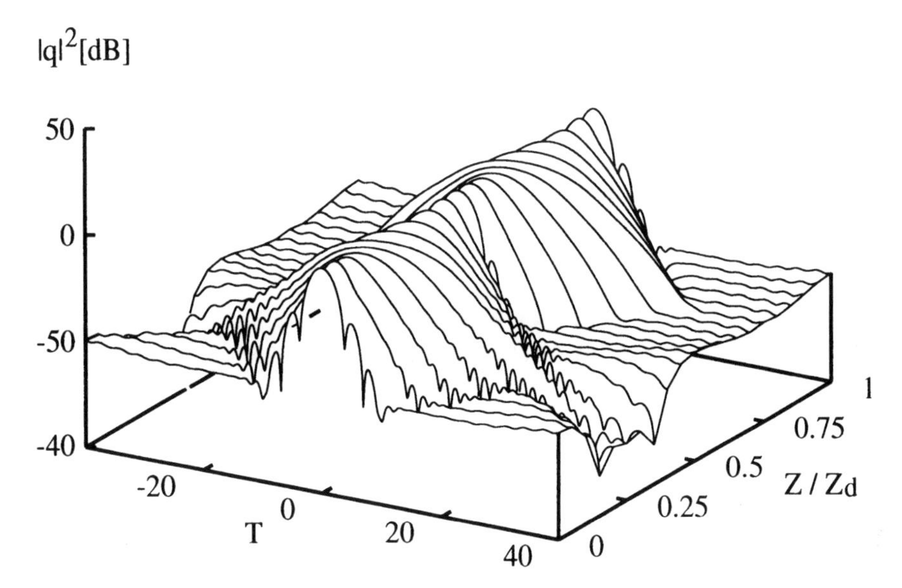

Figure 9. The stationary solution of equation (1) with $\Delta D = 100$ at the lower branch of the curve in Figure 6.

$f(\tau)$ and a real constant λ_0. The function $f(\tau)$ is a solution of

$$\frac{D_0}{2}\frac{d^2 f}{d\tau^2} + A_0 f^3 - \frac{K_0}{2}\tau^2 f = \lambda_0 f \ . \tag{31}$$

Because of the nontrapping term with $K_0 < 0$, one needs to have higher intensity of f^2 than the soliton case ($K_0 = 0$), which contributes to a part of the enhanced power [22]. Figure 10 shows the stationary (bound-state) solution of equation (31) corresponding to the parameters used in Figure 2. We here found that the enhanced peak power at $Z = nZ_d$ is 2.25, which is close to the value 2.24 obtained by the direct numerical simulation of equation (1). We also plot the pulse shape of the simulation at $Z = 200Z_d$. The difference is due to the corrections on the stationary solution f, and the core solution V may be obtained by a Fourier-like expansion in terms of the eigen-modes of equation (31). Then the oscillating tails in the stationary solution may be canceled by taking account of the higher order eigenmodes as in the usual case with the Fourier transform. Then the solution v of equation (12) can be constructed by using the guiding center theory. Note in Figure 10 that the pulse width of the stationary solution f is smaller than that of the simulation, and this contributes a reduction of the interaction between pulses [21]. The details of these results will be presented elsewhere.

Also note that the oscillating tails result that the total energy of the *stationary* solution, $\int_{-\infty}^{\infty} |V|^2 dT$, diverges. This implies that an input pulse

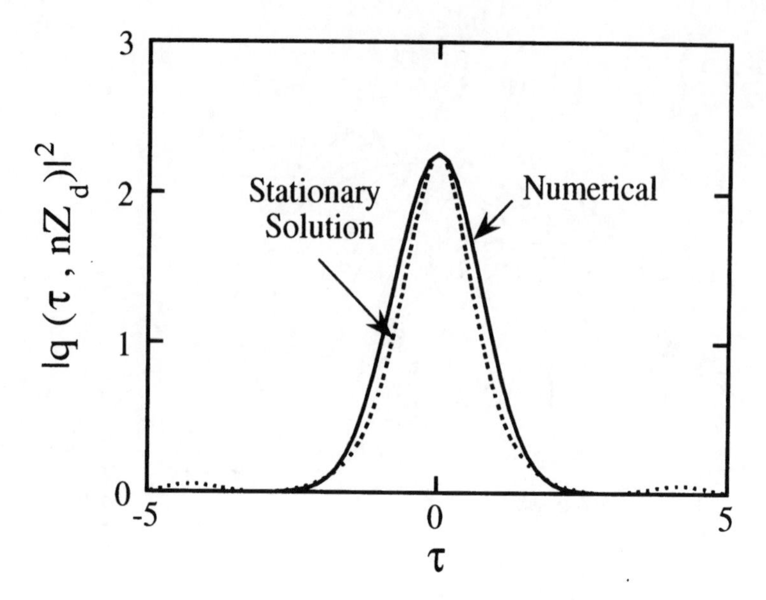

Figure 10. Stationary solution and numerical solution.

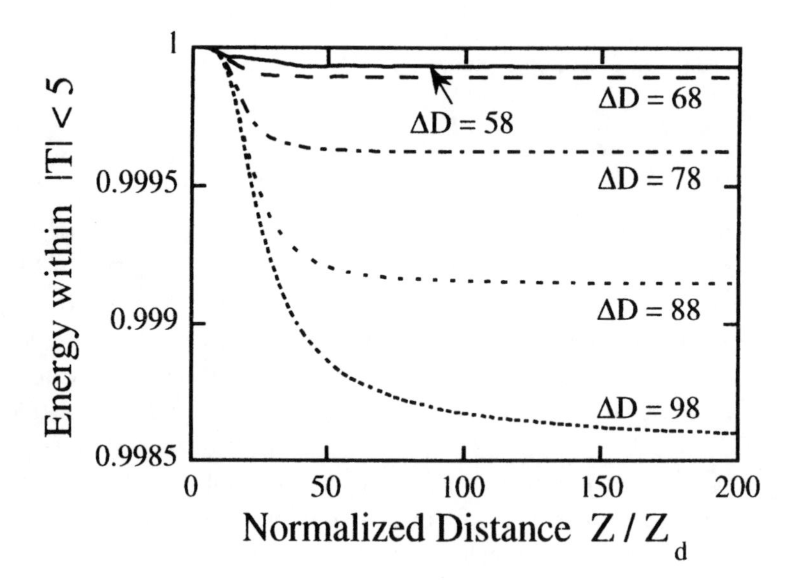

Figure 11. Energy leak due to tunneling effect.

having a finite energy eventually decays to dispersive waves due to the tunneling effect. However, in a practical situation with $\Delta D \cong 60$, the decay may be negligible. In fact, in Figure 11 the direct numerical simulation of equation (1) with Gaussian initial data shows a negligibly small decay of the energy within the window $|T| < 5$ after forming a quasi-stationary state having some oscillating tails. But for $\Delta D \geq 90$, one can observe a gradual decay of the energy due to the tunneling effect.

3. NRZ Pulse Propagation

The NRZ pulse propagation can also be described by the NLS equation,

$$i\frac{\partial q}{\partial Z} + \frac{\beta_2}{2}\frac{\partial^2 q}{\partial T^2} + |q|^2 q = 0 \,. \tag{32}$$

Unlike the soliton case, the NRZ pulse is non-stationary, and in order to reduce the pulse distortion the system should be designed in a small dispersion and low power regime. Also, because of the modulational instability for constant amplitude of the NRZ pulse, one has to operate the system at the regime $\beta_2 < 0$. In an ideal case, the best performance can be obtained in the zero dispersion and linear limit. However, the real system always includes a noise, and the signal pulse power should be kept higher than the noise level. Because of this limit in signal to noise ratio, the nonlinearity becomes an essential effect for a long distance transmission problem, and the zero dispersion causes a resonant interaction process between noise and signal, which is called the four wave mixing (FWM), and leads to a distortion of the pulse. The main purpose of this section is to present a mathematical theory to describe the pulse distortion in a weak dispersion limit of the NLS equation.

Since the limit $\beta_2 \to 0$ is singular in equation (32), we first rewrite q as

$$q(T, Z) = \sqrt{\rho(T, Z)}\exp\left(i\frac{1}{\beta}\sigma(T, Z)\right) \,. \tag{33}$$

Here we have set $\beta_2 = -\beta^2$. The quantities $\rho := |q|^2$ and $\sigma := \beta \mathrm{Arg}(q)$ represent the local intensity and phase of the electric field. Substituting equation (33) to equation (32), and rescaling the variables $Z \to Z/\beta$, we have

$$\frac{\partial}{\partial Z}\rho = \frac{\partial}{\partial T}(\rho u) \,, \tag{34}$$

$$\frac{\partial}{\partial Z}(\rho u) = \frac{\partial}{\partial T}\left(\rho u^2 + \frac{1}{2}\rho^2 - \frac{\beta^2}{4}\rho\frac{\partial^2}{\partial T^2}\ln\rho\right) \,, \tag{35}$$

where u represents the chirp defined by $u := \partial\sigma/\partial T$. In the limit $\beta \to 0$, if we assume that both ρ and u are smooth, then equations (34) and (35) for $\rho \neq 0$ can be approximated by a 2×2 quasi-linear system,

$$\frac{\partial}{\partial Z}\begin{pmatrix}\rho \\ u\end{pmatrix} = \begin{pmatrix}u & \rho \\ 1 & u\end{pmatrix}\frac{\partial}{\partial T}\begin{pmatrix}\rho \\ u\end{pmatrix} \,. \tag{36}$$

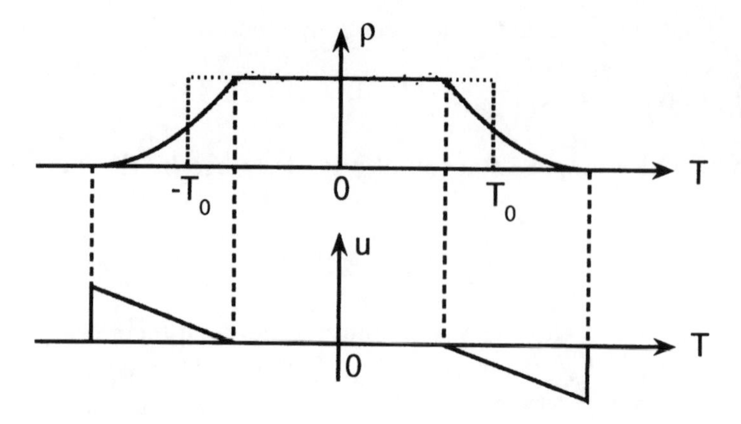

Figure 12. Deformation of NRZ pulse with no initial chirp.

The system (36) is known as the shallow water wave equation, and has been intensively discussed (for example, see reference [4]). It is then interesting to note that the distortion of the NRZ pulse may be understood as a deformation of the water surface.

For a demonstration of the NRZ pulse propagation, we consider the initial value problem of equation (36). As a simple example of an initial NRZ pulse, we take a square pulse having constant phase (zero-chirp),

$$\rho(T,0) = \begin{cases} \rho_0, & \text{for} \quad |T| < T_0 \\ 0, & \text{for} \quad |T| > T_0 \end{cases}, \tag{37}$$

$$u(T,0) = 0, \quad \text{for} \quad \forall T. \tag{38}$$

This is called “Dam-breaking problem”, since ρ and $-u$ represent the depth and the velocity of water which rests on the *spatial region* $|T| < T_0$ at the *time* $Z = 0$. We then expect to see a leakage of the water from the edges. In Figure 12, we plot the analytic solution, the pulse shape ρ and the chirp u. The thin dotted curve in Figure 12 shows the numerical result of the NLS equation with $\rho_0 = 1, T_0 = 10$ and $\beta_2 \equiv -\beta^2 = -0.1$ at $Z = 10\beta \approx 3.16$. Notice the good agreement with the analytical solution except some small oscillations on the top of ρ which disappear in the limit $\beta_2 \to 0$. As we have predicted from the hydrodynamic analogy, the deformation of the pulse induces the generation of chirp (or water velocity) at the edges, and the global nature of the solution is understood as an expansion of the water (the rarefaction wave). In order to reduce the expansion, it is natural to put initial chirp opposed to the chirp appearing in the edges. This is equivalent to give an initial velocity with a piston and to compress the water. Because of the quasi-linearity of the system (36), we then expect a shock formation in the solution, and therefore equation (36) is no longer valid as an approximate model of the NLS equation. In fact, if we start

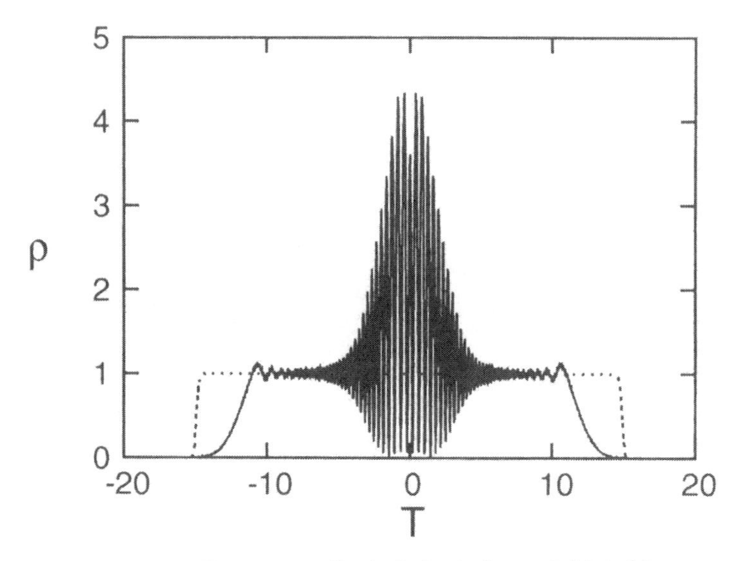

Figure 13. Optical shock due to initial chirp.

with the initial data with $\rho(T,0)$ in equation (37) and, for example, with $u_0 > 0$

$$u(T,0) = \begin{cases} -u_0, & \text{for} \quad T < 0 \\ u_0, & \text{for} \quad T > 0 \end{cases}, \tag{39}$$

the numerical solution of the NLS equation shows high oscillations starting from the discontinuous point of chirp, $T = 0$, as seen in Figure 13. Here we have set $\rho_0 = 1, u_0 = 2.85$ with $\beta_2 = -0.1$, and the solution $\rho(T, Z)$ is plotted for $Z = 0.95$. The main frequency of the oscillations is of order $1/\beta$, and usual optical filter can remove those oscillations as in the sense of *averaging*. So what we want to describe here is the average behavior of the solution. For this purpose, in the following section we extend the model (36) and show that the new approximate model (the NLS-Whitham equations) admits a global solution describing the average motion of the NLS equation (32) for several step initial data having different values of chirp.

3.1. THE NLS-WHITHAM EQUATIONS

Let us first give the NLS-Whitham equations defined on the genus g Riemann surface, and in particular the system (36) is just the simplest case ($g = 0$) of the NLS-Whitham equation. The two sheeted Riemann surface of genus g is defined by the algebraic (hyper-elliptic) curve $y^2 = R_g(\lambda)$ with

$$R_g(\lambda) = \prod_{k=1}^{2g+2} (\lambda - \lambda_k) \,, \tag{40}$$

where the branch points λ_k are real and are assumed to satisfy $\lambda_1 < \lambda_2 < \ldots < \lambda_{2g+2}$. The NLS-Whitham equation then describes the dynamics of $\lambda_k' s$ which corresponds to a slow modulation of quasi-periodic wavetrain of the NLS equation. Then the NLS-Whitham equation of genus g is defined by

$$\frac{\partial \omega_1}{\partial Z} = \frac{\partial \omega_2}{\partial T} , \tag{41}$$

where ω_1 and ω_2 are Abelian differentials of the second kind defined by

$$\omega_1 = \frac{1}{2}\left(1 + \frac{\hat{P}_g}{\sqrt{R_g}}\right) d\lambda \sim d\lambda + O\left(\frac{d\lambda}{\lambda^2}\right) , \tag{42}$$

$$\omega_2 = \frac{1}{2}\left(\lambda + \frac{\hat{Q}_g}{\sqrt{R_g}}\right) d\lambda \sim \lambda d\lambda + O\left(\frac{d\lambda}{\lambda^2}\right) . \tag{43}$$

Here $\hat{P}_g$, $\hat{Q}_g$ are the monic polynomials of degree $g+1$ and $g+2$ respectively. The coefficients of these polynomials are determined by their asymptotic properties and the normalizations on ω_1 and ω_2,

$$\oint_{a_i} \omega_1 = \oint_{a_i} \omega_2 = 0, \quad \text{for } i = 1, \cdots, g , \tag{44}$$

with the canonical a_i-cycle over the region $[\lambda_{2i+1}, \lambda_{2i}]$. In the simplest case of $g = 0$, the algebraic curve is given by $y^2 = (\lambda - \lambda_1)(\lambda - \lambda_2)$ with

$$\lambda_1 = u - 2\sqrt{\rho}, \quad \lambda_2 = u + 2\sqrt{\rho} , \tag{45}$$

and the $g = 0$ NLS-Whitham equation gives the Riemann invariant form of equation (36) whose Riemann invariants are λ_1 and λ_2. As shown in [4], the Whitham equation describes the conservation laws averaged over the high oscillations. Then one can introduce the fast and slow variables to describe the oscillation and the averaged behavior on the solution of the NLS equation. The average is taken over the fast variable so that the averaged equation describes the solution behaviour in the slow scale. Thus the average is given by

$$\langle f \rangle := \lim_{\beta \to 0} \frac{\beta}{2\pi} \int_{-\frac{\pi}{\beta}}^{\frac{\pi}{\beta}} f(T_{slow}, T_{fast}) dT_{fast} , \tag{46}$$

where T_{slow} and T_{fast} are the slow and fast variables, and we use T as the slow variable in the NLS-Whitham equation. Note that the rescaling $Z \to Z/\beta$ used to derive equation (36) defines the slow variable $Z = Z_{slow}$, i.e. $Z_{slow} = \beta Z_{fast}$. The Whitham equation leads to the averaged conservation laws,

$$\frac{\partial F_n}{\partial Z} = \frac{\partial G_n}{\partial T} \tag{47}$$

in which the cases for $n = 1$ and $n = 2$ correspond to the averages of (34) and (35). Here F_n and G_n are the averaged conserved densities and fluxies. In the particular case of $g = 0$, we have $\langle \rho \rangle = \rho$ and $\langle \rho u \rangle = \rho u$, which means there is no oscillations in these quantities.

3.2. CONTROL OF NRZ PULSE

We now study the initial value problem of the NLS-Whitham equation with the initial data given by equations (37) and (39). Here we vary the value of initial chirp u_0, and study its effect for a purpose of reducing the NRZ pulse broadening. The main objective is to determine the average behavior of the solution of the NLS equation (32) in the weak dispersion limit. In a practical situation, this corresponds to the NRZ pulse behaviour through an optical filter which averages out the high oscillations (shock) due to the initial chirp.

Before we give the results of analysis, let us first recall that the value of the u_0 corresponds to the initial velocity of the water, and the positive (negative) value of u_0 gives an compression (expansion). So we see a shock formation for the positive u_0 and a rarefaction wave for the negative u_0. We also note that the characteristic speed for zero chirp is given by $2\sqrt{\rho}$, so that we expect to see a different behavior of the solution for each u_0 value, (*i*) $u_0 > 2\sqrt{\rho_0}$, (*ii*) $0 < u_0 < 2\sqrt{\rho_0}$, (*iii*) $0 > u_0 > -2\sqrt{\rho_0}$, and (*iv*) $u_0 < -2\sqrt{\rho_0}$. This situation is quite similar to the case of Toda lattice equation discussed in reference [1]. We then show that by choosing an appropriate number of genus for a given initial data and by solving the corresponding NLS-Whitham equation, one can obtain a global solution. Below we give just the main theorems for those cases (see reference [17] for the details).

(i) The case where $u_0 > 2\sqrt{\rho_0}$: In this case, the chirp gives a compression of the pulse, and we expect a shock formation. Then we first identify the initial data as the $g = 1$ (*degenerated into* $g = 0$) data, and we have:

Theorem 1 *For* $u_0 > 2\sqrt{\rho_0}$, *the* $g = 1$ *NLS-Whitham equation with the initial data (37) and (39) has a global solution.*

The numerical solution $\rho(T, Z)$ at $Z = 0.95$ of the NLS equation (32) in this case is shown in Figure 13, where $\rho_0 = 1, u_0 = 2.85$ and $\beta_2 = -0.1$. We see here nearly steady oscillation in the center region, which indicates the $g = 1$ periodic solution of the NLS equation as predicted in the theorem.

(ii) The case where $2\sqrt{\rho_0} > u_0 > 0$:
This case also gives a compression, and we again need to use a nonzero

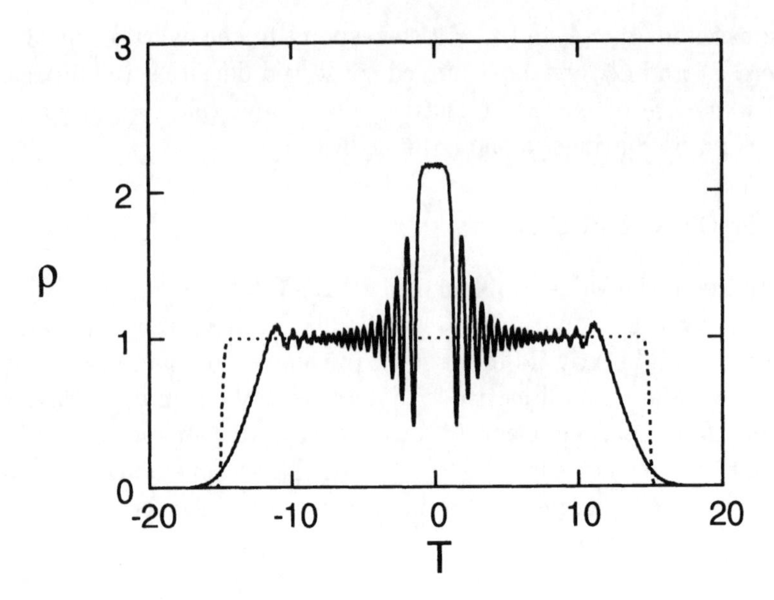

Figure 14. The solution $\rho(T, Z)$ for $2\sqrt{\rho_0} > u_0 > 0$.

genus NLS-Whitham equation for the regularization. In this case, the initial data can be considered as a $g = 2$ (*degenerated into* $g = 0$) data, and we obtain:

Theorem 2 *For* $2\sqrt{\rho_0} > u_0 > 0$, *the* $g = 2$ *NLS-Whitham equation with the initial data (37) and (39) admits the global solution.*

In Figure 14, we plot the numerical solution $\rho(T, Z)$ at $Z = 1.58$ of the NLS equation (32), where $\rho_0 = 1, u_0 = 0.95$ with $\beta_2 = -0.1$. Note in Figure 14 that the non-oscillating part with a constant level in the center region corresponds to the $g = 2$ (degenerated into $g = 0$) solution, and the oscillating parts in the side regions to the ones (degenertated into $g = 1$), as predicted in the theorem. One should compare Figure 13 with Figure 14 to see the difference in the center regions.

(iii) The case with $0 > u_0 > -2\sqrt{\rho_0}$:

In this case, the initial data gives a expansion of the pulse, and we have:

Theorem 3 *For* $0 > u_0 > -2\sqrt{\rho_0}$, *the* $g = 0$ *NLS-Whitham equation admits the global solution with the initial data (37) and (39).*

In Figure 15, we plot the numerical result $\rho(T, Z)$ at $Z = 3.16$ of the NLS equation (32) where $\rho_0 = 1, u_0 = -0.95, \beta_2 = -0.1$, which agrees quite well with the analytical solution shown as the dashed curve.

(iv) The case where $u_0 < -2\sqrt{\rho_0}$:

We first have that the initial data in this case can be considered as a $g = 1$

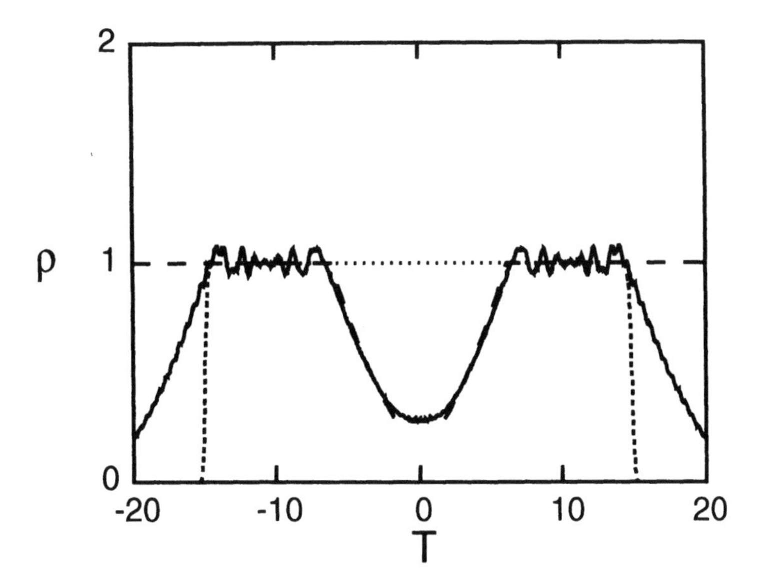

Figure 15. The solution $\rho(T, Z)$ for $0 > u_0 > -2\sqrt{\rho_0}$.

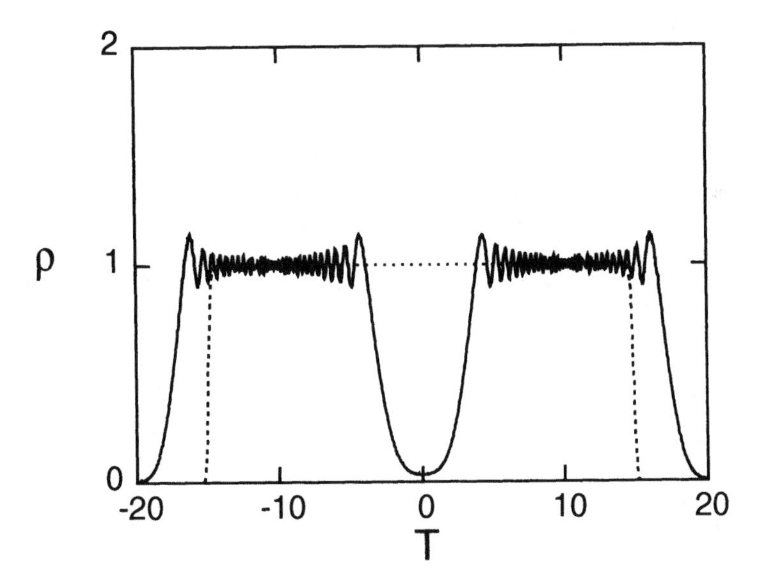

Figure 16. The solution $\rho(T, Z)$ for $u_0 < -2\sqrt{\rho_0}$.

(*degenerated into* $g = 0$) data. Then we obtain:

Theorem 4 *For* $u_0 < -2\sqrt{\rho_0}$*, the* $g = 1$ *NLS-Whitham equation with the initial data (37) and (39) has a global solution.*

In Figure 16, we plot the pulse profile $\rho(T, Z)$ at $Z = 0.95$ obtained by the numerical simulation of the NLS equation (32) with $\rho_0 = 1, u_0 = -2.85$

and $\beta_2 = -0.1$. The pulse level at $T = 0$ becomes zero right after the propagation, which can be predicted by the present theory. One should also note that the genus is actually degenerated into $g = 0$ for $\forall T$, even though we need the $g = 1$ regularized initial data for the global solution. This implies that the solution is indeed $g = 1$, but there is no oscillations in the solution. The small oscillations observed in Figure 16 are due to a nonzero value of β_2, and disappear in the limit $\beta_2 \to 0$.

References

1. Bloch, A. M. and Kodama, Y.: Dispersive regularization of the Whitham equation for the Toda lattice, *SIAM J. Appl. Math.*, **52**, (1992), p.909.
2. Cautaerts, V.: *Dispersion-managed soliton in normal averaged dispersion for small energy*, Osaka University, (December, 1997) in preparation.
3. Doran, N. J., Smith, N. J., Forysiak, W. and Knox, F. M.: Dispersion as control parameter in soliton transmission systems, in *Physics and Applications of Optical Solitons in Fibres '95*, Kluwer Academic Press, (1996), p.1.
4. Flaschka, H., Forest, M. G. and McLaughlin, D. W.: Multiphase averaging and the inverse spectral solution of the Korteweg-de Vries equation, *Comm. Pure Appl. Math.*, **33**, (1980), p.739.
5. Forysiak, W., Knox, F. M. and Doran, N. J.: Stepwise dispersion profiling of periodically amplified soliton systems, *J. Lightwave Technol.*, **12**, (1994), p.1330.
6. Forysiak, W., Blow, K. J. and Doran, N. J.: Reduction of Gordon-Haus jitter by post-transmission dispersion compensation, *Electron. Lett.*, **29**, (1993), p.1225.
7. Gabitov, I. R. and Turitsyn, S. K.: Breathing solitons in optical fiber links, *JETP Lett.*, **63**, (1996), p.814.
8. Georges, T. and Charbonnier, B.: Reduction of the dispersive wave in periodically amplified links with initially chirped solitons, *IEEE Photon. Technol. Lett.*, **9**, (1997), p.127.
9. Hasegawa, A. and Kodama, Y.: Guiding-centre soliton in fibres with periodically varying dispersion, *Opt. Lett.*, **16**, (1991), p.1385.
10. Hasegawa, A. and Kodama, Y.: *Solitons in optical communications*, Oxford Univ. Press, (1995).
11. Hasegawa, A., Kumar, S. and Kodama, Y.: Reduction of collision-induced time jitters in dispersion-managed soliton transmission systems, *Opt. Lett.*, **21**, (1996), p.39.
12. Hasegawa, A., Kodama, Y. and Maruta, A.: Recent progress in dispersion-managed soliton transmission technology *Opt. Fiber Tech.*, **3**, (1997), p.197.
13. Kodama, Y. and Wabnitz, S.: Analytical theory of guiding center NRZ and RZ signal transmission in normally dispersive nonlinear optical fibers, *Opt. Lett.*, **20**, (1995), p.2291.
14. Kodama, Y. and Wabnitz, S.: Compensation of NRZ signal distortion by initial frequency shifting, *Electron. Lett.*, **31**, (1995), p.1761.
15. Kodama, Y., Kumar, S. and Hasegawa, A.: Dispersion managements on soliton transmission in fibres with lumped amplifiers, in *Physics and Applications of Optical Solitons in Fibres '95*, Kluwer Academic Press, (1996), p.15; Kumar, S., Hasegawa, A. and Kodama, Y.: Adiabatic soliton transmission in fibers with lumped amplifiers: Analysis, *J. Opt. Soc. Am. B*, **14**, (1997), p.888.
16. Kodama, Y., Kumar, S. and Maruta, A.: Chirped nonlinear pulse propagation in a dispersion-compensated system, *Opt. Lett.*, **22**, (1997), p.1689.
17. Kodama, Y.: *The Whitham Equations for Optical Communications*, (1997), preprint.

18. Kumar, S. and Hasegawa, A.: Quasi-soliton propagation in dispersion managed optical fibers, *Opt. Lett.*, **22**, (1997), p.372.
19. Nakazawa, M. and Kubota, H.: Optical soliton communication in a positively and negatively dispersion-allocated optical fibre transmission line, *Electron. Lett.*, **31**, (1995), p.216.
20. Nijhof, J. H. B., Doran, N. J., Forysiak, W. and Knox, F. M.: Stable soliton-like propagation in dispersion managed systems with net anomalous, zero and normal dispersion, *Electron. Lett.*, **33**, (1997), p.1726.
21. Okamawari, T., Ueda, Y., Maruta, A., Kodama, Y. and Hasegawa, A.: Interaction between guiding center solitons in a periodically dispersion compensated optical transmission line, *Electron. Lett.*, **33**, (1997), p.1063.
22. Smith, N. J., Knox, F. M., Doran, N. J., Blow, K. J. and Bennion, I.: Enhanced power solitons in optical fibers with periodic dispersion management, *Electron. Lett.*, **32**, (1996), p.54.
23. Suzuki, M., Morita, I., Edagawa, N., Yamamoto, S., Taga, H. and Akiba, S.: Reduction of Gordon-Haus timing jitter by periodic dispersion compensation in soliton transmission, *Electron. Lett.*, **31**, (1995), p.2027.
24. Sugawara, H., Kato, H. and Kodama, Y.: Reduction of timing jitter in a soliton-WDM system with dispersion compensation, *Electron. Lett.*, **33**, (1997), p.1065.
25. Whitham, G. B.: *Linear and Nonlinear Waves*, John Wiley, New York, (1974).

(Received January 7, 1998)

CIRCULATING LOOP EXPERIMENTS ON SOLITONS WITH PERIODIC DISPERSION MANAGEMENT

Role of Band-Pass Filters and Path-Averaged Dispersion

J. M. JACOB, G. M. CARTER AND C. R. MENYUK
Department of Computer Science and Electrical Engineering
University of Maryland Baltimore County
1000 Hilltop Circle, Baltimore, Maryland 21250
The Laboratory for Physical Sciences
8050 Greenmead Drive, College Park, Maryland 20740

Abstract. We characterize the performance of long-haul 10 Gbit/s optical communication systems using stationary optical band-pass filters and periodic dispersion management. We have found for low values of path-averaged dispersion in a strong dispersion-managed system the pulse energy and dynamics are practically independent of path-averaged dispersion.

1. Introduction

Since the first discussion [1]-[3] of applying periodic dispersion management using alternating spans of opposite sign dispersion to soliton transmission, there has been tremendous progress in the experimental achievements and theoretical understanding of these systems. One of the promising aspects of this work is the ability to achieve long-distance, high data rate transmission without the need of active retiming or sliding frequency filter techniques. To date the best error-free transmission results using dispersion management and stationary filters are 10 Gbit/s×24,500 km, 20 Gbit/s×14,000 km, 40 Gbit/s×5,000 km and 20 Gbit/s×8 (WDM) × 4,000 km [4]-[7]. In all of the experimental work a wide variety of dispersion maps have been proposed. Soliton-like maps consist mostly of relatively low anomalous dispersion fiber with short spans of large normal dispersion fiber [1, 2]-[5]-[11]. NRZ-like maps consist mostly of relatively low normal dispersion fiber followed by short spans of large anomalous dispersion fiber [4], [12]-[14]. Other maps consist of large anomalous dispersion fiber followed by short spans of large normal dispersion fiber and have been proposed as soliton upgrades to

A. Hasegawa (ed.), New Trends in Optical Soliton Transmission Systems, 155–166.

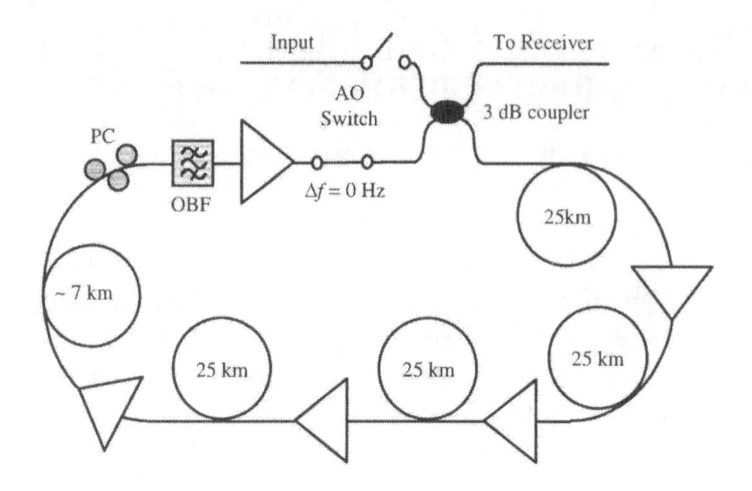

Figure 1. Experimental setup for circulating loop transmission.

existing terrestrial systems using standard non-dispersion shifted fiber [15]-[17].

Numerical and analytical studies have revealed that the pulse dynamics (breathing) of dispersion managed solitons results in an enhanced power when compared to solitons in fiber with uniform dispersion equal to the path-averaged dispersion, resulting in improved SNR performance and Gordon-Haus timing jitter reduction [18]-[20]. Dispersion managed solitons also differ from the standard soliton in the Gaussian-like pulse shapes which exists in these systems and can reduce pulse-pulse interactions [18, 21, 22]. In addition numerical results of soliton dispersion management for WDM applications are promising since dispersion management plays an important role in reducing the deleterious effects of collisions between solitons of neighboring channels [23]-[25]. The scope of this paper is to discuss our experimental work on dispersion managed solitons and illustrate the important role that stationary band-pass filters and path-averaged dispersion play in the error-free transmission of dispersion-managed solitons.

2. Experimental Setup

Our transmission system is a circulating dispersion-managed fiber optic loop shown schematically in Figure 1. The focus of our work in this system is the dispersion map and the stationary optical band-pass filter (OBF). We examine the impact of two different OBF's with a FWHM of 1.3 nm and 2.8 nm. The dispersion map consists of 100 km (4×25 km) of dispersion-shifted fiber ($\lambda_0 \cong 1565$ nm) with a normal dispersion of -1.1 ps/nm·km followed by ~7 km of standard fiber ($\lambda_0 \cong 1310$ nm) with an anomalous dispersion of +16.7 ps/nm·km at the operating wavelength of 1551 nm. By varying the length of the standard fiber, the path-averaged dispersion can be chosen to lie in the normal or anomalous dispersion regime at the operating wavelength. This dispersion-management scheme, which uses mostly normal dispersion fiber, is similar to that used in NRZ transmission systems [26]. We have successfully demonstrated error-free transmission of both NRZ and dispersion-managed soliton in this system with the appropriate sign of path-averaged dispersion [14]. The circulating loop has five variably spaced EDFA's pumped at 980 nm. The acousto-optic (AO) loop switch is followed by an AO frequency shifter such that a zero net optical frequency shift is obtained. A polarization controller is used to optimize the bit-error rate due to polarization- dependent effects in the loop. The receiver design is an unamplified dc–20 GHz O/E converter, 100 kHz–10 GHz low noise RF amplifier and a 5-pole Bessel filter with a 3 dB roll-off at 4.3 GHz. The optical front-end of the receiver is a low power EDFA followed by a 4 nm OBF.

3. Measurement Techniques

Bit-error rate (BER) measurements are made using a PRBS data pattern length equal to $2^{23}-1$. A cycle of our loop contains approximately 5.3 Mbits of data and is repeated every 100 ms, such that the error-counting duty cycle is fixed for all transmission distances up to 20,000 km. Due to transients arising from the 100 ns switching speed of the acousto-optic modulators, about 80 % of the loop data are gated for BER measurements. Therefore our experiments require approximately 250 seconds to accumulate 10 Gbits of data. Because of this duty cycle, we choose to define the amplitude and phase margins in our experiments to be at a level that has a constant BER $= 1\times10^{-6}$.

To determine the amplitude margin, the decision voltage is measured which yields a BER $= 1\times10^{-6}$ at the top and bottom of the eye at the center time position. The degradation in the amplitude as a function of transmission distance provides a distinct measure of the accumulated noise statistics for one-bits (marks) and zero-bits (spaces). This measurement determines the reduction of the amplitude and the increase of the noise in the one-bits and it determines the increase in the mean and standard deviation of the noise power in the zero-bits. Similarly, to determine the phase margin, the decision time position is measured which yields a BER $= 1\times10^{-6}$ at the leading and trailing edge of the eye at a constant optimum decision voltage. Assuming Gaussian statistics, the timing jitter is related to the phase margin by $\sigma = (\phi - \phi_0)$ / 9.5, where ϕ_0 is the initial phase margin.

Measuring the amplitude and phase margins in this way clearly identifies the cause of the error-free transmission distant limit arising from the accumulation of timing jitter and amplitude noise contributions. Therefore we do not present our data using the customary BER vs distance plot which emphasizes the combined effects of both amplitude noise and timing jitter beyond the error-free transmission distance limit.

We will demonstrate that our filtered dispersion-managed soliton transmission line has a particular steady state solution. Much like a standard filtered soliton system, launching a pulse that is different from the steady state solution gives rise to a transient regime where the pulses are transformed into the proper solution. The transient regime can be up to several thousand kilometers. The steady-state pulse dynamics within one map cycle are measured using a gated streak camera connected to a 90/10 tap coupler placed at various locations within the loop. The streak camera has a temporal resolution of 2 ps. The streak camera is synchronously scanned by a 100 MHz signal phase-locked to the transmitter clock. Therefore, at long transmission distances a low frequency jitter exists in the 10 Gbit/s data relative to the 100 MHz signal due to temperature dependent length fluctuations in the loop. Since the amplitude and frequency of this jitter increases with transmission distance, it is beneficial to quickly obtain a steady state pulse by pre-chirping the input data pulses [27]. Using a 7 ps pulse pre-chirped by 4 km of the standard anomalous dispersion fiber used in our map, we obtain a near steady-state solution within 4 Mm of transmission. It should be noted that the rate at which the steady state solution was formed has no effect on the error-free distance limit observed in our experiments. We have launched several pulse-widths in the range between 7–25 ps, including pre-chirped pulses and observed no change in the error-free transmission limit.

4. Results and Discussion

In soliton transmission systems, the nonlinear nature of the pulse together with optical band-pass filters is an effective way to reduce both amplitude a frequency fluctuations [28]-[30]. However, the disadvantage of using a filter is that noise can experience an exponential growth at the center of the band-pass filter. What this means for conventional intensity-modulated direct-detection (IMDD) is that the one bits can be very stable in amplitude and time, but noise can exponentially grow in the zero bits, thus degrading the overall SNR of the system. The following experimental results will focus on these issues as they apply to a stationary-filtered, dispersion-managed soliton transmission system. The results are presented in the following order; pulse stability and noise properties, timing jitter reduction due to enhanced

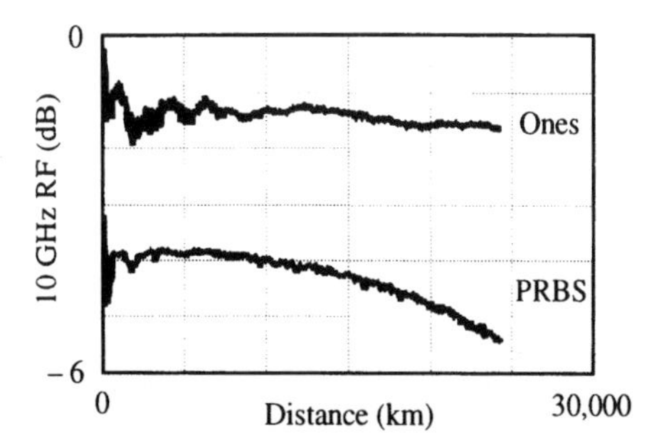

Figure 2. Measured 10 GHz fundamental RF component versus transmission distance for an unmodulated pulse train (ones) and a modulated 10 Gbit/s data stream (PRBS).

energy and periodic steady-state pulse dynamics.

4.1. PULSE STABILITY AND NOISE PROPERTIES

It is first necessary to establish that soliton-like pulses do exist in our system using the NRZ-style dispersion map. We observed stable propagation by monitoring a train of pulses as a function of transmission distance using a sampling oscilloscope and the streak camera [12]. Beyond the transient regime, no significant pulse distortions were observed up to our equipment limit of 28 Mm. Another measure of stable propagation can be inferred from the fundamental RF component of the optical pulse train versus distance. In Figure 2, the fundamental RF component of a 10 GHz pulse train (all ones) is measured at the receiver using the 1.3 nm OBF and a path-averaged dispersion (D_{avg}) of 0.15 ps/nm·km. The oscillations over the initial 5–10 Mm identifies the transient regime where the pulses are developing into a steady-state solution of the transmission line. This transient regime is characterized by significant nonlinear spectral evolution and temporal pulse compression and expansion which follows the high and low RF power respectively. At the distance where the RF component equilibrates ($z \cong 10$ Mm) a steady state solution is formed which is characterized by a Gaussian-like spectrum and a periodically stationary pulse which depends strongly on the band-pass filter's peak curvature. The stable 10 GHz RF component is further confirmation that stable, soliton-like long range propagation exists in our NRZ-style dispersion map.

The lower curve in Figure 2 is the fundamental RF component for a 10 Gbit/s data stream modulated with a PRBS $2^{23}-1$ pattern. The average power is 3 dB lower than the case with no data modulation (ones) to ensure that the peak power of the pulse is the same in both cases. As can be seen there is a significant roll-off of the 10 GHz RF component from the modulated data beyond 10 Mm. It will be shown that the roll-off is not due to pulse instability but rather an exponential growth of ASE optical power in the zero-bits originating from the excess gain, which accompanies stationary band-pass filters. This suggests that the 10 GHz RF component is a good measure of the overall SNR but does not distinguish between noise accumulation in the one or zero bits.

The best way to clearly characterize the transmission system performance is by means of BER measurements. In the first experimental configuration ($D_{avg} = 0.15$ ps/nm·km and 1.3 nm FHWM OBF) the error-free ($BER < 1\times10^{-9}$) transmission limit was measured to be 13,500 km, despite the fact that stable pulse propagation far exceeds this limit. The reason for this is clearly demonstrated in the amplitude margin measurement of Figure 3(a). The slow variation of the positive decision voltage level for a constant BER indicates a slow variation in the mean and variance of the pulse amplitude with transmission distance indicating stable pulse propagation. The rapid variation of the negative decision voltage level for a constant BER indicates the increase in the mean and variance of noise in the zero-bits and is due to the exponential growth of ASE noise which limits our error-free transmission distance to 13,500 km. From an SNR viewpoint, this clearly illustrates the advantage of pulse energy enhancement gained by using dispersion management. A simple method to reduce the rate of ASE noise growth is to reduce the curvature of the OBF. However, the amplitude and frequency stability given by the filter will also be reduced so the path-averaged dispersion must then be reduced. The next experimental configuration uses $D_{avg} = 0.02$ ps/nm·km and an OBF with a FWHM equal to 2.8 nm [4]. As indicated by the amplitude margin in Figure 3(b) we obtain very stable pulse propagation. The error-free transmission limit is extended to 24,500 km, limited again by the ASE noise growth in the zero-bits. The average optical power in the experiment with $D_{avg} = 0.15$ ps/nm·km and the 1.3 nm OBF was about -3 dBm. For the experiment with $D_{avg} = 0.02$ ps/nm·km and the 2.8 nm OBF at $D = 0.02$ ps/nm·km the average optical power was about 0 dBm. Thus the average optical power was increased by a factor of two as the path-averaged dispersion was reduced by nearly a factor of 8 and the filter bandwidth doubled. In addition, using the same 2.8 nm OBF and average optical power we varied the path-averaged dispersion from 0.02 ps/nm·km to 0.08 ps/nm·km and observed no change in the error-free transmission limit or in the amplitude margin measurements.

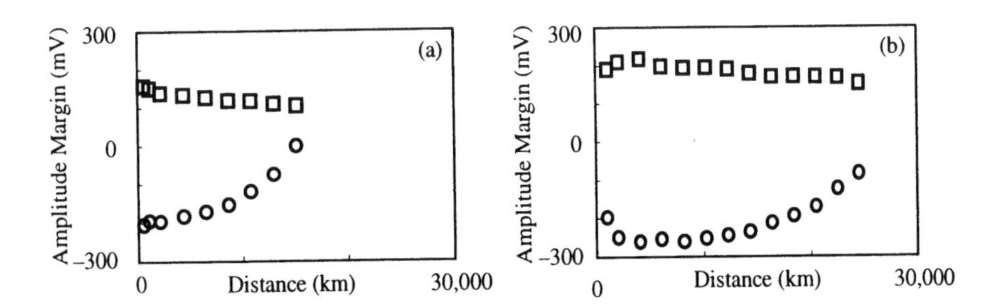

Figure 3. Amplitude margin defined by the decision level for a constant BER of 1×10^{-6} as a function of distance for experiments with (a) $D_{avg} = 0.15$ ps/nm·km, OBF = 1.3 nm and (b) $D_{avg} = 0.02$ ps/nm·km, OBF = 2.8 nm. The squares (circles) depict the amplitude noise of the one-bits (zero-bits).

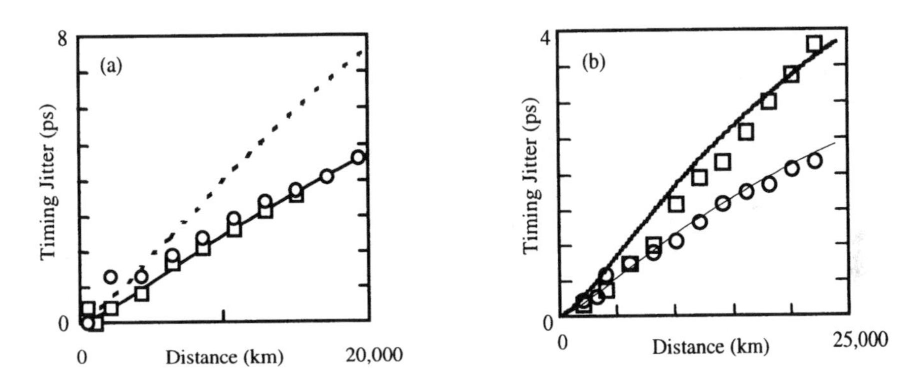

Figure 4. Gordon-Haus timing jitter measurements for dispersion managed solitons. (a) Experiment with $D_{avg} = 0.15$ ps/nm·km and 1.3 nm OBF. Dashed line is analytical prediction for solitons in uniform dispersion fiber. Solid line is the fit when the enhanced energy of dispersion-managed solitons is considered. Experimental data measured by BER phase margin (squares) and jitter statistics from sampling oscilloscope (circles). (b) Experiments with $D_{avg} = 0.04$ ps/nm·km (lower) and $D_{avg} = 0.07$ ps/nm·km (upper) and 2.8 nm OBF. Analytical fit used an enhancement factor of 5–6.

4.2. TIMING JITTER REDUCTION

Another advantage of dispersion-managed solitons is the reduction of Gordon-Haus timing jitter due to the enhanced pulse energy. This is demonstrated in Figure 4(a) using a $D_{avg} = 0.15$ ps/nm·km and the 1.3 nm OBF. The maximum measured timing jitter is about 4.5 ps at 20,000 km. Given a 65 ps timing window at the receiver, we could tolerate up to 5.6 ps of jitter and

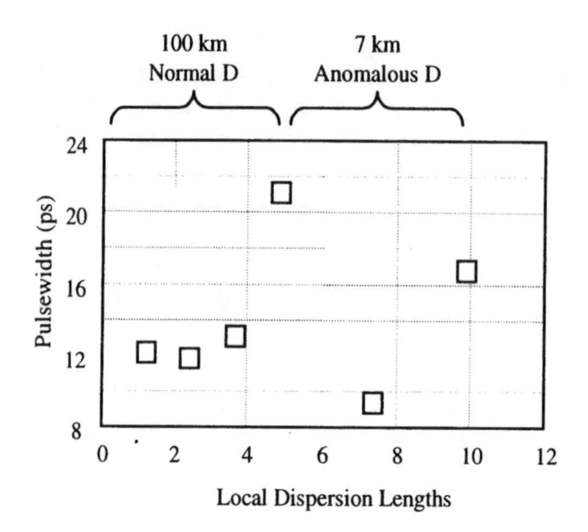

Figure 5. Steady state dynamics of dipsersion-managed soliton measured at several locations within one cycle of the map.

maintain a BER of 1×10^{-9}. Applying the analytical theory developed for timing jitter in a filtered soliton transmission system with uniform dispersion fiber would overestimate the measured jitter as shown by the dashed line. Modifying this fit by the square root of the enhancement energy, agrees well with the experimental data [13]. The enhancement factor of 2.6 used in the fit agrees quite well with experimental and numerical estimates of 2–3 [12, 31]. The timing jitter for experiments with the 2.8 nm OBF at $D_{avg} = 0.04$ ps/nm·km and 0.07 ps/nm·km is also shown with the modified fit using an enhancement factor between 5–6, which shows that the timing jitter scales as expected with the square root of D_{avg}.

4.3. STEADY-STATE PULSE DYNAMICS

The pulse dynamics measure in an experiment using a $D_{avg} = 0.02$ ps/nm·km and the 2.8 nm OBF are shown in Figure 5. Error-free operation over 24,500 km was confirmed before the measurements to insure that the pulse dynamics are measured in near identical conditions with the best communications performance. The FWHM of the pulses are measured after each EDFA following a 25 km span of the dispersion-shifted fiber, and in the middle(3.5 km) and end (7 km) of the standard fiber span. The map is normalized to the local dispersion length with the minimum pulse-width of 9 ps used as the characteristic time. In the 100 km normal dispersion span, the local dispersion, $D_{loc} = -1.1$ ps/nm·km such that the local dispersion length, $z_d \cong$ 20 km. In the 7 km anomalous dispersion span, $D_{loc} = +16.7$ ps/nm·km

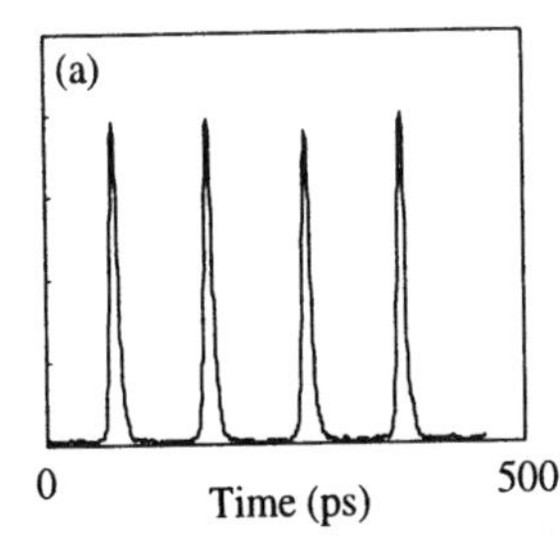

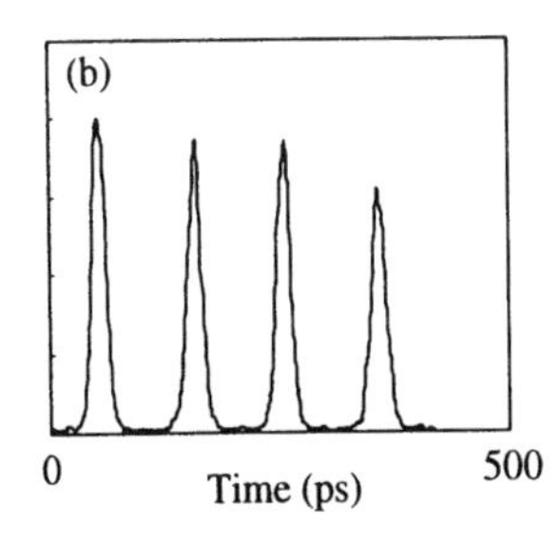

Figure 6. Streak camera traces of pulse dynamics within the anomalous dispersion span located at (a) 3.4 km mid-point and (b) 7 km end.

such that the local dispersion length, $z_d \cong 1.4$ km and the total map length is about 10 local dispersion lengths. The steady state dispersion-managed solitons compress to a minimum near the mid-section of each dispersion span as expected by the linear dynamics of a pulse propagation in a system with alternating signs of dispersion. The minimum value in the anomalous span is approximately 9 ps and the corresponding minimum in the normal dispersion span is about 11.5 ps. This asymmetry in the pulse-width minima can be attributed to the fiber nonlinearity. The location where the pulses experience maximum stretching occurs at the interface between the two dispersion regimes. The streak camera trace used to determine the minimum and maximum pulse-width in the anomalous span is shown in Figure 6. The spectrum taken at the minimum compression point in the anomalous dispersion span is shown in Figure 7.

The parabolic nature of the natural logarithmic plot depicts a near Gaussian pulse shape. The numerical Fourier transform of the spectrum results in a pulse-width of 8.7 ps and results in a time-bandwidth product of 0.54. The band-pass filter is the principal component of the transmission system which determines the shape of the nearly transform-limited pulse at the minimum compression point. It should be noted that the measured pulse dynamics over a range of D_{avg} from 0.02 ps/nm·km to 0.1 ps/nm·km varied by no more than 2 ps at the minimum compression point in the mid-span of the anomalous dispersion. Finally, it should be pointed out that the transient evolutions as described by the 10 GHz RF component in Figure 2 are largest at the interface between the dispersion regimes where the pulse experiences the most stretching. At the minimum compression points in either dispersion regime, the transient effects are much less pronounced.

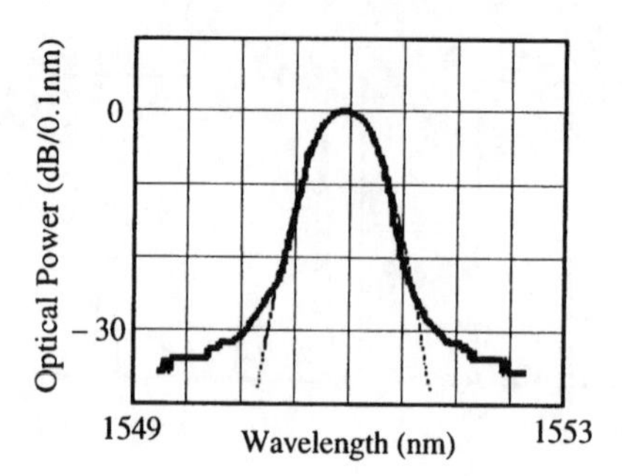

Figure 7. Optical spectrum measured at the mid-span of the anomalous dispersion regime. A Fourier transform of this spectrum predicts a pulse-width of 8.7 ps in good agreement with the measured redults. An ideal Gaussian shape (dashed-line) is presented for comparison.

5. Summary

In summary, we have demonstrated the benefits of using a filtered dispersion-managed soliton transmission system. The enhanced power arising from dispersion management together with the band-pass filter provides stable propagation of the solitons with reduced Gordon-Haus timing jitter at low values of path-averaged dispersion. The first experimental measure of the pulse dynamics in a periodically dispersion-managed soliton system are presented which are consistent with numerical and analytical results [27], [32]-[34]. However, we do observe a very weak dependence of the pulse energy on the path-averaged dispersion over the range of $D_{avg} = 0.02$ ps/nm·km to 0.08 ps/nm·km. The only expected change is the scaling of the Gordon-Haus timing jitter with the square root of the path-averaged dispersion. The relatively constant pulse energy does not agree with theoretical expressions which state that the pulse-energy is proportional to the path-averaged dispersion [27, 32].

Acknowledgement

The authors G.M. Carter and J.M. Jacob gratefully acknowledge the support of the National Science Foundation through the grant ECS-9610256.

References

1. Suzuki, M., Morita, I., Yamamoto, S., Edagawa, N., Taga, H. and Akiba, S.: Timing jitter reduction by periodic dispersion compensation in soliton transmission, *Conf. Optical Fiber Communications*, San Diego, CA, paper **PD-20**, (Feb. 1995).

2. Suzuki, M., Morita, I., Yamamoto, S., Edagawa, N., Taga, H. and Akiba, S.: Reduction of Gordon-Haus timing jitter by periodic dispersion compensation in soliton transmission, *Electron. Lett.*, **31**, (1995), pp.2027-2029.
3. Nakazawa, M. and Kubota, H.: Optical soliton communication in a positively and negatively dispersion-allocated optical fiber transmission line, *Electron. Lett.*, **31**, (1995), pp.216-217.
4. Jacob J. M. and Carter, G. M.: Error-free transmission of dispersion-managed solitons at 10 Gbit/s over 24,500 km without sliding-frequency filters, *Electron. Lett.*, **33**, (1997), pp.1128-1129.
5. Suzuki, M.: Soliton-based RZ transmission over transoceanic distances using periodic dispersion compensation, *Conf. Nonlinear Guided Waves and Their Applications*, Cambridge, England, paper **FA1**, (Aug. 1996).
6. Morita, I., Suzuki, M., Edagawa, N., Yamamoto, S. and Akiba, S.: Single-channel 40 Gbit/s, 5000 km straight-line soliton transmission experiment using periodic dispersion compensation, *Electron. Lett.*, **33**, (1997), pp.698-699.
7. Suzuki, M., Morita, I., Tanaka, K., Edagawa, N., Yamamoto, S. and Akiba, S.: 160 Gbit/s (8×20 Gbit/s) soliton WDM transmission experiments using dispersion flattened fibre and periodic dispersion compensation, *European Conf. Optical Communications*, Edinburgh, Scotland, paper **WE2B**, (Sept. 1997).
8. Suzuki, M., Morita, I., Edagawa, N., Yamamoto, S. and Akiba, S.: 20 Gbit/s-based soliton WDM transmission over transoceanic distances using periodic compensation of dispersion and its slope, *Electron. Lett.*, **33**, (1997), pp.691-692.
9. Morita, I., Suzuki, M., Edagawa, N., Yamamoto, S., Taga, H. and Akiba, S.: 20 Gb/s single-channel soliton transmission over 9000 km without inline filters, *Photon. Technol. Lett.*, **8**, (1996), pp.1573-1574.
10. Morita, I., Suzuki, M., Edagawa, N., Tanaka, K., Yamamoto, S. and Akiba, S.: Performance improvement by initial phase modulation in 20 Gbit/s soliton-based RZ transmission with periodic dispersion compensation, *Electron. Lett.*, **33**, (1997), pp.1021-1022.
11. Naka, A., Matsuda, T. and Saito, S.: Optical RZ signal straight-line transmission experiments with dispersion compensation over 5520 km at 20 Gbit/s and 2160 km at 2×20 Gbit/s, *Electron. Lett.*, **32**, (1996), pp.1694-1696.
12. Jacob, J. M., Golovchenko, E. A., Pilipetskii, A. N., Carter, G. M. and Menyuk, C. R.: Experimental demonstration of soliton transmission over 28 Mm using mostly normal dispersion fiber, *Photon. Technol. Lett.*, **9**, (1997), pp.130-132.
13. Carter, G. M., Jacob, J. M., Menyuk, C. R., Golovchenko, E. A. and Pilipetskii, A. N.: Timing-jitter reduction for a dispersion-managed soliton system: experimental evidence, *Opt. Lett.*, **22**, (1997), pp.513-515.
14. Jacob, J. M., Golovchenko, E. A., Pilipetskii, A. N., Carter, G. M. and Menyuk, C. R.: 10-Gb/s transmission of NRZ over 10,000 km and solitons over 13,500 km error-free in the same dispersion-managed system, *Photon. Technol. Lett.*, **9**, (1997), pp.1412-1414.
15. Yamada, E., Kubota, H., Yamamoto, T., Sahara, A. and Nakazawa, M.: 10 Gbit/s, 10,600 km, dispersion-allocated soliton transmission using conventional 1.3 μm singlemode fibres, *Electron. Lett.*, **33**, (1997), pp.602-603.
16. Favre, F., Le Guen, D., Moulinard, M. L., Henry, M., Michaud, G., Devaux, F., Legros, E., Charbonnier, B. and Georges, T..: Demonstration of soliton transmission at 20 Gbit/s over 2200 km of standard fibre with dispersion compensation and prechirping, *Electron. Lett.*, **33**, (1997), pp.511-512.
17. Harper, P., Knox, F. M., Kean, P. N., Bennion, I. and Doran, N. J..: 10-Gbit/s soliton propagation over 5250 km in standard fiber with dispersion compensation, *Conf. Optical Fiber Communications*, Dallas, TX, paper **THN-1**, (Feb. 1997) .
18. Smith, N. J., Knox, F. M., Doran, N. J., Blow, K. J. and Bennion, I..: Enhanced power solitons in optical fibres with periodic dispersion management, *Electron. Lett.*, **32**, (1996), pp.54-55.

19. Nakazawa, M., Kubota, H., Sahara, A. and Tamura, K..: Marked increase in the power margin through the use of a dispersion-allocated soliton, *Photon. Technol. Lett.*, **8**, (1996), pp.1088-1090.
20. Smith, N. J., Forysiak, W. and Doran, N. J.: Reduced Gordon-Haus jitter due to enhanced power solitons in strongly dispersion managed systems, *Electron. Lett.*, **32**, (1996), pp.2085-2086.
21. Yu, T., Golovchenko, E. A., Pilipetskii, A. N. and Menyuk, C. R.: Dispersion-managed soliton interactions in optical fibers, *Opt. Lett.*, **22**, (1997), pp.793-796.
22. Matsumoto, M. and Haus, H. A.: Stretched-pulse optical fiber communications, *Photon. Technol. Lett.*, **9**, (1997), pp.785-787.
23. Golovchenko, E. A., Pilipetskii, A. N. and Menyuk, C. R.: Collision-induced timing jitter reduction by periodic dispersion management in soliton WDM transmission, *Electron. Lett.*, **33**, (1997), pp.735-736.
24. Golovchenko, E. A., Pilipetskii, A. N. and Menyuk, C. R.: Periodic dispersion management in soliton WDM transmission with sliding filters, *Opt. Lett.*, **22**, (1997), pp.1156-1158.
25. Forysiak, W., Devaney, J. F. L., Smith, N. J. and Doran, N. J.: Dispersion management for wavelength-division-multiplexed soliton transmission, *Opt. Lett.*, **22**, (1997), pp.600-602.
26. Bergano, N.S., Davidson, C. R., Mills, M. A., Corbett, P. C., Evangelides, S. G., Pedersen, B., Menges, R., Zyskind, J. L., Sulhoff, J. W., Srivastava, A. K., Wolf, C. and Judkins, J.: Long-haul WDM transmission using optimum channel modulation: A 160 Gb/s (32×5 Gb/s) 9,300 km demonstration, *Conf. Optical Fiber Communications*, Dallas, TX, paper **PD-16**, (Feb. 1997).
27. Smith, N. J., Doran, N. J., Knox, F. M. and Forysiak, W.: Energy-scaling characteristics of solitons in strongly dispersion-managed fibers, *Optic. Lett.*, **21**, (1996), pp.1981-1983.
28. Hasegawa, A. and Kodama, Y.: *Solitons in Optical Communications*, Clarendon Press, Oxford, 1995.
29. Mecozzi, A., Moores, J. D., Haus, H. A. and Lai, Y.: Soliton transmission control, *Opt. Lett.*, **16**, (1991), pp.1841-1843.
30. Mollenauer, L. F., Mamyshev, P. V. and Neubelt, M. J.: Measurement of timing jitter in filter-guided soliton transmission at 10 Gbit/s and achievement of 375 Gbit/s-Mm, error-free, at 12.5 and 15 Gbit/s, *Opt. Lett.*, **19**, (1994), pp.704-706.
31. Golovchenko, E. A., Jacob, J. M., Pilipetskii, A. N., Menyuk, C. R. and Carter, G. M.: Dispersion-managed solitons in a fiber loop with in-line filtering, *Opt. Lett.*, **22**, (1997), pp.289-291.
32. Matsumoto, M.: Theory of stretched-pulse transmission in dispersion-managed fibers, *Opt. Lett.*, **22**, (1997), pp.1238-1240.
33. Kumar, S. and Hasegawa, A.: Quasi-soliton propagation in dispersion-managed optical fibers, *Opt. Lett.*, **22**, (1997), pp.372-374.
34. Grigoryan, V. S., Yu, T., Golovchenko, E. A., Menyuk, C. R. and Pilipetskii, A. N.: Dispersion-managed soliton dynamics, *Opt. Lett.*, **22**, (1997). To be published.

(Received November 17, 1997)

20 GBIT/S PDM SOLITON TRANSMISSION EXPERIMENT IN DISPERSION COMPENSATED STANDARD FIBER SYSTEMS

H. MURAI, M. SHIKATA AND Y. OZEKI
Semiconductor Technology Laboratories
Oki Electric Industry Co., Ltd.
550-5 Higashi-asakawa, Hachioji, Tokyo 193, Japan

Abstract. The behaviors of optical soliton propagation in standard single mode fiber (SMF) system with strongly dispersion compensated scheme using a dispersion compensating fiber (DCF) were studied by both the numerical simulation and the recirculating loop transmission experiment. The simulation revealed that the alternating average dispersion map arrangement changing between the normal and anomalous values and the use of an in-line optical bandpass filter (BPF) help stabilize the soliton pulse and extend the transmission distance over dispersion compensated standard single mode fiber (SMF). As predicted by the simulation, soliton transmissions at 10 Gbit/s over 6,100 km long SMF and 20 Gbit/s over 2,600 km long SMF were successfully demonstrated by employing the novel dispersion arrangement and suitable in-line optical BPFs.

1. Introduction

SMF, which has a large wavelength dispersion in 1.55 μm band of the erbium-doped fiber amplifier gain, constitutes the majority of already-installed fiber plants worldwide. To expand the SMF system's bandwidth, the combination of a dispersion compensation technique and optical soliton transmission has attracted much attention because of its potential for achieving transmission distance over several thousands km at high bit rates at and beyond 10 Gbit/s. Recently, several research groups have demonstrated a soliton propagation over 5,000 km of SMF at 10 Gbit/s with an inventive dispersion management [1]-[3]. And beyond 10 Gbit/s, 20 Gbit/s

A. Hasegawa (ed.), New Trends in Optical Soliton Transmission Systems, 167–181.

soliton transmission over 2,000 km of SMF has been realized by adapting a pre-chirp technique at the transmitter [4]-[6].

In this report, conditions to improve the dispersion compensated soliton transmission were examined in terms of parameters, such as dispersion map arrangement, types of in-line optical BPFs and SMF input optical power based on the numerical simulation. Accordingly, 10 and 20 Gbit/s soliton transmission over several thousands km of SMF with dispersion compensation were demonstrated using recirculating fiber-loop transmission experiment.

2. Characteristic of Optical Soliton in Strong Dispersion Compensated Fiber Systems

The conventional soliton transmission requires that the transmission fiber's wavelength dispersion satisfy the soliton condition along the entire section of the fiber. This is difficult condition to be met particularly for the already built fiber plant. The dispersion compensated soliton, on the other hand, only requires that a dispersion averaged over some distance, typically over an optical amplifier spacing, satisfy the soliton condition [7]- [8]. This flexibility in the dispersion arrangement makes the dispersion compensated soliton very attractive alternatives to the conventional soliton or NRZ transmission systems, as the longhaul transmission systems [1]-[6].

Figure 1 shows a schematic diagram of a dispersion compensated soliton system, where it consists of a succession of the transmission fiber such as SMF, a dispersion compensating element such as DCF having an appropriate opposite dispersion to offset the transmission fiber's large dispersion, and an optical amplifier. The dispersion, averaged over the amplifier spacing, is set to give a slightly anomalous dispersion, a requirement for the soliton formation. An averaged dispersion can be easily adjusted to satisfy the soliton condition, simply by adding or subtracting dispersion with dispersion compensating elements at every amplifier site.

In the SMF dispersion compensated soliton, the soliton pulse with the wavelength in 1550 nm band broadens in SMF section due to its large dispersion, and then narrows in the subsequent DCF section. As this process repeats itself along the transmission, a slightly anomalous average dispersion and cumulative Kerr effect over the many amplifier spacings help pulses to maintain their soliton characteristics. However, the large local dispersion of SMF section leads to the partial overlap between the neighboring pulses and the radiation of non-soliton component as dispersive wave. This is why soliton-soliton interaction is enhanced, leading to severe patterning effect, timing jitter and collision of pulses. These phenomena limit the system transmission distance in the dispersion compensated soliton.

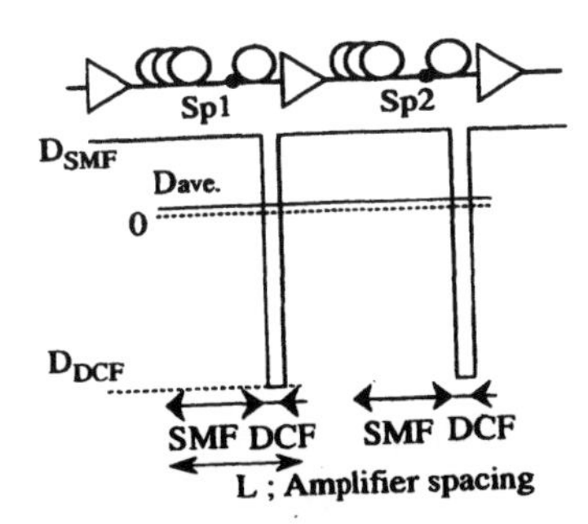

Figure 1. Dispersion map of conventional dispersion compensation scheme.

To investigate the details of the soliton propagation over the dispersion compensated SMF system, a numerical simulation was conducted, solving the non-linear Schrödinger's equation (NSE) with perturbation terms by the split-step-Fourier method [10]. The interaction distance, where solitons collide, depends on input optical power to the fiber, and span average dispersion. Calculated results, assuming DSMF = 17 ps/nm·km, DDCF varied to give various average dispersion ranging from 0.2 to 1.0 ps/nm·km and 10 Gbit/s 20 ps FWHM Gaussian pulse as input pulse, are shown in Figure 2. In the figure, the interaction distance of the DC soliton is plotted along with that of an ideal soliton. As the soliton order, N, increases, the interaction distance deteriorates because of further pulse broadening and the stronger radiation of non-soliton component.

Recently, it has been reported that initially pre-chirped pulses reduces the pulse broadening in the high dispersion transmission fiber so that the interaction distance increases [11].

Since the soliton-soliton interaction produces a spurious sideband component distorting the pulses, suppressing the growth of the side-band component by an in-line optical BPF with a suitable pass-band profile should stabilize the soliton pulse transmission.

To further improve the soliton transmission distance, the novel dispersion map arrangement is proposed, where an average dispersion of the neighboring span alternates between normal and anomalous value as shown in Figure 3, while maintaining the overall average dispersion slightly anomalous. With this type of alternating dispersion map arrangement, soliton pulse width stabilizes in shorter distance than with the uniform average dispersion map arrangement. Regarding the order of alternation, the improvement is obtained only when a normal average section precedes an anomalous one.

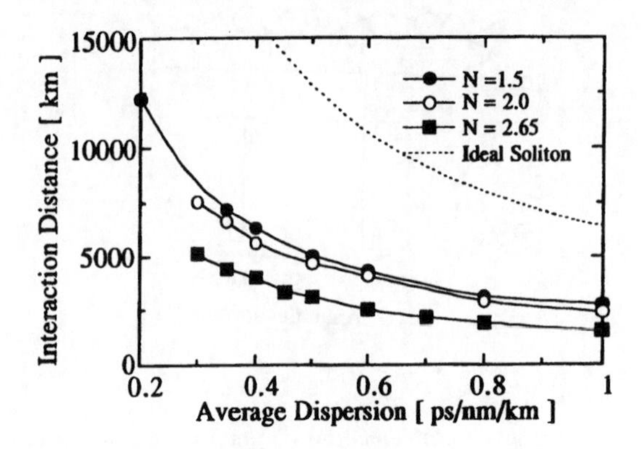

Figure 2. Calculated interaction distance of 10 Gbit/s 20 ps-FWHM Gaussian pulses in strongly dispersion compensated fiber line for N=1.5 (solid circles), 2.0 (open circles) and 2.65 (solid squares). Dotted line shows ideal soliton case.

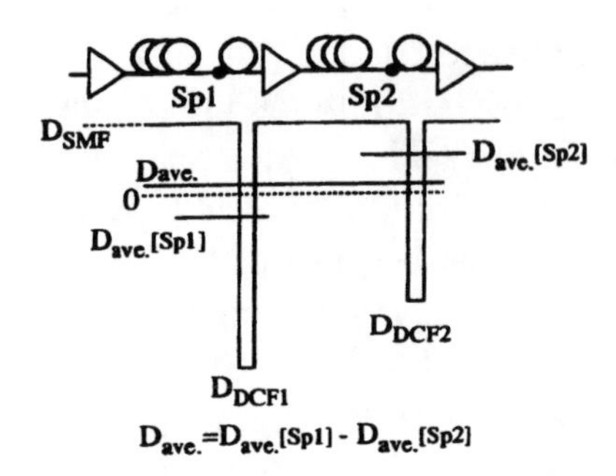

Figure 3. Dispersion map of alternative average dispersion scheme.

3. Optimization of Soliton Transmission

The optimization of system parameters was conducted numerically for the proposed system , with dispersion arrangement, which consists of repetition of 50 km-long SMF section, 10 km-long DCF section and an optical amplifier. An optical BPF was inserted after the optical amplifier. The ASE noise from the optical amplifier was calculated by adding a random number in the Fourier spectrum domain, commensurate with the noise figure of 5 dB. The detail of calculation parameters is shown in Table 1.

TABLE 1. System parameters for simulation

System parameter	
Wavelength	1559.5 nm
Pulse shape	Gaussian
FWHM	15 ps or 20 ps
D_{SMF}	17 ps/nm·km
D_{DCF}	——-
Loss	0.2 dB/km
Dispersion Slope	0.07 $ps/nm^2/km$ for both SMF and DCF
γ_{SMF} γ_{DCF}	$2.09 \times 10^{-3} W^{-1} m^{-1}$ $4.63 \times 10^{-3} W^{-1} m^{-1}$

3.1. EFFECT OF OPTICAL BANDPASS FILTER

In Figure 4, the interaction distance is plotted as a function of the bandwidth of two kinds of in-line BPFs ; Lorentzian and 2nd-order Butterworth filters. In the latter case, the interaction distance is almost constant, while the significant peaking characteristic is obtained in the former case. The use of appropriate optical BPFs mitigates soliton pulse degradation because spurious sideband components induced by the interaction are effectively suppressed. Using a 1 nm Lorentzian filter, the interaction distance can be extended to more than 5-fold. For the bandwidth less than 1.0 nm, the signal side-band spectrum is affected, incurring a more pulse broadening, and leading to the distortion of soliton pulses.

Figure 5 shows the interaction distance as an average dispersion is varied for various soliton orders; N, with and without in-line filters. The type of in-line filters was assumed to be Lorentzian of 1 nm-bandwidth.

The interaction distance for each soliton order is improved with in-line filters employed. For the higher order soliton of N = 2.65, significant improvement is obtained, while the result is less significant for lower order solitons. This is because in the lower order soliton, its spurious side-band generation by the nonlinear interaction is small and its reduction by in-line filter is less effective to the transmission performance.

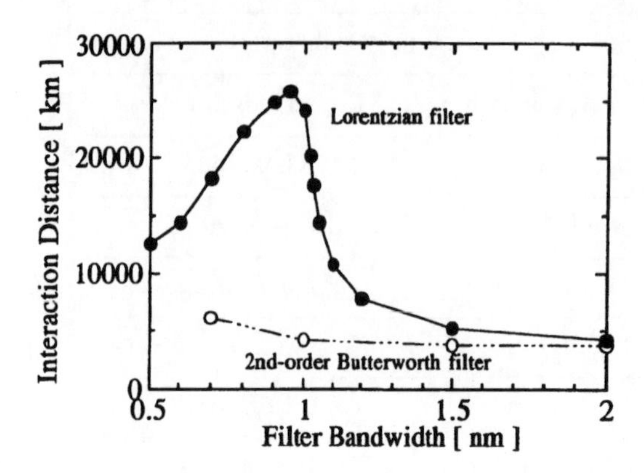

Figure 4. Calculated interaction distance for Lorentzian filter (solid circle) and 2nd-order Butterworth filter (open circle).

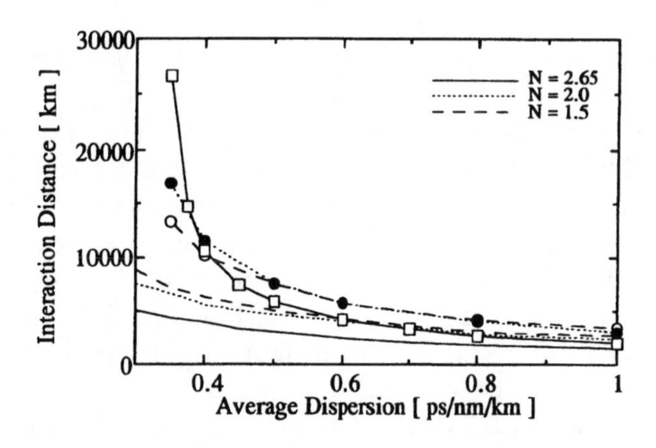

Figure 5. Calculated interaction distance for a paired soliton with soliton order of 1.5 (broken line), 2.0 (dotted line) and 2.65 (solid line). The line with symbols show the case of inserting 1 nm-Lorentzian in-filter.

3.2. STABILIZATION OF SOLITON PULSE TRANSMISSION WITH ALTERNATING AVERAGE DISPERSION SCHEME

In the strongly dispersion compensated soliton system, many studies so far assumed that the system have a uniform dispersion compensation map thoughout the entire length of transmission [1]-[6]. A novel dispersion com-

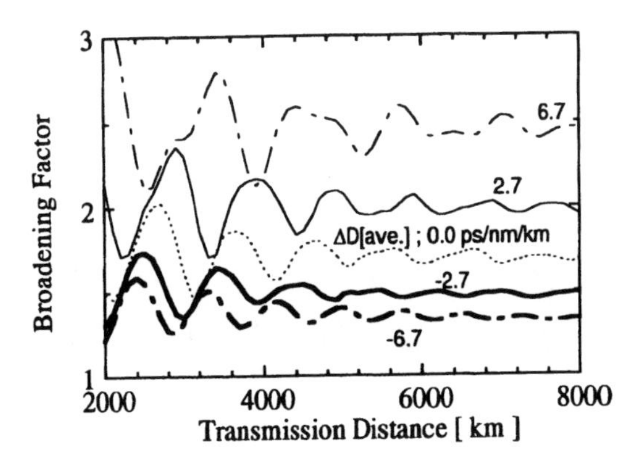

Figure 6. Broadening factor calculated every two spans for a soliton pulse. The factor 1 corresponds to 20 ps-FWHM for Gaussian pulse.

pensation scheme is proposed where a span average dispersion is arranged in such a way that normal and anomalous average dispersion spans come alternately along the transmission distance, while total average dispersion is kept in a slightly anomalous value. In Figure 6, broadening factor of pulse width calculated every two spans, normalized to initial launched pulse width of 20 ps, is plotted as a function of transmission distance for various dispersion arrangement. The parameter $\Delta D_{ave.}$ is defined as the difference between the first and second spans average dispersion in Figure 3. In Figure 6, the pulse widths increase asymptotically to various values, accompanied by underdamped oscillation depending on $\Delta D_{ave.}$. With the in-line filter, after $\sim$ 4,000 km, the pulse width fluctuation is well contained for each case. Broadening factor tends to approach unity, as the value of $\Delta D_{ave.}$ decreases further. Initially pulses are launched into normal average dispersion span, where they are up-chirped and their pulse widths broadens so that in next span of anomalous region they are linearly compressed to narrow. This chirping effect occurring in every normal average dispersion segment improves the pulse width stabilization and helps keep soliton pulse's original shape over longer distance as shown in Figure 6.

Additionally, temporal fluctuation of soliton pulses position is calculated with two runs of pseudo-random bit sequence signal of 6th stage and shown in Figure 7. The reduction of the temporal fluctuation was obtained by arranging the dispersion map with large negative value of $\Delta D_{ave.}$.

The main cause of the temporal fluctuation is attributed to the patterning effect and optical Kerr effect. In strongly DC soliton system such as dispersion compensated SMF soliton system, the patterning effect is

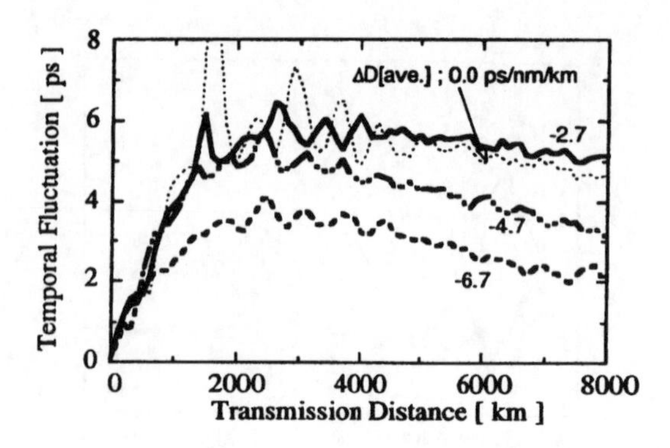

Figure 7. Temporal fluctuation calculated every two spans for various $\Delta D_{ave.}$.

larger due to large broadening. But with the systems having the dispersion arrangement of $\Delta D (< 0)$, pulse width fluctuation after every two spans reduces as shown in Figure 7.

3.3. 20 GBIT/S PDM SOLITON TRANSMISSION

As the transmission bit rate is doubled to 20 Gbit/s with the pulse width kept the same as before, the interaction between the neighboring soliton pulses becomes more acute. However, the polarization mode multiplexing (PDM) technique is effective to reduce soliton-soliton interaction [12]. In Figure 8, the comparison of transmission distance with 20 Gbit/s PDM soliton and single polarized soliton, giving $Q = 6$ corresponding to the bit error rate (BER) $= 10^{-9}$, is calculated for various system parameters [13]. Assumed overall average dispersion was 0.15 ps/nm·km. An optimized in-line BPF was also inserted after every amplifier. In the figure, the benefit of PDM soliton is clearly seen. At the soliton number below 1 where the effect of GVD is dominant, the transmission distance of each case asymptotically approaches the value limited by GVD. However, with increased soliton number (> 1) where the nonlinear effect is not negligible, the transmission distance of PDM is better than the single polarization case due to less soliton interaction.

To extend transmission distance further, alternating average dispersion scheme mentioned above is applicable. The maximum transmission distance is shown in Figure 9 as a function of $\Delta D_{ave.}$, with constant average dispersion value of 0.15 ps/nm·km. The transmission distance peaks at 3,500 km with an optimal value of $\Delta D_{ave.} = -16.3$ ps/nm·km for the soliton order of $N = 1.5$. The value of $\Delta D_{ave.}$ corresponds to $D_{ave.}[Sp1] = -8.0$ ps/nm·km

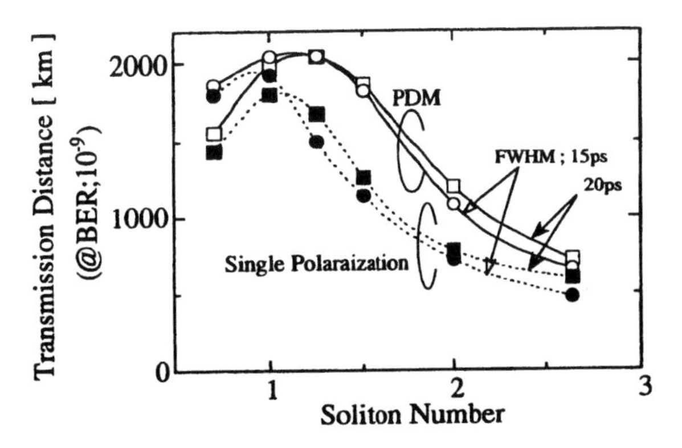

Figure 8. Calculated maximum transmission distance for PDM and single polarization soliton.

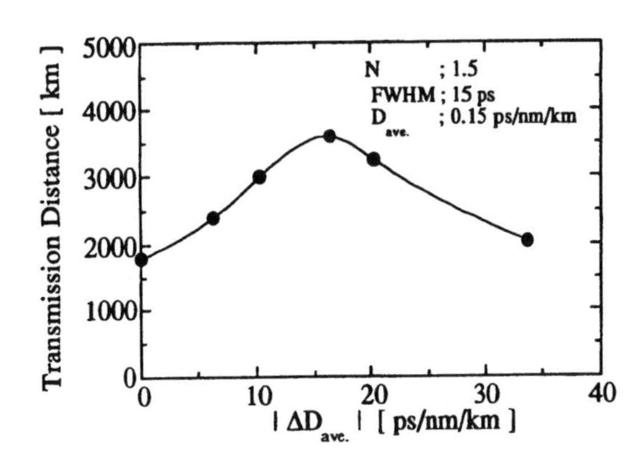

Figure 9. Calculated maximum transmission distance for the soliton of N = 1.5 to various $\Delta D_{ave.}$.

and $D_{ave.}[Sp2] = 8.3$ ps/nm·km.

4. Transmission Experiment

To verify the simulation results, the recirculating loop transmission experiment was conducted. The schematic diagram of the experimental setup is shown in Figure 10. The recirculating fiber loop circumference was set approximately 120 km. It consists of two optically amplified transmission segments, each of which consists of 50 km of SMF and 10 km of DCF sections, and erbium doped fiber amplifiers (EDFAs) with a noise figure (NF =

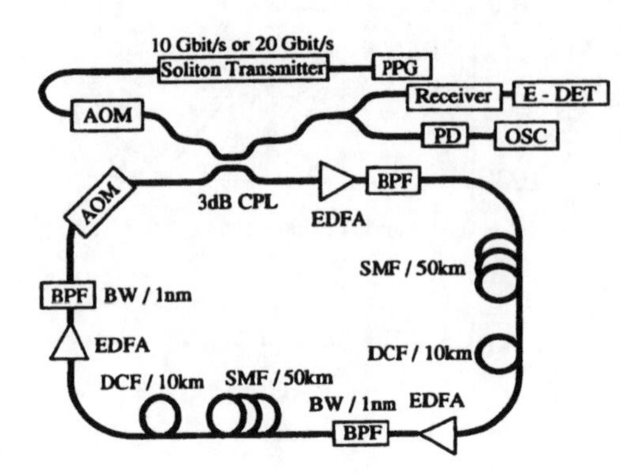

Figure 10. Experimental loop setup for 10 and 20 Gbit/s soliton transmission.

5.5 dB). An optical BPF of single cavity 1 nm bandwidth was inserted after each optical amplifier. In the transmitter, an optical soliton pulse stream was generated with a sinusoidally-driven electroabsorption (EA) modulator, and encoded by $LiNbO_3$ intensity modulator with 10 Gbit/s 2^7-1 pseudo-random binary sequence. The generated pulse had a Fourier-transform limited, nearly Gaussian shape and FWHM of 20 ps. The launched pulse power into SMF was varied around 6 dBm.

4.1. 10 GBIT/S-6,100 KM SOLITON TRANSMISSION

The average dispersion over the whole loop was set 0.3 ps/nm·km at the signal wavelength of 1559.5 nm. The two types of dispersion compensation scheme, based on the above numerical prediction, was examined. One was a nearly uniform dispersion compensation scheme as shown in Figure 1, with $\Delta D_{ave.} = -0.64$ ps/nm·km. The other was the alternating dispersion compensation scheme where the value of $\Delta D_{ave.}$ was -5.3 ps/nm·km.

Figure 11 indicates a BER characteristic as the function of the transmission distance, excluding DCF length. The maximum distance to obtain BER of 1×10^{-9} was measured to be approximately 4,700 km and 6,100 km for the uniform and alternating dispersion compensation scheme, respectively.

The optical eye pattern diagrams after 6,000 km transmission for both cases that of $\Delta D_{ave.}$ being -0.64 ps/nm·km and -5.32 ps/nm·km are shown in Figure 12 (a) and (b), respectively. In Figure 12 (b), the jitter was clearly seen to be reduced.

In Figure 13, optical input power dependence of the transmission dis-

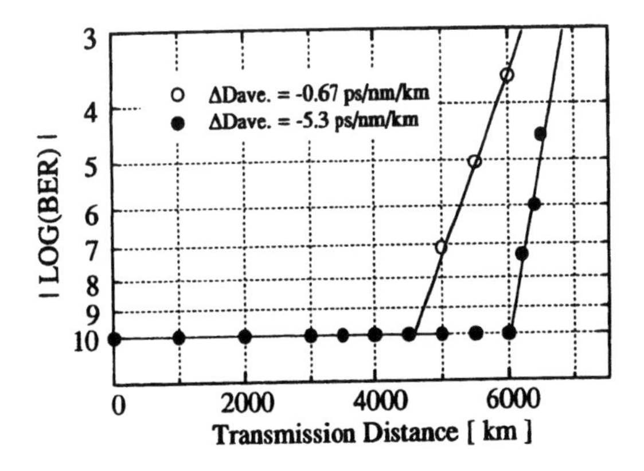

Figure 11. BER characteristic of 10 Gbit/s soliton transmission.

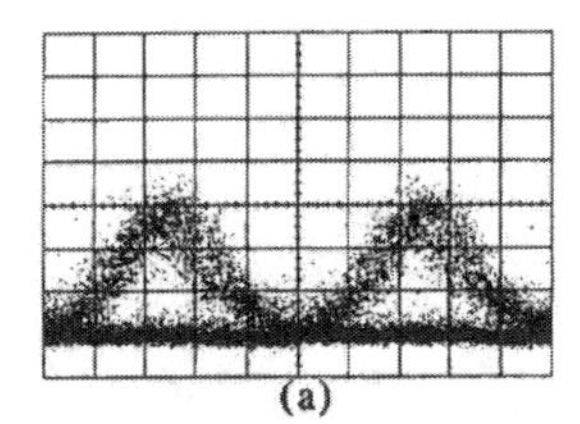

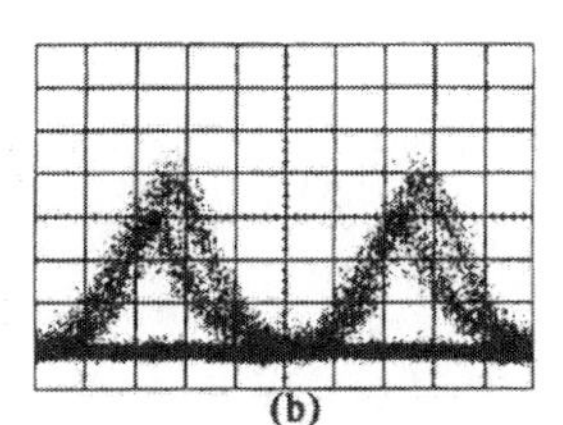

Figure 12. Detected optical eye pattern at 6,000 km. (a) $\Delta D_{ave.} = -0.67$ ps/nm·km. (b) $\Delta D_{ave.} = -5.3$ ps/nm·km.

tance is measured for both schemes. Although there exits an optimal power at 7 dBm for both cases, the transmission distance for the scheme (b) was improved by 10 to 30 % relative to the uniform scheme (a), in the power range of 4 through 8 dBm.

The limiting factor for the transmission distance in the case (a) was a timing jitter induced by patterning effect. So, the reduction of such an effect was realized by adapting the alternating dispersion compensation scheme as shown in Figure 13.

In the alternating dispersion compensated SMF system, soliton waveforms remain almost the same in the transmission range exceeding 2,000 km through 5,000 km after having experienced the transitional distortion of the pulse shape at 1,000 km. The pulse width is stabilized to the value of 30 to 35 ps as predicted in Figure 6. This is consistent with the results of the numerical simulation that removing the SPM-induced spectrum component

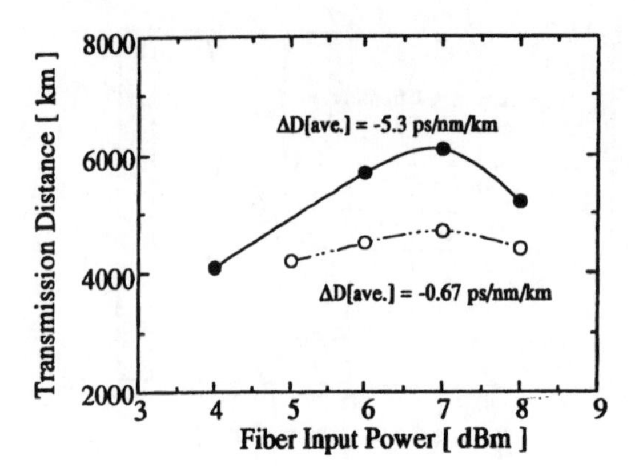

Figure 13. Input power dependence of obtained maximum transmission distance. Open circle ; $\Delta D_{ave.} = -0.67$ ps/nm·km, solid circle ; $\Delta D_{ave.} = -5.3$ ps/nm·km.

by an appropriate optical BPF effectively contains the soliton interaction and thus stabilize the waveform evolution.

4.2. 20 GBIT/S-PDM SOLITON TRANSMISSION

20 Gbit/s soliton transmission was conducted, using the similar arrangement of the dispersion compensation scheme. The average dispersion over the whole loop was set 0.06 ps/nm·km at the signal wavelength of 1559.5 nm. The FWHM of initial pulses was 15 ps. In the transmitter, 20 Gbit/s pulse train were generated by optically bit-interleaving 10 Gbit/s pulse trains with their state of polarization set orthogonal to each other. In the receiver, 20 Gbit/s pulse train is optically demultiplexed into 10 Gbit/s pulse train by employing a sinusoidally driven EA modulator which switches off every other bit of the input pulse train. Figure 14 shows the BER characteristic for transmission.

The circle symbols and square symbols correspond to the case of $\Delta D_{ave.} = -5.8$ ps/nm·km and $\Delta D_{ave.} = -16.8$ ps/nm·km, respectively. Open and close symbols represent one 10 Gbit stream and the other 10 Gbit/s stream, respectively. The maximum transmission distance was 2,600 km at input power of 1.0 dBm, while in the case of $\Delta D_{ave.} = -16.8$ ps/nm·km it was 2,200 km at input power of 2 dBm.

Figure 15 shows the optical eye pattern for (a) $\Delta D_{ave.} = -5.8$ ps/nm·km and (b) $\Delta D_{ave.} = -16.8$ ps/nm·km. The better eye-opening and BER for the case (b) are attributed to the reduction of soliton interaction.

Regarding the optimal input power, the transmission distance of the case (b) was above 2,200 km at the input power ranging from −1.0 dBm

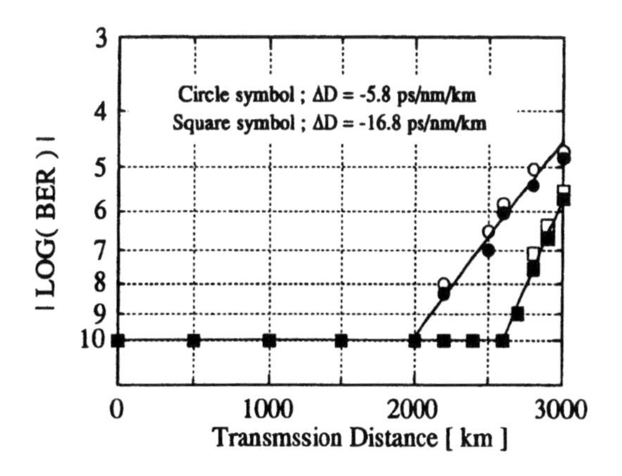

Figure 14. BER characteristic for 20 Gbit/s soliton transmission.

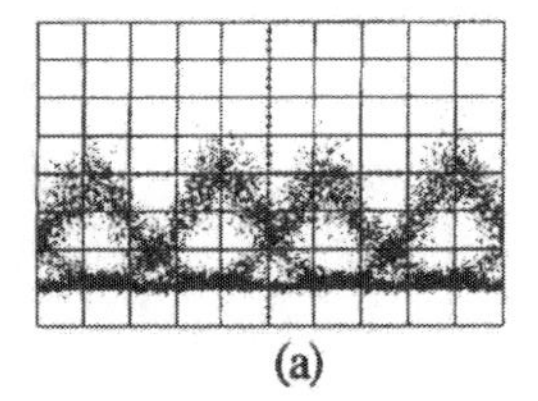

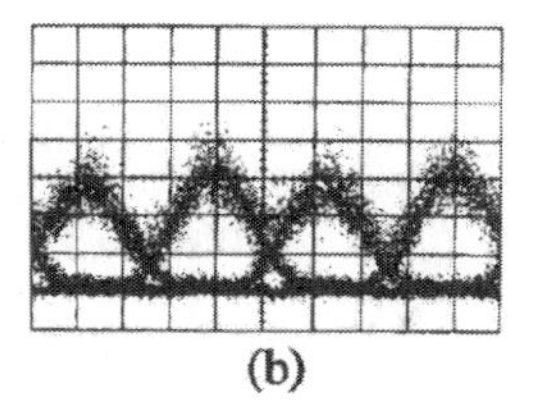

Figure 15. Detected eye pattern at 2,500 km. (a) $\Delta D_{ave.} = -5.8$ ps/nm·km. (b) $\Delta D_{ave.} = -16.8$ ps/nm·km.

to 3.0 dBm as shown in Figure 16.

5. Conclusion

To conclude, the behaviors of soliton propagation in the dispersion compensated SMF systems were investigated numerically, and experimentally. Based on the numerical results, the effectiveness of optical BPF and alternating average dispersion arrangement were proposed for extending the soliton transmission distance. These phenomena were experimentally confirmed by the recirculating fiber-loop transmission experiment, for both 10 Gbit/s and 20 Gbit/s PDM dispersion compensated soliton systems. As the system parameters were optimized, the transmission distance of 6,100 km and 2,600 km, excluding the length of DCF, were obtained for 10 Gbit/s and 20 Gbit/s PDM dispersion compensated soliton over SMF, respectively.

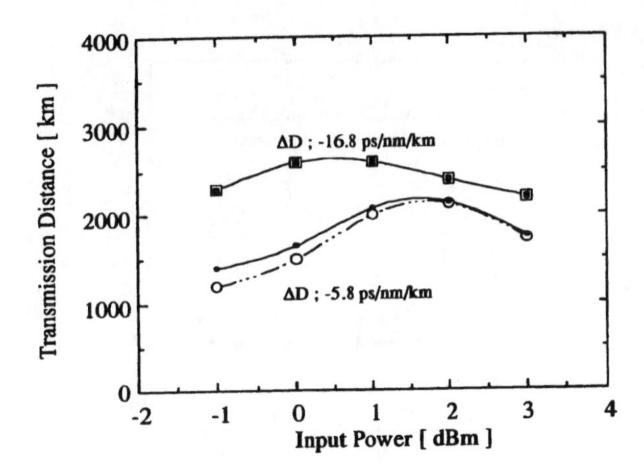

Figure 16. Measured maximum transmission distance for 20 Gbit/s PDM soliton with alternating average dispersion scheme of $\Delta D_{ave.} = -16.8$ ps/nm·km (open square symbols) and $\Delta D_{ave.} = -5.8$ ps/nm·km (open circle symbols). The transmission distances are also plotted to the orthogonally polarized signals (solid symbols).

References

1. Knox, F. M., Forysiak, W. and Doran, N. J.: 10-Gbit/s Soliton Communication Systems Over Standard Fiber at 1.55 μm and the Use of Dispersion Compensation, *J. Lightwave Technol.*, **13**, (1995), pp.1955-1962.
2. Harper, P., Knox, F. M., Kean, P. N., Bennion, I. and Doran, N. J.:10- Gbit/s soliton propagation over 5250 km in standard fiber with dispersion compensation, *Dig. OFC '97*, Dallas, Texas, **ThN1**, (1997).
3. Yamada, E., Kubota, H., Yamamoto, T., Sahara, A. and Nakazawa, M.: 10 Gbit/s 10,00 km dispersion-allocated soliton transmission using conventional 1.3 μm singlemode fibers, *Electron. Lett.*, **33**, (1997), pp.602-603.
4. Frave, F. D., Guen, Le. and G., Devaux.: 4×20 Gbit/s WDM transmission over 2000 km with 100 km dispersion-compensated spans of standard fibre, *Electron. Lett.*, **33**, (1997), pp.1234-1235.
5. Frave, F. D., Guen, Moulinnard, M. L., Henry, M., Michaud, G., Mace, L., Devaux, F., Charbonnier, B. and Geoges, T.: 200 Gbit/s 100 km-span soliton WDM Transmission over 1000 km of standard fiber with dispersion compensation and pre-chirping, *OFC'97*, Dallas, Paper **PD 17**.
6. Frave, F. D., Guen, Moulinnard, M. L., Henry, M., Michaud, G., Devaux, F., Legros, E., Charbonnier, B. and Geoges, T.: 20 Gbit/s over 2200 km of standard fiber with dispersion compensation and pre-chirping, *Electron. Lett.*, **33**, (1997), pp.511-512.
7. Hasegawa, A. and Kodama, Y.: Guiding-center soliton in fibres with periodically varying dispersion, *Opt. Lett.*, **16**, (1991), pp.1385- 1387.
8. Nakazawa, M. and Kubota, H.: Optical soliton communication in a positively and negatively dispersion-allocated optical fiber transmission line, *Electron. Lett.*, **31**, (1995), pp.216-217.
9. Nakazawa, M.: Dispersion-allocated soliton technologies and the stretched- pulse mode-locked fiber laser, *Dig. OFC '97*, Dallas, Texas, **ThN7**, (1997).
10. Agrawal, G. P.: Nonlinear fiber optics, *Academic Press*, San Diego, California, (1995), pp.28-59.
11. Georges, T. and Charbonnier, B.: Reduction of the dispersive wave in periodically

amplified links with initially chirped solitons, *Photon. Technol. Lett.*, **9**, (1997), pp.127-129.

12. Hasegawa, A. and Kodama, Y.: *Interaction between solitons in the same channel, Solitons in optical communications*, Clarendon press, Oxford, (1995), pp.151-172.
13. Bergano, N. S., Kerfoot, F. W. and Davidson, C. R. : Margin measurements in optical amplifier systems, *Photon. Technol. Lett.*, **Vol. 5**, (1997), pp.304-306.

(Received January 5, 1998)

DYNAMICS OF DISPERSION-MANAGED SOLITONS IN OPTICAL COMMUNICATIONS

A Geometric Interpretation

J. N. KUTZ
Hong Kong Polytechnic University
Hung Hom, Kowloon, Hong Kong, China

S. G. EVANGELIDES JR.
Submarine and Lightwave Systems International Inc.
101 Crawfords Corner Road, Holmdel, NJ 07733, USA.

AND

J. P. GORDON
Bell Laboratories, Lucent Technologies
101 Crawfords Corner Road, Holmdel, NJ 07733, USA.

Abstract. An analytic description is presented for the pulse dynamics in a dispersion managed communications systems. A variational formalism reduces the governing equations to a Hamiltonian system for which a geometrical interpretation of the pulse dynamics is given. The reduced model gives a simple method for calculating the ideal enhanced initial power for a dispersion managed breather. Periodic and quasi-periodic dispersion-managed solitons are observed in the mean-anomalous regime as well as the mean-zero and mean-normal regimes provided the map is sufficiently strong.

1. Introduction

In recent years, there have been increasing research efforts aimed toward understanding dispersion-managed communications systems. As a result, both experimental [1] and computational [2, 3] results have suggested the possibility of using return-to-zero (RZ) pulses in periodic dispersion maps where the average dispersion is in the anomalous (soliton) regime. A fundamental motivation for the dispersion mapping technique arises from the need to significantly suppress the phase-matchedfour-wave mixing processes which

A. Hasegawa (ed.), New Trends in Optical Soliton Transmission Systems, 183–195.

leads to substantial performance penalties in wavelength division multiplexing (WDM) systems. By allowing for large local dispersion, the efficiency of the four-wave mixing process can be substantially reduced [4, 5].

Although the four-wave mixing penalty is decreased, dispersion management imposes a strong perturbation to the nonlinear pulse propagation in the fiber. Despite this, original computational evidence by Smith *et al.* [2, 3] and later by Gabitov *et al* [6, 7, 8], Golovchenko *et al.* [9], and Kutz *et al.* [10] suggests that solitary waves are indeed robust and capable of propagating over ultra long distances without significant change in shape from their initial launch conditions when viewed stroboscopically after each dispersion map period. A key feature which enables the dispersion managed system to support these solitary waves is enhanced initial power, where the enhancement factor is relative to the one-soliton energy corresponding to the average dispersion of the system. It should be noted that during a given dispersion map period, the pulse can undergo $O(1)$ changes in amplitude, width and quadratic phase chirp. However, these fluctuations, sampled at appropriate locations of the dispersion map, can be periodic and thus a dispersion managed breather (soliton) is supported.

Although theory has been developed to describe the long-term dynamics due to the periodic dispersion-map perturbations [11, 12, 13, 14], only recently have analytic techniques by Gabitov *et al.* [6, 7, 8] and Kutz *et al.* [10] been developed to describe the amplitude and phase fluctuations which occur on the scale of the dispersion map period. By utilizing an averaged variational method and its corresponding Hamiltonian dynamics, the governing nonlinear Schrödinger (NLS) equation with a periodic dispersion map is reduced to a coupled set of nonlinear ordinary differential equations (ODEs) which accurately describe the full NLS dynamics. Although the Hamiltonian formulation renders the problem soluble in terms of a quadrature which can be explicitly integrated, we show that a *geometrical* analysis of the pulse dynamics in the amplitude–chirp phase plane, and the corresponding Poincaré section of the periodic problem, provide a natural and insightful framework in which to view the pulse dynamics [10].

2. Governing Equations

The evolution of a pulse propagating in a lossless optical fiber under the influence of the weak Kerr nonlinearity and which includes the effects of a periodically varying dispersion behaves according to the dispersion managed nonlinear Schrödinger equation

$$i\frac{\partial Q}{\partial Z} + \frac{\sigma(Z)}{2}\frac{\partial^2 Q}{\partial T^2} + |Q|^2 Q = 0, \tag{1}$$

where Q is the electric field envelope normalized by the peak field power $|E_0|^2$, and $\sigma(Z)$ is a periodic dispersion map. The variable T represents the physical time normalized by $T_0/1.76$, where T_0 is the full-width, half-max (FWHM) of the pulse (e.g. $T_0 = 50$ ps), and the variable Z represents the physical distance divided by the dispersion length Z_0 which corresponds to the average value $(\bar{D})$ of the dispersion map $\sigma(Z)$. Thus we find $Z_0 = (2\pi c/\lambda_0^2 \bar{D})\,(T_0/1.76)^2$, which gives $|E_0|^2 = (\lambda_0 A_{\text{eff}}/2\pi n_2 Z_0)$ for the peak field intensity of the one-soliton solution corresponding to the average dispersion. Here $n_2 = 2.6 \times 10^{-16}\text{cm}^2/\text{W}$ is the nonlinear coefficient of the fiber, $A_{\text{eff}} = 55\,\mu\text{m}^2$ is the effective cross sectional area of the fiber and $\lambda_0 = 1.55\,\mu\text{m}$ and c are the carrier's free-space wavelength and speed of light respectively. In the present analysis, we neglect the periodic intensity fluctuations due to fiber loss and amplification as they only introduce small perturbations to the average pulse dynamics [13]. Thus the peak power corresponds to the path-average peak power of the propagating pulse.

The parameters which are of greatest interest correspond to the particular dispersion map being used (see Figure 1). It is helpful therefore to consider the specific case of a piecewise constant (symmetrized) step function dispersion map which yields a minimal amount of pulse deformation [11, 12]

$$\sigma(Z) = \frac{1}{\bar{D}} \begin{cases} D_- & 0 < Z < \frac{1}{2}\frac{Z_-}{Z_0} \\ D_+ & \frac{1}{2}\frac{Z_-}{Z_0} < Z < \frac{1}{2}\frac{Z_-}{Z_0} + \frac{Z_+}{Z_0} \\ D_- & \frac{1}{2}\frac{Z_-}{Z_0} + \frac{Z_+}{Z_0} < Z < P = \frac{Z_- + Z_+}{Z_0}\,, \end{cases} \tag{2}$$

where $\sigma(Z) = \sigma(Z+P)$, $D_+ > 0 > D_-$ are the dispersion values in each segment of fiber (in ps/km·nm), $Z_\pm$ is the length of each fiber segment (in km) and the average dispersion is simply given by $\bar{D} = (D_- Z_- + D_+ Z_+)/(Z_- + Z_+)$. Recall that for $\sigma > 0$, the governing NLS equation (1) supports bright hyperbolic secant solitons while for $\sigma < 0$ nonlocal dark solitons are supported. Although we have utilized the standard soliton scalings of equation (1), solitons in the strict mathematical sense are in fact not supported due to the rapid and $O(1)$ dispersion changes experienced by the pulse.

3. Variational Method

We can restate the governing NLS equation (1) in terms of its variational form by defining the Lagrangian [15]

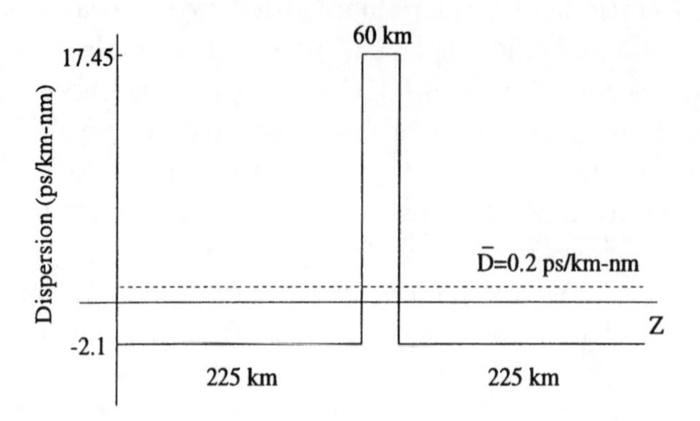

Figure 1. Typical dispersion map used in experiments. Note here that the average dispersion is $\bar{D} = 0.2$ ps/km·nm with $Z_- = 450$ km, $Z_+ = 60$ km, $D_- = -2.1$ ps/km·nm and $D_+ = 17.45$ ps/km·nm.

$$L = \int_{-\infty}^{\infty} \mathcal{L}(Q, Q^*) dT = \int_{-\infty}^{\infty} i \left(Q \frac{\partial Q^*}{\partial Z} - Q^* \frac{\partial Q}{\partial Z} \right) + \sigma(Z) \left| \frac{\partial Q}{\partial T} \right|^2 - |Q|^4 dT \,, \tag{3}$$

where the Lagrangian density $\mathcal{L}$, is determined by requiring $\delta\mathcal{L}/\delta Q^* = \delta\mathcal{L}/\delta Q = 0$ reproduce equation (1) and its complex conjugate respectively.

Given then an initial ansatz with appropriately chosen free parameters (i.e., amplitude, chirp, etc.), the Lagrangian L can be evaluated via equation (3). In the reduced model developed here, we assume that the fundamental dynamics occurs due to changes in amplitude and quadratic phase chirp. Further, we take as our initial condition a Gaussian pulse rather than the usual hyperbolic secant shape. This gives an ansatz of the form

$$Q(Z,T) = A \left(\frac{8\kappa^2}{\pi} \right)^{1/4} \sqrt{\eta} \exp \left[-(\kappa\eta T)^2 + i \left(\beta\kappa^2 T^2 + \frac{\phi(Z)}{2} \right) \right] \,, \tag{4}$$

where η and β are free parameters which depend upon Z and correspond to the amplitude and quadratic chirp dynamics respectively. Here we take $\kappa = 1.18/1.76$ in order that the Gaussian ansatz be of the same width (FWHM) as the one-soliton (hyperbolic secant). To make direct comparison with Smith *et al.* [2], we choose the Gaussian pulse energy ($\int |Q|^2 dT$) when $A = 1$ to be equal to that of the one-soliton hyperbolic secant pulse with $\bar{D}$ (i.e. $\int |Q|^2 dT = \int \operatorname{sech}^2 T dT = 2$). In general, *any* localized ansatz for which the amplitude and width are related as in equation (4) can be utilized [10]. However, the variational method is only valid provided the pulse evolution maintains a nearly-Gaussian pulse shape during propagation. Further, it is incapable of describing or capturing the slight growth of continuous spectra which is generated by the dispersion management perturbations [16].

To follow up previous work of Smith *et al.* [2], the amplitude enhancement factor A is defined as

$$A^2 = 1 + \alpha \left[\frac{\lambda_0^2}{2\pi c T_0^2} \left\{ (D_+ - \bar{D}) Z_+ - (D_- - \bar{D}) Z_- \right\} \right]^2 . \tag{5}$$

In Section 6 we vary A (or equivalently α) in order to characterize the pulse dynamics. Note that the pulse ansatz of equation (4) allows us to evaluate the Lagrangian L via equation (3). And although variations with respect to the free parameters η and β yield the appropriate ODE equations [6, 7, 8], we instead develop the Hamiltonian formulation.

4. Hamiltonian Dynamics

The Lagrangian system is brought to Hamiltonian form via a Legendre transformation by defining the canonical coordinate β and its conjugate momentum variable $\mathrm{p}_\beta = \partial L / \partial (d\beta/dZ) = A^2/\eta^2$ along with the Hamiltonian $H(\beta, \mathrm{p}_\beta) = \mathrm{p}_\beta (d\beta/dZ) - L(\beta, \mathrm{p}_\beta)$. The evolution equations for η and β, determined from Hamilton's equations $d\beta/dZ = \partial H/\partial \mathrm{p}_\beta$ and $d\mathrm{p}_\beta/dZ = -\partial H/\partial \beta$, yield:

$$\frac{d\eta}{dZ} = -2\kappa^2 \sigma(Z) \beta \eta \tag{6a}$$

$$\frac{d\beta}{dZ} = 2\kappa^2 \sigma(Z) \left(\eta^4 - \beta^2 \right) - \frac{2\kappa}{\sqrt{\pi}} A^2 \eta^3 . \tag{6b}$$

This system of equations describes the nonlinear amplitude and width fluctuations along with the quadratic chirp variations as the pulse propagates through a given dispersion map. Loss and gain fluctuations can also be easily incorporated into this formalism [6, 7, 8].

Since the resulting nonlinear ODE system of equation (6) is Hamiltonian and $\sigma(Z)$ is piecewise constant, we can utilize the fact that $dH/dZ = 0$ and construct level sets of a scaled version of H rewritten in terms of η and β:

$$\gamma \eta - \frac{d\phi}{dZ} - \kappa^2 \sigma(Z) \left(\eta^2 + \frac{\beta^2}{\eta^2} \right) = C , \tag{7}$$

where $\gamma = 2\kappa A^2/\sqrt{\pi}$. Solving for β and substituting into equation (6a) yields a first order ODE in terms of η alone

$$\frac{d\eta}{dZ} = \mp 2\kappa^2\sigma(Z)\eta^2 \left(\frac{\gamma\eta - C - \frac{d\phi}{dZ}}{\kappa^2\sigma(Z)} - \eta^2\right)^{1/2} . \tag{8}$$

Upon integrating this equation with initial conditions $\eta = \eta_0$ at $Z = 0$, we find

$$Z = \mp\frac{1}{2\kappa^2\sigma}\int_{\eta_0}^{\eta} \frac{dx}{x^2\left[\frac{\gamma x - C - d\phi/dZ}{\kappa^2\sigma} - x^2\right]^{1/2}}, \tag{9}$$

where σ is a fixed value, i.e. $\sigma = D_-/\bar{D}$ or $D_+/\bar{D}$ depending on whether the pulse is propagating in the normal or anomalous regime respectively. Although in principle we have an analytic solution for the pulse dynamics via equation (9), this equation proves to be unwieldy for practical purposes since the integration gives a complicated (closed form) expression which must be inverted to give $\eta = \eta(Z)$.

5. Phase-Plane Dynamics

It is often convenient and insightful to plot the phase-plane dynamics of a planar dynamical system such as equation (6). This geometric representation of the solution treats the distance Z as a parameter while plotting the η (amplitude) and β (chirp) variables on the x- and y-axes respectively. It elucidates some fundamental aspects of the breather dynamics which will be clearly shown in what follows. The phase-plane dynamics are markedly different depending on the sign of the dispersion. Thus we consider separately the anomalous and normal dispersion regimes. However, there are common features which are discussed initially.

We begin by seeking fixed points of equation (6) by setting the right hand sides equal to zero. The two fixed points are:

$$\text{I. } \beta = 0, \eta = 0 \qquad \text{and} \qquad \text{II. } \beta = 0, \eta = (A^2/\kappa\sqrt{\pi}\sigma(Z)) = \eta_0 . \tag{10}$$

Note that for the moment, we have included $\sigma(Z)$ as a variable parameter. In the analysis which follows, special care will be taken to look in the anomalous and normal dispersion regimes where the sign of $\sigma(Z)$ is positive and negative respectively.

A linear stability calculation about each of the fixed points gives insight into the general dynamics of the system. In particular, the fixed point $(\eta, \beta) = (0, 0)$ is degenerate so that its linear stability cannot be determined. However, the linear stability of the fixed point II can be determined. Linearizing about this point by setting $\eta = \eta_0 + \tilde{\eta}$ and $\beta = 0 + \tilde{\beta}$ yields:

$$\frac{d\tilde{\eta}}{dZ} = -2\kappa^2\sigma(Z)\eta_0\tilde{\beta} \quad \text{and} \quad \frac{d\tilde{\beta}}{dZ} = \left(8\kappa^2\sigma(Z)\eta_0^3 - \frac{6\kappa A^2}{\sqrt{\pi}}\eta_0^2\right)\tilde{\eta}. \quad (11)$$

The eigenvalues of this *linear* system are $\lambda_\pm = \pm 2i\kappa^2\eta_0^2\sigma(Z)$. Thus fixed point II is a center, regardless of the sign of the dispersion.

5.1. NORMAL DISPERSION: $\sigma = D_-/\bar{D} < 0$

We begin by considering the phase-plane dynamics in the normal dispersion regime. In this case, we can rewrite $\sigma = -|D_-|/\bar{D} < 0$. We then consider the location and stability of the fixed points of equation (6). Fixed point I has already been determined to be degenerate, and fixed point II lies in the left half plane since $\eta_0 < 0$. Because we only consider values of $\eta \geq 0$, only the fixed point at the origin is relevant. The phase-plane dynamics is shown in Figure 2 (left) along with the clockwise flow (arrows) in phase space. Note that the right half plane is filled with homoclinic orbits which emanate and terminate in the origin [17].

5.2. ANOMALOUS DISPERSION: $\sigma = D_+/\bar{D} > 0$

In the anomalous dispersion regime, the phase-plane dynamics is significantly different than that of the normal dispersion regime since fixed point II is located at

$$\beta = 0, \; \eta = \frac{A^2\bar{D}}{\kappa\sqrt{\pi}D_+} > 0 \quad (12)$$

with the pair of eigenvalues $\lambda_\pm = \pm 2i\kappa^2\eta_0^2 D_+/\bar{D}$. The phase portrait and its associated counter-clockwise phase flow is depicted in Figure 2 (right).

In this case, the phase-flow for $\eta \geq 0$ is partially determined by both the critical points I and II. It can be more clearly understood by making use of equation (7) for $C = -d\phi/dZ$. This gives rise to the separatrix

$$\beta = \pm\eta\sqrt{\frac{\gamma\eta\bar{D}}{\kappa^2 D_+} - \eta^2}, \quad (13)$$

which has a cusp at the origin ($\eta = \beta = 0$), depicted by the bolded curve in Figure 2 (right). Solutions outside this separatrix eventually flow into the origin (homoclinic orbits) [17], while those inside the separatrix are periodic. More generally, solution curves depicted in Figure 2 are all level sets of equation (7) for distinct C values.

The dynamic behavior of equation (6) is therefore governed by the two phase flows shown in Figure 2. From them, we can construct a geometric

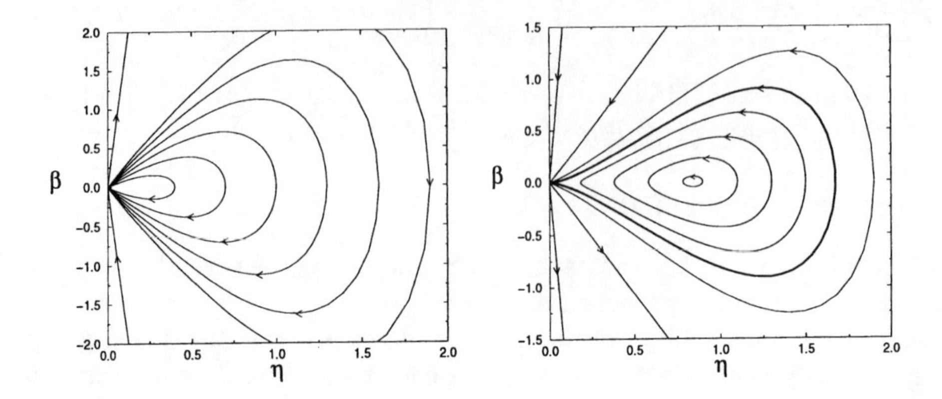

Figure 2. Phase-plane dynamics for $\sigma = \mp 1$ (left and right figures respectively) and $A = 1$ plotted for values of $\eta \geq 0$. The origin is a degenerate fixed point while the fixed point at $\eta = \eta_0 \geq 0$, $\beta = 0$ for $\sigma = 1$ is a center.

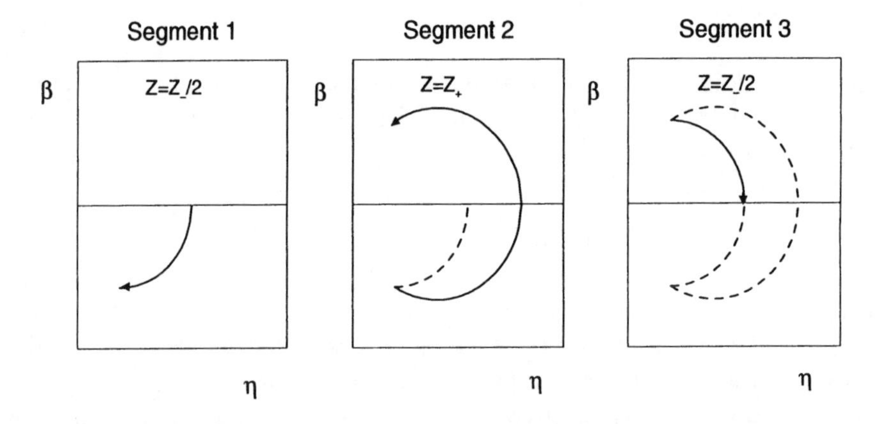

Figure 3. Qualitative depiction of the pulse dynamics on the phase-plane. The left figure represents the flow through the first part of the normal dispersion segment (see Figures 1 and 2). The middle figure represents the flow under anomalous dispersion (the dotted line is the flow of the previous fiber segment). The right figure is the final segment of the normal dispersion fiber. Note that if the parameters are chosen appropriately, a periodic orbit can be constructed.

representation of the solution and its associated dynamics. Furthermore, for a particular critical value of A, we can construct a strictly periodic dispersion managed breather solution from the dispersion map in Figure 1 as depicted in Figure 3. In this periodic solution, the pulse first evolves according to the first segment of the dispersion map for which $\sigma < 0$.

The dynamics then reverses its flow in the anomalous fiber (the dispersion switches sign) and then finally is again subject to the dynamics of the normal regime. Note the reflection symmetry about $\beta = 0$, which derives from the choice of dispersion map and the resulting invariance of equation (6) under the reversibility transformation $(\eta, \beta, Z) \mapsto (\eta, -\beta, -Z)$. Although it is only a qualitative description, the geometric representation of Figure 3 provides an insightful method for understanding the pulse dynamics. This periodic solution will be illustrated in the next section via direct numerical simulations.

6. Numerical Simulations

This section aims to demonstrate the geometric aspects of the dispersion-managed evolution. A thorough investigation and comparison of the PDE and ODE models is given in reference [10]. We can easily construct the phase-plane dynamics by plotting the solution of equation (6) as a function of Z. It is also insightful to plot the position in phase space at the end of each dispersion map period (thus defining a Poincaré map [17]). These results are plotted in Figure 4 for both the ODE model and the full governing NLS equation for propagation through 40 dispersion map periods of Figure 1. The light gray curves are the actual amplitude and chirp fluctuations in the phase-plane as a function of Z (see Figures 2 and 3), while the bolded points represent the solution at the end of each dispersion map period. For $\alpha = 0.4$ and 1.5, the Poincaré map shows a closed loop (an invariant set [17]) which corresponds to a long term quasi-periodic behavior. For $\alpha = \alpha_c = 0.723$ we construct a numerical analog of Figure 3 and find that the Poincaré map has a fixed point located at $(\eta, \beta) = (1, 0)$. The PDE dynamics are observed to be in good agreement with the ODE model with deviations due again to the ansatz assumption and shedding of dispersive radiation [16]. Figure 4 also shows that the ODE model gives a useful method for predicting the maximum variations in the phase and amplitude fluctuations.

From Figure 4 it should be pointed out that given a fixed dispersion map and propagation distance, different channels may see different enhancement factors. However, if the desire is to receive an unchirped pulse at the end of the transmission line, one needs to simply put on the right amount of initial chirp ($\beta \neq 0$) so that the solution in the Poincaré section (at the end of the transmission line) ends on the η-axis ($\beta = 0$). Thus the role of pre-chirping of pulses has a simple geometric interpretation through Figure 4.

We also utilize the reduced ODE system to understand the system performance for various alterations of the dispersion map. Thus, given a dispersion map, we can calculate the appropriate initial pulse energy (α_c) so

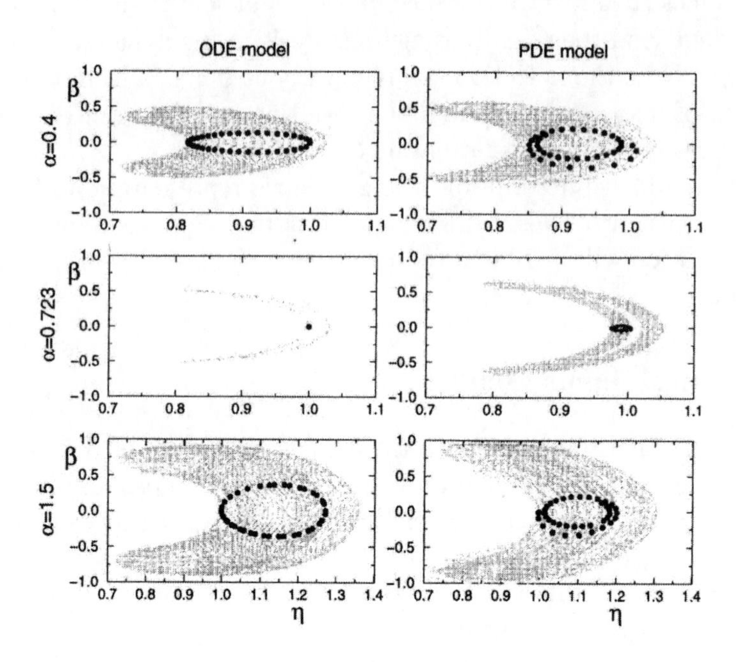

Figure 4. Phase-plane dynamics of the ODE model (left) and governing dispersion managed NLS (right). The light gray curves represent the η and β fluctuations as a function of Z while the dots represent the solution after every dispersion map period, i.e. the Poincaré map. The spiraling in the PDE model is due to energy loss to the continuum.

that the performance is optimized for the given map. A simple shooting method can be utilized to calculate the value of α_c by noting that for values of $\alpha < \alpha_c$ the Poincaré map flows to the left, while values of $\alpha > \alpha_c$ flow to the right. Hence α_c can be found with any predetermined accuracy. This allows for a complete investigation of the optimal performance of the system given a fixed dispersion map. In Figure 5 we plot the relevant performance characteristics for the dispersion map of Figure 1 and modifications thereof. In particular, we show the value of α_c as a function of the pulse width (top left), the magnitudes of dispersion (top right), the dispersion period length (bottom left) and the average dispersion (bottom right). These results can be calculated rapidly using a shooting method in conjunction with equation (6).

Finally, we note that these results are not limited to the mean-anomalous regime. Rather, for a dispersion map above a critical strength, dispersion-managed solitons can be supported with a mean-normal dispersion (this is considered in detail elsewhere [18]). The periodic phase-plane (ODE) and evolution (PDE) dynamics is illustrated in Figure 6. Although the mean-

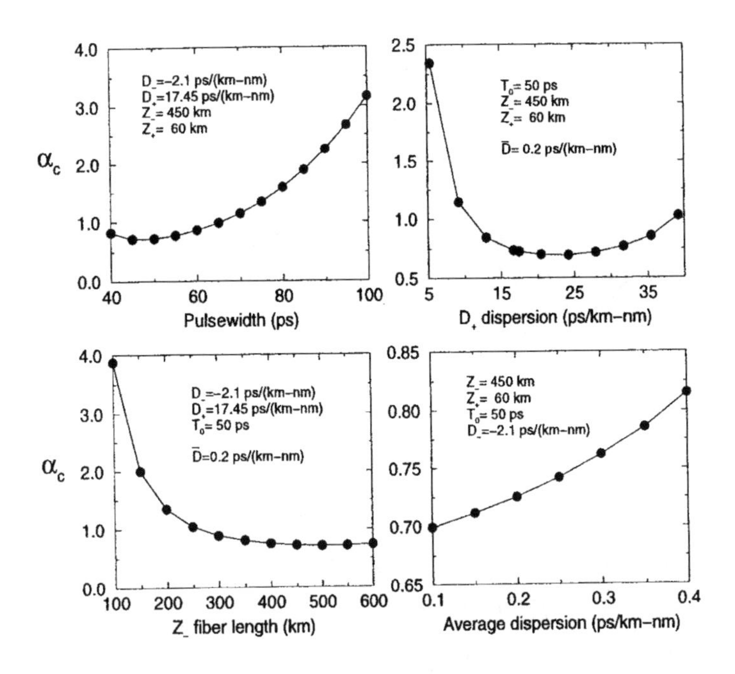

Figure 5. Dependence of α_c on various dispersion map configurations and pulse widths. Note that the parameter α_c can be much different than ≈ 0.7 for certain dispersion maps.

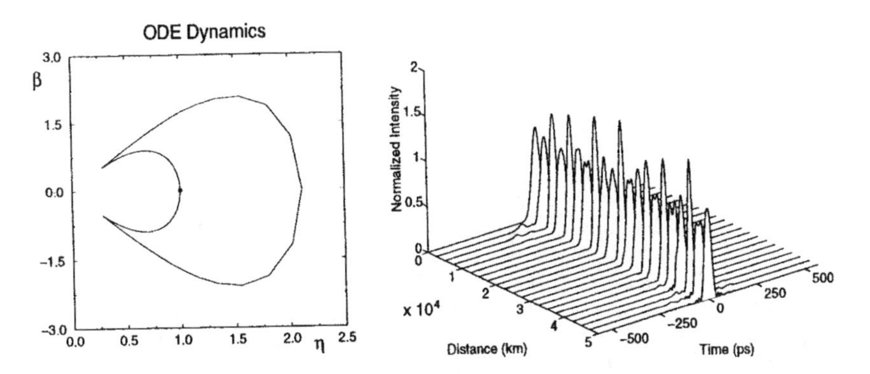

Figure 6. Phase-plane (ODE) and evolution (PDE) dynamics for $\bar{D} = -0.04$ ps/km·nm, $Z_+ = 53$ km, $T_0 = 30$ ps and launch power $|E_0|^2$=10.251 mW depicting the stabilized and localized evolution with a mean-normal dispersion.

zero regime also supports periodic solutions in the ODE setting, the PDE is subject to a long-wavelength instability which is discussed in detail in Bronski and Kutz [11, 12].

7. Summary and Discussion

Through the use of a variational method, the pulse dynamics in a dispersion managed optical system can be reduced to a pair of coupled, nonlinear ODEs. Numerical simulations show that the ODE model accurately captures the pulse dynamics as governed by the dispersion managed NLS equation (1). Thus a useful tool is provided for the modeling and design of dispersion managed systems, since the computational efforts are reduced by two to three orders of magnitude over direct simulations of equation (1). We have also found that the reduced ODE system supports periodic solutions in the mean-normal, mean-zero and mean-anomalous dispersion regimes. This is confirmed with simulations of equation (1) and suggests the possibility of using mean-normal WDM channels near the zero-dispersion wavelength for sending stable, dispersion-managed pulses. In addition, the Poincaré map reveals a long term periodicity which is present in the full equation (1) and elucidates the role or pre-chirping of pulses. Finally, with minor modifications, the variational method may be used capture the Gordon-Haus jitter dynamics with dispersion-management [19]

Acknowledgements

We thank P. Holmes for many helpful discussions related to this work. J. N. Kutz acknowledges support from a NSF University-Industry Postdoctoral Fellowship (DMS-9508634) while at the Program in Applied and Computational Mathematics, Princeton University, and Bell Laboratories, Lucent Technologies, and an International Research Fellow Award (INT-9703769) while at Hong Kong Polytechnic University.

References

1. Naka, A., Matsuda, T. and Saito, S.: Optical RZ signal straight line transmission with dispersion compensation over 5220 km at 20 Gb/s and 2160 km at 2X20 Gb/s, *Elec. Lett.*, **32**, (1996), pp.1694-1696.
2. Smith, N. J., Knox, F. M., Doran, N. J., Blow, K. J. and Bennion, I.: Enhanced power solitons in optical fibers with periodic dispersion management, *Elec. Lett.* , **32**, (1996), pp.54-55.
3. Smith, N. J., Doran, N. J., Knox, F. M. and Forysiak, W.: Energy-scaling characteristics of solitons in strongly dispersion-managed fibers, *Opt. Lett.*, **21**, (1996), pp.1981-1983.
4. Kurtzke, C.: Suppression of fiber nonlinearities by appropriate dispersion management, *IEEE Pho. Tech. Lett.*, **5**, (1993), pp.1250-1253.
5. Tkach, R. W., Chraplyvy, A. R., Forghieri, F., Gnauck, A. H. and Derosier, R. M.: 4-Photon mixing and high-speed WDM systems, *J. Light. Tech.*, **13**, (1995), pp.841-849.
6. Gabitov, I. and Turitsyn, S. K.: Breathing solitons in optical fiber links, *Pisma v JETP*, **63**, (1996), pp.814-819.

7. Gabitov, I., Shapiro, E. G. and Turitsyn, S. K.: Optical pulse dynamics in fiber links with dispersion compensation, *Opt. Comm.*, **134**, (1997), pp.317-329.
8. Gabitov, I., Shapiro, E. G. and Turitsyn, S. K.: Asymptotic breathing pulse in optical transmission system with dispersion compensation, *Phys. Rev. E*, **55**, (1997), pp.3624-3633.
9. Golovchenko, E. A., Jacob, J. M., Pilipetskii, A. N., Menyuk, C. R. and Carter, G. M.: Dispersion-managed solitons in a fiber loop with in-line filtering, *Opt. Lett.*, **22**, (1997), pp.289-291.
10. Kutz, J. N., Holmes, P., Evangelides Jr., S. G. and Gordon, J. P.: Hamiltonian dynamics of dispersion managed breathers, *J. Opt. Soc. Am. B*, **15**, (1998), pp.87-96.
11. Bronski, J. C. and Kutz, J. N.: Guiding-center pulse dynamics in nonreturn-to-zero (return-to-zero) communications systems with mean-zero dispersion, *J. Opt. Soc. Am. B*, **14**, (1997), pp.903-911.
12. Bronski, J. C. and Kutz, J. N.: Asymptotic behavior of the nonlinear Schrödinger equation with a rapidly-varying, mean-zero dispersion, *Physica D*, **108**, (1997), pp.315-329.
13. Hasegawa, A. and Kodama, Y.: Guiding-center soliton in optical fibers, *Opt. Lett.*, **15**, (1990), pp.1443-1445 .
14. Gabitov, I. R. and Turitsyn, S. K.: Averaged pulse dynamics in a cascaded transmission system with passive dispersion compensation, *Opt. Lett*, **21**, (1996), pp.327-329.
15. Anderson, D.: Variational approach to nonlinear pulse propagation in optical fibers, *Phys. Rev. A*, **27**, (1983), pp.3135-3145.
16. Gordon, J. P.: Dispersive perturbations of solitons of the nonlinear Schrödinger equation, *J. Opt. Soc. Am. B*, **9**, (1992), pp.91-97
17. Guckenheimer, J. and Holmes, P.: *Nonlinear Oscillations, Dynamical Systems and Bifurcation of Vector Fields*, Springer-Verlag, New York, (1983).
18. Kutz, J. N. and Evangelides Jr., S. G.: Dispersion managed breathers with average normal dispersion, (submitted).
19. Kutz, J. N. and Wai, P. K. A.: Gordon-Haus timing jitter reduction in dispersion-managed soliton communications, (submitted).

(Received December 25, 1997)

SUPERB CHARACTERISTICS OF DISPERSION-ALLOCATED SOLITON TRANSMISSION IN TDM AND WDM SYSTEMS

M. NAKAZAWA, H. KUBOTA, K. SUZUKI, E. YAMADA, AND A. SAHARA

NTT Optical Network Systems Laboratories
Tokai, Ibaraki-ken 319-11, Japan

Abstract. Recent progress on soliton transmission is described, in which dispersion management plays an important role in increasing the power margin and the dispersion tolerance in TDM and WDM systems. The quality of the dispersion-allocated soliton is compared with RZ and NRZ pulses.

1. Introduction

It has long been said that a serious drawback to soliton transmission is that a special fiber must be prepared which has an anomalous group velocity dispersion (GVD). However, we have recently developed a new soliton transmission technology which enables us to use an already installed fiber cable. This is called the "dispersion-allocated (D-A) soliton" technique, by which a soliton can propagate even in fibers with normal GVDs as long as the average GVD is kept in the anomalous region [1]-[5].

The idea of employing dispersion compensation for soliton transmission was first reported in 1992 with a view to minimizing pulse broadening and expanding the amplifier spacing [6]. When a D-A soliton is used, the sideband instability [7] between the soliton and the dispersive waves has less influence because phase matching for the four wave mixing is unlikely to occur. Thus, a D-A soliton has a much larger power margin than an ordinary soliton with a uniform dispersion. In this paper, we describe superb characteristics of the D-A soliton in TDM and WDM systems. It is also shown that the physics of the D-A soliton is similar to that of stretched pulse mode-locking [8, 9].

A. Hasegawa (ed.), New Trends in Optical Soliton Transmission Systems, 197–224.

The dispersion-allocated soliton does not experience a large nonlinear phase change when the dispersion perturbation is large. Here the soliton behaves like a linear pulse such that the pulse broadening due to positive GVD can be compensated for by negative GVD. However, the dispersion-allocated soliton retains its soliton nature over the average soliton period. The important point is that when the dispersion perturbations are small, the system is fully described by a perturbed nonlinear Schrödinger equation, but when the dispersion perturbation is large, the steady-state pulse deviates from being a soliton and is closer to being a Gaussian pulse with chirp. The physics of this mechanism is similar to that of FM mode-locking , in which a phase modulation is generated by an averaged self phase modulation (SPM).

2. Power Margin and Waveform of D-A Soliton

We show first how the D-A soliton can carry a larger energy compared with conventional solitons. Figure 1 shows how an N = 3 ideal soliton, a dynamic (average) soliton, and D-A solitons propagate over 3,000 km. Soliton self frequency shift (SSFS) with a coefficient of 5.9 fs and a third order dispersion (D_λ) of 0.07 ps/km·nm^2 are taken into account. The initial pulse width is 12 ps and the average GVD is 0.15 ps/km·nm. A split-step Fourier method is used for the calculation. Figure 1(a) is an ideal soliton propagating in a lossless fiber, where SSFS = 0, $D_\lambda = 0$, the GVD is uniform, and no optical filters are installed. Figure 1(b) is a dynamic (average) soliton with a uniform GVD of 0.15 ps/km·nm. Figure 1(c) is a D-A soliton where the GVD changes from normal to anomalous. The fiber with a normal GVD of −2 ps/km·nm is 80 km long and the succeeding fiber with an anomalous GVD of 34.55 ps/km·nm is 5 km long. Figure 1(d) is a D-A soliton with a dispersion change from anomalous to normal GVD. The fiber with an anomalous GVD of 2 ps/km·nm is 80 km long and the fiber with a normal GVD of −29.45 ps/km·nm is 5 km long. In Figure 1(b), (c) and (d), the amplifier spacing is 85 km, a 2 nm optical filter is installed every 85 km, the fiber loss is 0.2 dB/km and no ASE is taken into account.

We have shown that the waveform evolution in high-order soliton propagation is initiated by the modulational instability that occurs at the parabolic top of the soliton pulse [10]. This occurs without noise because the parabolic top of the soliton pulse causes phase-matched four wave mixing which becomes the origin of the modulational instability. This is the principal reason for the breaks which occur in ideal high order solitons. These breaks stop occurring when the phase matching is disturbed.

By comparing Figure 1(a) and (b), it is clearly seen in Figure 1(b) that the N = 3 soliton breaks into multiple pulses during the propagation. This

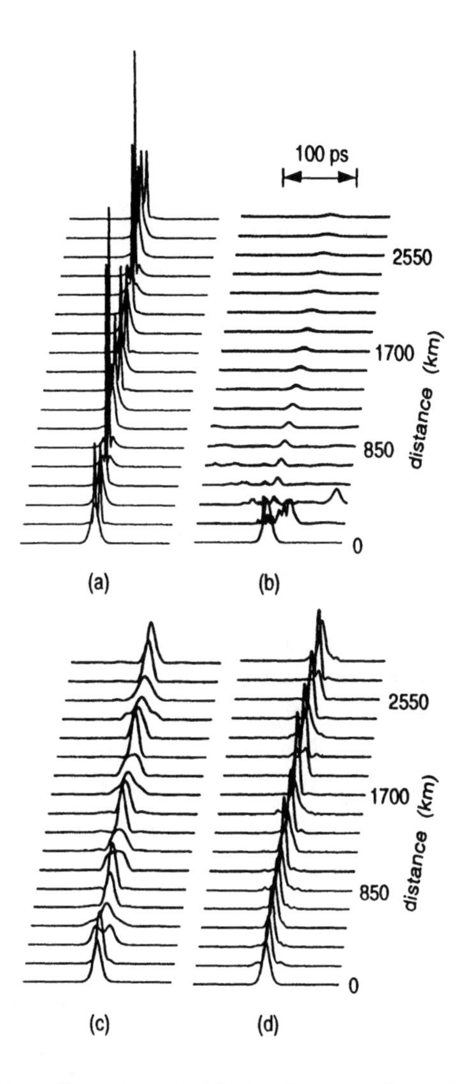

Figure 1. Propagation of an N = 3 ideal soliton, a dynamic soliton and dispersion-allocated soliton over 3,000 km. The GVD=0.15 ps/km·nm, SSFS=5.9 fs, the third order dispersion $D_\lambda = 0.07$ ps/km·nm^2. (a) Ideal soliton propagation in a lossless fiber, SSFS=0, $D_\lambda = 0$ and uniform GVD, in which no optical filters are installed. (b) Dynamic (average) soliton with uniform GVD. 2 nm optical filters are installes. (c) A dispersion-allocated soliton with a dispersion change from normal to anomalous GVD. The fiber with a normal GVD of −2 ps/km·nm is 80 km-long and the succeeding fiber with an anomalous GVD of 34.55 ps/km·nm is 5 km-long. (d) A dispersion-allocated soliton with a dispersion change from anomalous to normal GVD. The fiber with an anomalous GVD of 2 ps/km·nm is 80 km-long and the fiber with a normal GVD of −29.45 ps/km·nm is 5 km-long.

is due to the fact that the SSFS and the third order dispersion break the N = 3 soliton into N = 1 solitons with different amplitudes. On the other hand, no such break occurs in Figure 1(c) and (d) and a single pulse can propagate over 3,000 km. The N = 3 D-A soliton in Figure 1(c), which can be obtained by launching a sech pulse with an N = 3 average soliton amplitude, has a broader pulse width than that in Figure 1(d). This is because the soliton pulse in Figure 1(c) experiences a normal GVD fiber first which gives rise to nonlinear pulse broadening. That is, the soliton effect in Figure 1(c) is weaker than that in Figure 1(d). The soliton pulse in Figure 1(d) passes through an anomalous GVD fiber first and so propagates as an ideal soliton without broadening. These results suggest that the normal GVD fiber prevents the soliton from breaking into N = 1 solitons as the phase matching is not completely achieved. In other words, the soliton does not evolve ideally and a single-like pulse with a higher amplitude is maintained. This high order (high intensity) soliton pulse is very useful for increasing the power margin by as much as 3^2, that is 9.5 dB, compared with power of the N = 1 average soliton. It is important to note that a 2 nm optical filter installed every 85 km plays an important role in stabilizing the single-like pulse.

Figure 2 shows the propagation characteristics of D-A solitons with N = 1, 2 and 3 amplitudes. The parameters are the same as those in Figure 1(d) except that one span consists of a 50 km anomalous GVD fiber and a 35 km normal GVD fiber. Figure 2(a) shows how the pulse width changes in one dispersion-allocated span. Figure 2(b) shows the waveforms of the transmitted D-A soliton fitted to sech and Gaussian pulses. It is interesting to see in Figure 2(a) that the D-A soliton changes drastically in one span [2, 5]. In the steady-state, an input pulse in the anomalous GVD fiber has a linear chirp due to SPM. Thus, the pulse is compressed and rebroadened up to the input of the fiber with normal GVD. At this point, the sign of the chirping becomes opposite. Thus the transmitted pulse is recompressed and then rebroadened up to the next input of the fiber with anomalous GVD. That is, after propagation over one span, the chirp has the same value as at the beginning. This feature was first reported in reference [2]. The pulse width for higher N values changes more than that for N = 2 because the intensity dependent spectral change is larger.

As seen in Figure 2(b), the transmitted N = 2 and N = 3 solitons are similar to a Gaussian pulse than a sech pulse. These pulses are extracted at the minimum pulse width position at which the pulse distortion is minimized. The results suggest that the nature of the propagating soliton can be described as a linear pulse transmission with a small soliton effect (Kerr nonlinearity). When the SPM as a perturbation can be treated as a parabolic phase modulation, a steady-state pulse is a Gaussian pulse with

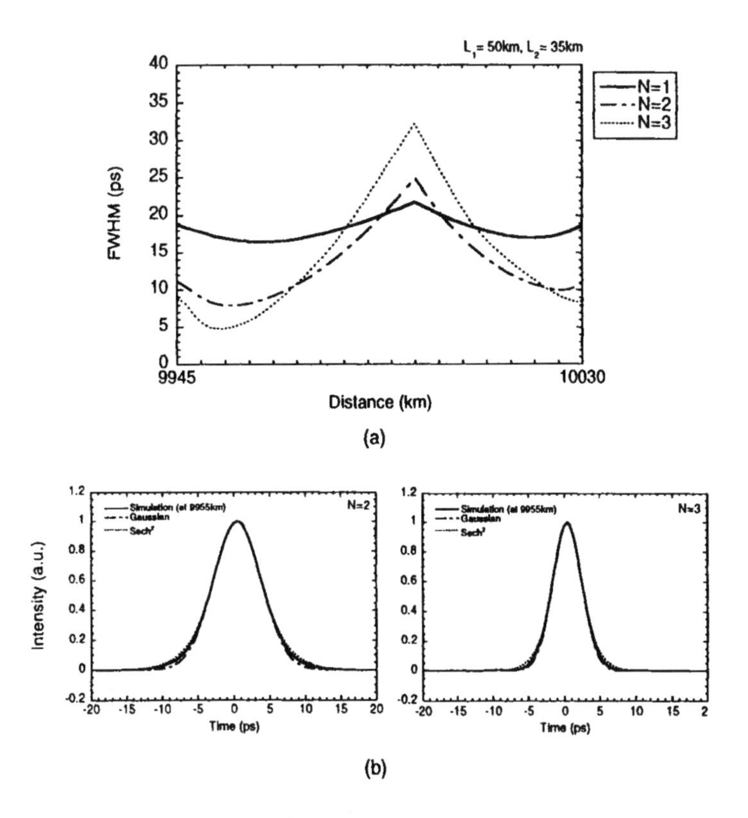

Figure 2. Propagation characteristics of the N=1, 2 and 3 dispersion-allocated solitons. All parameterts are the same as those in Figure 1(d) except that one span consists of 50 km anomalous fiber and a 35 km normal fiber. (a) Pulse width (FWHM) changes in one amplifier spacing, (b) Transmitted pulse waveform with N=2 and N=3 amplitudes. Dotted line: a sech pulse, dashes line: a Gaussian pulse.

chirp according to FM mode-locking theory. A larger dispersion allocation results in a pulse which is more nearly Gaussian pulse with chirp. For a small dispersion-allocation, however, the pulse is close to a sech pulse as the soliton formation mechanism should be fully taken into account.

3. D-A Soliton and the Stretched-pulse Mode-locking

We describe here the relationship between the D-A soliton and the stretched-pulse mode-locked fiber laser [8, 9]. The pulse propagation in a soliton laser corresponds to an ideal or average soliton transmission. It has an anomalous GVD, and the pulse maintains a constant width and remains transform-limited over a single-pass. By contrast, the pulse propagation in a stretched-pulse mode-locked laser corresponds to a D-A soliton transmission, in which the net GVD is close to zero, and the pulse undergoes large dispersive changes in width and peak power in a single pass.

It is important to note that there is a difference between the operation GVD in the stretched-pulse mode-locked laser and the D-A soliton. With stretched-pulse mode-locking, pulse compression occurs due to a nonlinear polarization rotation in the fiber cavity, which means that there is no need to keep the average GVD anomalous. Even when the cavity has a net normal GVD, it still works because of nonlinear wing shaping (intensity dependent switching). However, there is no such switching effect in the D-A soliton transmission, and therefore an anomalous GVD is needed to shorten the transmitting pulse through the use of SPM. When the average GVD is normal, the pulse is eventually dispersed.

With the pulse propagation in a soliton laser, which corresponds to a uniform soliton transmission, the master equation is written in the following form with a perturbed nonlinear Schrödinger equation.

$$\frac{du}{dz} = \left\{ g - l + j\Psi + \left(jD + \frac{f_0}{\omega_f^2} \right) \frac{d^2}{dt^2} + (\gamma - j\delta)|u|^2 \right\} u = 0 \,. \tag{1}$$

It is well known that the solution of equation (1) is a chirped sech pulse as shown by

$$u = U_0 \mathrm{sech} \left(\frac{t}{\tau} \right)^{1+j\beta} \,. \tag{2}$$

On the other hand, the master equation for a stretched pulse mode-locked laser or a D-A soliton, is described with a linear Schrödinger equation as

$$\frac{du}{dz} = \left\{ g - l + j\Psi + \left(jD + \frac{f_0}{\omega_f^2} \right) \frac{d^2}{dt^2} + (\gamma - j\delta)|U_0|^2(1 - \mu t^2) \right\} u = 0 \,. \tag{3}$$

In the stretched-pulse laser, the pulse width is greatly modified as described in Section 2 and the average pulse width is broader than the minimum width. Therefore, the field u experiences average nonlinearity, in which we can approximate the nonlinear potential with a parabola. Thus we obtain the above equation. The solution of equation (3) is also well known and is given by

$$u = U_0 \exp\left(-Qt^2 \right) \,. \tag{4}$$

Here Q is the complex and the D-A soliton has a Gaussian rather than a sech pulse shape. In the case of the D-A soliton, g is zero in equation (3).

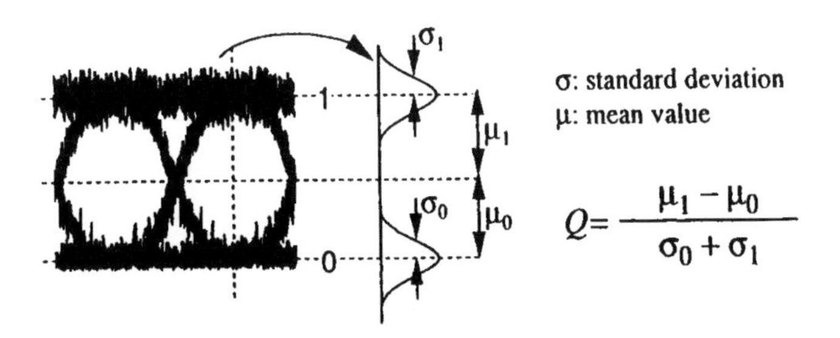

Figure 3. Definition of the Q value.

4. Q Value of D-A Soliton

The Q value is the signal-to-noise ratio of the received data eye and is directly related to the bit error rate of the transmission system [11]. A Q-value is calculated from fluctuations in the transmitted waveform (eye pattern) as shown in Figure 3. If we assume the fluctuation has a Gaussian distribution, the bit-error rate (BER) can be directly calculated from the Q value. For example, Q $= 7$ corresponds to a BER of 1×10^{-12}. Although the Q value is a measure of the signal-to-noise ratio, it can also include a degradation in the transmission characteristics due to a nonlinear interaction of the pulse when a pseudo random bit stream (PRBS) is used as the input data stream. Therefore, we can evaluate the transmission characteristics of a complicated transmission system which has a nonlinear interaction and noises simply from the Q value. The pulse propagation in an optical fiber is described by a nonlinear Schrödinger (NLS) equation and is numerically solved by the split-step Fourier method. The NLS equation with a fiber loss and dispersion change as a function of the distance is given by

$$-i\frac{\partial u}{\partial z} = \frac{1}{2}\frac{k''(z)}{|k''|}\frac{\partial^2 u}{\partial t^2} - \frac{i}{6}\frac{k'''}{|k_0''|t_0}\frac{\partial^3 u}{\partial t^3} + |u|^2 u + i\Gamma u - \frac{t_n}{t_0}\frac{\partial |u|^2}{\partial t}\,, \quad (5)$$

where $k''(z)$ is the normalized local GVD value and Γ is the normalized fiber loss. k_0'' corresponds to the average GVD. The last term indicates the self Raman effect where t_n is assumed to be 5.9 fs. The ASE power P_{ASE}, generated by the EDFA is given by

$$P_{\mathrm{ASE}} = \mu(G-1)h\nu B\,, \quad (6)$$

where $h\nu$ is the photon energy, B is the bandwidth, G is the gain and μ is an inversion parameter. The ASE is assumed to be white noise and added to the signal after each amplification process. An optical filter is also applied to the transmitted waveform after the ASE is added. The transmitted signal is detected by a quadrature detector. A base band electrical filter of $0.65 \times$ bandwidth is applied to the detected signal to reduce the noise and then an eye pattern is drawn to measure a Q value. Since this method is applicable to any waveform and any dispersion configuration, we can compare the transmission characteristics of various transmission systems from the Q value.

In Figure 4, we show how the Q value changes as a function of propagation distance. The dispersion allocation corresponds to (d) in Figure 1. Figure 4(a) shows the maximum Q vs. propagation distance and Figure 4(b) shows jitter evolution vs. propagation distance. The data bit rate is 20 Gbit/s with a soliton pulse width of 12 ps. The fiber loss, noise figure of the optical amplifier, baseband filter width and length of the random data pattern are 0.22 dB/km, 7 dB, 13 GHz and $2^7 - 1$, respectively. For an $N = 1$ D-A soliton, the propagation distance is approximately 2,000 km for $Q = 7$ as shown in Figure 4 (a), which increases to as much as 6,000 km when an $N = 2$ D-A soliton is used. When an $N = 3$ D-A soliton is adopted, the propagation distance is also increased to 4,600 km which is however, shorter than that of the $N = 2$ D-A soliton. This is because a small amount of ripple appears on the wing of the $N = 3$ D-A soliton as seen in Figure 1 (d).

Figure 4(b) shows how the timing margin for maintaining $Q = 7$ signal quality is reduced as a function of propagation distance. For an $N = 2$ D-A soliton, the maximum transmission distance at which a phase margin of ± 5 ps can be maintained is 5,000 km, which is much longer than the distance of 1,800 km for an $N = 1$ D-A soliton. These results indicate that the power and phase margins of soliton transmission can be greatly increased through the use of the D-A soliton. When the pulse is shortened to 10 ps, the transmission distances of $N = 1$, 2 and 3 D-A solitons for $Q = 7$ change to 3,600 km, 6,000 km and 4,900 km, respectively.

5. Q Contour Mapping and System Evaluation

In the next step we compare the characteristics of the D-A soliton with those of conventional NRZ and RZ pulse transmission at zero GVD. Figure 5 shows a new method for evaluating the performance of optical transmission systems at 20 Gbit/s. This is called "Q factor contour mapping" which can immediately evaluate the power margin and dispersion tolerance of a system [12]. With this map, we can compare the transmission performance of different optical systems. Figure 5(a) and (b) show Q contour maps of the transmitted peak optical power (EDFA output power) and the GVD

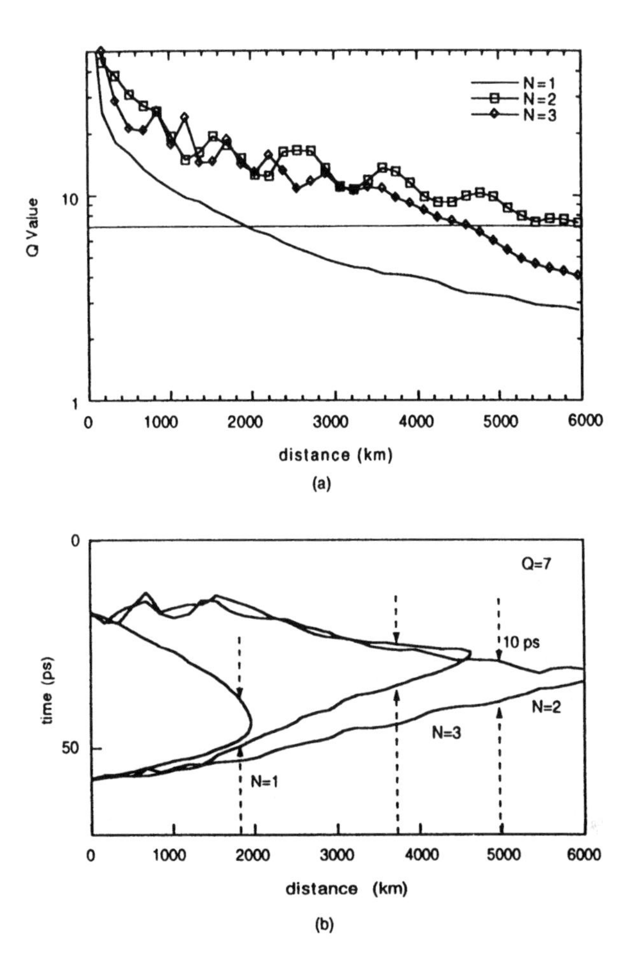

Figure 4. Changes in the Q value as a function of propagation distance. The bit rate =20 Gbit/s, soliton width =12 ps. The dispersion allocation corresponds to Figure 1(d). (a) Q value vs. propagation distance. (b) Timing jitter evaluation for Q=7 vs. propagation distance.

of a transmission fiber. The transmission distance is set at 2,560 km. This Q mapping enables us to evaluate the power margin from the vertical axis and the dispersion tolerance from the horizontal axis. Figure 5(a) and (b) correspond to the soliton and the NRZ transmission, respectively. In both cases, the GVD is uniform.

The contours of the Q map in Figure 5(b) are entirely different from those in Figure 5(a), in which the Q = 7 contour is much smaller than that in Figure 5(a). It is interesting to note that the Q map is not symmetric in relation to zero GVD, which would be expected from a linear transmission property, but the results suggest that better transmission quality is obtained in the anomalous GVD region. This means that some soliton effect

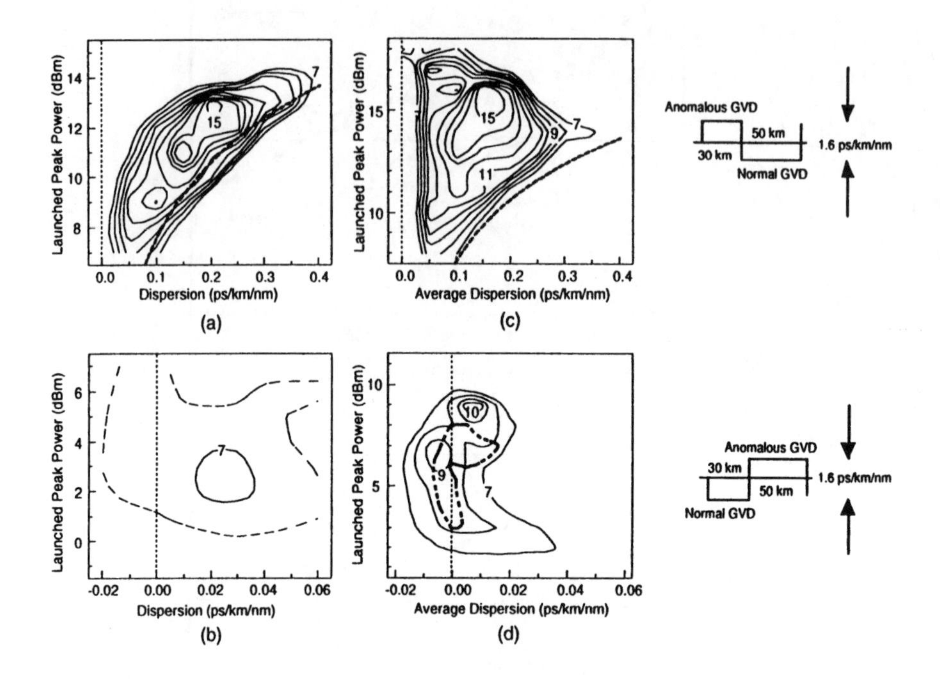

Figure 5. Q contour maps for various optical transmission systems. (a) Soliton with uniform GVD, (b) NRZ with uniform GVD, (c) D-A soliton and (d) D-A NRZ.

improves the transmission quality even for NRZ pulses. These results imply that a soliton transmission with uniform GVD is of better quality than an NRZ transmission at exactly zero GVD.

Figure 5(c) and (d) show how the contours of the Q map can be improved through the use of dispersion allocation. The transmission distance is set at 5,120 km in Figure 5(c) and at 2,560 km in Figure 5(d). As seen in Figure 5(c), the D-A soliton can drastically extend the transmission distance to over 5,120 km which is much further than that of a soliton with uniform GVD. In addition the maximum Q can be obtained at a much higher transmission power which agrees well with the content of Figure 1. These improvement are realized because phase matching is unlikely to occur. The transmission distance of an NRZ pulse using dispersion allocation is also further than that in Figure 5(b), but it is limited to 3,200 km (dotted line with $Q = 7$), which is much shorter than that of the D-A soliton. As clearly seen in Figure 5(c), D-A soliton transmission has a much larger power margin and dispersion tolerance than with D-A RZ and NRZ pulse transmissions at zero GVD. The contours in Figure 5(c) also show that the dispersion tolerance is much larger than that of the NRZ system.

These results suggest that dispersion-allocated soliton transmission is a very powerful optical transmission technique. D-A solitons are useful not only to compensate for the GVD but also to eliminate other nonlinear effects which occur as a result of the phase mismatching caused by the large GVD in each fiber segment.

Figure 6 numerically compares the performance of three optical transmission systems (D-A soliton, NRZ and RZ transmission zero GVD). Two-step dispersion allocation is used in these simulations. Figure 6 (a) and (b) correspond to 10 and 20 Gbit/s systems, respectively. The horizontal axis ΔD_1, is the amount of dispersion allocation from the average GVD in the first fiber segment with anomalous GVD and the vertical axis is the transmission distance. The maximum transmission distance is defined as the point at which the Q factor becomes 7. For the 10 Gbit/s evaluation, we used a 60 km span consisting of a 50 km-long anomalous GVD fiber and a 10 km dispersion allocation fiber. The amplifier noise figure was 8 dB, PRBS was $2^7 - 1$ and the pulse width was 15 ps. The FWHM of the filter width was 3 nm for both the D-A soliton and NRZ at 10 Gbit/s, and the band width of the baseband filter was 6.5 GHz. For the 20 Gbit/s evaluation, we employed different parameters. Here one span was extended to 80 km. The first fiber with anomalous GVD was 30 km and the succeeding fiber with a normal GVD was 50 km. The pulse width was 10 ps and the baseband filter width was 13 GHz. The FWHM of the installed optical filter was 2 nm for the D-A soliton and 10 nm for the NRZ pulse.

Our intention was to derive a fundamental principle for two-step dispersion allocation even under different conditions. The various transmitted powers were chosen to obtain the maximum transmission distance for each dispersion allocation. As it is known that modulational instability degrades the quality of an NRZ transmission [13], the dispersion allocation was reversed, that is the pulse train was transmitted from a normal to an anomalous GVD fiber.

It is clear from Figure 6 that the D-A soliton has the longest transmission distance in both 10 and 20 Gbit/s systems. It is interesting to note that the transmission distance is the most sensitive to dispersion allocation in the soliton system. For example the maximum transmission is obtained at a dispersion allocation of approximately 4–6 ps/km·nm for 10 Gbit/s and 2–3 ps/km·nm for 20 Gbit/s. This means that there is an optimum allocation at which the average Kerr effect can be stably balanced with the average GVD. Even for RZ transmission at zero GVD (open square), the maximum distance depends on the dispersion allocation. This is because the soliton nature can be maintained in the anomalous GVD fiber although the overall soliton effect is weak since the average GVD over one span is zero. For NRZ transmission, the maximum transmission distance

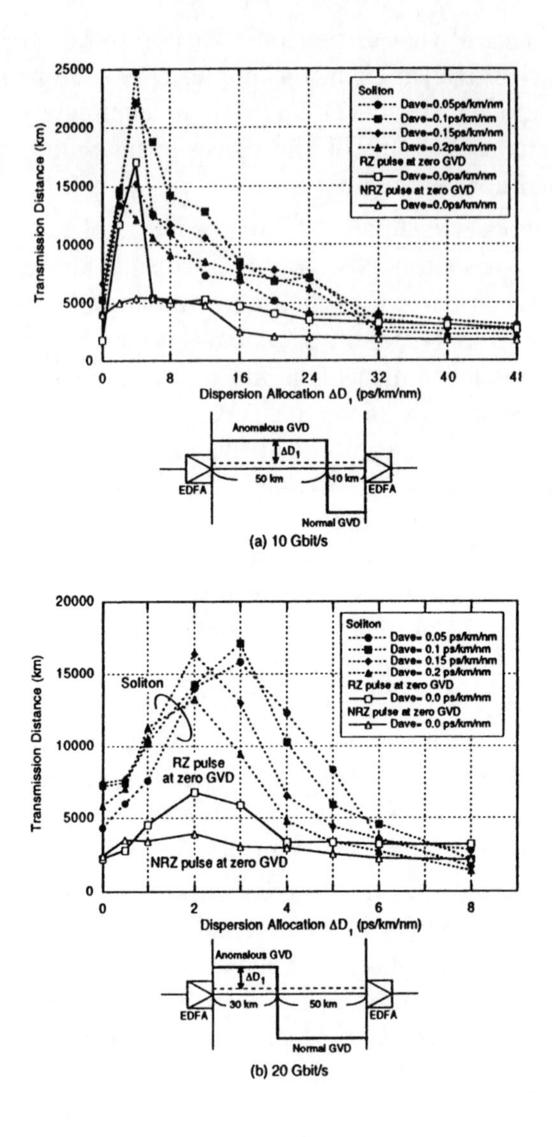

Figure 6. The maximum transmission distance vs. two-step dispersion allocation, ΔD_1 is the amount of dispersion allocation from the average GVD in the fiber segment. (a) 10 Gbit/s system, (b) 20 Gbit/s system.

is almost independent of dispersion allocation. This indicates that NRZ is most stable in the linear transmission region under relatively low transmitted power. From Figure 6 (a) and (b), the typical transmitted powers needed to obtain the maximum transmission distance for D-A soliton, RZ pulse and NRZ pulse transmission at 10 Gbit/s were $+6 \sim +8$ dBm, $+8$ dBm and $+2$ dBm, respectively and those at 20 Gbit/s were $+4 \sim +8$ dBm, $0 \sim +4$ dBm and $-4 \sim +5$ dBm, respectively. These results indicate that the D-A soliton has the largest transmitted power and suppresses four wave mixing with the aid of the dispersion allocation.

Finally we compare between the D-A soliton and NRZ transmission how the transmission distance is decreased when the average dispersion is changed. If the change in the transmission distance is very sensitive with respect to average dispersion, it maybe difficult to realize a commercial system. Hence the present analysis is very important. The result is shown in Figure 7 for 1.3 μm SMF D-A soliton. The bit rate is 10 Gbit/s. The conditions are the same as those of Figure 6. The launched power was optimized for each dispersion. The arrows in the figure indicate the transmission distance at a dispersion tolerance of 0.4 ps/km·nm. The distance is about 5,000 km for a D-A soliton pulse and 2,000 km for an NRZ signal. Here again, the transmission distance of the soliton signal is about 2.5 times greater than that of the NRZ signal against the change in the average GVD. These results indicate that a high quality with a larger margin is possible with the D-A soliton. It is also very interesting to note that even in the normal GVD region ($-0.5-0.0$ ps/km·nm), the transmission distance is still longer than the NRZ case. This indicates that a small amount of Kerr nonlinearity still works at the anomalous GVD fiber in the D-A transmission line to shorten the pulse due to the soliton effect.

6. Experimental Comparison of D-A Soliton, NRZ and RZ Pulse Transmission

We confirmed the results reported in the previous sections with transmission experiments using a fiber loop as shown in Figure 8. The experiments comprised a 10 Gbit/s data transmission using 1.3 μm single-mode fibers (SMF) and dispersion compensation fibers (DCF) [14]. The amplifier spacing (L_a) was 59 km, where a 50 km SMF was succeeded by a 9 km DCF, and the loop length was 472 km. The average GVD was 0.1 ps/km·nm. This condition corresponds to the filled squares on a dotted line in Figure 6(a). The pulse width was 16 ps and the corresponding spectral width was 0.21 nm. The time-bandwidth product was 0.42. The signal wavelength was 1.553 μm, and the data length was 2^7-1 PRBS. The fiber loop consisted of eight amplifier spans, with eight 0.98 μm pumped erbium-doped fiber amplifiers (EDFAs) and one 1.48 μm pumped EDFA. In Figure 6(a), the dispersion allocation with the 1.3 μm SMF corresponds to as much as 16 ps/km·nm in the 1.5 μm region, nevertheless this still contributes to extending the transmission distance. This is the reason why we used conventional 1.3 μm fiber. This technique is applicable to dispersion-shifted fibers (DSFs) with which much better results can be obtained as expected from Figure 6 because the dispersion allocation is much smaller than that for 1.3 μm SMF.

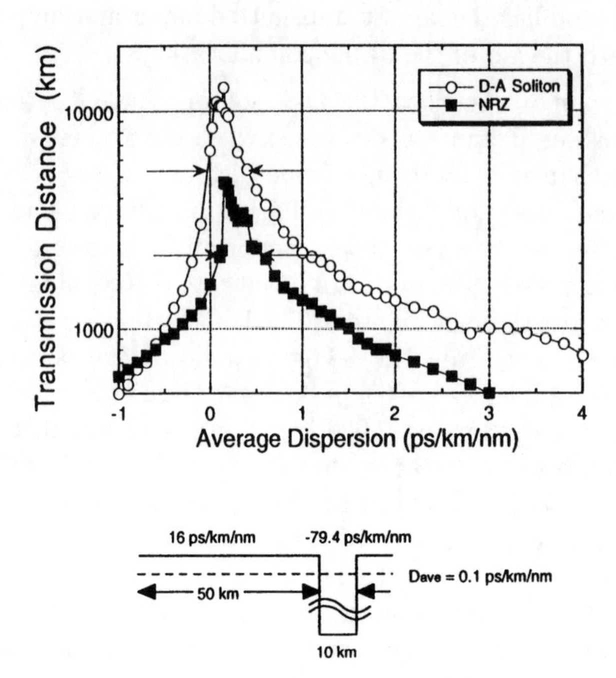

Figure 7. Dispersion tolerance of the dispersion-allocated soliton and NRZ systems.

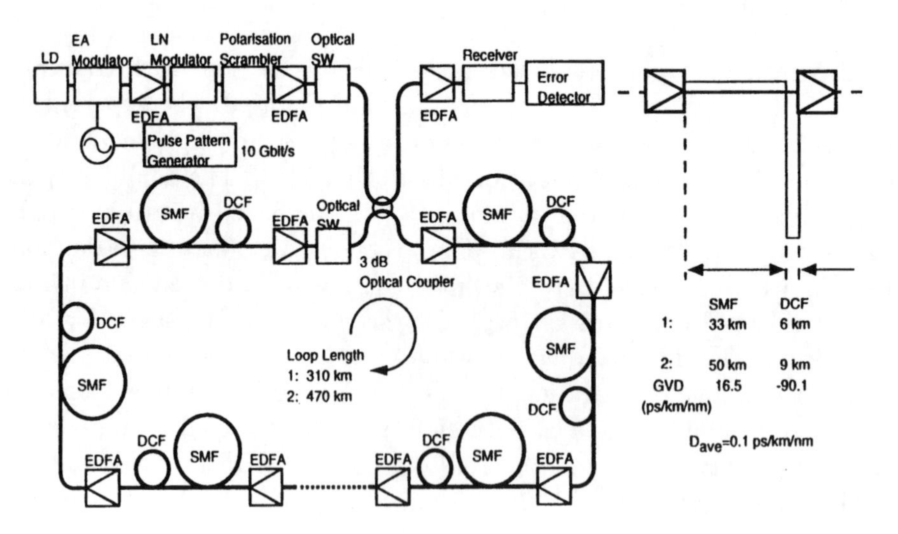

Figure 8. Experimental setup for system comparison using 1.3 μm single-mode fibers and dispersion compensation fiber.

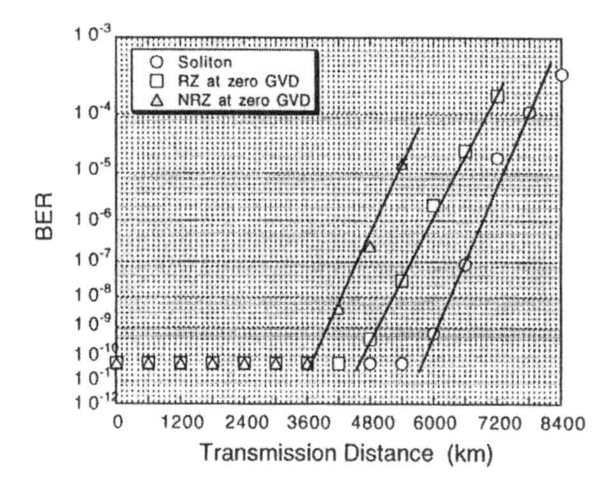

Figure 9. BER vs. transmission distance for the D-A soliton, RZ and NRZ pulses. O: D-A soliton, □: RZ pulse at zero GVD, △: NRZ pulse at zero GVD.

Figure 9 shows BER results for solitons and RZ and NRZ pulses at zero GVD. The transmission distances for the NRZ, RZ and D-A soliton systems which guarantee a BER of 1×10^{-10} were 3,800 km, 4,800 km and 6,000 km, respectively. As clearly seen, the longest transmission distance was achieved with the D-A soliton. This result agrees well with the numerical analysis (see at 16 ps/km·nm in Figure 6 (a)). In Figure 6 (a) the D-A soliton has the longest transmission distance of 8,000 km, next is the RZ pulse at zero GVD (5,000 km) and then the NRZ pulse (2,500 km), where the BER is set at 1×10^{-12} in all cases. These results agree well with the numerical results. The difference between the experiment and the numerical analysis may be due to polarization dependence of the gain and loss in the system and small parameter differences. It is interesting to note that the input pulses were immediately broadened to as much as 180 ps by the dispersion, nevertheless the results indicate that there is an accumulation of a small amount of Kerr nonlinearity which balances the average dispersion giving a value of as small as 0. 1 ps/km·nm.

We could further extend the maximum transmission distance for 1.3 μm D-A soliton transmission, by incorporating a slow speed polarization scrambler to eliminate the polarization hole burning effect in the EDFAs. BER curves before and after the 10,000 km transmission are shown in Figure 10. A BER below 1×10^{-9} was achieved with a received optical power of -27 dBm. The power penalty at that point was 5.2 dB. The inset photograph shows the eye diagram after a 10,000 km transmission,

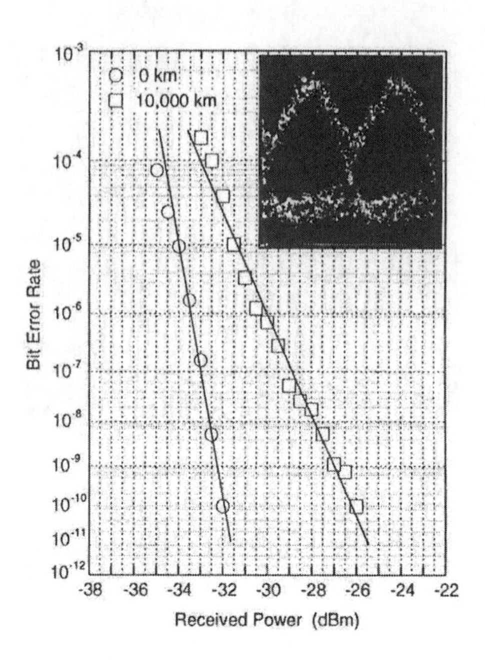

Figure 10. Bit error rate curves before and after 10,000 km transmission. ○: 0 km, □: 10,000 km.

which indicates a clear eye opening. Figure 11 shows the BER obtained as a function of the transmission distance. A bit error rate of 10^{-9} was achieved at 10,600 km. This result agrees with our numerical simulation which gives a maximum transmission distance of more than 11,000 km. Even when the amplifier spacing was extended to 59 km (50 km SMF and 9 km DCF), a BER below 1×10^{-9} was achieved at a transmission distance of 9,200 km.

7. 40 Gbit/s Single Channel D-A Soliton Transmission over 70,000 km Using Soliton Control

Advantage of D-A soliton transmission over a WDM system is that a capacity of a single channel becomes much larger than that of a WDM system. In this section, we describe a single channel 40 Gbit/s soliton transmission over 70,000 km with the use of in-line synchronous modulation [15, 16]. The experiment was performed in a 250 km dispersion-shifted fiber loop, as shown in Figure 12. The optical soliton source was a 10 GHz harmonically mode-locked erbium fiber laser operating at 1550.5 nm [17]. The pulse and spectral widths were 5 ps and 0.50 nm, respectively ($\Delta\nu\Delta\tau = 0.31$). The pulse was modulated at 10 Gbit/s with a $2^{11}-1$ PRBS using a Lithium

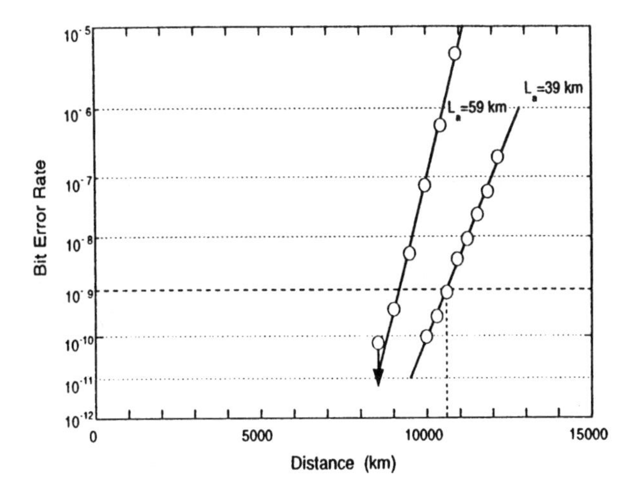

Figure 11. Bit error rate as a function of the transmission distance.

Niobate (LN) intensity modulator. A planar lightwave circuit (PLC) was used to optically multiplex the 10 Gbit/s signal into a 40 Gbit/s data train. The PLC multiplexer consisted of a two stage Mach-Zehnder interferometer with different arm lengths, which corresponded to time differences of 25 and 50 ps. In order to obtain a 10 GHz clock signal from the transmitted 40 Gbit/s signal, 10 GHz soliton units were superimposed on each other with slightly different amplitudes. This technique is also useful for reducing soliton-soliton i nteraction. The amplifier spacing was 50 km and the average fiber loss for the span was 12.5 dB. We used a four-segment dispersion-decreasing configuration to reduce the influence of the dispersive waves. The four 12.5-km long dispersion shifted fibers (DSFs) we used had GVDs of 0.24, 0.06, −0.04 and −0.10 ps/km·nm. We set the average GVD at approximately 0.04 ps/km·nm to maximize the transmission distance when soliton control was not employed. The average launched power was +7.0 dBm and the peak power into the first segment was 50 mW, which corresponded to an almost N = 1 soliton.

After a 250 km transmission through the loop, we applied in-line soliton control using synchronous modulation and narrowband optical filtering, in which the filter bandwidth was 0.80 nm. The 40 GHz clock signal was extracted from part of the transmitted soliton pulses with an ultrahigh-speed photodetector which had a 50 GHz bandwidth, and a high Q dielectric filter. The LN modulator for soliton control was driven by the extracted 40 GHz clock.

The transmitted 40 Gbit/s soliton data signal was demultiplexed into a

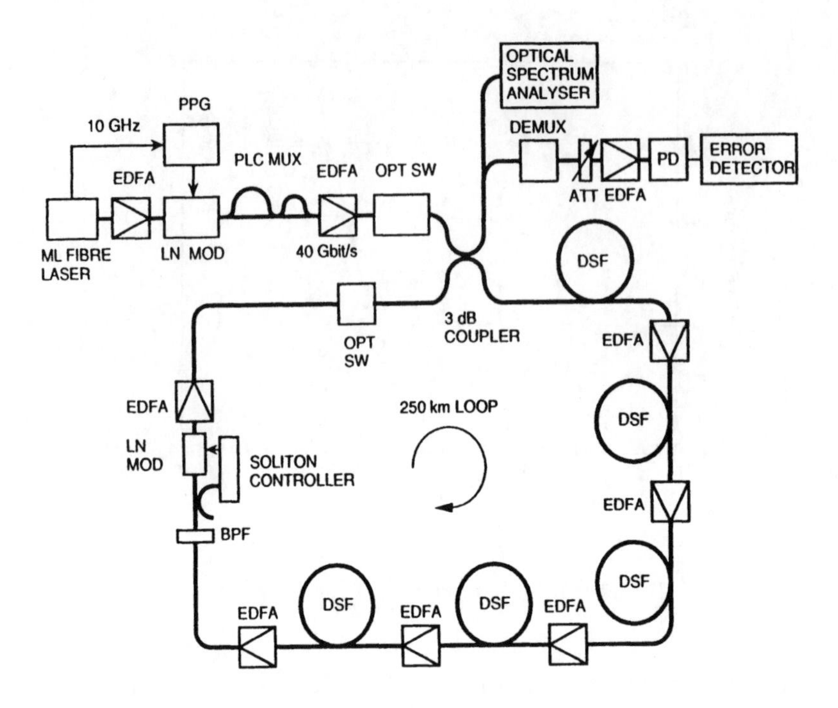

Figure 12. Experimental setup for 40 Gbit/s single channel soliton transmission using in-line synchronous modulation and optical filtering.

10 Gbit/s signal by using an electroabsorption (EA) intensity modulator. The 10 GHz clock signal was extracted from the transmitted signal, where a high S/N clock was obtained because the unequal amplitude soliton train retains the 10 GHz clock component. Then, by adjusting the DC bias voltage and the amplitude of the 10 GHz clock signal for the EA modulator, the gate width of the EA modulator was set at 20 ps to extract a 10 Gbit/s signal from the 40 Gbit/s signal. The demultiplexed signal was detected with a 10 Gbit/s optical receiver and BER was measured.

Figure 13(a) and (b) show the extracted 40 GHz clock signal and the optical spectrum after a 70,000 km transmission, respectively. As shown by Figure 13(a) the 40 GHz signal for the soliton control was extracted with a high S/N ratio. In Figure 13(b), small peaks in the broad spectrum show the 40 GHz spectral component. Figure 14 shows the measured BERs. The open circles indicate the BER with soliton control after a transmission of 70,000 km. The filled circles show the BER without soliton control after a transmission of 4,500 km, and the triangles show the BER at 0 km. When the soliton control was not applied, the maximum transmission distance was 4,500 km. The BER beyond 4,500 km increased rapidly with an increase in the transmission distance due to the accumulation of the noise component

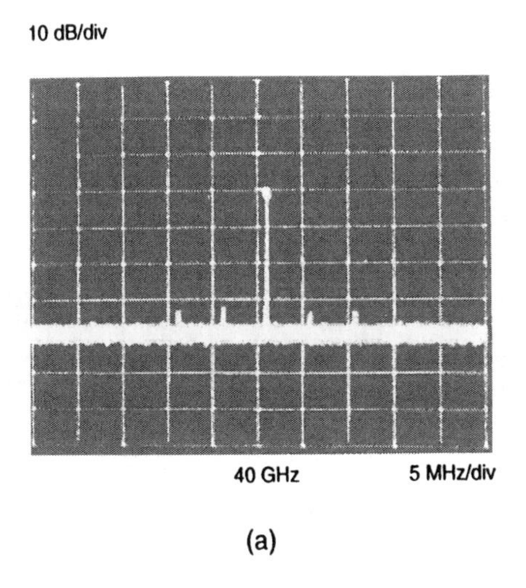

(a)

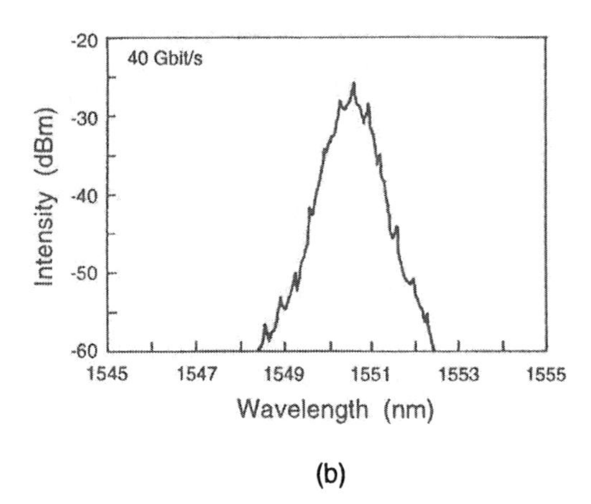

(b)

Figure 13. The 40 GHz extracted clock signal and optical spectrum after the 70,000 km transmission. (a) 40 GHz clock signal, (b) Optical spectrum.

and timing jitter. In contrast, when we employed the soliton control [15], the maximum transmission distance was greatly extended up to 70,000 km. When the received power was larger than −31.0 dBm, no error appeared at a 10^{-8}-order error counter setting, which indicates that the BER was less than 1×10^{-8}. The power penalty was 1.2 dB for the transmission with control and 2.9 dB without control. This shows that the degradation due to the accumulation of noise and timing jitter is larger than when soliton control is used. The inset photograph shows an eye pattern after a 70,000 km transmission, in which the eye is clearly open.

These results indicate the possibility of ultra-high capacity single chan-

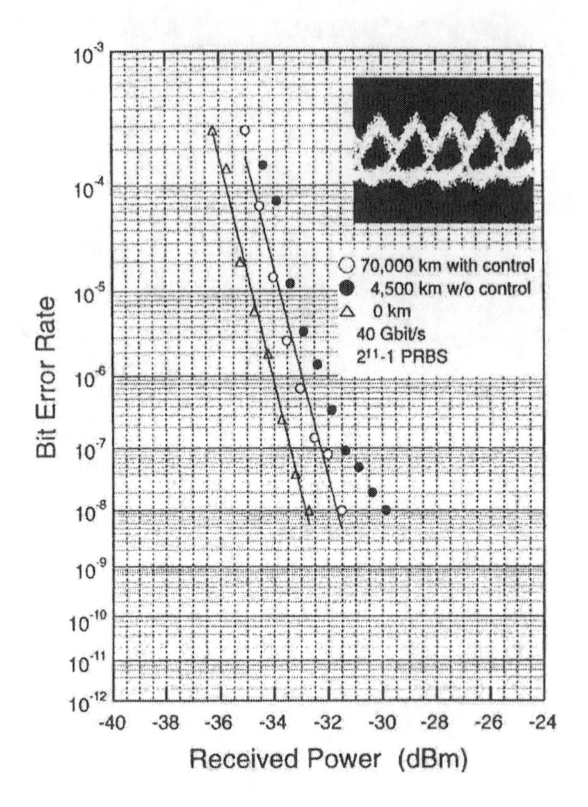

Figure 14. Bit error rate (BER) characteristics of 40 Gbit/s soliton data. ○: BER after 70,000 km transmission with soliton control, ●: BER after 4,500 km transmission without soliton control, △: BER at 0 km.

nel soliton transmission over several tens of thousands of km through the use of the in-line soliton control technique.

8. High Capacity WDM D-A Soliton Transmission

D-A soliton transmission is very advantageous not only for TDM but also for WDM in realizing a large-capacity long-distance system. The biggest advantage in WDM soliton transmission is that one channel can carry a much greater capacity than one NRZ channel, thus enabling us to reduce the number of channels. In addition, a combination of TDM and WDM becomes very flexible. It has been thought that WDM soliton is not so advantageous because of soliton collision between different wavelengths. However, dispersion allocation of the transmission line works very well to reduce the soliton collision effect and even 200 Gbit/s (20 Gbit/s$\times$10 channels) transmission over 1,000 km with a span of 100 km has been reported by using a dispersion-compensated 1.3 μm single-mode fiber [18]. There have been several reports of WDM soliton transmissions of 80 Gbit/s (10 Gbit/s$\times$8 channels) over 9,000 km using the sliding filter method [19] and 100 Gbit/s (20 Gbit/s$\times$5 channels) over 10,000 km with the synchronous modulation method [20]. In addition, a 40–60 Gbit/s (20 Gbit/s$\times$ 2–3 channels) transmission has also been demonstrated without soliton controls [21]. The dispersion-tuned synchronous modulation technique was also reported as an alternative to delay time control [22]. When the in-line modulation method is used, it is highly advantageous to increase the bit rate as it is similar to laser mode-locking at an ultrahigh repetition rate.

By contrast, NRZ transmission experiments with a total capacity of as high as 100–160 Gbit/s have been successfully reported in which the single channel capacity was 5 Gbit/s [23]-[25]. In this section, we first describe that in which condition the WDM D-A soliton is advantageous compared with the WDM NRZ transmission. Then, we show that it is possible to send 160 Gbit/s WDM soliton data (20 Gbit/s$\times$8 channels) over 10,000 km by using in-line soliton control.

Figure 15 shows a "Q map"for WDM NRZ and soliton systems. In the Q map for a WDM system, the Q value is represented by that of the worst channel so that the best performance is given by the worst channel. Figure 15 (a) and (b) correspond to uniform dispersion and large dispersion allocation cases, respectively. There are a total number of 10 channels and the bit rate is 10 Gbit/s (a pulse width of 25 ps). The channel separation is 1 nm. In Figure 15 (a), one span is an 80 km-long fiber with uniform GVD. In Figure 15 (b), the 80 km fiber has a GVD of 16 ps/km·nm for the soliton and the GVD of the 5 km DCF was varied according to the average GVD. With NRZ transmission, the 80 km fiber has a normal GVD of 16 ps/km·nm to reduce the modulational instability. A dispersion slope compensation (DSC) fiber is also installed every 320 km, where the third order dispersion is zero and the GVD is constant in each channel. In both figures, the WDM Q map for NRZ is shown on the l.h.s. and that for the soliton is shown on the r.h.s.

As shown in Figure 15 (a), WDM soliton transmission has characteristics comparable to those of WDM NRZ transmission. Although severe soliton collision occurs in a fiber with a uniform GVD, it can be reduced by adopting dispersion slope compensation. That is by making the soliton power in each channel the same. Without DCF, the soliton transmission is worse than NRZ transmission. However, as shown in Figure 15 (b), the situation is completely reversed. That is, the WDM D-A soliton transmission has much better Q value characteristics than the WDM NRZ transmission. This means that soliton collisions at different wavelengths are reduced due

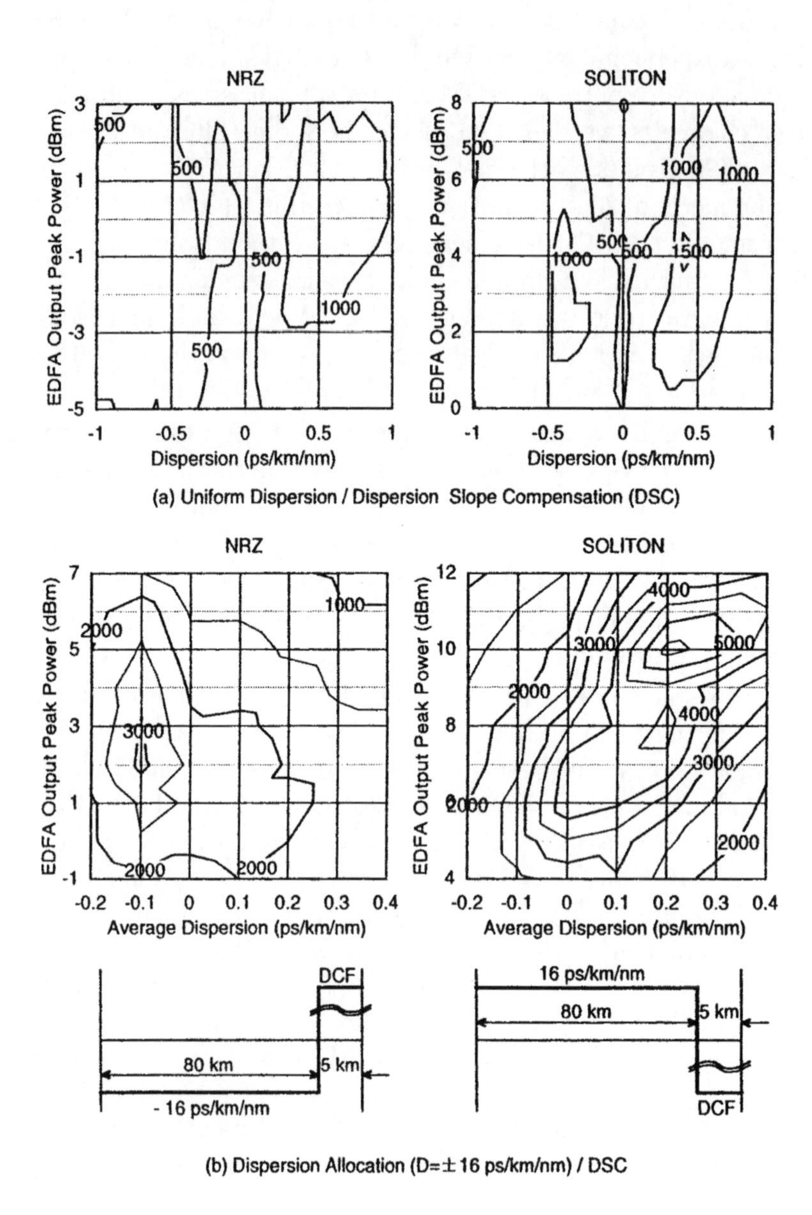

Figure 15. Q contour maps for evaluation of 10 Gbit/s×10 channel WDM systems. (a) Uniform dispersion, (b) Large dispersion allocation. The Q map for NRZ pulse is in l.h.s. and that for D-A soliton is in r.h.s.

to a process of pulse broadening in D-A soliton transmission. These results indicate that the D-A soliton is also applicable to WDM and provides a better result than the conventional NRZ WDM system.

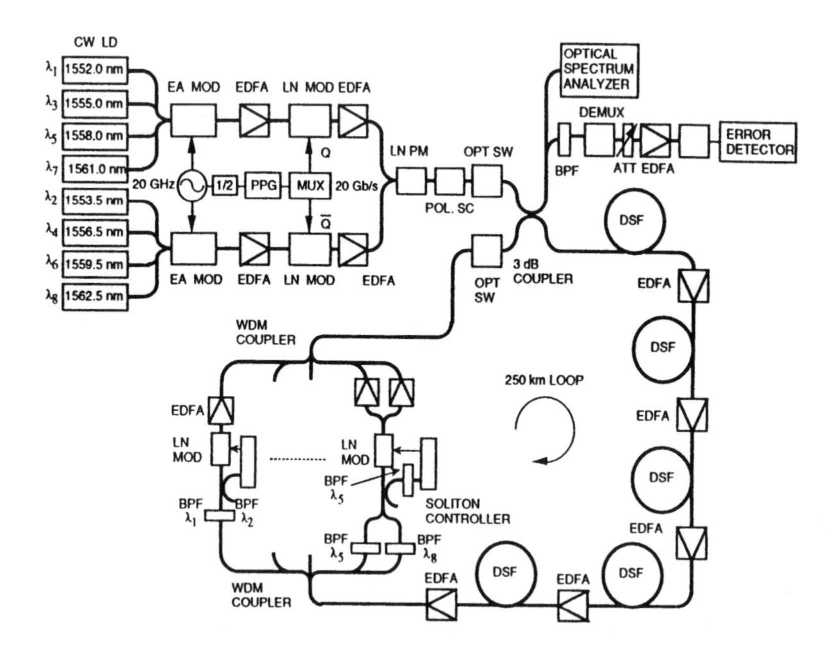

Figure 16. Experimental setup for a 160 Gbit/s (20Gbit/s×8 channels) WDM soliton transmission in a 250 km dispersion-shifted fiber loop.

We undertook a 160 Gbit/s WDM soliton transmission experiment in a 250 km dispersion-shifted fiber loop, as shown in Figure 16. The key to success to increase the system performance from 100 Gbit/s to 160 Gbit/s is adoption of a polarization scrambler and a phase modulator at the input [26]. The amplifier spacing was 50 km and the average fiber loss for one span was 12.5 dB. We used eight wavelength-stabilized cw LDs which were equally spaced at 1552.0 nm, 1553.5 nm, 1555.0 nm, 1556.5 nm, 1558.0 nm, 1559.5 nm, 1561.0 nm and 1562.5 nm as optical sources. A four-segment dispersion decreasing configuration was employed. The use of dispersion-decreasing configuration is a very powerful way of reducing collision effects and dispersive waves although it requires a rather more complicated selection procedure than when we use fibers with uniform dispersion [27]. Two polarization-insensitive electro-absorption modulators were used to convert the cw beams into 11–13 ps optical pulses at 20 GHz. One modulator was for signals at 1552.0 nm, 1555.0 nm, 1558.0 nm and 1561.0 nm (group 1) and the other for 1553.5 nm, 1556.5 nm, 1559.5 nm and 1562.5 nm (group 2). Two LN modulators were independently used for the data coding with a $2^{11}-1$ pseudorandom signal at 20 Gbit/s. Data signals for group 1 were modulated with a $\overline{Q}$ format and those for group 2 were modulated with a format in order to obtain uncorrelated neighboring channels. After amplifying each of the pulses to the corresponding soliton power level, all signals were polarization scrambled at low speed and synchronously phase-modulated at 20 GHz, and were introduced into the 250 km loop. The EDFAs were pumped by 1.48 μm LDs and had a relatively large bandwidth for WDM use. The total transmitted power was approximately +10 dBm.

After the 250 km transmission, the eight channel soliton signals were separated into their respective wavelengths with a WDM coupler. To keep the same GVD at each channels, a DCF was installed to each channel after a 250 km four-step dispersion-decreasing configuration. In this case, the average GVD for all channels was set to 0.12 ps/km·nm by connecting a different DCF to each channel. The DCFs we used have GVDs of 21.3 ps/nm, 47.0 ps/nm, 72.8 ps/nm, 98.8 ps/nm, 124.9 ps/nm, 151.1 ps/nm, 177.7 ps/nm and 204.5 ps/nm corresponding to signals at 1552.0 nm, 1553.5 nm, 1555.0 nm, 1556.5 nm, 1558.0 nm, 1559.5 nm, 1561.0 nm and 1562.5 nm, respectively. Since the average GVD is kept at 0.12 ps/km·nm in each channel, the average power for each channel was kept at the same value (+1.0 dBm).

Then we applied in-line soliton control using synchronous modulation and narrowband filtering [15]. For three modulators, each one controlled two channels and for other two modulators, each modulator controlled one channel. The bandwidth of the optical filter was approximately 0.35 – 0.45 nm. To obtain polarization insensitive synchronous modulation characteristics, two LN modulators were connected orthogonally. This modulator was driven by an extracted 20 GHz clock for each channel. Then while keeping the soliton power at an appropriate level, the pulses were multiplexed again through the WDM coupler and fed back to the input of the loop.

The eight clock signals for in-line modulation were cleanly extracted. When each channel was not well controlled, the wing of the 20 GHz clock signal had a broad and high pedestal which does not appear when the solitons are under control. This means that even in the presence of strong interactions between solitons with different wavelengths, stable solitons can be transmitted over long distances.

BERs after a 10,000 km transmission are shown in Figure 17. When the received power was larger than −28.0 dBm, the BER was less than 1×10^{-8}. The maximum power penalty difference was 1 dB. The inset photograph shows an eye pattern after a 10,000 km transmission at 1555.0 nm. When the polarization scrambler was installed the number of the channels can be increased and when the phase modulation is also applied the more stable transmission with a lower penalty was achieved.

Figure 18 shows how the BERs are degraded as the propagation dis-

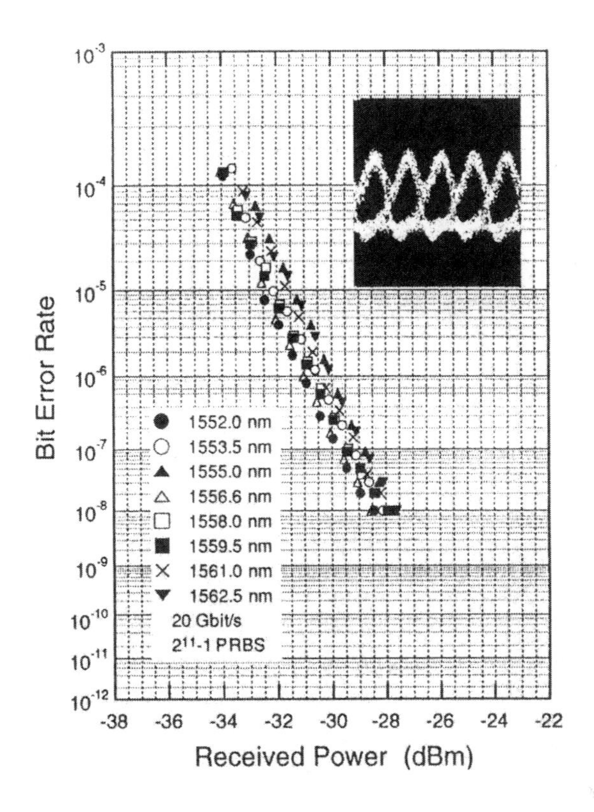

Figure 17. BERs after a 10,000 km transmission. The inset photographs show the eye patters after a 10,000 km transmission at 1555.0 nm.

tance is increased. The BERs started to be worse after a transmission of 12,000 km. This indicates that even when soliton transmission control is employed in each channel, it is not possible to send WDM soliton over unlimited distances. This would be attributed to the fact that there exist many collisions between different channels. However, it was possible to send 160 Gbit/s (20 Gbit/s×8 channels) WDM soliton at least 10,000 km. These results indicate that ultra high capacity WDM transmission over 10,000 km is also feasible through the use of soliton technology.

9. Summary

We have described recent progress on the dispersion-allocated soliton. The D-A soliton has a large power margin and dispersion tolerance compared with the conventional NRZ or RZ pulse transmission at zero GVD. This is because the dispersion allocation enables us to keep the transmitted pulse energy at a high level as the GVD in each fiber segment is much larger

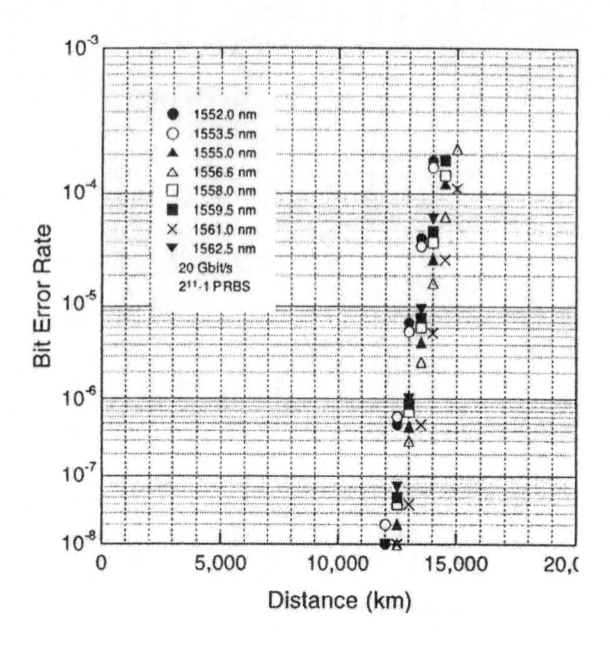

Figure 18. Change in the BERs as a function of the transmission distance.

than the average GVD. That is, pulse broadening occurs due to the large GVD, so that a larger energy is needed to maintain its intensity. In other words, this is a power enhancement factor. It is also important to note that simply increasing the power is not sufficient to improve the transmission characteristics. Because of the dispersion allocation, nonlinear efects such as a four wave mixing and modulation instability cannot evolve smoothly. Therefore, even when the input power is large, the signal-to-noise ratio can be maintained at a high level.

D-A soliton transmission using a conventional 1.3 μm SMF seems to be a very promising method as it can improve a capacity of 600 Mbit/s to at least 10 Gbit/s. There are two possible ways of realizing the potential of the D-A soliton transmission. One is TDM with soliton control, which enables us to send an ultrahigh speed signal channel transmission over 10,000 km. The other is WDM which also enables us to construct a high capacity system with high speed single channels. The difference between conventional WDM and D-A soliton WDM systems is the channel capacity of a signal channel. The D-A soliton can offer a much higher bit rate per channel.

Acknowledgment

The authors would like to express their thanks to Dr. I. Yamashita and Dr. I. Kobayashi for their fruitful comments and unceasing encouragement.

References

1. Nakazawa, M. and Kubota, H.: Optical soliton communication in a positively and negatively dispersion-allocated optical fiber transmission line, *Electron. Lett.*, **31**, (1995), pp.216-217.
2. Nakazawa, M. and Kubota, H.: Construction of a dispersion-allocated soliton transmission line using conventional dispersion-shifted nonsoliton fibers, *Jpn. J. Appl. Phys.*, **34**, (1995), pp.L681-683; see also Nakazawa, M. and Kubota, H.: Analyses of the dispersion-allocated bright and dark solitons, *Jpn., J. Appl. Phys.*, **34**, (1995), pp.L889-891.
3. Nakazawa, M., Kimura, Y., Suzuki, K., Kubota, H., Komukai, T., Yamada, E., Sugawa, T., Yoshida, E., Yamamoto, T., Imai, T., Sahara, A., Yamauchi, O. and Umezawa, M.: Soliton transmission at 20 Gb/s over 2,000 km in Tokyo metropolitan optical network, *Electron. Lett.*, **31**, (1995), pp.1478-1479.
4. Smith, N. J., Knox, F. M., Doran, N. J., Blow, K. J. and Bennion, I.: Enhanced power solitons in optical fibres with periodic dispersion management, *Electron. Lett.*, **32**, (1996), pp.54-55.
5. Nakazawa, M., Kubota, H., Sahara, A. and Tamura, K.: Marked increase in the power margin through the use of a dispersion-allocated solitons, *IEEE Photonics. Tech. Lett.*, **8**, (1996), pp.1088-1090.
6. Kubota, H. and Nakazawa, M.: Partial soliton communication systems, *Opt. Commun.*, **87**, (1991), pp.15-18.
7. Kelly, S. M. J.: Characteristic sideband instability of periodically amplified average soliton, *Electron. Lett.*, **28**, (1992), pp.806-807.
8. Tamura, K., Ippen, E. P. and Haus, H. A.: Pulse dynamics in stretched-pulse fiber lasers, *Appl. Phys. Lett.*, **67, 10**, (1995), pp.158-160.
9. Haus, H. A., Tamura, K., Nelson, L. E. and Ippen, E. P.: Stretched-pulse additive pulse mode-locking in fiber ring lasers: theory and experiment, *IEEE J. Quantum Electron.*, **31, 3**, (1995), pp.591-598.
10. Nakazawa, M., Suzuki, K., Kubota, H. and Haus, H. A.: High-order solitons and the modulational instability, *Phys. Rev. A.*, **39**, (1989), pp.5768-5776.
11. Bergano, N. S., Kerfoot, F. W. and Davidson, C. R.: Margin measurements in optical amplifier systems, *Photo. Tech. Lett.*, **5**, (1993), pp.304-306.
12. Sahara, A., Kubota, H. and Nakazawa, M.: Q-factor contour mapping for evaluation of optical transmission systems:soliton against NRZ against RZ pulses at zero group velocity dispersion, *Electron. Lett.*, **32**, (1996), pp.915-916.
13. Tkach, R. W., Gnauck, A. H., Chraplyvy, A. R., Derosier, R. M., Giles, C. R., Nyman, B. M., Ferguson, G. A., Sulhoff, J. W. and Zyskind, J. L.: One-third terabit transmission through 150 km of dispersion-managed fiber, *ECOC'94*, Firenze, Italy, (1994), pp.45-48.
14. Yamada, E., Kubota, H., Yamamoto, T., Sahara, A. and Nakazawa, M.: 10 Gbit/s, 10,600 km dispersion-allocated soliton tarnsmission using convnetional 1.3 μm singlemode fibres, *Electron. Lett.*, **33**, (1997), pp.602-603.
15. Nakazawa, M., Yamada, E., Kubota, H. and Suzuki, K.: 10 Gbit/s soliton transmission over one million kilometers, *Electron. Lett.*, **27**, (1991), pp.1270-1272.
16. Suzuki, K., Kubota, H., Sahara, A. and Nakazawa, M.: 40 Gbit/s single-channel optical soliton transmission over 70,000 km using in-line synchronous modulation and optical filtering, *Electron.lett.*, (1997), to be published.
17. Nakazawa, M., Yoshida, E. and Kimura, Y.: Ultrastable harmonically and regener-

atively modelocked polarisation-maintaining erbium fibre ring laser, *Electron. Lett.*, **30**, (1994), pp.1603-1604.
18. Le Guen, D. et al., 200 Gbit/s 100 km-span soliton WDM transmission over 1,000 km of standard fibre with dispersion compensation and prechirping, *OFC'97*, Dallas, U.S.A., (Feb. 1997).
19. Mollenauer, L. F., Mamyshev, P. V. and Neubelt, M. J.: Demonstration of soliton WDM transmission at up to 8×10 Gbit/s, error-free over transoceanic distances, *OFC'96*, San Jose, U.S.A., **PDP 22**, (Feb. 1996).
20. Nakazawa, M., Suzuki, K., Kubota, H., Sahara, A. and Yamada, E.: 100 Gbit/s WDM (20 Gbit/s×5 channels) soliton transmission over 10, 000 km, *OFC'97*, Dallas, U.S.A., **PDP 21**, (Feb. 1997) .
21. Suzuki, M., Morita, I., Edagawa, N., Yamamoto, S. and Akiba, S.: 20 Gbit/s-based soliton WDM transmission over transoceanic distances using periodic compensation of dispersion and its slope, *ECOC'96*, Oslo, **ThB3.4.**, (Sept. 1996).
22. Desurvire, E., Leclerc, O. and Audouin, O.: Synchronous in-line regeneration of wavelength-division multiplexed soliton signals in optical fibers, *Opt. Lett.*, **21, No. 4**, (1996), pp.1026-1028.
23. Bergano, N. S. et al., 100 Gbit/s error free transmission over 9,100 km using twenty 5 Gbit/s WDM data channels, *OFC'96*, San Jose, U.S.A., **PDP 23**, (Feb. 1996).
24. Taga, H., Takeda, N., Imai, K., Yamamoto, S. and Akiba, S.: 110 Gbit/s (22×5 Gbit/s), 9,500 km transmission experiment using 980 nm pump EDFA 1R repeater without forward error correction, *OA&A'96*, Monterey, CA, **PDP 5**, (July 1996).
25. Bergano, N. S. et al.: Long-haul WDM transmission using channel modulation: A 160 Gb/s (32×5 Gb/s) 9,300 km demonstration, *OFC'97*, Dallas, U.S.A., **PDP 16**, (Feb. 1997).
26. Morita, I. et al.: Performance improvement by initial phase modulation in 20 Gbit/s soliton-based RZ transmission with periodic dispersion compensation, *Electron. Lett.*, **33**, (1997), pp.1021-1022.
27. *Physics and Applications of Optical Solitons in Fibers '95*, Ed. Hasegawa, A., Kluwer Academic Publishers, **Chapt. 1-2**, (1996).

(Received January 13, 1998)

THEORY OF GUIDING-CENTER BREATHING SOLITON PROPAGATION IN OPTICAL COMMUNICATION SYSTEMS WITH STRONG DISPERSION MANAGEMENT

S. K. TURITSYN
Institut für Theoretische Physik I
Heinrich-Heine-Universität Düsseldorf
40225 Düsseldorf, Germany.

Abstract. It is presented a theory of a chirped breathing soliton propagation in the dispersion-managed fiber links. It is shown that a fast (over one period) dynamics of the pulse is governed by a system of two ordinary differential equations for the pulse width and chirp. Dispersion-managed pulses can propagate stably not only in the region of the anomalous average dispersion, but also with zero average path dispersion and even in the region of normal residual dispersion. Under some assumptions a slow (average) evolution of the central part of the peak can be approximated by the nonlinear Schrödinger equation (NLSE) with additional parabolic potential. If an effective parabolic potential is of a nontrapping type, the radiation is tunneling from the main peak. Asymptotic dispersion-managed (DM) soliton forming in such systems presents a central peak surrounded by oscillatory tails. It is demonstrated that using additional gratings after compensation cell it is possible to produce a pulse with strong confinement. Such pulse has all attractive features of the quasi-soliton suggested by Kumar and Hasegawa, but can be produced with simple dispersion management. Fast decaying tails and a rotation of the relative phase (due to chirp) between neighboring solitons reduce soliton interaction and allow for denser information packing.

1. Introduction

Recent fast developments toward multimedia service have stimulated investigations of the high- and ultrahigh-bit-rate optical communications. Both the traditional methods and new approaches to optical data transmis-

A. Hasegawa (ed.), New Trends in Optical Soliton Transmission Systems, 225–243.

sion are under intensive research due to growing demand for large capacity communication systems. Recent years have seen tremendous progress in soliton-based ultra-long high-bit-rate optical communication. In the experiment of L. Mollenauer *et al.* [1] it has been demonstrated a feasibility of 80 Gbit/s transmission over 9,000 km by using 8-channel 10 Gbit/s soliton WDM signals. Studies of dispersion-managed optical communication systems have demonstrated a great potential of the dispersion compensation technique both for ultralong transmission using dispersion-shifted fiber (DSF) and for upgrade of the installed terrestrial links mostly based on standard monomode fibers (SMFs). In the experiments of M. Nakazawa *et al.* [2] 60 Gbit/s has been transmitted 10,000 km using in-line synchronous modulation and optical filtering with three unequally spaced channels. In the experiment of M. Suzuki *et al.* 160 Gbit/s (8×20 Gbit/s) transmission over 4,000 km has been achieved using dispersion flattened fiber [3]. 320 Gbit/s soliton WDM transmission over 1,100 km of standard monomode fiber with large amplifier spacing of 100 km using dispersion compensating fiber (DCF) to equalize dispersion has been demonstrated in the work D. Le Guen *et al.* [4]. Soliton-based communication systems can carry higher bit rate per channel than NRZ format systems and thus, soliton WDM transmission is a promising way to realize future fiber communication links with ultra-large transmission capacity. This field continues to progress rapidly and the achieved results are expected to be improved continuously.

Transmission of a breathing soliton in optical fiber links with nonuniform dispersion has been put into the focus of intensive research recently [14], [2]-[39]. Large variations of dispersion (strong dispersion management) strictly modify the soliton propagation, inducing breathing-like oscillations of the pulse width during amplification period. This dynamics differs substantially from guiding-center soliton propagation. Nevertheless, numerical simulations and experiments have demonstrated that it is possible to reach extremely stable propagation of a breathing soliton in the fiber links with strong dispersion management. Pulse dynamics presents rapid oscillations of the power and width on the amplification distance, and slow evolution on the larger scales due to the fiber nonlinearity and residual dispersion [13]. Numerical simulations and experiments have revealed the following features of the dispersion-managed soliton:

- the pulse width experiences large oscillations during compensation period leading to "breathing-like"soliton dynamics;
- the shape of the forming asymptotic pulse (at least in the central part) is not always sech as for the NLSE soliton, but varies from sech to a Gaussian shape and sinclike profile;
- a forming pulse is chirped;

- energy of the stable breathing pulse is well above that of the NLSE soliton with the same pulse width and of the corresponding average dispersion.
- interaction between dispersion-managed solitons is suppressed in comparison with corresponding interaction between fundamental solitons.

The chirp is one of the most important feature of the dispersion-managed soliton [13, 20]. Soliton chirp leads to a fast rotation of the relative phase shift between neighboring soliton resulting into suppression of interaction. Important consequence of the chirp of the dispersion-managed soliton is that input signal launched into the transmission line should be either chirped [20, 4] or launched at some specific points of the dispersion map (for the case of transform-limited input pulse) [19, 36, 35]. As a matter of fact, the average balance can be provided even without anomalous residual dispersion [27, 28, 36] or even with normal average path dispersion [28]. These observations make clear a difference between soliton-like pulse in a system with dispersion compensation and the soliton of the NLSE. This indicates that an average model describing evolution of the breathing pulse should differ from the NLSE. In other words, the dispersion management imposes such a strong perturbation that a carrier pulse in this case is not more than NLSE soliton. Of course, a modification of the carrier pulse, for instance, by the transmission control elements like sliding filters is rather well-known. Indeed, mathematically, effect of the filtering can be considered as a formation of the autosoliton (solitary wave with parameters determined by the system parameters) with chirp from the input pulse in the form of the NLSE soliton. The energy of such autosoliton is enhanced in comparison with the energy of the corresponding fundamental soliton. This change of the features of the carrier pulse can be used for further improvement of the transmission capacities of the soliton-based communication systems. The dispersion-managed soliton is a new kind of the information carrier whose features differ significantly from that of the fundamental soliton. Due to the chirp of such pulse the interaction between two neighboring dispersion-managed solitons is suppressed as a result of a fast rotation of the relative phase. Channel bit rate of 20 Gbit/s in ultra-long transmission has been achieved using dispersion-managed soliton. Problem of realization of optical communication system with the ultra-large capacity around 100 Gbit/s stimulates investigation of new types of the information carrier with the tails decaying faster than exponential tails of the NLSE soliton [24, 33, 31, 38]. Using of such pulses for transmission would result in a substantial suppression of the soliton interaction and, consequently, in a possibility of a more dense information packing per channel.

One of the major factors limiting transmission capacity achievable by the modern optical soliton-based communication systems is the interaction between two neighboring solitons. Overlap of the exponential tails of the closely spaced pulses leads to the interaction of the solitons and the information loss. To provide for a stable transmission, a separation between two neighboring fundamental solitons should be not less than five soliton widths. This is a principal limitation for a transmission based on the soliton with sech shape described by the NLS equation. To overcome this limitation one can use information carrier different from the NLSE soliton. In general, soliton interaction can be suppressed, for instance, if a phase difference of the interacting pulses will rapidly vary with propagation or providing a carrier pulse with the tails decaying faster than exponential tails of the NLSE soliton. This would result in a suppression of the soliton interaction and, consequently, in a possibility to increase transmission rate. A new information carrier in communication systems with relatively complex dispersion management has been suggested recently in reference [31]. As it has been found in reference [31] a combination of programmed chirp and dispersion profiles allows to produce a quasi-soliton optical pulse with many attractive features. Such a pulse has Gaussian tails and as a result soliton interaction is significantly reduced. It has been shown in reference [31] that quasi-soliton (having Gaussian tails) practically does not interact with neighboring pulses if spacing is about three soliton widths. It has been predicted in reference [31] that single-channel transmission of 100 Gbit/s could be achievable with such carrier pulse. Relative disadvantage of this approach is a rather special dispersion profile that provides propagation of quasi-soliton.

In this paper it is presented analytic and numerical analysis of the average soliton propagation in periodically amplified system with dispersion management. A fast (over one period) dynamics of the pulse is governed by a system of two ordinary differential equations for the pulse width and chirp. This system of equations have been derived previously in our works by using variational approach [13, 21] (see also references [27, 34, 23, 39]). Variational approach is an effective tool that can describe with good enough accuracy dynamics of the central part of the dispersion-managed pulse. Variational approach is used to optimize and to design dispersion-managed transmission systems [36, 27, 34, 23, 39]. As it was shown in reference [33] the basic system of ODEs used in the variation approach can be derived directly from the basic model. Using developed method and direct numerical simulations it can be shown that the dispersion-managed pulse can propagate stably not only in the region of the anomalous average dispersion, but also with zero average path dispersion [27, 28, 36] and even in the region of normal residual dispersion [28].Under some assumptions (that are not so direct as in the guiding-center concept), in the leading order an average equation describing a central part of the breathing soliton propagation in the transmission systems with strong dispersion management is the NLS equation with additional quadratic potential [24, 33, 32]. In the case of transmission links with dispersion maps studied in references [16, 19] a carrier pulse presents a central peak with oscillatory tails that occur due to a tunneling part of pulse energy through potential barrier formed by the nonlinear term and effective parabolic potential of nontrapping type. As was shown in reference [32] for true periodic solutions quadratic potential is always nontrapping. However, using of the additional grating makes possible to form potential that is always trapping in the section as for quasi-soliton. As a particular limit of this general theory, for a specific dispersion profile suggested in reference [31] pulse (quasi-soliton) propagation is described exactly by the NLS equation with additional parabolic trapping potential. In the present work, we have linked quasi-soliton that has attractive properties, but requires relatively complex dispersion management with general case of the DM soliton. We show that quasi-soliton described in reference [31] presents a limiting case of the usual DM soliton and all remarkable properties of quasi-soliton can be obtained with simple dispersion management realized by gratings. Note also that solitons with the Gaussian tails occur in the systems with in-line phase modulation [37].

2. Basic Equations

Optical pulse propagation down the cascaded transmission system with varying dispersion is governed by

$$iA_z + d(z)A_{tt} + c_0|A|^2A = iL(-\gamma + r\sum_{k=1}^{N}\delta(z - z_k))A = i\,G(z)A. \qquad (1)$$

Here $d(z) = -\beta_2 L/(2t_0^2)$ and $c_0 = \sigma P_0 L$; β_2 is the first order group velocity dispersion, σ is the nonlinear coefficient; Z_k $(i = k, .., N)$ are the amplifier locations, amplification period is Z_a, and $r = [\exp(\gamma Z_a) - 1]$ is an amplification coefficient. We introduce dimensionless variables in the following way: the coordinate along the fiber z is normalized to the dispersion compensation length $z = Z/L$; time $t = T/t_0$ is normalized by the parameter t_0 proportional to the initial pulse width; an envelope of the electric field $E = E(T, Z)$ is scaled by the power parameter P_0: $|E|^2 = P_0|A|^2$. There are few distinctive lengths in the problem: $Z_{NL} = 1/(\sigma P_{in})$ is the characteristic nonlinear length (P_{in} is the input pulse power), $Z_{dis} = t_0^2/|\beta_2|$ — the local dispersion length corresponding to the transmission fiber. Amplification distance in general can be different from the compensation period. Chromatic dispersion $d(z) = \tilde{d}(z) + \langle d\rangle$ presents a sum of a rapidly (over one compensating period) varying high local dispersion and a constant residual dispersion ($\langle d\rangle \sim L/Z_{RD} \ll \tilde{d}$), here $Z_{RD} = t_0^2/|\beta_{2RD}|$ is a dispersion length corresponding to the average dispersion. Term strong dispersion management means that a variation of the dispersion on the compensation period is large. Consequently, not only pulse power, but also pulse width experiences substantial variations during the compensation period. Formally this can be formulated as a condition $R(z) = \int_0^z d(z')dz' \geq 1$ (moderate and strong management), in contrast to the weak management, when $R(z) = \int_0^z d(z')dz' \ll 1$.

3. Fast Dynamics of the Dispersion-managed Pulse

In this section we consider the fast dynamics of the dispersion-managed pulse. It is worthy to recall well-known exact solution of the linear problem (see, for instance, reference [13]). Neglecting nonlinear term in equation (2) fast oscillations of the linear pulse amplitude and width are given by

$$A(z,t) = \int_{-\infty}^{+\infty} d\omega A(\omega,0) \exp\left[i\omega t - i\frac{\omega^2}{2}\int_0^z d(\xi)d\xi\right] \exp\left(\int_0^z G(z')dz'\right) . \tag{2}$$

Here $A(\omega,0)$ is a Fourier transform of the initial pulse. For a Gaussian input signal $A(0,t) = N\exp(-t^2)$ linear oscillations are described by

$$A(z,t) = \frac{N}{\sqrt{\tau(z)}} \exp[-(1+iC)t^2/\tau^2(z) + i\Phi] \exp\left(\int_0^z G(z')dz'\right), \tag{3}$$

here $\tau^2(z) = 1 + 16R^2$, $dR/dz = d(z)$, $C = 4R$ and $\Phi = -0.5\arctan[4R]$. In the first approximation, on the distances of the order of the compensation period, the pulse dynamics in the system with strong dispersion management is given by this linear solution. If the dispersion d is exactly compensated the pulse recovers its power and width periodically. Important observations that can be made from equation (3) are that the dispersion-managed pulse is chirped and that it has self-similar structure. These features must be held (at least in the central, energy containing part) by the full theory that includes nonlinearity and residual dispersion. Evolution of the central part is the most important for the optimization of the system design. Note that self-similar structure of the breathing pulse allows to describe most of the observed features of the dispersion-managed soliton including power spectrum evolution over one period. It should be noted, however, that the oscillatory tails of the dispersion-managed pulse are likely not self-similar and this should be taken into account by a more precise theory.

It has been shown in reference [21] by numerical simulations of equation (2) that arbitrary input pulse typically evolves into an asymptotic structure that presents oscillating main peak and a dispersive pedestal. By a proper choice of the parameters of the input pulse this radiation can be significantly suppressed. However, oscillations around the main peak cannot be entirely eliminated for typical dispersion maps. Our first goal now is to describe the fast dynamics of the main peak. Following standard approach we make transformation $A(z,t) = B(z,t)\exp(\int_0^z G(z')dz')$.

Let us consider evolution of the following integral quantities [33] related to the pulse width

$$b(z) = \left[\frac{\int t^2|B|^2 dt}{\int |B|^2 dt}\right]^{1/2}, \tag{4}$$

and the pulse chirp

$$\mu(z) = \frac{i}{4}\frac{\int t(BB_t^* - B^*B_t)dt}{\int t^2|B|^2 dt}. \tag{5}$$

It is easy to check that the evolution of $b(z)$ and $\mu(z)$ is given by

$$b_z = 4d(z)\, b(z)\, \mu(z), \tag{6}$$

$$\mu_z + 8d(z)\mu^2 = \frac{4d(z)\int |B_t|^2 dt - c(z)\int |B|^4 dt}{4\int t^2|B|^2 dt}. \tag{7}$$

To express integrals $\int |B_t|^2 dt$ and $\int |B|^4 dt$ in terms of b and μ we should make additional assumptions about the structure of the solution. Let us make the following transformation of the function B

$$B(z,t) = \frac{Q[\frac{t}{T(z)}, z]}{\sqrt{T(z)}}\exp[i\frac{M(z)}{T(z)}t^2]. \tag{8}$$

It is easy to check that

$$b(z) = T(z)\left[\frac{\int x^2|Q(x,z)|^2 dx}{\int |Q(x,z)|^2 dx}\right]^{1/2}, \tag{9}$$

$$\mu(z) = \frac{M(z)}{T(z)} + \frac{i}{4T(z)^2}\frac{\int x(QQ_x^* - Q^*Q_x)dx}{\int x^2|Q|^2 dx}, \tag{10}$$

here $x = t/T(z)$. Closed system of equations for T and M (or for related quantities b and μ) can be obtained if we assume that a propagating pulse has a self-similar structure close to the solution of the linear problem presented above. Namely, let us assume that the function Q satisfies the requirement $\partial(\arg Q)/\partial x = 0$. In other words, we assume that in the transformation equation (8) the fast oscillations of the pulse chirp are almost accounted by the term proportional to $M(z)$. Under this assumption, straightforward calculations yield

$$T_z = 4d(z)\, M\,, \tag{11}$$

$$M_z = \frac{d(z)C_1}{T^3} - \frac{c(z)C_2}{T^2}\,. \tag{12}$$

Here

$$\begin{aligned} C_1(z) &= \int |Q_x(x,z)|^2 dx / (\int x^2 |Q(x,z)|^2 dx)\,, \\ C_2(z) &= \int |Q(x,z)|^4 dx / (4 \int x^2 |Q(x,z)|^2 dx)\,. \end{aligned}$$

If a solution is close to self-similar structure given by equation (3), in other words, if in the leading order one can neglect dependence of the function $|Q|$ on z (more precise we assume that $Q(x,z) = Q(x)\exp(i\lambda(z))$ for the central energy-bearing part of the asymptotic pulse), C_1 and C_2 become constants. For instance, for the soliton profile $f(x) = \mathrm{sech}(x)$, $C_1 = 2C_2 = 4/\pi^2$ and for the Gaussian shape $f(x) = \exp(-x^2)$, $C_1 = 4$, $C_2 = 1/\sqrt{2}$. It should be pointed out that even if the above assumption is not correct for the tails of the pulse, it still can be used for approximate description of the evolution of the central, energy bearing part of the dispersion-managed soliton.

Relations between integral and local characteristics of the pulse width and chirp are

$$\frac{b(z)}{b(0)} = \frac{T(z)}{T(0)}, \quad \mu(z) = \frac{M(z)}{T(z)}. \tag{13}$$

Of course, made above assumption must be justified by direct numerical simulations of the master equation (2). As it has been shown in reference [33], indeed, asymptotically the central part of the solution of equation (1) is described by a self-similar oscillating structure close to the solution given by equation (8). Numerical investigation of an asymptotic pulse formed in the system after input signal passes many compensation sections shows that $T(z)$ and $M(z)$ defined by equations (4), (5) and (13) vary in accordance with equations (11) and (12) with constant C_1, C_2 (also found numerically). Note also that equations (11) and (12) have been derived in reference [13] using variational approach, but here these equations describe exact transformation of the equation (2). To keep at the beginning of each section pulse width normalized as above, we use the following initial conditions: $T(0) = T_0 = 1$. As it has been shown in references [34, 39, 36] the considered method gives a reasonably good approximation of the energy enhancement. A general self-consistent procedure to describe dynamics and shape of dispersion-managed soliton is beyond the scope of this paper and will be published elsewhere.

Though the main results of the paper will be formulated in a general form and can be used for arbitrary dispersion map, in the illustrations without loss of generality we consider a symmetrical dispersion map studied in reference [16]. Piece of a dispersion-shifted fiber (DSF) with the anomalous dispersion $\beta_2^{(1)} < 0$ and length Z_c is followed by the DSF with the normal dispersion $\beta_2^{(2)} > 0$ and length $L - 2Z_c$ in the center and at the end of the section it is placed symmetrically the same fiber as at the beginning. Input pulse in the numerical simulations using the master equation (2) has a Gaussian form with T_{FWHM} =16.65 ps that corresponds to t_0 =10 ps and C_1 =1. Parameters of the system are the following: L = 400 km, Z_c = 100 km. Taking amplification period as Z_a = 40 km one can average over fast oscillations of the power and obtain lossless model considered in reference [16]. As was pointed out in reference [19] the inclusion of periodic amplification and dispersion compensation can be handled as separate problems, provided that amplification distance is substantially different from the period of dispersion map. Note that this is a typical situation for ultra-long transoceanic optical communication systems [34].

In Figure 1 it is compared fast dynamics (over one compensation cycle) obtained by direct numerical simulations of the partial differential equation (2) – crosses both for $T(z)$ and for $M(z)$ and by means of the solving of the ordinary differential equations (11) and (12) – solid lines. Dispersion map is shown above. The fast dynamics has been considered after pulse has passed many compensation sections. It has been checked that the fast dynamics for dispersion-managed soliton practically is not changed with evolution along the line. The same dimensionless picture can describe few practical dispersion-managed systems due to evident scaling.

Investigation of the fundamental equations (11) and (12) allows to reveal new very interesting features of the dispersion-managed pulses. As it has been shown recently in reference [27, 28] the stable dispersion-managed pulse can exist with zero average path dispersion (and even with normal average path dispersion [28]).

In Figure 2 it is shown predictions (made using equations (11) and (12)) on the dependence of the pulse width (taken at the boundary between fibers and in the chirp-free points of fiber pieces with anomalous and normal dispersion, respectively) of such pulses (compare with reference [28]). Here we have considered as an example (to make comparison with the numerical results of reference [28]) fibers with larger local dispersions than in the previous figure.

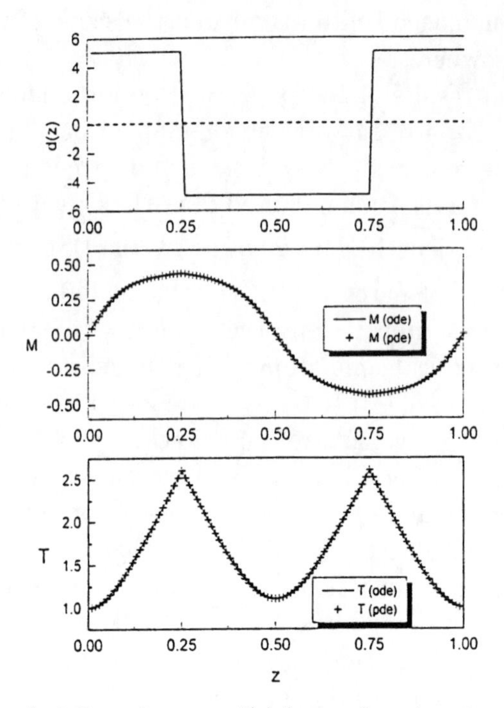

Figure 1. Symmetrical dispersion map $d(z)$ (upper figure) and typical periodic solutions $T(z)$ (lower figure) and $M(z)$ (middle) for this map (initial conditions are $T(0) = 1$, $M(0) = 0$). Both for $T(z)$ and $M(z)$: solid lines – solution of equations (11) and (12) (ODE), crosses –direct simulations of equation (2) (PDE). Here map parameters are: local dispersion $d = \pm 5(\beta_2^{(1,2)} = \pm 2.5\ \text{ps}^2/\text{km})$, average dispersion $\langle d \rangle = 0.05$, $C_1 = 1$, $C_2 = 1.03$.

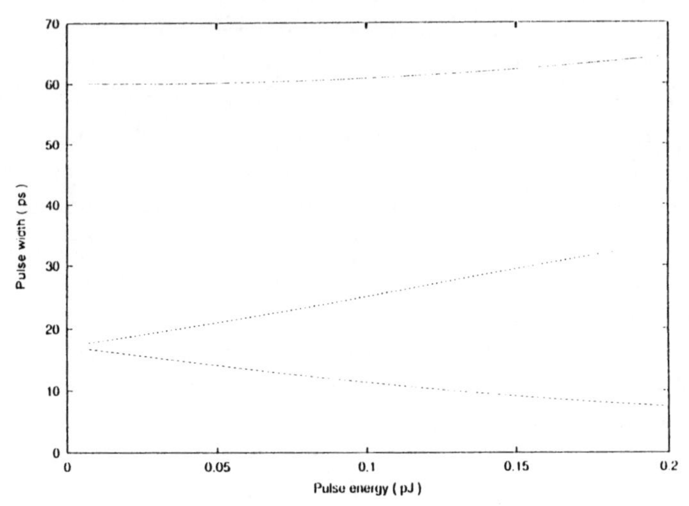

Figure 2. Pulse width at the boundary between fibers (solid line), and in the chirp-free points of the pieces of fibers with anomalous (dashed line) and normal (short-dashed line) dispersion as a function of the pulse energy. Parameters of the map are: local dispersion $d = \pm 1.785$, average dispersion $\langle d \rangle = 0$.

4. Slow (Average) Pulse Dynamics in the Systems with Strong Dispersion Management

In the guiding-center (average) soliton concept that has been developed in references [6, 5, 7] for cascaded transmission systems with constant dispersion or with weak variations of the dispersion, the average pulse dynamics is given in the leading order by the NLS equation. Using a powerful method of Lie transformation [6] it has been shown that higher-order corrections to NLSE appear in the second order of a small parameter $\epsilon = Z_a/Z_0$ (here Z_a is the amplification distance and Z_0 is a dispersion length). Large variation of the dispersion on the compensation period is the main obstacle to the direct application of the Lie-transform method [6] to obtain averaged (slow) dynamics in equation (2). The main technical idea of the approach suggested in reference [24] (see also recent paper [32]) is to use first a transformation that accounts for a fast pulse dynamics and to apply averaging procedure to the transformed equation.

Let us make the following exact transformation that is similar to the so-called "lens"transform first suggested by Talanov [40] in the theory of self-focusing

$$\begin{aligned} A(t,z) &= \frac{Q(\xi,z)}{\sqrt{\tau(z)}}\exp[i\frac{\nu(z)}{\tau(z)}t^2]\exp(\int_0^z G(z')dz') \\ &= \sqrt{\frac{c(z)}{\tau(z)}}Q(\xi,z)\exp[i\frac{\nu(z)}{\tau(z)}t^2]\,, \end{aligned} \tag{14}$$

here $\xi = t/\tau$ and functions $\tau(z)$ and $\nu(z)$ are periodic and will be specified later. Our goal now is to average equation (1) keeping general structure of the rapid oscillations given by equation (3). To specify $\tau(z)$ and $\nu(z)$ we use results of the previous section. Let us fix the first equation for τ, ν as

$$\tau_z = 4d(z)\,\nu, \tag{15}$$

equation (2) is then transformed by equation (14) to

$$iQ_z + \frac{d(z)}{\tau^2}Q_{\xi\xi} + \frac{c(z)}{\tau}|Q|^2Q - \tau\nu_z\xi^2Q = 0. \tag{16}$$

We fix an equation for ν_z using analysis above as

$$\nu_z = \frac{C_1 d(z)}{\tau^3} - \frac{C_2 c(z)}{\tau^2}. \tag{17}$$

Here $d(z)$ and $c(z)$ as defined above. Note that equations (15) and (17) are nothing more but equations (11) and (12) have first been derived in reference [13] using variational approach. However, here these equations describe the exact transformation of the equation (2).

In the primitive consideration of the evolution of the central (energy-bearing) part of the pulse under above assumptions one can average equation (16) assuming that a function Q varies slowly on the compensation distance and it can be placed outside the averaging integrals. Recall that we consider here only systems with strong dispersion management. After straightforward calculations, in the leading order, the averaged equation describing slow evolution of the chirped pulse due to nonlinearity and residual dispersion reads

$$i\frac{\partial U}{\partial z}+r_1\frac{\partial^2 U}{\partial \xi^2}-a\xi^2 U+r_2|U|^2U=0;\ r_2=\langle\frac{c}{T}\rangle;\ r_1=\langle\frac{d}{T^2}\rangle;\ \ a=(r_1C_1-r_2C_2). \tag{18}$$

Here $\langle f\rangle$ denotes averaging over one compensation period in z. Obviously, properties of steady-state solutions of equation (18) strictly depend on the sign of the parameter a. The positive a would provide well confined solutions with Gaussian tails in the main part, while the negative a corresponds to the solutions with oscillatory tails.

5. A Chirped Soliton with Strong Confinement

Now we link results obtained for dispersion-managed soliton in system with dispersion compensation and quasi-soliton found in reference [31]. We show that incorporation of grating to reverse periodically a pulse chirp allows to produce carrier pulse with strong confinement even without using rather complex dispersion profile considered in reference [31].

As it has been shown above the slow evolution of the central (energy-containing) part of the pulse propagating in the system with strong dispersion management is described under some assumptions by the NLSE with additional parabolic potential $\propto a\ t^2$ with a defined above. As it was proved in [32] a is always negative for periodic solutions of equations (11) and (12). This corresponds to the nontrapping effective potential and to the tunneling of the radiation from the central part of the dispersion-managed soliton. Note that even for quasi-soliton a defined as above is negative. However, in the section instant potential $\tilde{a}(z)$ is always positive (for quasi-soliton) and this provides strong confinement of the quasi-soliton and its superior characteristics related to soliton interaction. Though, for periodic solutions of equations (11) and (12) the parameter a is always negative, by simple dispersion management it is possible to make $\tilde{a}(z)$ to be always positive within the compensation section. This results into formation of a stable chirped pulse with strong confinement and suppressed interaction.

We point out that in our scheme attractive features of the quasi-soliton can be obtained for simple dispersion management in contrast to relatively complex dispersion maps considered in reference [31]. The field decay in the central region of the pulse where almost all energy is concentrated is faster than exponential. Far-field tails are yet exponential, but due to fast decay of the intensity in the central part, their amplitude is much less than the amplitude of the tails of the fundamental soliton with the same peak power (corresponding to the same local dispersion). As a matter of fact, we show also that quasi-soliton found in reference [31] is a particular limiting case of the DMS.

Lumped action of the amplifiers and gratings are accounted through pulse parameters transformations at junctions corresponding to locations of amplifiers and gratings. It is worth to recall that the both devices are linear. Point action of the grating can be described in the Fourier space [43] as $\Psi_{out}(t) = \int_{-\infty}^{\infty} \exp(-ig\omega^2 - i\omega t)\Psi(\omega)d\omega$, where $\Psi(\omega)$ is a Fourier image of the pulse before grating. The effect of grating can be easily understood for a chirped Gaussian pulse $\Psi_{in}(t) = A_{in}\exp[-t^2/(2T_{in}^2) + it^2 M_{in}/T_{in}]$. When the condition $g = -2M_{in}T_{in}^3/(1 + 4M_{in}^2 T_{in}^2)$, holds then $T_{out} = T_{in}$, $M_{out} = -M_{in}$, $|A_{out}| = |A_{in}|$ and, therefore, the grating is "tuned" to hold the pulse shape and chirp unchanged. In other words under certain conditions the effect of grating is equivalent to complex conjugation of the entering Gaussian pulse. Under these conditions the in-line grating transforms the parameters T and M as $T_{out} = T_{in}$ and $M_{out} = -M_{in}$. For a large chirp magnitude the similar formula can be derived for arbitrary pulse shape by means of the method of stationary phase. Though the main results of this work will be formulated in a general form and can be used for many dispersion maps, to make the idea the most clear, in the illustrations without loss of generality we use the simplest possible dispersion-managed system using chirped gratings (see e.g. references [44, 45, 46]). Line is built from DSF with anomalous dispersion $D = 1$ ps/nm·km; gratings are placed with the period $L = 100$ km.

Results of numerical simulations are presented in Figures 3-6. In Figure 3 it is compared evolution of the soliton chirp (above) and pulse width found by direct simulations (rombs and squares) and by solving equations (11) and (12) (solid lines). Coefficients C_1 and C_2 are determined through the pulse shape as described above.

Slow (average) dynamics is shown stroboscopically (at the end of the compensation cells) in Figure 4. Straight lines are for the chirped soliton and oscillations are for the Gaussian input signal with optimized parameters.

Figure 5 displays the shape of the chirped soliton (solid lines) compared with the sech-shaped solitons with the same peak power (short-dashed line) and the same pulse width (dashed line) corresponding to the local disper-

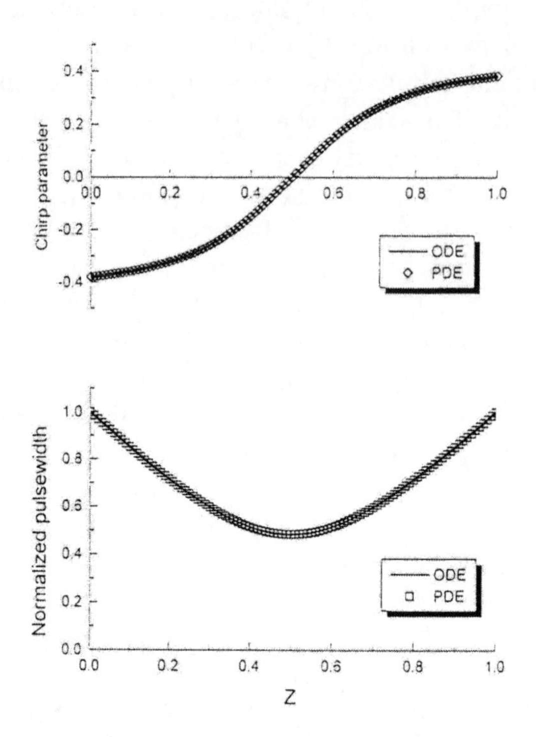

Figure 3. Comparison of the evolution of the soliton chirp (above) and pulse width in the system with in-line gratings found by direct simulations (rombs and squares) and by solving equations (11) and (12).

sion in the section. Right close-up picture shows corresponding shapes in the conventional scale. In the left inset it is shown that, as expected, found chirped pulse has enhanced power in comparison with the sech-soliton of the same width corresponding to the same path average dispersion. Interaction of chirped solitons is substanially reduced. For instance, solitons initially separated by 2.5 soliton width can propagate practically unlimited. Conventional solitons (corresponding to the same local dispersion) with the same pulse width and with the same input separation collapse after passing 10 sections.

Thus, we have demonstrated that utilization of the gratings at the end of a compensation cell allows to produce chirped carrier with strong confinement. Interaction of such solitons are suppressed similar to reduction of interaction of quasi-solitons found in reference [31] for special dispersion profile. Reduction of the interaction is both due to fast oscillations of the relative phase shift induced by the chirp and due to strong confinement of the pulse. The important issue is that in our scheme attractive features of the quasi-soliton can be obtained for simple dispersion management in

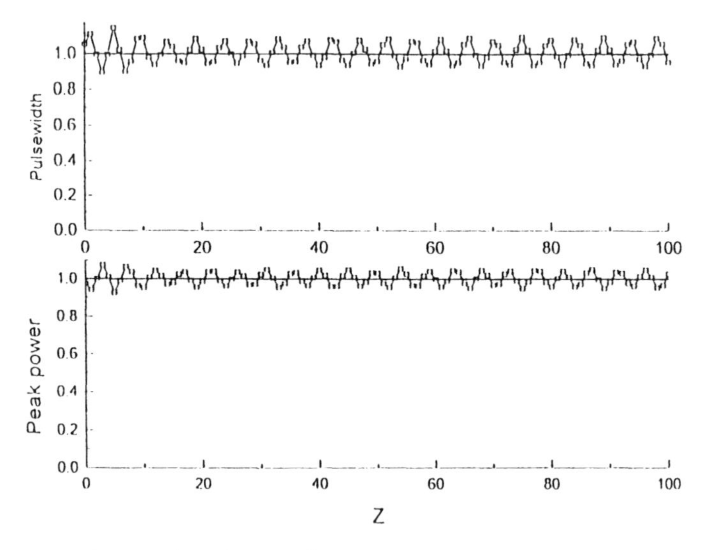

Figure 4. Slow (average) evolution of the dispersion-managed pulse in the system with gratings. Pulse peak power shown stroboscopically at the end of the compensation sections. Straight lines are for the chirped soliton and oscillations are for the Gaussian input signal with optimized parameters.

contrast to relatively complex dispersion maps considered in reference [31]. Such a pulse is an attractive candidate to be used as an information carrier in the future optical transmission systems with ultra-large capacity around 100 Gbit/s. We have shown that quasi-soliton with Gaussian tails found in reference [31] is a particular limiting case of the DM soliton.

6. Conclusions

It is studied the structure and dynamics of the chirped breathing soliton in the optical communication systems. It is shown that a "fast"(over one compensation period) dynamics of the dispersion-managed pulse is described by the system of two ordinary differential equations (11) and (12). This system of equation is capable to describe most of the features of the dispersion-managed soliton such as energy enhancement and pulse chirping. Our method predicts also that the dispersion-managed chirped breathing soliton can be formed not only in the region of anomalous dispersion, but also with zero and normal average path dispersion. An average (slow) dynamics of the central part of the chirped soliton in the transmission systems with strong dispersion management in the leading order is given (under some assumptions) by the NLS equation with additional quadratic potential. Coefficients of this equation are determined by the averaged characteristics of the periodic functions describing fast dynamics. We have found

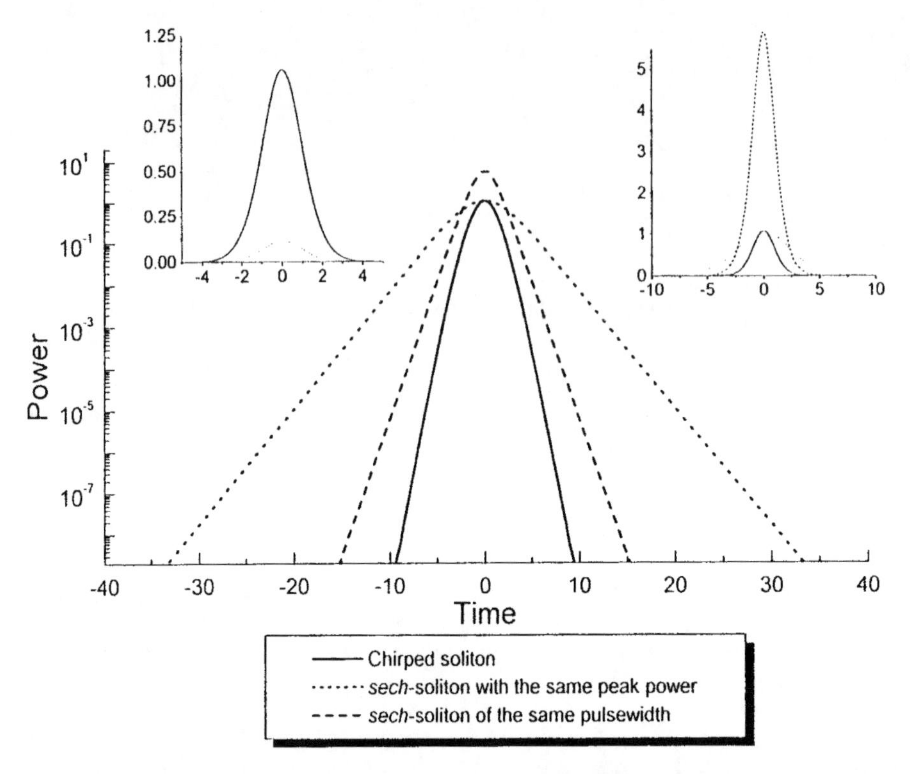

Figure 5. Shape of the asymptotic pulse (shown in logarithmic scale) forming in the system with gratings is compared with the sech solitons of the same peak power (short-dashed line) and of the same pulse width (dashed line). Right close-up picture shows these solitons in the conventional scale. In the left inset it is shown the power enhancement: it is plotted found chirped pulse and the sech-soliton with the same width corresponding to the same path-average dispersion.

that incorporation of a gratings for a periodical reverse of the pulse chirp makes possible to produce carrier pulses with strong confinement even without using rather complex dispersion profile considered in reference [31]. Dispersion-managed soliton propagating in such transmission system has all remarkable properties of quasi-soliton described in reference [31], but can be produced with more simple dispersion management. Strong confinement and the chirp of a obtained pulse results into substantial suppression of the soliton interaction and a possibility of a very dense information packing. Such pulse is an attractive candidate to be used as an information carrier in the future optical transmission systems with ultra-large capacity around 100 Gbit/s. As a particular limit of a general theory, for specific dispersion profile suggested by Kumar and Hasegawa, a pulse propagation is described exactly by the NLS equation with additional parabolic trapping potential. Analytical results are supported by direct numerical simulations.

Acknowledgements

I would like to thank V. Mezentsev and E. G. Turitsyna for their help and assistance. Valuable discussions with I. Gabitov, P. Mamyshev and A. Hasegawa are greatly appreciated.

References

1. Mollenauer, L. F., Mamyshev, P. V. and Neubelt, M. J.: Demonstration of soliton WDM transmission at up to 8×10Gbit/s, error-free over transoceanic distances, *OFC'96*, San Jose, Post Deadline presentation, **PD22-1**, (1996).
2. Nakazawa, M., Suzuki, K., Kubota, H. and Yamada, E.: 60G/bits WDM (20G/bits×3 unequally spaced channels) soliton transmission over 10000 km using in-line synchronous modulation and optical filtering, *Electron. Lett.*, **32**, (1996), p.1686.
3. Suzuki, M., Morita, I., Tanaka, K., Edagawa, N., Yamamoto, S. and Akiba, S.: 160 Gbit/s (8×20 Gbit/s) soliton WDM transmission experiments using dispersion flattened fibre and periodic dispersion compensation, *ECOC'97*, Edinburgh, **3**, (1997), p.99.
4. Le Guen, D., Favre, F., Moulinard, M. L., Henry, M., Devaux, F. and Georges, T.: 320 Gbit/s soliton WDM transmission over 1100 km with 100 km dispersion-compensated spans of standard fibre, *ECOC'97*, Edinburgh, Postdeadline paper, **5**, (1997), p.25.
5. Mollenauer, L. F., Evangelides Jr., S. G. and Haus, H. A.: Long-distance soliton propagation using lumped amplifiers and dispersion-shifted fiber, *IEEE J. Lightwave Tech.*, **9**,, (1994), pp.194-201.
6. Hasegawa, A. and Kodama,Y.: Guiding-center soliton in optical fibers, *Opt. Lett.*, **15**, (1990), pp.1443-1445 ; Guiding-center soliton, *Phys. Rev. Lett.*, **66**, (1991), pp.161-164.
7. Blow, K. J. and Doran, N. J.: Average soliton dynamics and the operation of soliton systems with lumped amplifiers,*IEEE Photon. Technol. Lett.*, **3**, (1991), pp.369-371 .
8. Lin, C., Kogelnik, H. and Cohen, L. G.: Optical-pulse equalization of low-dispersion transmission in single-mode fibers in the 1.3-1.7 μm spectral region,*Opt. Lett.*, **5**, (1980), pp.476-479.
9. Ellis, A. D. and Spirit, D. M.: Unrepeated transmission over 80 km standard fibre at 40 Gbit/s, *Electron. Lett.*,**30**, (1994), pp.72-74.
10. Breuer, D., Juergensen, K., Kueppers, F., Mattheus, A., Gabitov, I. and Turitsyn, S. K.: Optimal schemes for dispersion compensation of standard monomode fiber based links, *Opt. Commun.*, **140**, (1997), pp.15-18.
11. Mattheus, A. and Turitsyn, S. K.: Pulse interaction in nonlinear communication systems based on standard monomode fibres, *Proc. of European Conference on Optical Communications, ECOC'93*, Montreux, Switzerland, Regular papers **2**, MoC2.3, (September 12-16, 1993), pp.37-38.
12. Knox, F. M., Forysiak, W. and Doran, N. J.: 10 Gbit/s soliton communication systems over standard fibre at 1.55 μm and the use of dispersion compensation, *IEEE J. Lightwave Technol.*,**13**, (1995), pp.1955-1963.
13. Gabitov, I. and Turitsyn, S. K.: Averaged pulse dynamics in a cascaded transmission system with passive dispertion compensation, *Opt. Lett.*, **21**, (1996), pp.327-330 ; Breathing solitons in optical fiber links, *JETP Letters*, **63**, (1996), pp.861-865 ; Averaged Pulse Dynamics in the Cascaded Transmission Systems with a Passive Dispersion Compensation, Preprint Los Alamos, LAUR-95-3633, (October 13, 1995).

14. Suzuki, M., Morita, I., Edagawa, N., Yamamoto, S., Taga, H. and Akiba, S.: Reduction of Gordon-Haus timing jitter by periodic dispersion compensation in soliton transmission, *Electron. Lett.*, **31**, (1995), pp.2027-2028.
15. Nakazawa, M. and Kubota, H.: Construction of a dispersion-allocated soliton transmission line using conventional dispersion-shifted nonsoliton fibers, *Japan J. Appl. Phys.*, **34**, (1995), L681.
16. Smith, N., Knox, F. M., Doran, N. J., Blow, K. J. and Bennion, I.: Enhanced power solitons in optical fiber transmission line, *Electron. Lett.*, **32**, (1996), pp.54-55.
17. Nakazawa, M., Kubota, H., Sahara, A. and Tamura, K.: Marked increase in the power margin through the use of a dispertion-allocated soliton, *IEEE Photon. Technol. Lett.*, **8**, (1996), pp.452-454.
18. Nakazawa, M. and Kubota, H.: Optical soliton communication in a positively and negatively dispertion-allocated optical fiber transmission line, *Electron. Lett.*, **31**, (1995), pp.216-217.
19. Smith, N. J., Doran, N. J., Knox, F. M. and Forysiak, W.: Energy scaling characteristics of solitons in strongly dispertion-managed fibers, *Opt. Lett.*, **21**, (1997), pp.1981-1984.
20. Georges, T. and Charbonnier, B.: Reduction of the dispersive wave in periodically amplified links with initially chirped solitons, *IEEE Photon. Techn. Lett.*, **9**, (1997), pp.127-128.
21. Gabitov, I., Shapiro, E. G. and Turitsyn, S. K.: Optical pulse dynamics in fiber links with dispertion compensation,*Opt. Commun.*, **134**, (1996), pp.317-329 ; Asymptotic breathing pulse in optical transmission systems with dispertion compensation, *Phys. Rev. E*, **55**, (1997), pp.3624-3633.
22. Jacob, J. M.,Golovchenko, E. A., Pilipetskii, A. N., Carter, G. M. and Menyuk, C. R.: Experimental demonstration of soliton transmission over 28 Mm using mostly normal dispertion fiber, *IEEE Photon. Technol. Lett.*, **9**, (1997), pp.130-132.
23. Matsumoto, M. and Haus, H. A.: Stretched-Pulse Optical Fiber Communications, *IEEE Photon. Technol. Lett.*, **9**, (1997), pp.785-788.
24. Turitsyn, S. K.: Theory of averaged pulse propagation in high-bit-rate optical transmission systems with strong dispersion management, *JETP Letters*, **65**, (1997), pp.845-851.
25. Golovchenko, E. A., Pilipetskii, A. N. and Menyuk, C. R.: Dispersion- managed soliton interactions in optical fibers, *Opt. Lett.*, **22**, (1997), pp.793-795.
26. Yang, T. S. and Kath, W. L.: Analysis of enhanced-power solitons in dispersion-managed optical fibers, *Opt. Lett.*, **22**, (1997), pp.985-987.
27. Breuer, D., Kueppers, F., Mattheus, A., Shapiro, E. G., Gabitov, I. and Turitsyn, S. K.: Symmetrical dispersion compensation for standard monomode-fiber-based communication systems with large amplifying spacing, *Opt. Lett.*,**22**, (1997), pp.546-549.
28. Nijhof, J. H. B., Doran, N. J., Forysiak, W. and Knox, F. M.: Stable soliton-like propagation in dispersion managed systems with net anomalous, zero and normal dispersion, *Electron. Lett.*, **33**, (1997), pp.1726-1727.
29. Matsumoto, M.: Theory of stretched-pulse transmission in dispersion-managed fibers, *Opt. Lett.*, **22**, (1997), pp.1238-1240.
30. Sahara, A., Kubota, H. and Nakazawa, M.: Optimum Fiber Dispersion for Two-Step Dispersion-Allocated Optical Soliton, RZ at Zero GVD and NRZ Systems, *IEEE Photon. Techn. Lett.*, **9**, (1997), pp.1179-1181.
31. Kumar, S. and Hasegawa, A.: Quasi-soliton propagation in dispersion managed optical fibers, *Opt. Lett.*, **22**, (1997), pp.372-375.
32. Hasegawa, A., Kodama, Y. and Maruta, A.: Recent Progress in Dispersion-Managed Soliton Transmission Technologies, *Opt. Fiber Techn.*, **3**, (1997), pp.197-213.
33. Shapiro, E. and Turitsyn, S. K.: Theory of guiding-center breathing soliton propagation in optical communication systems with strong dispersion management, *Opt. Lett.*, **22**, (1997), pp.1544-1547.

34. Kutz, J. N., Holmes, P., Evangelides Jr., S. G., Gordon, J. P.: Hamiltonian dynamics of dispertion managed breathers, *JOSA B*, (1997), to be published.
35. Yang, T.S., Kath, W. L. and Turitsyn, S. K.: Optimal dispersion maps for wavelength-division-multiplexed soliton transmission, submitted to *Opt. Lett.*, (1997).
36. Turitsyn, S. K., Shapiro, E. G.: Variational approach to the design of optical communication systems with dispersion management, submitted to *Opt. Fiber Techn.*, (1997).
37. Mezentsev, V. K. and Turitsyn, S. K.: Soliton with Gaussian tails as carrier pulse in optical communication systems with in-line phase modulators, *Opt. Commun.*, (1997).
38. Mezentsev, V. K. and Turitsyn, S. K.: Dispersion-managed soliton with Gaussian tails in fibre links with in-line chirped fibre gratings, submitted to *Opt. Lett.*, (1997).
39. Shapiro, E. and Turitsyn, S. K.: Enhanced power breathing soliton in communication systems with dispersion management, *Phys. Rev. E*, **56**, (1997).
40. Talanov, V. I.: Focusing of light in cubic media, *JETP Letters*, **11**, (1970), pp.199-201; *Pis'ma Zh. Eksp. Teor. Fiz.*, **11**, (1970), pp.303-306.
41. Haus, H. A., Tamura, K., Nelson, L. E. and Ippen, E. P.: Stretched-pulse additive pulse mode-locking in fiber ring lasers: Theory and Experiment, *IEEE J. Quantum Electronics.*, **31**, (1995), pp.591-603.
42. Turitsyn, S. K.: Stability of an optical soliton with Gaussian tails, *Phys. Rev. E*, **56**, (1997), pp.3784-3788.
43. Kashyap, R.: Photosensitive Optical Fibers: Devices and Applications, *Opt. Fiber Techn.*, **1**, (1994), pp.17-43.
44. Kashyap, R., Chernikov, S. V., McKee, P. F. and Taylor, J. R.: 30 ps chromatic dispersion compensation of 400 fs pulses at 100 Gbit/s in optical fibers using an all photoinduced chirped reflection grating, *Electron. Lett.*, **30**, (1994), pp.1078-1080.
45. Kashyap, R., Ellis, A. D., Malyon, D., Froehlich, H. G., Swanton, A. and Armes, D. J.: Eight wavelength x 10 Gbit/s simultaneous dispersion compensation over 100 km singlemode fibre using a single 10 nm bandwidth, 1.3 metre long, super-step-chirped fibre Bragg grating with a continuous delay of 13.5 ns, *ECOC'96*, Oslo, Norway, Postdeadline paper **Th. 3.2**, (1996), pp.7-10.
46. Grudinin, A. B. and Goncharenko, I. A.: *Electron. Lett.*, **32**, (1996), p.1602.

(Received January 26, 1998)

ANALYSIS OF GORDON-HAUS JITTER IN DISPERSION COMPENSATED OPTICAL FIBERS

T. OKAMAWARI, A. MARUTA AND Y. KODAMA
Graduate School of Engineering
Osaka University
2-1 Yamadaoka, Suita, Osaka 565, Japan

Abstract. We theoretically analyze the Gordon-Haus timing jitter in an optical transmission system with a dispersion compensation. The enhanced power obtained in the system reduces the frequency shift induced by the noise, and a choice of an appropriate position of the receiver further reduces the timing jitter. Based on these results, we obtain the reduction of the Gordon-Haus timing jitter for the pulse in the dispersion compensated system. The reduction is more when compared with a soliton having equal pulse width and path-averaged dispersion.

1. Introduction

In a conventional soliton transmission system, an anomalous uniform dispersion optical fiber has been used for balancing the nonlinear effect with the dispersion effect everywhere in the transmission system. In this system, the Gordon-Haus jitter is one of the most effective factor to determine the transmission distance or the bit-rate [1]. To reduce the Gordon-Haus jitter, a technique with a periodic dispersion compensation has been developed [2]-[4]. Inserting the normal dispersion fiber (which is called the dispersion compensation fiber : DCF) into the anomalous dispersion fiber, we can reduce the path-averaged dispersion of the system. Therefore, the conversion from the frequency shift induced by the noise into the timing jitter is suppressed [3]. The observed pulse propagating in such a system has a Gaussian-like shape with the enhanced power [4]. This property of the pulse improves the signal-to-noise ratio at the receiver. Moreover, it has been numerically observed that there is an extra timing jitter reduction in such a system compared with a soliton system with a uniform anomalous

A. Hasegawa (ed.), New Trends in Optical Soliton Transmission Systems, 245–260.

dispersion fiber having equal pulse width and path-averaged dispersion [5]-[7]. Such properties of the pulse enable us to make a new transmission line with higher bit-rate and/or longer transmission distance than a conventional soliton transmission system. Another application of this technology is an upgrade of a pre-installed system. The pre-installed line will be economically reformed to be a high speed line by inserting the DCF. For the optimum design of these newly developed systems, it is required to evaluate the Gordon-Haus timing jitter. In this paper, we theoretically explain the reduction of the Gordon-Haus jitter in a periodically dispersion compensated system. According to the original theory given by Gordon and Haus [1], the noise added to the pulse at each amplifier induces the frequency shift of the pulse, and this frequency shift is converted into the timing jitter through the dispersion. Therefore, we have to examine the dynamics of the frequency and position of the pulse under the influence of the nonlinear interaction between the noise and the pulse. We use the variational method, in which a Gaussian pulse with a linear chirp is used as a trial function.

This paper is organized as follows. Section 2 gives a basic property of the pulse in a dispersion compensated transmission line. We consider an ideal transmission line having periodic change of the dispersion. Then, we numerically analyze the propagation property of the pulse in the dispersion compensated system. In Section 3, based on the result of the previous section, we apply the variational method to the system and derive the equations of the motion for the pulse parameters describing its amplitude, width, chirp, frequency, center position and phase. We also discuss the validity of the variational method. Section 4 gives a theoretical explanation of the Gordon-Haus jitter in the dispersion compensated system. By using the variational method, we derive the formulas which give both the frequency shift and the conversion to the timing jitter. In Section 5, we consider an upgrade of a pre-installed system by using the dispersion compensation. Applying the formulas given in Section 4, we analyze the Gordon-Haus jitter in the system. Section 6 is devoted to the conclusion.

2. Basic Property of the Pulse in the Dispersion Compensated System

In this section, we analyze the basic property of the pulse propagating in a dispersion compensated transmission line. Here, we consider the normalized dispersion porfile shown in Figure 1. Here, we take the local dispersions D_1 and D_2 are anomalous and normal respectively. Z_d is the period of the dispersion compensation. The normalized path-averaged dispersion assumed to be $\bar{D} = 1$. Here, we consider an ideal transmission line without both fiber loss and amplification. In this case, the model equation in the normalized

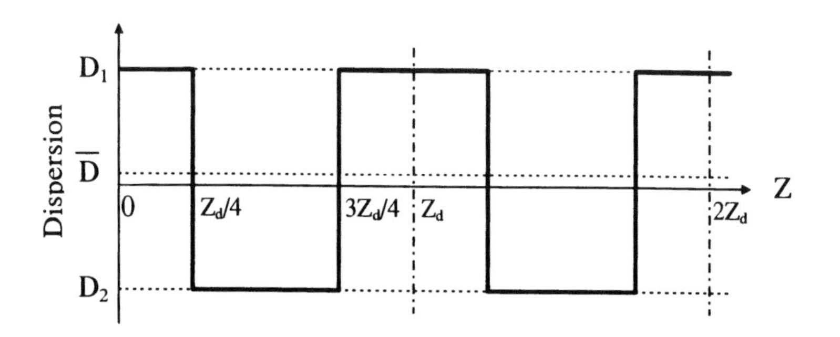

Figure 1. Dispersion profile

unit is given by

$$iu_Z + \frac{1}{2}D(Z)u_{TT} + |u|^2u = 0 \ . \tag{1}$$

The subscripts represent the partial derivatives. T, Z and u are normalized time, distance and complex envelope of the electric field, respectively. $D(Z)$ gives the dispersion profile normalized by the path-averaged dispersion $\bar{D}$.

The obserbed pulse in the dispersion compensated system has a Gaussian-like shape with the periodic change of the amplitude, pulse width and chirp [4, 8]. We then choose a Gaussian pulse having the same pulse width as soliton ($T_s = 1.763$) for the input. In this case, the initial waveform is given by $u(T,0) = A_0 \exp[-0.446T^2]$. At first, we analyze the relation between the peak power A_0^2 and the dipsersion difference $\Delta D \equiv D_1 - D_2$. For each ΔD, we can find optimum A_0 which makes the propagation of the pulse the most stable. Here, we calculate the difference between the maximum and the minimum value of the pulse width ΔT_s at each end of the dispersion compensation period ($Z = mZ_d$, for integer m) over 60 dispersion compensation period. We then assume that the smallest ΔT_s gives the optimum A_0. In Figure 2, we show the relation between A_0 and ΔT_s for various ΔD. Here, the dispersion compensation period is chosen to be $Z_d = 0.156$. One can see that there is an optimum A_0 for each ΔD, and the most stable propagation is obtained in the case of $\Delta D = 58$ and $A_0 = 1.52$. One can also see that the larger ΔD requires the larger A_0 [4]. In Figure 3, we show the pulse propagtaion over 60 dispersion compensation period for $\Delta D = 58$, where the sampling point is chosen to be $Z/Z_d = m$, for integer m. In Figure 4, we show the pulse propagation in a dispersion compensation period for Figure 3. One can see that the pulse is stably propagating over a long transmission distance. However, in the dispersion compensation period, the waveform of the pulse is oscillating with the period [4].

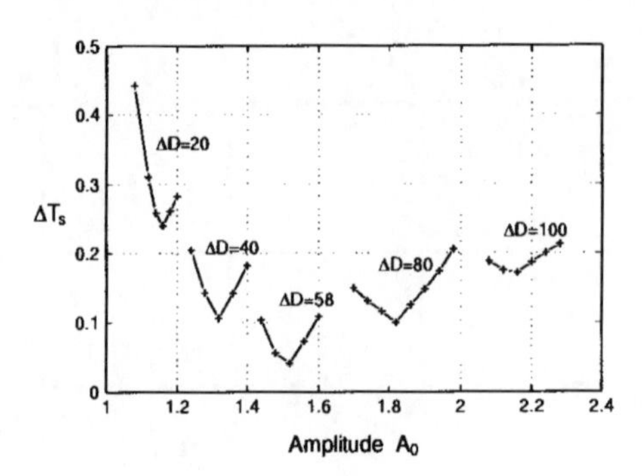

Figure 2. The stability of the pulse in the dispersion compensated system.

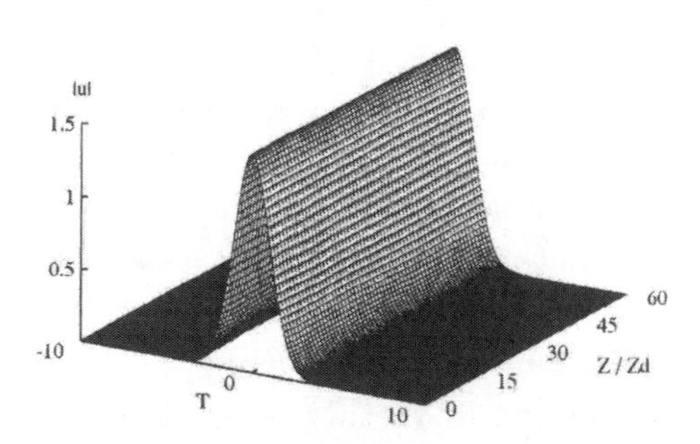

Figure 3. Pulse propagation in the dispersion compensated system (over 60 dispersion compensated period).

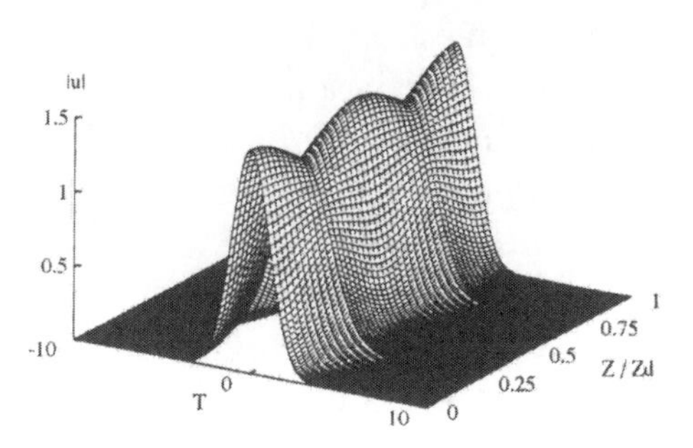

Figure 4. Pulse propagation in the dispersion compensated system (in a dispersion compensation period).

It has been reported that the pulse in the dispersion compensation system can be characterized by the amplitude, width and chirp [9]. We then write the solution for equation (1) in the form,

$$u(T, Z) = a(Z)v[Z, p(Z)T] \exp\left[\frac{i}{2}C(Z)T^2\right] . \tag{2}$$

Equation (2) represents the solution having the core v with moving its amplitude a, inverse of pulse width p, and chirp C periodically. Here, we do not specify the waveform of the core $|v|$ as a Gaussian, because the waveform of the core may depend on the dispersion profile. Substituting equation (2) into equation (1), we obtain,

$$iv_Z + \frac{1}{2}Dp^2 v_{\tau\tau} + a^2|v|^2 v - \frac{1}{2}K\tau^2 v = 0 , \tag{3}$$

$$\frac{da}{dZ} = -\frac{1}{2}DaC , \tag{4}$$

$$\frac{dp}{dZ} = -DpC , \tag{5}$$

where

$$K = \frac{1}{p^2}\left[\frac{dC}{dZ} + C^2 D\right] , \tag{6}$$

$$\tau = p(Z)T . \tag{7}$$

Using equations (4) and (5), we obtain the relation $a^2(Z) = \beta p(Z)$ with a constant β. Here, if dC/dZ were given, the motion of the parameters $a(Z)$, $p(Z)$ and $C(Z)$ are determined with the cannonical coordinate $p(Z)$ and $C(Z)$ [9]. If the motion were periodic one, we can obtain a closed orbit in a phase plane ($p - C$ plane). dC/dZ will be determined by the variational method given in Section 3. Here, we can obtain $p(Z)$ and $C(Z)$ by the direct numerical simulation of equation (1). In Figure 5, we show the phase diagram obtained by the numerical simulation for Figure 1 with $\Delta D = 40$, 58 and 80. In the case of $\Delta D = 58$, the orbit is almost closed in a period and the motion is almost a periodic one. However, in the cases of $\Delta D = 40$ or $\Delta D = 80$, the orbit is not closed in a period and the motion will be a quasi-periodic one. These results correspond to the result of Figure 2 and show that the Gaussian core is the most suitable for $\Delta D = 58$ in Figure 1.

3. Variational Method

As we have shown in the previous section, the pulse in the dispersion compensated system is almost stationary with an oscillation in its amplitude,

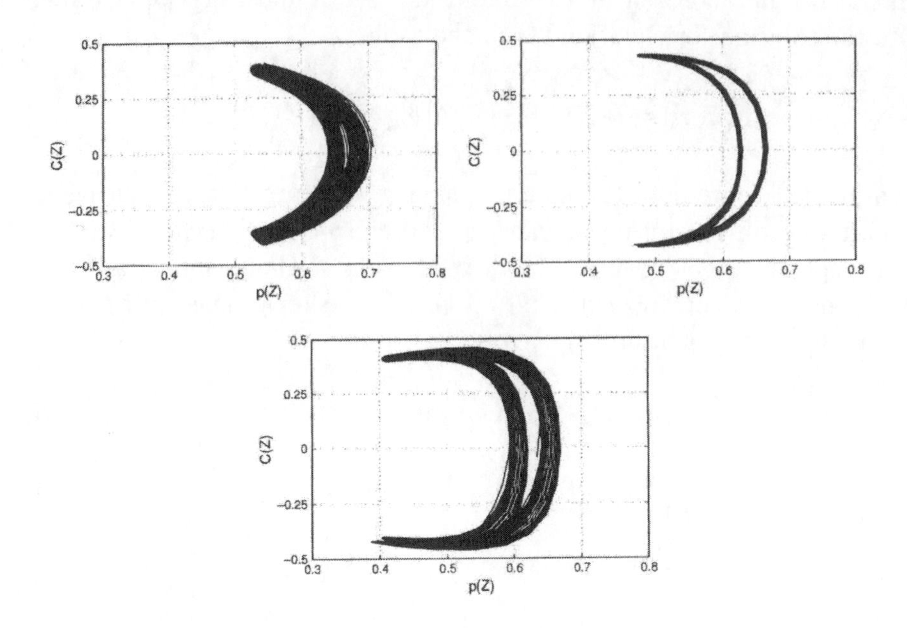

Figure 5. The orbit on the $p - C$ phase plane. $\Delta D = 40$(upper left), $\Delta D = 58$(upper right), $\Delta D = 80$(lower)

width and chirp. For this system, we apply the variational method. In this section, we give a brief summary of the variational method. The pulse propagation in a transmission line with the periodic dispersion compensation and the lumped amplification can be described by

$$iq_Z + \frac{1}{2}D(Z)q_{TT} + |q|^2 q = iG(Z)q + i\epsilon R \ , \tag{8}$$

in the normalized units. The term $G(Z) \equiv -\Gamma + G_A \sum_{m=1}^{N} \delta(Z - mZ_a)$ represents the normalized fiber loss (Γ) and the effect of a lumped amplifier ($G_A \equiv \Gamma Z_a$) which is periodically installed with the spacing Z_a. The term ϵR represents a perturbation with $\epsilon R \ll 1$. Using the transformation $q(Z,T) = \sqrt{A(Z)}u(Z,T)$, equation (8) can be written in

$$iu_Z + \frac{1}{2}D(Z)u_{TT} + A(Z)|u|^2 u = i\epsilon \frac{R}{\sqrt{A(Z)}} \ , \tag{9}$$

where $A(Z) \equiv A(0) \exp\left[2\int_0^Z G(Z')dZ'\right]$, and the averaged values of $A(Z)$ and $D(Z)$ are assumed to be both unity in the normalized unit. The Lagrangian for equation (9) with $\epsilon R = 0$ is given by

$$L_0 = \int_{-\infty}^{\infty} \left[-\frac{D(Z)}{2}|u_T|^2 + \frac{A(Z)}{2}|u|^4 + \frac{i}{2}(u_Z u^* - u_Z^* u) \right] dT \ , \qquad (10)$$

As was discussed in Section 2, the pulse in the dispersion compensated system oscillates in its amplitude, width and chirp with the compensation period. We then assume that such a solution of equation (9) takes the form,

$$u(Z,T) = a(Z) f(\tau) \exp[i\varphi(Z,T)] \ , \qquad (11)$$

$$\left\{ \begin{array}{lcl} \tau(Z,T) & = & p(Z)(T - T_0(Z)) \ , \\ \varphi(Z,T) & = & \frac{C(Z)}{2}(T - T_0(Z))^2 + \kappa(Z)(T - T_0(Z)) + E(Z) \ . \end{array} \right. \qquad (12)$$

The parameters $a(Z)$, $p(Z)$, $C(Z)$, $\kappa(Z)$, $T_0(Z)$ and $E(Z)$ represent the amplitude, inverse of width, chirp, frequency, center position and phase of the pulse respectively. Note here that equation (11) corresponds to equation (2) with $T_0(Z) = 0$, $\kappa(Z) = 0$ and $v(Z,\tau) = f(\tau)\exp[iE(Z)]$. Substituting equation (11) into equation (10) and applying the variational method [10, 11], the equations of the motion for the parameters are given by,

$$\frac{da}{dZ} = -\frac{1}{2}DaC + \frac{\epsilon}{2\sqrt{A}} \int_{-\infty}^{\infty} \mathrm{Re}[Re^{-i\varphi}] \left(\frac{3}{I_L} - \frac{\tau^2}{I_C} \right) f d\tau \ , \qquad (13)$$

$$\frac{dp}{dZ} = -DpC + \frac{\epsilon p}{\sqrt{A}a} \int_{-\infty}^{\infty} \mathrm{Re}[Re^{-i\varphi}] \left(\frac{1}{I_L} - \frac{\tau^2}{I_C} \right) f d\tau \ , \qquad (14)$$

$$\frac{dC}{dZ} = D\left(\frac{I_D}{I_C}p^4 - C^2 \right) - \frac{I_N A a^2 p^2}{2 I_C} - \frac{\epsilon p^2}{I_C \sqrt{A} a} \int_{-\infty}^{\infty} \mathrm{Im}[Re^{-i\varphi}](f + 2\tau f_\tau) d\tau \ , \qquad (15)$$

$$\frac{d\kappa}{dZ} = \frac{2\epsilon}{I_L \sqrt{A} a p} \int_{-\infty}^{\infty} \{ C\mathrm{Re}[Re^{-i\varphi}]\tau f - p^2 \mathrm{Im}[Re^{-i\varphi}] f_\tau \} d\tau \ , \qquad (16)$$

$$\frac{dT_0}{dZ} = D\kappa + \frac{2\epsilon}{I_L \sqrt{A} a p} \int_{-\infty}^{\infty} \mathrm{Re}[Re^{-i\varphi}]\tau f d\tau \ , \qquad (17)$$

$$\frac{dE}{dZ} = D\left(\frac{\kappa^2}{2} - \frac{I_D}{I_L}p^2 \right) + \frac{5 I_N A a^2}{4 I_L} + \frac{\epsilon}{2 I_L \sqrt{A} a p} \int_{-\infty}^{\infty} \{ p\mathrm{Im}[Re^{-i\varphi}](3f + 2\tau f_\tau) + 4\kappa \mathrm{Re}[Re^{-i\varphi}]\tau f \} d\tau \ . \qquad (18)$$

where I_L, I_D, I_C, and I_N are constants defined by

$$I_L = \int_{-\infty}^{\infty} f^2 d\tau, \ I_D = \int_{-\infty}^{\infty} (f_\tau)^2 d\tau, \ I_C = \int_{-\infty}^{\infty} (\tau f)^2 d\tau, \ I_N = \int_{-\infty}^{\infty} f^4 d\tau.$$

Using equations (13) to (18), we can describe the pulse dynamics in a dispersion compensated line.

To analyze the basic property of the pulse in a dispersion compensated system, we assume unperturbed ($\epsilon R = 0$) and lossless ($A(Z) = 1$) case in equation (9), which corresponds to equation (1). In this case with initial condition $\kappa(0) = 0$, $T_0(0) = 0$, equations (13)~(18) are simplifyed as follows,

$$\frac{da}{dZ} = -\frac{1}{2}DaC \,, \tag{19}$$

$$\frac{dp}{dZ} = -DpC \,, \tag{20}$$

$$\frac{dC}{dZ} = D\left(\frac{I_D}{I_C}p^4 - C^2\right) - \frac{I_N a^2 p^2}{2I_C} \,, \tag{21}$$

$$\frac{dE}{dZ} = -D\frac{I_D}{I_L}p^2 + \frac{5I_N a^2}{4I_L} \,. \tag{22}$$

Note here that equations (19) and (20) are the same as equations (4) and (5). Substituting $a^2 = \beta p$ into equation (21), we obtain

$$\frac{dC}{dZ} = D\left(\frac{I_D}{I_C}p^4 - C^2\right) - \frac{I_N}{2I_C}\beta p^3 \,. \tag{23}$$

As was discussed in Section 2, the pulse propagation in the dispersion compensated system is determined by equations (20) and (23) with the cannonical coordinates $p(Z)$ and $C(Z)$. We solve equations (20) and (23) numerically for Figure 1 with $\Delta D = 58$ and Gaussian core ($f(\tau) = e^{-\tau^2}$), and compare with the direct numerical simulation of equation (1). Figure 6 shows that the pulse propagating in the dispersion compensated system can be accurately analyzed by the variational method. We then solve equation (3) with $p(Z)$ and $C(Z)$ given by Figure 6. The result is shown in Figure 7. Comparing Figure 7 with Figure 4, one can see that the core of the pulse $|v|(= |f|)$ is propagating stably. Figure 7 also shows the validity of the variational method.

4. Evaluation of the Gordon-Haus Jitter

Noise is added to the pulse at each periodically located amplifier and the interaction between the noise and the pulse causes the frequency shift of the pulse. This frequency shift is converted into a timing jitter throuth the dispersion. This effect is called the Gordon-Haus effect [1]. In this section, we theoretically examine the Gordon-Haus jitter in the dispersion compensated system. Let us first consider the frequency shift induced by the noise.

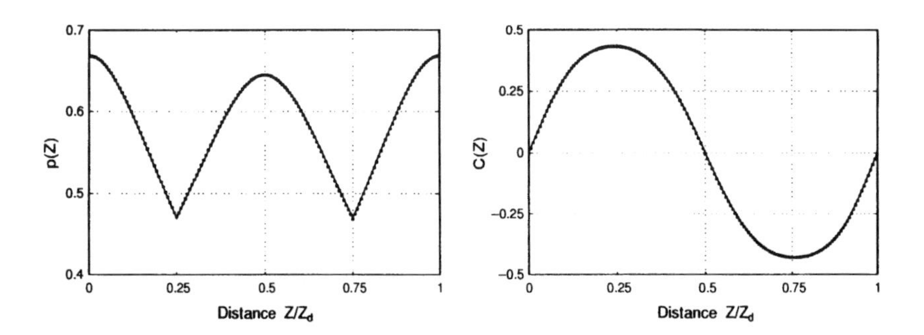

Figure 6. The comparison of the results obtained by the variational method (solid curve) and the numerical simulation (symbol).

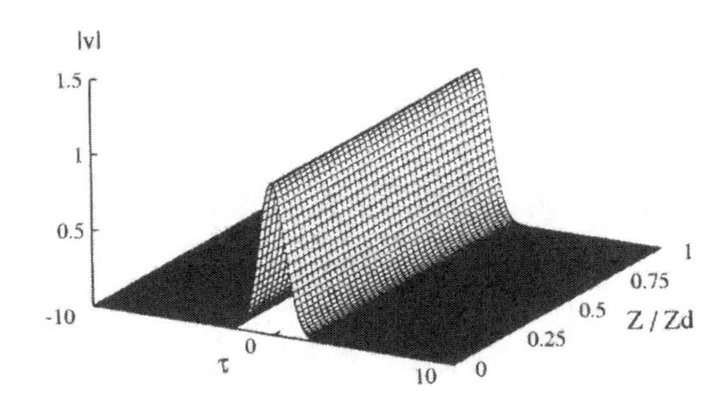

Figure 7. The propagation of the core in a dispersion period.

The noise having equal frequency and phase to the pulse may influence the pulse dynamics. Then, we assume that the perturbation in equation (8) caused by the noise at the m-th amplifier can be written in the form,

$$\epsilon R = n \exp[i\kappa(Z)(T - T_0(Z)) + iE(Z)]\delta(Z - mZ_a) , \tag{24}$$

where $n = n_R + in_I$ is the complex amplitude of the noise. By using equations (16), (24) and the Gaussian trial function ($f(\tau) = e^{-\tau^2}$), the evolution of the parameter for the normalized frequency κ is given by,

$$\frac{d\kappa}{dZ} = \frac{2}{ap}\sqrt{\frac{2}{\pi A}}\delta(Z - mZ_a) \tag{25}$$

$$\times \int_{-\infty}^{\infty} \left[(Cn_R + 2p^2 n_I)\cos\left(\frac{C\tau^2}{2p^2}\right) + (Cn_I - 2p^2 n_R)\sin\left(\frac{C\tau^2}{2p^2}\right)\right] \tau e^{-\tau^2} d\tau.$$

The correlation for n_R and n_I is assumed to be

$$\langle n_i n_j \rangle = n_0 \delta_{ij} \delta(T - T') , \qquad i, j = R \text{ or } I , \tag{26}$$

where

$$n_0 = \frac{2\pi^2 c^2 h n_{sp} n_{nl} t_s}{1.763 \bar{d} \lambda^4 A_{eff}} (G_0 - 1) . \tag{27}$$

Here, c is the velocity of light in vacuum, h is the Planck's constant, n_{sp} is the spontaneous emission coefficient of the amplifier, n_{nl} is the nonlinear coefficient of the fiber, t_s is the full width half maximum of the pulse, $\bar{d}$ is the path-averaged dispersion, λ is the wavelength of the carrier, A_{eff} is the effective core area of the fiber, and $G_0 = \exp(2\gamma z_a)$ is the amplifier gain. Here, we assume that the average values of the parameters a, p, C, κ and T_0 will be the same as those of unperturbed case (without noise). Integrating equation (25) and using the relation equation (26), we can obtain the mean square of the frequency shift in the normalized unit,

$$\langle \kappa_m^2 \rangle = \sqrt{\frac{2}{\pi}} \left(\frac{4p_m^4 + C_m^2}{a_m^2 p_m} \right) n_0 S(G_0) . \tag{28}$$

where $x_m = x(mZ_a)$ for $x = a$, p and C, and $S(G) = (G-1)/(G \ln G)$. This formula has been also obtained by Kumar and Lederer, who have used somewhat different perturbation method [12]. From equation (28), we find that the frequency shift is inversely proportional to the peak power of the pulse [5]. Thus the enhanced power reduces the frequency shift.

We now compute the conversion from the frequency shift into the timing jitter. This conversion is given by the first term in equation (17). Here, we neglect the second term in equation (17) which is much smaller than the first term for longer propagation distance [1]. Integrating equation (17), we obtain the mean square of the timing jitter induced by m-th amplifier in the system in following form reference [13]

$$\langle (T_0)_m^2 \rangle = \langle \kappa_m^2 \rangle \left[\int_{mZ_a}^{NZ_a} D(Z) dZ' \right]^2 . \tag{29}$$

Equation (29) shows that the timing jitter depends on the cumulative dispersion between the m-th amplifier and the receiver. The total timing jitter in the system is given by

$$\langle t_0^2 \rangle = \left(\frac{1.763\lambda^2}{2\pi c \tau_s} \right)^2 \sum_{m=1}^{N} \langle \kappa_m^2 \rangle \left[\int_{mz_a}^{Nz_a} d(z') dz' \right]^2 , \tag{30}$$

in the real unit. In equation (30), N represents the total number of the amplifier in the system.

Here, we summarize how to evaluate the Gordon-Haus jitter in the dispersion compensated system. At first, the frequency shift at each amplifier is calculated by using equation (28). In the calculation of equation (28), we use $a(Z)$, $p(Z)$ and $C(Z)$ given by the variational method with unperturbed case. Then, we calculate the total timing jitter by using equation (30).

5. Upgrade of the Pre-installed Transmission Line

One of the application of the dispersion compensation technique is an upgrade of a pre-installed transmission line. Inserting the DCF periodically into the pre-installed line, the bit-rate of the system can be increased economically. Here, we consider the upgrade of the transmission line with 3.0 ps/nm·km fiber dispersion and 39.8 km amplifier spacing. By inserting the DCF into this line, the path-averaged dispersion $\bar{d}$ will be reduced to 0.1 ps/nm·km as shown in the models (a) and (b) in Figure 8. We assume that the length of the DCF is so short that the nonlinear effect and the fiber loss in the DCF are negligible (this is equivalent to the case with the grating). In this case, $D(Z)$ is given by

$$D(Z) = D_1 + \hat{\alpha} \sum_{m=1}^{M} \delta\left(Z - mZ_d\right) . \tag{31}$$

The parameters using here are $n_{nl} = 3.2 \times 10^{-20}$ m^2/W, $A_{eff} = 50$ μm^2, $\lambda = 1.55$ μm, $n_{sp} = 1.5$, $t_s = 20$ ps and $\gamma = 0.023$ /km (0.2 dB/km).

To evaluate the frequency shift induced by the noise, we calculate $a(Z)$, $p(Z)$ and $C(Z)$ for Figure 8 (a) and (b) by using equations (19)~(21) with the Gaussian trial function ($f(\tau) = e^{-\tau^2}$) and without both amplification ($A(Z) = 1$) and perturbation ($\epsilon R = 0$). The result is shown in Figure 9. Here, we neglect the effects of the amplification and the fiber loss. However, these effects are averaged out by using the guiding center theory because of the shorter amplifier distance [14]. So, the result shown in Figure 9 is applicable to the calculation of the frequency shift. By inserting the result of Figure 9 into equation (28), we evaluate the frequency shift induced by the noise for Figure 8(a) and (b) as shown in Table 1. We also evaluate the frequency shift for the uniform dispersion soliton transmission system having equal pulse width and the path-averaged dispersion. One can see that the frequency shift is reduced by using the dispersion compensated technique. One can also see that the enhanced power obtained in model (b) further reduces the frequency shift [5, 15].

Next, we consider the conversion from the frequency shift into the timing jitter. Here, the question is to find which dispersion profile in Figure 8 is more effective to reduce the timing jitter. Since the frequency shift depends on the dispersion profile in the real system as shown in Table 1,

Installed system

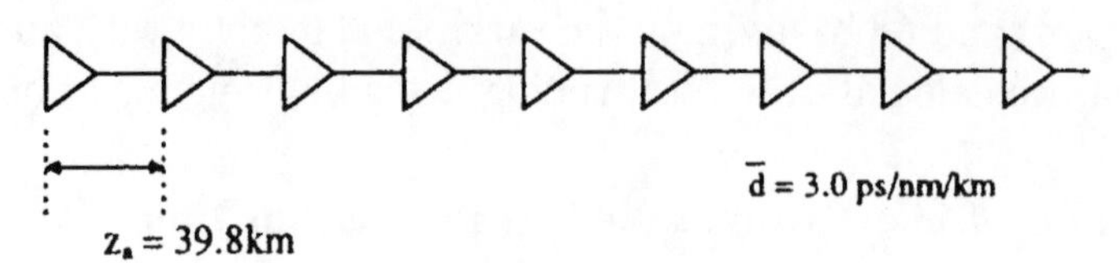

Model. (a)

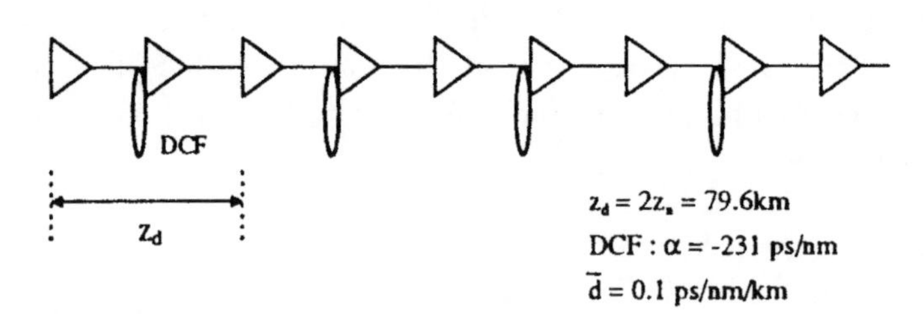

Model. (b)

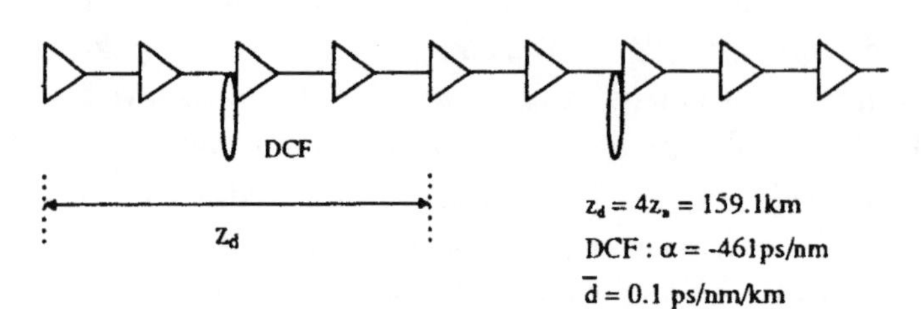

Figure 8. The upgrade of the pre-installed system.

TABLE 1. The frequency shift induced by m-th amplifier $\langle\kappa_m^2\rangle/(n_0 S(G_0))$ given by the variational method.

	$m = 4n$	$m = 4n + 1$	$m = 4n + 2$	$m = 4n + 3$
Soliton	0.667	0.667	0.667	0.667
model (a)	0.417	0.433	0.417	0.433
model (b)	0.171	0.146	0.144	0.156

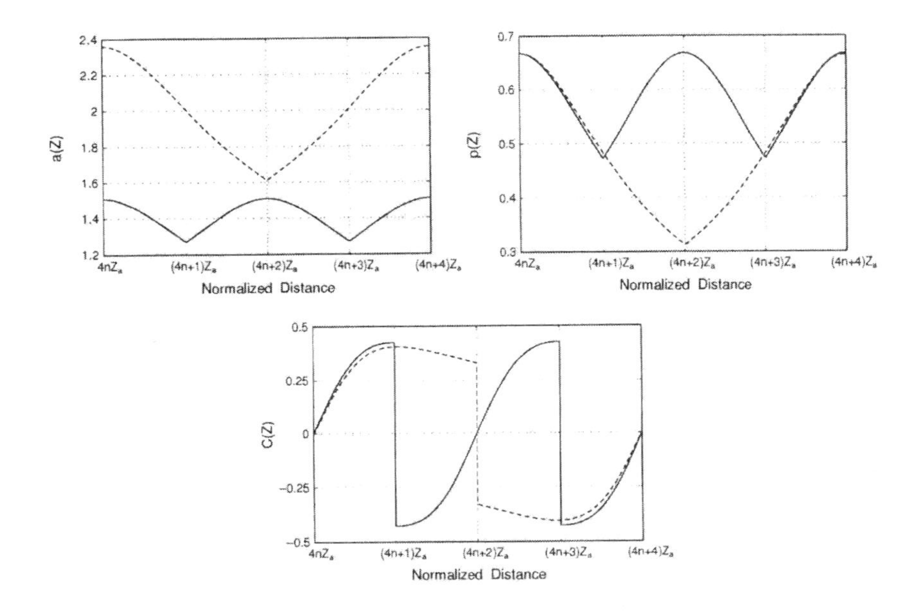

Figure 9. The change of $a(Z)$, $p(Z)$, $C(Z)$ for Figure 8(a) and (b) (solid curve : model (a), dashed curve : model (b)).

it is difficult to evaluate the pure effect of the conversion. We then assume all the frequency shifts to be equal and normalized to $\langle \kappa_m^2 \rangle = 1$, and calculate equation (30) for Figure 8(a), (b) and the soliton. The result is shown in Figure 10. The timing jitter of the soliton is monotonically increasing because of the uniform dispersion. However, the timing jitter for the dispersion compensated system is increasing with the oscillation because of the periodic change of the dispersion. Since the cummulative dispersion between each amplifier and the receiver is minimized by placing the receiver at the point right after the DCF, we can obtain about 16 % and 38 % reduction for the model (a) and (b) compared with the soliton. Otherwise, if the receiver were misplaced, we may degrade the performance of the system. In addition, we can obtain smaller cumulative dispersion by using the DCF with larger $\hat{\alpha}$. By this reason, the conversion is further reduced in model (b) than (a).

By using equations (28) and (30) we can evaluate the total timing jitter. In Figure 11, we compare our calculation (left) with the numerical simulation (right) for Figure 8 (a), (b) and the soliton case. It can be seen from Figure 11 that the theoretical analysis is in agreement with the numerical simulation. We can also find that the model (b) gives better performance

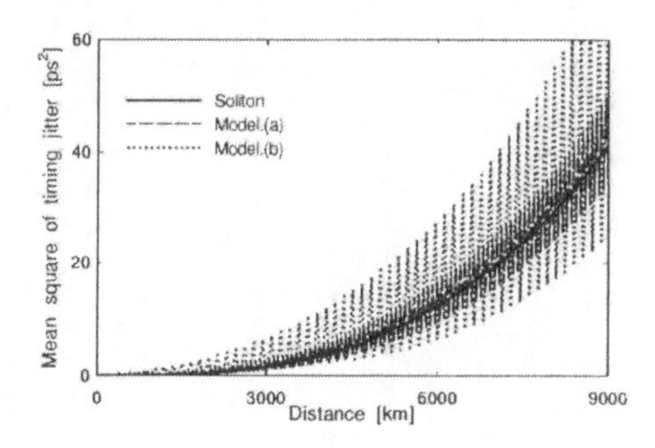

Figure 10. Conversion from the frequency shift to the timing jitter.

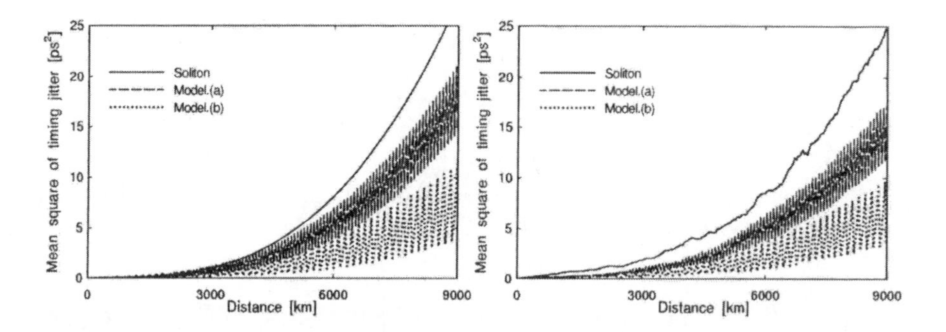

Figure 11. The total timing jitter calculated by the variational method. Variational method (left) and numerical simulation (right).

for suppressing the Gordon-Haus jitter. This indicates that the larger step of the dispersion value or the longer peirod of the dispersion compensation is more effective to reduce the Gordon-Haus jitter.

6. Conclusion

In this paper, we have theoretically shown the reduction of the Gordon-Haus jitter in the dispersion compensated line.

To analyze the pulse propagation in the dispersion compensated system, at first, we examine the basic property of the pulse. We have shown that the amplitude, width and chirp of the pulse is periodically changing with the dispersion compensation period. We then make sure of the stability of the Gaussian core of the pulse. By using these properties of the pulse in the dispersion compensated system, we apply the variational method with the Gaussian shape trial function and show the validity of the method by comparing with the numerical simulation.

Next, we apply the variational method to evaluate the Gordon-Haus jitter. Then, we show that the frequency shift is inversely proportional to the peak power of the pulse. In order to enhance the peak power, it is required to make larger dispersion step and/or elongate the dispersion compensation period [15]. We also show that the conversion of the frequency shift into the timing jitter is reduced by placing the receiver right after the strong DCF. Based on these results, we can reduce the Gordon-Haus timing jitter in newly developed optical transmission line or the upgrade of the pre-installed system by using the DCF. However, as shown in reference [9], the larger dispersion step or the longer compensation period introduce oscillating tails in the pulse, and the interaction between neighboring pulses generates an additional timing jitter. The optimum dispersion step or the compensation period is then determined by minimizing the timing jitter caused by the Gordon-Haus effect and the pulse-pulse interactions. We can also say that the peak power of the pulse at the point right after the DCF is the smallest in the dispersion compensation period (see Figure 9). So, we have to optimize the position of the receiver by cosidering both the Gordon-Haus jitter and the signal to noise ratio.

References

1. Gordon, J. P. and Haus, H. A.: Random walk of coherently amplified solitons in optical fiber transmission, *Opt. Lett.*, **11**, (1986), pp. 665-667.
2. Nakazawa, M. and Kubota, H.: Optical soliton communication in a positively and negatively dispersion-allocated optical fibre transmission line, *Electron. Lett.*, **31**, (1995), pp.216-217.
3. Suzuki, M., Morita, I., Edagawa, N., Yamamoto, S., Taga, H. and Akiba, S.: Reduction of Gordon-Haus timing jitter by periodic dispersion compensation in soliton transmission, *Electron. Lett.*, **31**, (1995), pp.2027-2029.
4. Smith, N. J., Knox, F. M., Doran, N. J., Blow, K. J. and Bennion, I.: Enhanced power solitons in opitcal fibres with periodic dispersion management, *Electron. Lett.*, **32**, (1996), pp.54-55.
5. Smith, N. J., Forysiak, W. and Doran, N. J.: Reduced Gordon-Haus jitter due to enhanced power solitons in strongly dispersion managed systems, *Electron. Lett.*, **32**, (1996), pp.2085-2086.
6. Carter, G. M., Jacob, J. M., Menyuk, C. R., Golovchenko, E. A. and Pilipetskii, A. N.: Timing-jitter reduction for a dispersion-managed soliton system : experimental evidence, *Opt. Lett.*, **22**, (1997), pp.513-515.
7. Matsumoto, M. and Haus, H. A.: Stretched-pulse optical communications, *IEEE Photon. Technol. Lett.*, **9**, (1997), pp. 785-787.
8. Georges, T. and Charbonnier, B.: Reduction of the dispersive wave in periodically amplified links with initially chirped solitons, *IEEE Photon. Technol. Lett.*, **9**, (1997), pp.127-129.
9. Kodama, Y., Kumar, S. and Maruta, A.: Chirped nonlinear pulse propagation in a dispersion compensated system, *Opt. Lett.*, **22**, (1997), pp.1689-1691.
10. Anderson, D.: Variational approach to nonlinear pulse propagation in optical fibers,

Phys. Rev. A, **27**, (1983), pp.3135-3145.

11. Wabnitz, S., Kodama, Y. and Aceves, A. B.: Control of optical soliton interactions, *Optical Fiber Technol.*, **1**, (1995), pp.187-217.
12. Kumar, S. and Lederer, F.: Gordon-Haus effect in dispersion managed soliton systems, *Opt. Lett.*, **22**, (1997), pp.1870-1872.
13. Doran, N. J., Smith, N. J., Forysiak, W. and Knox, F. M.: Dispersion as control parameters in soliton transmission systems, in *Physics and Applications of Optical Solitons in Fibres '95*, Ed. A. Hasegawa (Kluwer Academic Publishers, Dordrecht, 1996), pp.1-14.
14. Hasegawa, A. and Kodama, Y.: Guiding center soliton, *Opt. lett.*, **15**, (1990), pp.1443-1445.
15. Smith, N. J., Doran, N. J., Knox, F. M. and Forysiak, W.: Energy-scaling characteristics of solitons in strongly dispersion-managed fibers, *Opt. Lett.*, **21**, (1996), pp.1981-1983.

(Received January 14, 1998)

FOUR-WAVE MIXING IN FIBER LINKS WITH DISPERSION MANAGEMENT

S. BURTSEV

Corning Incorporated, Science & Technology Process Engineering & Modeling
SP-PR-02-2, Corning, NY 14831, USA.

AND

I. GABITOV

L.D. Landau Institute for Theoretical Physics
Russian Academy of Sciences
2 Kosygin St., Moscow, Russia

1. Introduction

Wavelength division multiplexing (WDM) is currently one of the most effective ways to increase the transmission capacity of optical lines [1, 2, 3]. Four-wave mixing (FWM) [4] is the major factor which limits the potential of the WDM approach. This limitation is due to the nonlinear interaction of neighboring frequency channels which leads, in particular, to the generation of Stokes and anti-Stokes sidebands (FWM-components), that play the role of an extra noise source and lead to signal deterioration [5, 6]. Dispersion management is a powerful technique of fiber optic telecommunication. There is strong evidence that dispersion management helps also to suppress the FWM [5, 6, 7] and, hence, makes it possible to realize the true potential of WDM. The goal of this paper is to study theoretically FWM-characteristics of systems based on the return-to-zero signal format in the presence of a dispersion map. Based on this study, we propose a way to minimize FWM-imposed limitations on data streams by choosing an optimal dispersion map.

A. Hasegawa (ed.), New Trends in Optical Soliton Transmission Systems, 261–270.

2. Basic Model

The group velocity of a wave packet in an optical fiber depends on the signal wavelength. Thus, in WDM systems, in different channels pulses propagate with different velocities and, as a result, collide with each other. Such collisions, for a general signal format, lead to signal degradation due to cross-phase modulation and FWM [1]. An adequate model to describe pulse propagation and interaction in WDM systems is the nonlinear Schrödinger (NLS) equation [1, 3].

It is well-known that for the ideal case of "pure" NLS soliton pulses, these pulses are stable and results of such collisions are negligible: solitons collide elastically [8] and emerge from a collision with their shape, amplitude, and frequency unchanged. The only net effect is a shift in the soliton position/phase. FWM components (and, also, the transient frequency shift) appear only temporarily – they are re-absorbed by solitons after the collision. In an optical line with loss/gain variations there is a net frequency shift whose value depends on both the channel frequency spacing and the amplifier distance spacing. This shift leads to pulse jitter, and hence limits system performance by setting the limit on the maximum system bandwidth. This frequency shift may be suppressed [9] by choosing the collision distance of more than double the amplifier spacing. Nevertheless, impairments caused by FWM are still not resolved by this approach. The source of these impairments is the resonant growth of FWM components during 2-soliton collision due to loss/gain variations. This fact was demonstrated and studied numerically in reference [5]. The corresponding simple and effective analytical solution of this problem for a constant dispersion fiber was presented in reference [7]. It was also suggested in reference [5] and successfully verified in the experiment [10] that the use of exponentially tapered fibers helps to suppress FWM effect strongly. Even though such an approach is very effective to suppress FWM, the use of a dispersion map with the piece-wise constant dispersion profile represents an alternative which is more practical and may be as effective as the use of exponentially tapered fibers.

The dynamics of the electric field envelope $q(z,t)$ (normalized by $\sqrt{P_0}$, where P_0 is the peak pulse power) is governed by the perturbed nonlinear Schrödinger (NLS) system with fast periodic dispersion and loss/gain variations [2]

$$iq_z + \frac{1}{2}d(z)q_{tt} + q^2q^* = Rq\,, \tag{1}$$

$$Rq = i\gamma q + iG(z)q\,.$$

Here $d(z)$ measures dispersion variations, the distance z is normalized by the nonlinear length ($Z_{nl} = 1/(\alpha P_0)$, α is the Kerr nonlinearity), and the retarded time t is normalized by the pulse width. In addition to the dispersion variations that cause pulse "breathing",there exist fast and strong pulse-amplitude oscillations due to the losses/gain variations, which are modeled by the right-hand side Rq. Here $\gamma < 0$ is the loss constant; the amplification, necessary to compensate losses, is given by

$$G(z) = G_0 \sum_{n=0}^{N} \delta(z - nz_a) , \quad G_0 = \exp(\gamma z_a) , \tag{2}$$

a periodic sequence of δ-functions.

We study the case when the amplifier spacing and the dispersion-cycle length are small compared to the nonlinear length. Hence, the corresponding normalized values of these parameters, z_a, z_c, are small. For soliton-based systems, these two scales usually coincide, $z_a \sim z_c$, while for NRZ systems the dispersion scale z_c contains $10 \div 20$ amplification scales z_a.

We apply a standard transformation [2]

$$q(t, z) = a(z)u(t, z) , \qquad da/dz = Ra , \tag{3}$$

and reduce equation (1) to the "pure"NLS with a variable nonlinear coefficient, $a^2(z)$,

$$iu_z + \frac{1}{2}d(z)u_{tt} + a^2(z)u^2u^* = 0 . \tag{4}$$

The rapidly varying coefficients $d(z)$ and $a^2(z)$ can be split in two parts $d(z) = \langle d\rangle + \tilde{d}$, $a^2(z) = \langle a^2\rangle + \tilde{a}$, where the angle brackets denote averaging over the fast dispersion scale z_c.

$$\langle f\rangle = \frac{1}{z_c}\int_0^{z_c} f(z)\, dz . \tag{5}$$

Here, the mean values $\langle d\rangle$, $\langle a^2\rangle$ and the periodic function $\tilde{a}$ are of order one, while the variable part $\tilde{d}$ of the dispersion may be of order $1/z_c$ if the local dispersion is high.

Following reference [11], we apply the Fourier transform

$$u(t, z) = \frac{1}{2\pi}\int_{-\infty}^{\infty} d\omega\, v_\omega(\omega, z) \exp[-i\omega t - i\omega^2 \tilde{d}_1/2 + i\omega^2 < a^2\tilde{d}_1\rangle/\langle 2a^2\rangle] , \tag{6}$$

where

$$\tilde{d}_1 = \int_0^z d(z)\, dz + d_{10} , \tag{7}$$

and d_{10} is integration constant (over z). In the frequency domain, we obtain the integro-differential equation

$$i\frac{d}{dz}v_\omega - \frac{1}{2}\langle d\rangle\omega^2 v_\omega + M = 0\,. \tag{8}$$

Thus, variations of both the dispersion and loss/gain are incorporated into the integral term M given by

$$\begin{aligned} M(\omega, z) &= \int_{-\infty}^{\infty}\int_{-\infty}^{\infty}\int_{-\infty}^{\infty} d\omega_1 d\omega_2 d\omega_3 v_{\omega_1} v_{\omega_2} v^*_{\omega_3} \times \\ & \quad \delta(\omega_1 + \omega_2 - \omega_3 - \omega) S(\omega_1, \omega_2; \omega_3, \omega; z)\,. \end{aligned} \tag{9}$$

Here, the kernel S depends on all the frequencies $\omega_1, \omega_2, \omega_3, \omega$ and the distance z as

$$S(\omega_1, \omega_2; \omega_3, \omega; z) = \frac{1}{2\pi^2} a^2(z) \times \exp[-i\beta\tilde{d}_1/2 + i\beta\langle a^2\tilde{d}_1\rangle/\langle 2a^2\rangle]\,, \tag{10}$$

$$\beta = \omega_1^2 + \omega_2^2 - \omega_3^2 - \omega^2\,, \tag{11}$$

where the angle brackets denote averaging over the fast dispersion scale z_c and $\tilde{d}_1{}' = d(z)$.

The use of the integration constant (over z) d_{10} (and, also, the 3rd last term in the exponent of expressions (6) and (10)) is necessary because it helps to make the function $\tilde{d}_1$ mean-free and, hence, to symmetrize the kernel S.

The role of all transformations of the original system (1) so far is to separate the "fast" scales (due to the losses/amplification and dispersion variations) from the "slow" scales (due to the nonlinearity plus the averaged dispersion), which simplifies the model system.

The averaging of the equation (8) leads to the effective model system: an integro-differential equation that describes, in principle, among other things, the pulse spectrum supported by a given dispersion map [11]. If $S = 1$, then equation (8) is reduced to the "pure" NLS with the averaged dispersion value $\langle d\rangle$, obtained by taking the inverse Fourier transform. Thus, the use of the transformed equation (8) makes possible the correct comparison between an optical system with a dispersion map and a system built from a homogeneous fiber with the dispersion parameter chosen to equal the averaged dispersion $\langle d\rangle$.

Note that the kernel S is a periodic function of the distance z. Therefore, we can expand it in a Fourier series

$$S = \sum_{n=-\infty}^{\infty} S_n \exp(-ink_c z)\,, \quad k_c = 2\pi/z_c\,, \tag{12}$$

where

$$S_n = \langle S \exp(ink_c z)\rangle\,, \qquad k_c = 2\pi/z_c\,. \tag{13}$$

For simplicity, we consider the nonlinear interaction of two channels only. If these channels are spaced sufficiently far apart in the frequency domain, then, with a good accuracy, the FWM generation is modeled by decomposing the total wave-field into the sum

$$q = q_1 + q_2 + q_{fwm} \tag{14}$$

of the channels q_1 and q_2 and the FWM-term q_{fwm}. The terms q_1 and q_2 are leading, while the FWM-term q_{fwm} is small provided pulse collision is fast enough. This is true because the WDM channels are far apart in the frequency domain. The envelopes q_1 and q_2 of these channels are located around the frequencies Ω_1 and Ω_2, correspondingly. As a result, the FWM-part q_{fwm} contains two sidebands. The first sideband's (Stokes) peak frequency is around $\Omega_{221} = 2\Omega_2 - \Omega_1$, while the second sideband's (anti-Stokes) peak frequency is around $\Omega_{112} = 2\Omega_1 - \Omega_2$, so $q_{fwm} = q_{221} + q_{112}$. In the linear approximation, the Stokes and anti-Stokes sidebands are independent of each other and can be treated separately. Thus, by substituting formulae (14) into the model equation (1) and considering the FWM-term about, say, the frequency Ω_{112}, one can obtain a single simplified equation to determine the growth and saturation of the FWM-component. Such an approach has been already successfully used for the case of a homogeneous fiber [7]. In our paper, we demonstrate that the approach can be generalized and applied to the case when a dispersion map is used.

At first glance, the derived equation (8) appears more complicated than the original system (1). In reality, however, it is very useful for modelling WDM problems. To demonstrate this fact, we apply the decomposition (14) to (8) and derive a simple 1st-order inhomogeneous ordinary differential equation (ODE) to study the 112 FWM component, which is described by $(v_{fwm})_\omega$. This equation is

$$i\frac{d}{dz}(v_{fwm})_\omega - \frac{1}{2}\langle d\rangle \omega^2 (v_{fwm})_\omega + \sum_{n=-\infty}^{\infty} M_n \exp(-ink_c z) = 0\,, \tag{15}$$

where

$$M_n = \int_{-\infty}^{\infty}\int_{-\infty}^{\infty}\int_{-\infty}^{\infty} d\omega_1 d\omega_2 d\omega_3\, \delta(\omega_1+\omega_2-\omega_3-\omega)\times$$
$$(v_1)_{\omega_1}(v_1)_{\omega_2}(v_2^*)_{\omega_3} S_n(\omega_1,\omega_2,\omega_3,\omega). \tag{16}$$

The source term M is determined by the spectrum of the pulse which plays the role of the carrier of information. We can consider the spectrum of this pulse as a known function. Further analysis can be performed on both the analytical approximation of such a pulse and the spectrum of the pulse obtained numerically. Below we illustrate it on the simplest examples of continuous waves and soliton approximation which is valid in the case of a weak dispersion map.

Continuous Waves

We begin by applying the above-derived general system (15) to the simplest possible case: FWM by two continuous waves. The channels $(v_{1,2})_\omega$ are represented by δ-functions. We choose zero frequency in the middle between the channels separated by Ω

$$(v_1)_\omega = 2\pi v_0\, \delta(\omega-\Omega/2)\exp(ik_1 z),$$
$$(v_2)_\omega = 2\pi v_0\, \delta(\omega+\Omega/2)\exp(ik_1 z), \tag{17}$$

where the wave number $k(\omega)$ is estimated at the leading order by $k(\omega)\simeq -\langle d\rangle\omega^2/2$, k_1 is the value of the wave number at the channel frequency $(\Omega/2)$ $k_1 = k(\Omega/2)$, and v_0 is the amplitude. The corresponding solution of system (15) is the sum of all the Fourier harmonics with different wave numbers but the same frequency $3\Omega/2$,

$$(v_{fwm})_\omega = \delta(\omega-3\Omega/2)\exp(ik_{fwm}z)\sum_{n=-\infty}^{\infty} v_n,$$
$$k_{fwm} = k(3\Omega/2), \tag{18}$$

where the expansion functions $v_n(z)$ are given by

$$v_n = S_n v_0^3 \frac{\exp[i(\Delta k - nk_c z)]-1}{\Delta k - nk_c}, \tag{19}$$

such that $v_n(z=0)=0$. The phase mismatch vector Δk is given by

$$\Delta k = k_1 + k_1 - k_2 - k_{fwm} = \langle d\rangle\Omega^2. \tag{20}$$

The expression for the coefficient S_n should be evaluated at the corresponding frequencies of the signal and FWM-component

$$\begin{aligned} S_n &= S_n(\omega_1 = \Omega/2, \omega_2 = \Omega/2, \omega_3 = -\Omega/2, \omega = 3\Omega/2) \\ &= \langle a^2(z) \exp(-i\tilde{d_1}\Omega^2 + i\Omega^2 \langle a^2 \tilde{d_1}\rangle / \langle 2a^2 \rangle + ink_c z) \rangle \,. \end{aligned} \quad (21)$$

Hence, to obtain the expression (19) for the FWM-component created by CW's in the presence of a dispersion map, all one needs to do is to take the FWM-expression for the well-known case of a homogeneous fiber with a dispersion parameter $\langle d \rangle$ and multiply each of its harmonics by a certain renormalization coefficient S_n which is due to the dispersion map. Even though the expressions for the S_n are unwieldy, they have a very simple meaning: they describe exactly the suppression of the FWM by the given dispersion map. For simplicity, we present the expression for the zeroth order coefficient S_0 only in the case when only dispersion variations are present and there are no loss/gain variations. This expression is

$$S_0(\Omega/2, \Omega/2, -\Omega/2, 3\Omega/2) = \frac{sin(2\mu\Omega^2)}{2\mu\Omega^2} \,, \quad (22)$$

where μ measures the maximum pulse width change during the "breathing" cycle z_c. For CW's there is no "breathing "and the parameter μ denotes how many FWM-lengths $Z_{fwm} = 2\pi/\Delta k$ are contained in the 1st fiber link, provided one assumes the dispersion map is composed of two different fiber pieces only. So, even if the value of the parameter μ is rather small, say, $\mu = 0.1$ (weak dispersion management), the fact that the practical value of the normalized channel-frequency spacing Ω is rather large, say, $\Omega = 5$, leads to a small value of the coefficient $S_0 \sim 1/5$. Moreover, at certain values of the parameter $\mu\Omega^2$: $\mu\Omega^2 = n\pi$, the renormalization coefficient S_0 is identically zero, which implies complete suppression of the zeroth component of the FWM-term in the framework of our model. Note that each coefficient S_n picks up the factor $1/(\mu\Omega^2)$, which is responsible for the FWM suppression. Hence, the dispersion map effectively re-normalizes the Kerr effect. In the presence of the dispersion management, the effective "nonlinear response"in general depends on the specific characteristic of the map itself and also becomes frequency dependent. Moreover, this dependence is such that the nonlinear response tends to zero for high frequencies (or large frequency spacing between channels). This effect furnishes a simple physical explanation of why the FWM is suppressed by the dispersion map. The presence of the map does not really change the positions of the resonances

$$\Delta k - nk_c = 0 \,, \quad (23)$$

contained in the expression for FWM (19). To define the positions of the resonances in the presence of the map, all one has to do is to take the "old" expression for the case of the homogeneous fiber and insert the mean value $\langle d \rangle$ instead of the dispersion parameter d.

3. Sech-pulse

The standard soliton of the "pure" NLS with a constant dispersion value d (no dispersion map) has the following form

$$\begin{aligned} u_s &= \frac{2d^{1/2}\,\eta\, e^{i\phi}}{\cosh(2\eta\theta)}\,, \\ \theta &= t + 2\xi d\, z\,, \\ \phi &= -2\xi\theta + 2(\eta^2+\xi^2)d\, z\,. \end{aligned} \tag{24}$$

Here, η and ξ parameterize the soliton's amplitude and frequency, respectively. In the presence of a dispersion map, the physical system no longer supports a soliton mode similar to equation (24). In general, when the dispersion variations are strong due to the high local dispersion

$$\int_0^z d(z)dz \geq 1\,, \qquad 0 < z < z_c\,, \tag{25}$$

then stable RZ-pulses propagating through a fiber with a dispersion map no longer have the familiar $1/\cosh$-spectrum and the effective model equation is "far" from the NLS [11]. During their propagation, these pulses "breathe" by changing their width ("breathing" is due to the linear dispersion; as a result it does not affect the pulse spectrum over relatively short distances such as the dispersion cycle z_c).

When dispersion variations are moderate, one has

$$\int_0^z d(z)dz \ll 1\,, \qquad 0 < z < z_c\,. \tag{26}$$

If the "breathing" is rather strong, then the corresponding spectrum is close to the Gaussian. So far, it is not known how to present the pulse spectrum analytically in the general case. When the breathing is rather small and the corresponding stable mode is not that different from soliton (24), then with a good accuracy it may be described by the Fourier transform (6), whose spectrum $(v_s)_\omega$ looks exactly like that of a standard soliton (24)

$$(v_s)_\omega = \frac{\pi d^{1/2}}{\cosh[\frac{\pi}{4\eta}(\omega - 2\xi)]}\exp[2i(\eta^2+\xi^2-2\xi\omega)\,d\,z]\,. \tag{27}$$

To study the *resonant* growth of FWM due to the breathing solitons collisions we use exactly the same tool, equation (15), as we used to analyze the case of CW's. The only new thing is that now instead of δ-functions (17), one has to substitute the pulse-spectrum instead of $v_{1,2}$ in the kernel S_n. If the breathing is small, then, we substitute formulae (27) into S_n. As a result, we get an unwieldy triple integral. If there is no dispersion map $S_n = 1$, this triple integral can be exactly calculated (see reference [7]). If $S_n \neq 1$, then to the best of our knowledge, it is not clear how to take this triple integral exactly. Nevertheless, we can simplify things by making an approximation: the spectrum of a breathing pulse is rather thin. As a result, instead of the full expression for the kernel S_n one can use its value at the signal- and FWM-frequencies $\omega_1 = \Omega/2 = \omega_2 = -\omega_3 = \omega/3$.

Omitting the intermediate calculations, we present the final result

$$v_{fwm}(\omega, z) = F(\underline{\omega} = \omega - 3\Omega/2, z)\exp(ik_1 z)\,, \tag{28}$$

where we shifted the frequency $\omega \to \underline{\omega}$ around $3\Omega/2$ and "switched"from the FWM-field v_{fwm} to its envelope F by factoring out the fast exponent $\exp(ik_1 z)$

$$\begin{aligned} F &= i\pi\frac{(2\eta\sqrt{\langle d\rangle})^3}{\Omega}\mathrm{sech}(\frac{\pi\underline{\omega}}{4\eta})\exp[-i(\underline{\omega}^2 + 3\underline{\omega}\Omega + 2\Omega^2)\langle d\rangle\, z] \times \\ & \sum_{n=-\infty}^{\infty} S_n \int_{-\infty}^{2\eta\langle d\rangle\Omega z} d\zeta \exp[i\gamma_n(\omega)\zeta] W(\zeta, \omega)\,, \end{aligned} \tag{29}$$

$$\gamma_n = (\underline{\omega}^2 + 2\underline{\omega}\Omega + 2\Omega^2 - \frac{2\eta k_c}{\langle d\rangle} + 4\eta^2)\frac{\langle d\rangle}{2\Omega}\,, \tag{30}$$

$$\begin{aligned} W(\zeta, \omega) &= \frac{1}{\sinh^2(2\eta\langle d\rangle\zeta)} \times \\ & \left[\cosh(2\eta\langle d\rangle\zeta) + \frac{i\underline{\omega}}{2\eta}\sinh(2\eta\langle d\rangle\zeta) - \exp(i\underline{\omega}\langle d\rangle\zeta)\right]\,. \end{aligned} \tag{31}$$

The result (28) describes the generation and saturation of the FWM-component. Basically, it means that when the approximation of a "thin"pulse spectrum holds, the FWM in the presence of a dispersion map is presented by taking the "old" FWM-expression, derived in reference [7], and multiplying its every harmonic by the corresponding renormalization coefficient S_n, which describes the suppressive action by the dispersion map.

4. Conclusion

The result of this article is that the negative influence of the four-wave mixing in the WDM transmission can be reduced by the optimization of the dispersion map. This gives additional evidence of the power of the dispersion management technique for high-bit-rate optical telecommunications.

Acknowledgment

This research in part (I.Gabitov) was supported by the NSF Group Infrastructure Grant DMS-9510728.

References

1. Kaminow, I.P. and Koch, T.L. (1997) *Optical Communications IIIa*, Academic Press.
2. Hasegawa, A and Kodama, Y. (1995) *Solitons in Optical Communication*, Oxford University Press.
3. Agrawal, G.P. (1994) *Nonlinear Fiber Optics*, Academic Press.
4. Butcher, P.N. and Cotter, D. (1993) *The Elements of Nonlinear Optics*, Cambridge Studies in Modern Optics, **Vol. no. 9.**
5. Mamyshev, P.V. and Mollenauer, L.F. (1996) *Optics Letters* **Vol. no. 21,** pp. 396.
6. N.S. Bergano and C.R. Davidson, (1996) *Jornal of Lightwave Technology*, **14**, pp.1299.
7. Ablowitz, M.J., Biondini, G., Chakravarty, S., Jenkins, R.B. and Sauer, J.R., (1996) *Optics Letters*, **Vol. no. 21**, pp.1646.
8. Zakharov, V.E. and Shabat, A.B. (1972) *Soviet Physics. JETP*, **Vol. no. 34**, pp.62 .
9. Mollenauer, L.F., Evangelides, S.G. and Gordon, J.P. (1991) *Journal of Lightwave Technology*, **Vol. no. 9**, pp.362, .
10. Mollenauer, L.F., Mamyshev, P.V. and Neubelt, M.J. (1996) *Electronics Letters,* **Vol. no. 32**, pp.471.
11. Gabitov, I.R. and Turitsyn, S. K.(1995) *Optics Letters,* **Vol. no. 21.**

(Received December 25, 1997)

STATISTICAL PROPERTIES OF SIGNALS TRAVELLING IN NOISY NONLINEARLY DISPERSIVE OPTICAL CHANNELS

M. MIDRIO
Dipartimento di Elettronica e Informatica
Università degli Studi di Padova
via G. Gradenigo 6/A, 35131 Padova, Italy

1. Introduction

In recent years, request for availability of continuously increasing transmission bandwidth has led to a great interest in the study and the analysis of optical communication systems. Research efforts have been spent to understand how the system performance can be characterized. In particular, the main target of a number of study has been the computation of the transmission bit error rate (BER), which is the main quantity to be evaluated in order to decide whether a system may be used or not. Studies, most of the time numerical, have been carried out on several different transmission schemes, either using dispersion shifted or step index fibers, possibly along with some dispersion compensators, and soliton or non return to zero (NRZ) pulses.

However, it should be noticed that, although these system may severely differ from one another, they all suffer from the same impairment causes, related to the presence of fiber group velocity dispersion (GVD) and Kerr nonlinearity, and to the noise which is introduced by the optical amplifiers that make up the communication system. In addition, one further feature is common to all the above systems. As a matter of fact, they all resort to an Intensity–Modulation Direct–Detection (IM–DD) scheme, i.e. they all foresee use of a transmission system in which a photodiode closes the optical path, and the decision on the transmitted symbol is taken by neglecting the received pulse phase, and by considering its amplitude only.

Apart from the evolution of the pulses themselves in the nonlinearly dispersive noisy channel, the evaluation of the system performance may then be done in a quite general and unified way, by using standard techniques of stochastic system analysis. In particular, it was proved that reliable results

A. Hasegawa (ed.), New Trends in Optical Soliton Transmission Systems, 271–283.

on the system performance may be obtained by using a gaussian approximation for the signal constituting the decision variable, and by resorting to the BER evaluation method based on the study of bit subsequences [1, 2, 3].

This yields, for sequences of $p = 2^{n_B}$ bits, the following approximation for the BER

$$\mathrm{BER}(S) = \sum_{k=1}^{p} \frac{n_k}{n'} \mathrm{BER}_k(S) , \quad \mathrm{BER}_k(S) = \mathrm{Erfc}\left(\frac{|S - \mu_k|}{\sigma_k}\right) . \tag{1}$$

Here S is the decision threshold, n_k the number of occurences of bit pattern k, and $n' = \sum_{k=1}^{p} n_k$. Moreover, μ_k and σ_k represent the mean and the variance of the bit pattern k, as evaluated in the sampling time instant.

Equation (1) then shows that evaluation of the system performance basically requires computation of the mean and of the standard deviation values that pulses received at the end of the optical transmission link have in the sampling time instant. This means, on the one hand, that the description of the evolution of pulses undergoing fiber nonlinearity and dispersion is needed. On the other hand, that the characterization of the statistical properties of the noise that accumulates down the link is essential as well.

Up to now, reasearch has been mainly devoted to the first of the above two points, and led to elegant and useful results both for solitons and for NRZ pulses [4, 5]. On the contrary, a complete characterization of the accumulated noise statistics is not fully developed yet. As a matter of fact, it has been studied only in purely linear noisy optical channels, or in the presence of fiber nonlinearity, but at the zero dispersion wavelength [6].

The aim of the present communication is that of illustrating the closed–form theory we have developed, allowing statistical description of the received pulse mean value and variance in the presence of *both* fiber GVD and nonlinearity. Through equation (1), this eventually allows to evaluate the system performance in a closed form and, in particular, this will be done here for an NRZ transmission system.

2. Governing Equations.

The theory we are presenting focuses on the performance of an optical transmission system. As shown by equation (1), this requires knowledge of the statistical properties of the signal

$$y(t) = u * h(t) = |s|^2 * h(t) \tag{2}$$

which is detected after N amplified spans of a nonlinearly dispersive noisy optical channel by a receiver like the one depicted in Figure 1.

This is constituted by a quadratic photodiode, followed by an electrical filter with impulse response $h(\cdot)$, and by a threshold decision element. The

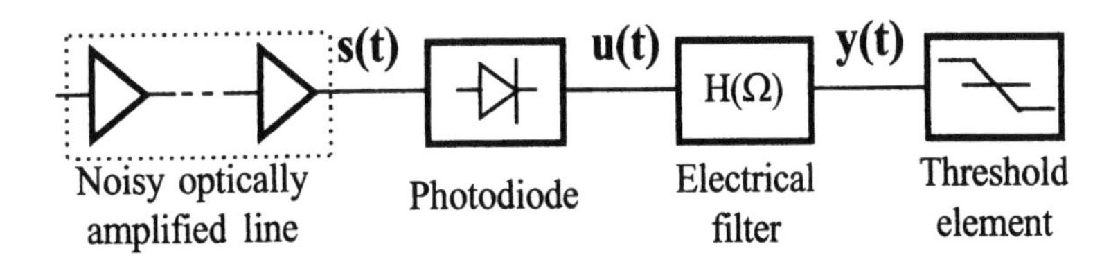

Figure 1. Schematic diagram of the transmission system and detection stage.

electrical filter is supposed to be ideal over a bandwidth B_{el}, e.g. its transfer function reads as

$$H(f) = \begin{cases} 1 & , \quad |f| < B_{el}/2 \\ 0 & , \quad |f| > B_{el}/2 \,. \end{cases}$$

The aim of this section is that of stating the equation giving the mathematical model for the problem at hand, and representing the starting point of our analysis.

Signal s travels along a chain of optically amplified fiber spans and its evolution is hence described by the nonlinear Schrödinger equation [4, 7]

$$i\frac{\partial s}{\partial z} - \frac{k''}{2}\frac{\partial^2 s}{\partial t^2} + \gamma_{NL}|s|^2 s = -i\alpha s + i\left[\sqrt{G}s + \zeta(t)\right]\sum_n \delta(z - nz_A)\,, \quad (3)$$

where $t = t' - z/V_g$ is the retarded time, V_g and k'' are the fiber mode group velocity and its dispersion, respectively, and $\gamma_{NL} = 2\pi n_2/\lambda A_{\text{eff}}$ is the fiber nonlinearity coefficient, with n_2 the nonlinear refractive index, A_{eff} the core effective area and λ the wavelength of the carrier optical wave. Moreover, α is the fiber loss, z_A the amplifier spacing, $G = \exp\{2\alpha z_A\}$ the amplifier power gain and $\zeta(t)$ the noise term accounting for the spontaneous emission introduced by each amplifier. As customary in the study of the NLS equation, it is convenient to introduce two further quantities, namely the pulse nonlinear and dispersion lengths that respectively read as $L_{NL} = 1/(\gamma_{NL}P_{REF})$ and $L_D = T_{REF}^2/|k''|$. In these quantities, T_{REF} and P_{REF} are arbitrary time and power units.

In the following, we will show that analytical results may be obtained whenever the fiber dispersion is suitably small, so that the amplifier spacing will always be much shorter than the dispersion length. As known, in these cases the propagation may be considered linear between any couple of adjacent amplifiers, that is the nonlinearity and the dispersion introduce a pulse distortion over length scale of several amplifier distances. This allows to average out the short-length variations of the field to describe the propagation in terms of average (or guiding-center) pulse [8, 9, 10]. Moreover, the amplifier noise becomes equivalent to a distributed white noise term that may be assumed delta correlated in space [11]. Finally, it proves convenient to recast the propagation equation in dimensionless form. Within the range of validity of the average model, this read as

$$i\frac{\partial q}{\partial Z} - \frac{\beta}{2}\frac{\partial^2 q}{\partial T^2} + |q|^2 q = n(Z,T)\ , \tag{4}$$

with $T = t/T_{REF}$, $Z = z/L_{NL}$ the normalized time and distance coordinates, respectively. Moreover, $\beta = L_{NL}/L_D$ and $q(t,z) = s(t,z)/A_0\sqrt{P_{REF}}$ being $A_0^2 = 2\Gamma Z_A/(1-\exp\{-2\Gamma Z_A\}) = \log G/(1-1/G)$ the average soliton preemphasis factor, with $\Gamma = \alpha L_{NL}\log(10)/20$ the normalized loss. Furthermore, $Z_A = z_A/L_{NL}$ is the normalized amplifier spacing and $n(Z,T)$ the spatially δ-correlated normalized noise term. This is supposed to be stationary and white over a band B, and we denote its power spectral density as $W_n = W_\zeta(f)/(Z_A A_0^2 P_{REF}) = n_{\text{sp}}(G-1)hf/(Z_A\ A_0^2 P_{REF})$ for $|f| < B/2$ with n_{sp} the spontaneous emission factor accounting for noncomplete inversion of the active medium [11]. The average power of the noise is $\sigma_n^2 = W_n B$.

3. System Analysis in the Presence of Noise

In this section the behaviour of the optical communication system is analyzed in the presence of the noise introduced by the in-line amplifiers making up the link. We remind that our target is the evaluation of the system performance, and that this requires decription of two topics, the statistics of the pulses received from a photodiode at the end of the noisy channel, and, related to this, the statistical properties of the noise undergoing nonlinearly dispersive propagation.

3.1. STATISTICAL PROPERTIES OF DETECTED PULSES

Let us then begin by describing under a general point of view how the statistics of a noisy signal detected by a photodiode and electrically filtered after that look like. To this end, we consider the receiver stage depicted in Figure 1 and we suppose that the signal exiting the propagation through the nonlinearly dispersive optical channel may be approximated as

$$s(t) = A + \eta_R(t) + i\eta_I(t)$$

i.e. it is assumed to be the sum of a continuous wave radiation with power $|A|^2$, plus a noise η representing the accumulation of the noise contributions ζ introduced by each amplifier at the end of each amplified fiber span. Note that results from this model, where the average part of the received field is a constant, will hold as long as the noise bandwidth is much broader than the transmitted pulse spectrum.

Concerning the noise, it may be assumed to be statistically independent of A, with zero mean and the correlation matrix

$$\left\langle \begin{bmatrix} \eta_R(t)\eta_R(t+\tau) & \eta_R(t)\eta_I(t+\tau) \\ \eta_I(t)\eta_R(t+\tau) & \eta_I(t)\eta_I(t+\tau) \end{bmatrix} \right\rangle = \begin{bmatrix} r_R(\tau) & r_{RI}(\tau) \\ r_{IR}(\tau) & r_I(\tau) \end{bmatrix} .$$

The average value of the detected signal is then

$$m_u = m_y = \langle u(t) \rangle = \langle |s(t)|^2 \rangle = |A|^2 + \sigma_R^2 + \sigma_I^2 = |A|^2 + \sigma_\eta^2 . \qquad (5)$$

Whereas, the correlation function at the photodiode output is easily computed to be

$$r_u(\tau) = \langle u(t)u(t+\tau) \rangle = m_u^2 + 4|A|^2 r_R(\tau) + r_\eta^2(\tau) , \qquad (6)$$

where $r_\eta(\tau)$ denotes the autocorrelation function of the squared modulus of the noise. Then, in the presence of a white noise having flat power spectral density W_η within the band $|f| < B/2$, and if $H(\cdot)$ is assumed to be an ideal electrical filter with bandwidth B_{el}, the variance of the decision variable $y(t)$ turns out to be

$$\sigma_y^2 = (\sigma_\eta^2)^2 \frac{B_{el}}{B} \left(1 - \frac{B_{el}}{4B}\right) + 4|A|^2 \int_{-B_{el}/2}^{B_{el}/2} R_R(f)\, \mathrm{d}f \qquad (7)$$

and hence it is given by the sum of two terms. The first is independent of the transmitted pulse, and is only due to the noise-noise beating in the photodiode. On the contrary, the second term depends on the transmitted optical power, through the *only in-phase component of the noise.* This is not surprising, since we are dealing with the signal exiting from a photodiode, where the squared modulus of the optical signal is processed. Then, since the in-quadrature component of the noise does not produce any beating with the signal, it does not appear in the pulse variance.

Although obvious, this however is the key point in the determination of the system performance. As a matter of fact, we will show in the following that the noise content which is in phase with the signal at the end of a nonlinearly dispersive optical link *strongly depends on the value of the group velocity dispersion.* In particular, we will find that, in the presence of anomalous dispersion, modulation instability causes enhancement of the noise spectrum with respect to an ideal system where dispersion and nonlinearity are neglected. On the contrary, propagation in the presence of normal dispersion will be proved to lead to a *reduction* of $R_R(f)$ with respect to the ideal case, eventually reflecting in a reduced variance σ_y^2. This fact had been observed in numerical simulations reported in reference [12], and in experiments as well (see references [13, 14]), and will be here theoretically addressed.

However, before going through the properties of the pulses detected at the end of the nonlinearly dispersive link, it is worth computing the signal variance for an ideal system where both the nonlinearity and the dispersion are neglected. In such a case, at the output of the N-th amplifier, the power spectral density of the accumulated noise is $R_R(f) = R_I(f) = W_\eta/2 = NW_\zeta/2$ for $|f| < B/2$ and since $\sigma_\eta^2 = W_\eta B$,

$$\sigma_y^2 = (NW_\zeta)^2 \frac{B_{el}}{B}\left(1 - \frac{B_{el}}{4B}\right) + 2|A|^2 N\sigma_\zeta^2 \frac{B_{el}}{B} . \tag{8}$$

3.2. STATISTICAL PROPERTIES OF THE NOISE

Let us now come to describe the behaviour of the noise when dispersive and nonlinear effects are included in the model. For simplicity of notation, we refer to the averaged normalized equation (4), and we consider the behaviour of a continuous wave radiation plus a *small* broadband noise propagating between two adjacent amplifiers. The governing equation hence is

$$i\frac{\partial q}{\partial Z} - \frac{\beta}{2}\frac{\partial^2 q}{\partial T^2} + |q|^2 q = 0 .$$

We write the solution as

$$q(T, Z) = [\sqrt{V} + n(T, Z)] \exp\{iVZ\}$$

and linearize about V. The following evolution equation for the small noise component n is found [7, 15]

$$i\frac{\partial n}{\partial Z} - \frac{\beta}{2}\frac{\partial^2 n}{\partial T^2} + V(n + n^*) = 0 \tag{9}$$

so that, letting $\hat{n}_R = \mathcal{F}[n_R]$, $\hat{n}_I = \mathcal{F}[n_I]$, the differential equations

$$\frac{\partial}{\partial Z}\begin{bmatrix} n_R(\hat{\Omega}, Z) \\ n_I(\hat{\Omega}, Z) \end{bmatrix} = \begin{bmatrix} 0 & -\beta/2\Omega^2 \\ \beta/2\Omega^2 + 2V & 0 \end{bmatrix}\begin{bmatrix} \hat{n}_R(\Omega, Z) \\ \hat{n}_I(\Omega, Z) \end{bmatrix} \tag{10}$$

are found, and they may be solved to yield

$$\begin{aligned} \begin{bmatrix} \hat{n}_R(\Omega, Z) \\ \hat{n}_I(\Omega, Z) \end{bmatrix} &= \mathbf{M}\begin{bmatrix} \hat{n}_R(\Omega, Z=0) \\ \hat{n}_I(\Omega, Z=0) \end{bmatrix} \\ &= \begin{bmatrix} \cos(\chi_p) & -\sin(\chi_p)/\sqrt{K} \\ \sqrt{K}\sin(\chi_p) & \cos(\chi_p) \end{bmatrix}\begin{bmatrix} \hat{n}_R(\Omega, 0) \\ \hat{n}_I(\Omega, 0) \end{bmatrix} \end{aligned} \tag{11}$$

with

$$\chi_p = \sqrt{\frac{\beta}{2}\Omega^2\left(\frac{\beta}{2}\Omega^2 + 2V\right)}\,, \quad K = 1 + \frac{4V}{\beta\Omega^2}\,. \tag{12}$$

These are quite general results, that do not depend on peculiar statistical properties of the noise n, since they simply resort to linearization of the nonlinear Schrödinger equation. We may now specialize the results to the case of a noise being described by a complex gaussian random process with uncorrelated real and imaginary components. In this case, the power spectrum at the end of the propagation may be computed by evalutaing the matrix product $\mathbf{M}\mathbf{M}^T$, and it results

$$R_R = \frac{W_n}{2}\left(\cos^2(\psi) + \frac{\sin^2(\psi)}{K}\right)\,, \quad R_I = \frac{W_n}{2}\left(\cos^2(\psi) + K\sin^2(\psi)\right) \tag{13}$$

with $\psi = \chi_p Z_A$. Consider first the case of the anomalous dispersive regime: $\beta < 0$. In this case, χ_p is imaginary for $|\Omega| < 2\sqrt{V/|\beta|}$, and the harmonic functions appearing in the above expressions actually are hyperbolic functions. This is the well known phenomenon of the modulational instabilty, as driven by the cw pump with power V (note that this is the average value of the cw power along the fiber span) [7]. The broadband noise grows by substracting energy to the cw pump. Accordingly, the power spectrum of both the in-phase and in-quadrature components of the noise grow larger.

On the contrary, a completely different behaviour is observed in the normally dispersive regime. In this case, χ_p is always real, and an exchange of energy between the noise components takes place. By inspecting equations (10)-(13) it may be verified that the noise components become mutually correlated and that an energy transfer occurs from the in-phase to the in-quadrature and mutual power spectra. In fact, as soon as the nonlinear effects are taken into account (i.e. letting $V > 0$), it turns out that $K > 1$ in equations (13). Hence, the power spectrum of the in-phase component is attenuated with respect to case of linear propagation, whilst the out of phase component grows larger. Through equation (7) this reflects in the fact that a pulse propagating in the presence of normal GVD and Kerr may exhibit a lower variance than it would do if the propagation were either in the linear regime, or in the anomalous dispersion region. Moreover, equation (1) also shows that this may eventually leads to find that the performance of a system operating in the presence of normal dispersion may become comparable or even better than that of an *ideal system* where nonlinearity and dispersion are neglected.

Before using equation (7) to derive explicit expressions for the received pulse variance, we now evaluate the power spectra that are obtained when the propagation through N amplifiers is considered. In this case, since the noise of each amplifier may be supposed to be statistically independent of all the others, the linearized theory we have developed yields the following expressions for the in-phase and in-quadrature power spectra

$$R_R(\Omega) = \frac{W_n}{2} \sum_{\ell=1}^{N} \left\{ \cos^2(\psi\ell) + \frac{\sin^2(\psi\ell)}{K} \right\} = N \frac{W_n}{2} \left[1 - G_R(V, N, \Omega)\right] , \tag{14}$$

$$R_I(\Omega) = \frac{W_n}{2} \sum_{\ell=1}^{N} \left\{ \cos^2(\psi\ell) + K \sin^2(\psi\ell) \right\} = N \frac{W_n}{2} \left[1 + G_I(V, N, \Omega)\right] , \tag{15}$$

where, as usual, W_n is the power spectral density of the squared modulus of the noise introduced by each amplifier, and

$$G_R(V, N, \Omega) = \frac{2V}{4V + \beta\Omega^2} \left[1 - \frac{\cos[(N+1)\chi_p Z_A] \sin[N \chi_p Z_A]}{N \sin[\chi_p Z_A]} \right] , \tag{16}$$

$$G_I(V, N, \Omega) = \frac{4V + \beta\Omega^2}{\beta\Omega^2} G_R(V, N, \Omega) \tag{17}$$

represent the spectral gain that the in-phase and in-quadrature components of the noise respectively experience in the nonlinear propagation, with respect to the linear case.

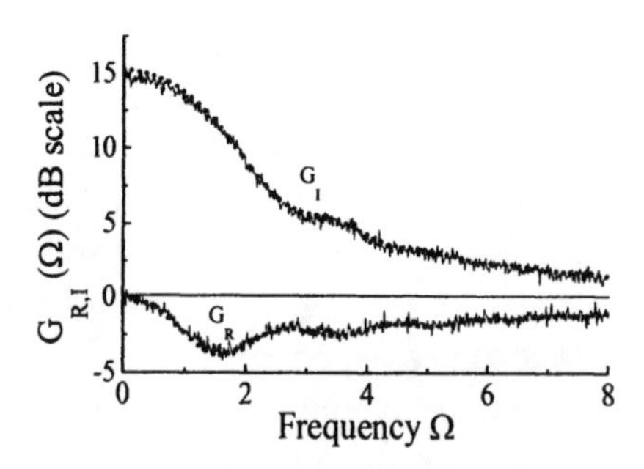

Figure 2. Power spectral densities for the in-phase and in-quadrature components of noise experiencing nonlinear propagation in the presence of normal fiber dispersion.

In particular, it is interesting to evaluate the spectral gains for a noisy cw radiation experiencing nonlinear propagation in the presence of normal fiber dispersion. Those are illustated in Figure 2, referring to the case of propagation of a continuous wave radiation with the power $P_{Peak} = 2.56$ mW through a normally dispersive noisy optical channel with overall lenght $L = 9,000$ km, and amplifier spacing $z_A = 50$ km. The fiber dispersion is

$k'' = 0.05$ ps/nm·km, the nonlinearity coefficient $\gamma_{NL} = 2.54$ W^{-1}km^{-1}, and the fiber loss $\alpha_{dB} = 0.2$ dB/km, so that $G = 10$ and $A_0^2 = 2.56$. The bandwidth of the optical noise and of the electrical filter are $B = 400$ GHz and $B_{el} = 40$ GHz, respectively. In the framework of the normalized units used in equation (4), this would correspond to the following set of parameters: $\beta = 0.2$, $Z_A = 0.16$, $N = 180$ and $V = 1$.

In the figure the zero level is the level of the in-phase and in-quadrature noise spectra for propagation in an ideal fiber link, where both the dispersion and the nonlinearity are neglected, and the noise is simply added at each amplification stage. As a result, both the in-phase and out of phase power spectral densities are flat over their bandwidth, and they have the same value. On the contrary, solid lines, which are results of numerical simulations, and dotted lines which are theoretical results, equations (16) and (17), describe the power spectral densities obtained at the end of the normally dispersive fiber propagation. Note that, as we previously announced, nonlinear propagation in the absence of normal fiber dispersion causes *reduction* of the in-phase noise spectrum (G_R), and enhancement of the in-quadrature component (G_I).

3.3. STATISTICS OF THE RECEIVED PULSES

We may now proceed to give the statistical characterization for the pulses detected at the end of the nonlinearly dispersive noisy optical channel. We again suppose the received field to be a continuous wave radiation with normalized power $V = P_{peak}/P_{REF}A_0^2$, and with a superimposed noise having the above statistical properties. With these hypotheses, and with reference to the dimensional units, we get from eq.(5) the mean received power at the distance $L = Nz_a$

$$\begin{aligned} m_y(t,L) &= |s(t,L)|^2 + \sigma_\eta^2 = |s(t,L)|^2 + N\sigma_\zeta^2 \\ &= |s(t,L)|^2 + N n_{sp}(G-1)h\nu B\, 10^{12}\,. \end{aligned} \tag{18}$$

Whereas, the pulse variance is given by equation (7) and it reads as

$$\sigma_y^2(t,L) = (N\sigma_\zeta^2)^2 \frac{B_{el}}{B}\left(1 - \frac{B_{el}}{4B}\right) + 2|U|^2 N\sigma_\zeta^2 \frac{B_{el}}{B}\left[1 - D(P_{peak}, N)\right]\,, \tag{19}$$

where σ_ζ^2 is, as usual, the power of the squared modulus of the noise which is introduced by each amplifier and which is supposed to have flat power spectral density W_ζ within the band $-B/2 < f < B/2$. Moreover B_{el} is the bandwidth of the receiving photodiode, and

$$D(P_{peak}, N) = \frac{1}{2}\left\{1 - \frac{1}{Nw}\left(\mathrm{Si}\left[\left(N + \frac{1}{2}\right)w\right] - \mathrm{Si}\left[\frac{1}{2}w\right]\right)\right\} \tag{20}$$

with

$$w = \frac{z_A}{A_0} 2\pi\, 10^{-3} \sqrt{10^{-3} \gamma_{NL} P_{peak} |k''|}\,, \quad \mathrm{Si}(z) = \int_0^z \frac{\sin(t)}{t}\, dt\,.$$

In these expressions dimensional quantities are used. In particular, $|s|^2$ is the received pulse power, as expressed in mW, $h\nu$ is in J, and B in GHz, so that m_y is measured in mW. Moreover, z_A is expressed in km, γ_{NL} in W^{-1} km^{-1}, P_{peak} in mW and k'' in ps^2/km.

Let us now discuss the results we have been obtained. Equation (18) shows that the mean value of the received pulses is the same as in a linear system, i.e. *nonlinearity and dispersion do not alter the average received pulse power.* On the contrary, equation (19) shows that the variance carries the "sign"' of the presence of fiber GVD and nonlinearity through the parameter D. As a matter of fact, it results $D = 0$ in the absence of fiber GVD and nonlinearity, and equation (19) reduces to the usual expression for the variance of the squared modulus of a random gaussian process, namely equation (8). Whereas, it results $D < 0$ for $k'' < 0$, i.e. in the anomalous dispersion regime. Note that $D < 0$ means an increased pulse variance, resulting from the modulation instability process, which drains energy from the pulse to enhance both the noise spectral component in phase with the useful signal, and the one in quadrature with it. In the framework of the above analytical expressions, this reflects in the fact that $G_R > 1$ and $G_I > 1$. Finally, it results $0 < D < 1$ for $k'' > 0$, e.g. in the normal dispersion regime. As a consequence, the pulse variance turns out to be even lower than in the ideally nondispersive linear case. As shown in Figure 2, this is related to the fact that, in the normal dispersion regime, nonlinear dispersive propagation causes reduction of the noise component in phase with the useful signal, along with an enhancement of the in-quadrature component.

4. An Example of Application: Evaluation of NRZ Transmission System Performance.

We said in the introduction that, in order to analytically compute the performance of an optical transmission system, knowledge of two items is required. Those are the evolution of bit patterns down the nonlinearly dispersive channel, and the statistics of the accumulated noise. Then, we presented a theory which allows to characterize the noise in a quite general way. Now, we intend to combine the results so far obtained with some analytical description of bit pattern evolution to show how closed-form computation of system performance can be made. In particular, we will do this for a system using NRZ modulation format, and operating in the presence of weak normal dispersion. As a matter of fact, it was proved by

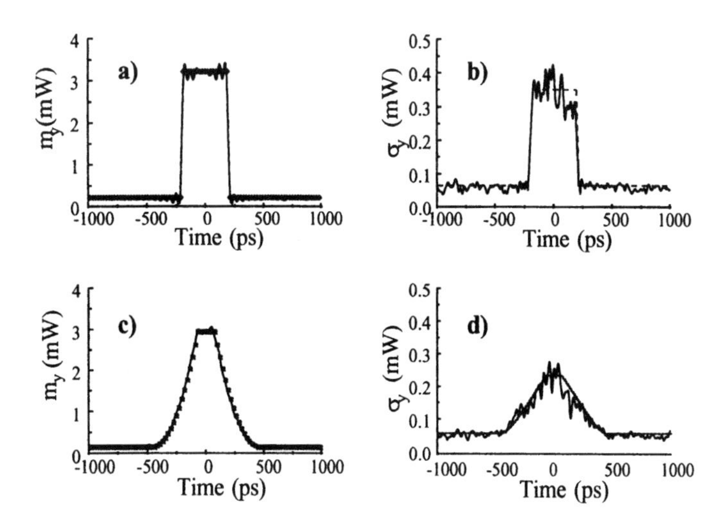

Figure 3. Mean value and standard deviation of an NRZ pulse detected after propagation in an ideal system (a) and (b), and in a normally dispersive nonlinear system (c) and (d).

Kodama and Wabnitz that the evolution of pulses of such a transmission system may be analytically predicted through a set of equations that is formally identical to those describing the breaking of a dam [5]. This means that, on the one hand, one may know which is the power received in the sampling time instant when one of the bit patterns to be used in equation (1) is launched inside the transmission link. On the other hand, the variance of this quantity may be computed with the help of the above theory, so that everything is known in equation (1).

Figure 3 illustrates the numerical simulation, and the comparison with theoretical predictions we have obtained by studying the behaviour of one particular subsequence, namely the one having one mark only among a number of spaces. In paticular, in insets a) and b) of Figure 3 the shape of an initially 400 ps NRZ pulse is shown after propagation through a 9,000 km long ideal link, i.e. in the absence of fiber nonlinearity and dispersion. Whereas, insets c) and d) refers to the propagation in the presence of nonlinearity and weak normal dispersion. The input pulse peak power is 2.56 mW, and the link parameters are the same as in Figure 2. In Figure 3 solid lines are results form numerical simulations, and dotted lines are theoretical predictions. Note that, quite obviously, propagation in the ideal system preserves the initially squared pulse shape, and gives rise to a variance due to the noise which linearly accumulates along the link. On the contrary, presence of normal GVD induces pulse distortion as described by the Kodama–Wabnitz theory of reference [5], but it also allows reduction of the received pulse variance. Moreover, note that an excellent agreement is found between theory and numerics. This suggests that computation of the system performance through equation (1) can effectively be made by considering a suitable number of bit subsequences, and it may lead to reliable results. This in fact is confirmed by Figure 4 showing comparison between results from equation (1) and numerics we have obtained for a 2.5 Gbit/s NRZ transmission operating in the same link as above. In the figure solid lines and discrete points are numerical and theoretical results, respectively. Moreover, circles are for the system operating in the presence of normal dispersion, and squares for the ideal system. It may be noticed that, for relatively low peak powers, i.e. until nonlinearity does not cause severe pulse distortion, the performance of the nonlinearly dispersive link is *even better* than in the ideal system. This is related to the reduction of the detected pulse variance we have discussed above.

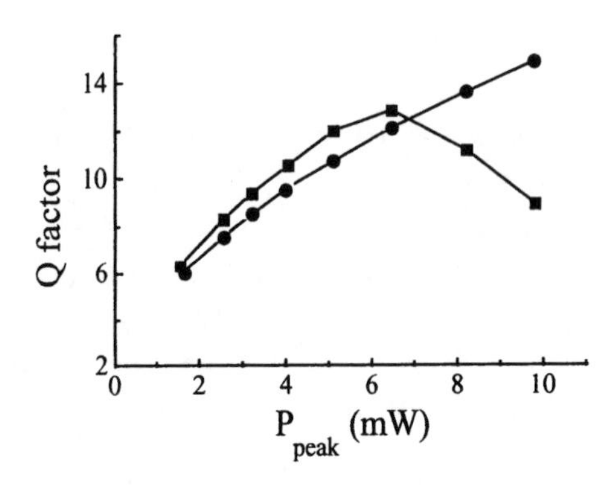

Figure 4. Standard deviation of the detected signal after propagation in an ideal system (circles), and in a normally dispersive nonlinear system (squares).

References

1. Humblet, P. A. and Azizoglu, M.: Derivation of analytical expressions for the bit-error probability of lightwave systems with optical amplifiers, *IEEE Journal of Lightwave Technol*, **9**, (1991), pp.1576-1582.
2. Anderson, C. J. and Lyle, J. A.: (1994) Technique for evaluating system performance using Q in numerical simulations exhibiting intersymbol interference, *El. Lett.*, **30**, (1994), pp.71-72.
3. Matera, F. and Settembre, M.: (1995) Performance evaluation of optical systems operating on long fibre links. *Opt. Quantum Electron.*, **27**, (1995), pp.831-846.
4. Hasegawa, A. and Kodama, Y.: *Solitons in Optical Communications*, Oxford University Press, New York, (1995).

5. Kodama, Y. and Wabnitz, S.: (1995) Analytical theory of guiding center nonreturn to zero and return to zero signal transmission in normally dispersive nonlinear optical fibers, *Opt. Lett.*, **20**, (1995), pp.2291-2293.
6. Mecozzi, A.: *J. Opt. Soc. Am.*, **B11**, (1995), pp.462-470.
7. Agrawal, G. P.: *Nonlinear Fiber Optics*, Academic Press, San Diego, (1989).
8. Hasegawa, A. and Kodama, Y.: Guiding center soliton, *Phys. Rev. Lett.*, **66**, (1991), pp.161-164.
9. Mollenauer, L. F.,Evangelides, S. G., and Haus, H. A.: (1991) Long-distance soliton propagation using lumped amplifiers and dispersion shifted fiber, *IEEE J. Lightwave Technol.*, **9**, (1991), pp.194-197.
10. Blow, K. J. and Doran, N. J.: (1991) Average soliton dynamics and the operation of soliton systems with lumped amplifiers, *IEEE Photon. Technol. Lett.*, **3**, (1991), pp.369-371.
11. Haus, H. A.: (1991) Quantum noise in a solitonlike repeater system, *J. Opt. Soc. Am.*, **B8**, (1991), pp.1122-1126.
12. Midrio, M., Matera, F. and Settembre, M.: (1997) Reduction of the detrimental ASE noise effect in optical communication systems by means of unusual modulation instability, *Proc. OFC '97*, Dallas, (February 16-21, 1997).
13. Ryu, S.: *Electron. Lett.*, **26**, (1990), pp.424-425.
14. Hui, R., O'Sullivan, M.: *Electron. Lett.*, **32**, (1996), pp.2001-2003.
15. Kikuchi, K. Enhancement of optical amplifier noise by nonlinear refractive index and group velocity dispersion of optical fibers, *IEEE Photon. Technol. Lett.*, **5**, (1993), pp.221-223.

(Received November 17, 1997)

DISPERSION MANAGEMENT ISSUES IN LONG AMPLIFIER-SPACING FREQUENCY-GUIDING MULTIWAVELENGTH SOLITON SYSTEMS

J.-P. HAMAIDE, B. BIOTTEAU, F. PITEL, O. AUDOUIN, AND E. DESURVIRE

Alcatel Corporate Research Center
c/o Alcatel Alsthom Recherche
route de Nozay, F - 91460 Marcoussis, France

Abstract. The performance of 4×10 Gbit/s soliton transoceanic links based on sliding filtering and exponential dispersion management is evaluated both numerically and experimentally. Considering discrete N-step implementation, we show that the management rule is determinative for optimal performance. We also show that the required number N of steps is related to the amplifier spacing. Experimental WDM soliton transmission in a single-span recirculating loop with large 60 km-amplifier spacing and 4-step dispersion management is then described. Error-free distances over 15, 7 and 3.9 Mm have been obtained in 2, 3 and 4×10 Gbit/s WDM transmission, in close agreement with numerical results. The agreement validates other numerical predictions, showing substantial improvement in realistic straight-line implementation and error-free transmission over more than 9 Mm. In this last case, tolerance on actual fibre dispersion is also discussed.

1. Introduction

Soliton transmission has shown promising application to ultra-long fibre-optic amplified systems. Soliton format is compatible with wavelength-division- multiplexing (WDM) providing system capacity increase and routing functionality, while allowing high channel-data rates at or above 10 Gbit/s. Such high-capacity systems were made possible by exploiting passive or active control techniques, in order to reduce both timing instability (or jitter) and signal-to-noise degradation.

A. Hasegawa (ed.), New Trends in Optical Soliton Transmission Systems, 285–301.

Several effects determine soliton transmission performance. In single-channel systems, amplified spontaneous emission (ASE) noise generated by the optical amplifiers induces timing jitter and amplitude fluctuations [1]. In addition, periodic power variation produces continuum generation, albeit very weak in path-averaged soliton systems (i.e. amplifier spacing short compared to soliton period), that may lead to neighbouring soliton interaction. These effects can be efficiently limited by passive or active control techniques. As for passive system experiments, examples are: transmission at 10 Gbit/s over 35 Mm [2] by use of fixed or sliding narrowband filtering that simultaneously reduces jitter and amplitude fluctuations; 20 Gbit/s over 14 Mm [3] by periodic dispersion compensation and filtering that takes advantage of the enhanced soliton power and of the low average chromatic dispersion that reduces timing jitter [4, 5]. Active control techniques involve periodic synchronous amplitude and/or phase modulation, which, in conjunction with soliton propagation, provides full signal regeneration (for instance, 10 Gbit/s transmission over unlimited distances [6]) and allows very large amplifier spacing (for instance, 20 Gbit/s transmission over 20 Mm with 140 km amplifier span [7]).

In amplified WDM soliton systems, collisions between pulses of the different channels causes an additional jitter. This additional jitter can also be kept under control by narrowband filtering, through in-line Fabry-Pérot filters [8]. For instance, two channels at 10 Gbit/s have been successfully transmitted over 19 Mm, using sliding filters and 30 km amplifier spacing [9]. However, larger amplifier spacing and/or larger number of channels lead to asymmetrical collisions and pseudo-phase matched four-wave-mixing (FWM) between channels [10]. Additional timing jitter and amplitude fluctuations result, which severely degrade the transmission performance. In order to overcome this difficulty, different dispersion management (DM) schemes have been proposed: exponential dispersion-decreasing fibre (DDF) between amplifiers [11] that removes all power-variation induced effects (asymmetric collisions, pseudo-phase-matched FWM and continuum generation), and dispersion management between anomalous and normal dispersion fibres [12] that reduces FWM by high local dispersion and residual jitter by low average dispersion. The former technique has been exploited with passive sliding narrowband filters making possible, for instance, to achieve 8×10 Gbit/s transmission over 9 Mm [13] and with active regeneration to achieve a transmission of 5×20 Gbit/s over 10 Mm [14]. The latter technique has been applied to 3×20 Gbit/s transmission over 8 Mm in conjunction with MUX/DEMUX devices to individually adjust the residual dispersion cumulated by the different channels [12], and to 4×20 Gbit/s transmission over 2 Mm with management between standard single-mode fibre and negative dispersion compensating fibre [15]. A recent and extensive review of dispersion management techniques and soliton properties can be found in reference [16].

The aforementioned results make it relevant to investigate the potential of DM in transoceanic systems having significantly longer amplifier spacings. In this paper, we shall focus on the numerical and experimental study of a 4 $\times$10 Gbit/s WDM system with long (60 km) amplifier spacing and using passive control technique. The system under consideration is based on a pratical N-step DM approaching the ideal exponential-like dispersion-decreasing fibre (DDF) to remove power-variation induced effects. Sliding narrowband filters are used to further reduce residual jitter and amplitude fluctuations. These filters also provide amplitude stabilisation that may partially compensate for EDFA gain variation over the multiplex range. Comparison between dispersion management rules, choice of the number N of steps and impact of dispersion fluctuation are key issues that will be addressed. In the next Section 2, we present the general link configuration and list the value of the main system parameters. Section 3 is devoted to the numerical design of the considered system, paying particular attention to the practical N-step DDF-implementation and on the relation between the required number of steps and amplifier spacing. Next, Section 4 describes experimental performance obtained using a recirculating loop setup. Careful comparison between experimental and numerical results is made, showing a very good agreement. Finally, we emphasise in Section 5 the unfavorable conditions of the loop configuration in contrast to that related to an actual straight-line configuration. Impact of dispersion fluctuation and uncertainty in a practical system is also assessed.

2. General Link Configuration of Passively-controlled DDF WDM Systems

It is known that the propagation equation of pulses through an actual fibre with loss reduces to the ideal lossless equation when chromatic dispersion exponentially decreases along its length like the optical power. Then, ideal solitons can be transmitted as the exact balance between exponentially- decreasing dispersion and also exponentially-decreasing Kerr nonlinearity is kept along all the line. The dispersion-decreasing configuration is schematically presented in Figure 1, assuming that loss is periodically compensated for by erbium-doped fibre amplifiers (EDFA). The figure also shows how DDF can be implemented in practice by use of N discrete steps following the ideal profile [11, 17].

The system also involves four transmitters at 10 Gbit/s assuming ideal sech^2-pulses and Fabry-Pérot (FP) filters after each amplifier. At the receiver side, a narrow filter selects one channel and the signal is detected by

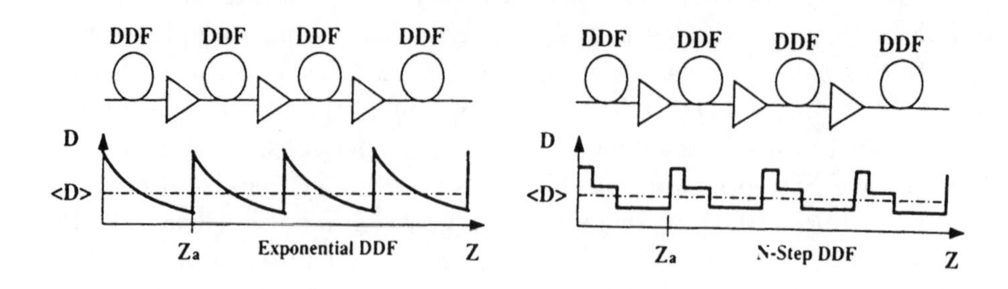

Figure 1. General link configuration and principle of dispersion-decreasing management, with both continuous and practical N-step implementation.

a square-law detector. The main values of the different parameters used in the following numerical study are listed in Table 1.

These values are as close as possible to realistic system. The performance of WDM transmission is computed numerically through the scalar nonlinear Schrödinger equation. After demultiplexing at the link output, amplitude and timing Q-factors are evaluated for each channel, while the minima between all these Q-factors are taken as the final value (Q = 6 corresponds to a BER equal to 10^{-9}). Performing several simulations with different amplifier noise seeds, we estimate a Q-factor accuracy equal to 0.5. These parameter values have been optimised through extensive simulations [18, 19], based on the ideal exponentially-decreasing dispersion case. Optimum in-line F-P filter FWHM width, defined as the ratio between free spectral range (FSR) and Finesse, is strongly coupled to filter spacing (assumed to be equal to the amplifier spacing). For filter spacing equal to 30, 40, 50, 60 and 70 km, the optimum filter width is 0.93, 0.80, 0.72, 0.66 and 0.61 nm respectively. Corresponding optimum amplifier output power is 4.5, 5, 6, 6.5 and 7 dBm, respectively, while amplifier output power is optimised with a 0.5 dB resolution.

3. Numerical Design of the Step-wise Exponential Dispersion Management

3.1. CHOICE OF A DISPERSION MANAGEMENT RULE

Let us consider the transmission of 4×10 Gbit/s WDM soliton channels up to 9 Mm and with an amplifier spacing Za = 60 km. First, one must note that without dispersion management, Q-factors rapidly decay [19]; this results from the relatively short collision length Zc ($= 2T_{FWHM}/D\Delta\lambda$),

TABLE 1. List of the main parameters and their values used in the numerical simulations.

Soliton transmitters :	Bit rate	10 Gbit/s
	Number of channels	4, temporally interleaved
	Pulse shape and width T_{FWHM}	$Sech^2$ pulses, 20 ps wide
	Channel spacing ($\Delta\lambda$)	0.8 nm
	Central multiplex wavelength	1557.3 nm
	Pulse sequence length	128 bits
In line fibre :	Average dispersion, D	0.4 ps/nm·km
	Dispersion slope	0.08 ps/nm^2·km
	Attenuation	0.22 dB/km
	Kerr coefficient n_2	2.7×10^{-20} m^2/W
	Effective area	50 μm^2
Amplifiers :	Amplifier type	1480 nm-pumped EDFA
	n_{sp}/C_1	3 dB
	Gain shape and bandwidth	Lorentzian; 25 nm FWHM
Guiding filters :	Filter type	Fabry-Pérot, 0.8 nm FSR (phase response included)
	Filter FWHM width	0.61 to 0.93 nm
	Sliding rate	+7 GHz/Mm
Receiver :	Optical filter and bandwidth	Rectangular shape; 0.8 nm
	Electrical filter and bandwidth	5th-order Bessel; 5.6 GHz
	Receiver acceptance window	70 ps

equal to 40 km for the outermost channels and lower than the 2×Za limit [20], even for the shorter studied amplifier spacing Za = 30 km. Conversely, with the ideal exponential profile, the quality factor is seen to reach 9.5 at 9 Mm. This is illustrated in Figure 2, showing the evolution of the worst-channel Q-factor against propagation distance.

Concerning the N-step approach, different DM rules are available in the literature. The "HAS"rule derived by Hasegawa *et al.* [11] is obtained by minimising the residual frequency shift produced by asymmetrical collisions. On the other hand, the "FOR"rule derived by Forysiak *et al.* [17] is obtained by minimising an empirical error function between the ideal lossless propagation equation and the actual lossy one. Both "HAS"and "FOR "rules converge toward the ideal exponential profile for large number of steps. Table 2 gives the values of the step length and dispersion according to both rules in the case of N = 4 DM, 60 km amplifier spacing and 0.4 ps/nm·km average dispersion. Q-factor evolution against propaga-

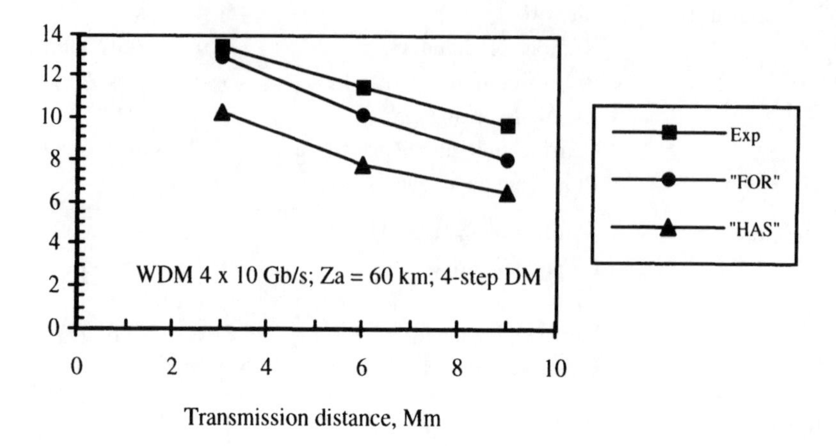

Figure 2. Evolution of the numerical worst-channel Q-factor against propagation distance, in case of 4×10 Gbits/s WDM transmission, and comparison between performance of ideal exponential dispersion management and 4-step DM following "HAS"and "FOR "DM rules.

TABLE 2. Step-length and step-dispersion in case of 4-step 60 km amplifier spacing system according to the dispersion management rule. Average dispersion is 0.4 ps/nm·km in both cases.

	HAS-like management		FOR-like management	
i	Zi [km]	Di [ps/nm·km]	Zi [km]	Di [ps/nm·km]
1st step	5.3	1.12	8.6	1.03
2nd step	7.4	0.81	11.0	0.63
3rd step	12.0	0.50	15.4	0.33
4th step	35.3	0.17	25.0	0.12

tion distance using these 4-step management rules is also plotted in Figure 2. This shows that the 4-step approach already induces a penalty on the Q-factors with respect to the ideal exponential DM. However, we shall consider that this penalty is acceptable in the FOR-type management ($Q = 8$ at 9 Mm). For larger number of steps (≥ 8), we did not observe any performance difference between the management rules, but when the number of steps becomes relatively small we always obtained best performance with management following the FOR-type rule.

Let us note however that this does not demonstrate that the FOR-type rule is the optimum one, nor that it does not exist cases where HAS-type rule gives better performance. In the following, we shall use the FOR-

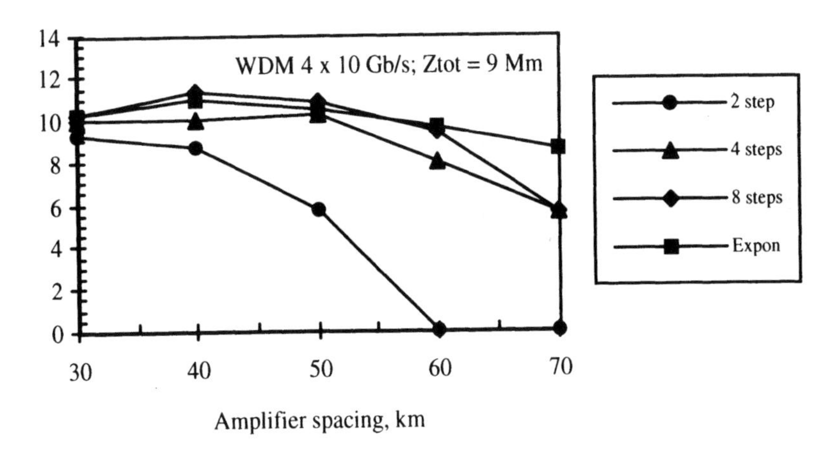

Figure 3. Theoretical worst-channel Q-factors of 4×10 Gbit/s system over 9 Mm versus amplifier spacing and number of DM-steps. Ideal exponential management case is also plotted.

type rule for assessing the relation between amplifier spacing and required minimum number of steps.

3.2. RELATION BETWEEN AMPLIFIER SPACING AND NUMBER OF STEPS

Next, we consider the impact of the amplifier spacing and number N of dispersion-management steps, in the case of 4×10 Gbit/s transmission over 9 Mm. Results are plotted in Figure 3, showing the evolution of the worst-channel Q-factor against amplifier spacing. Exponential DM is also reported in Figure 3, and may be considered as the optimum case, within the 0.5 uncertainty range of the Q-factors.

It appears that with 2-step DM, Q-factor up to $Q = 9$ can be achieved, providing the amplifier spacing does not exceed 40 km. For larger 60 km amplifier spacing, 4-step DM gives acceptable performance ($Q = 8$) with only a small penalty with respect to the ideal DM ($Q = 9.5$) as already deduced from Figure 2. Above 60 km amplifier spacing, Q-factor degradation becomes noticeable, even with 8-step DM, and results from the discrepency between N-step and exponential DM as well as from the noise increase in larger amplifier-spacing systems. One also observes that 8-step DM yields performance nearly equal to the ideal exponential management case, up to 60 km amplifier spacing.

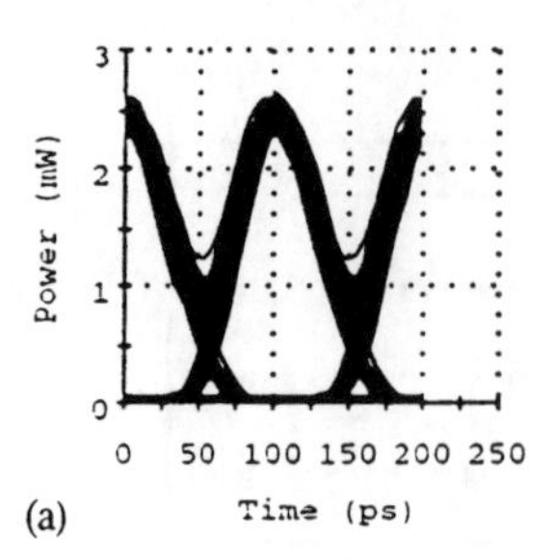

(a)

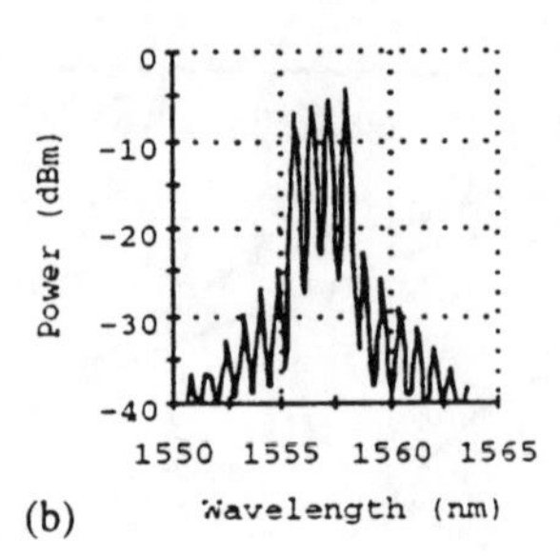

(b)

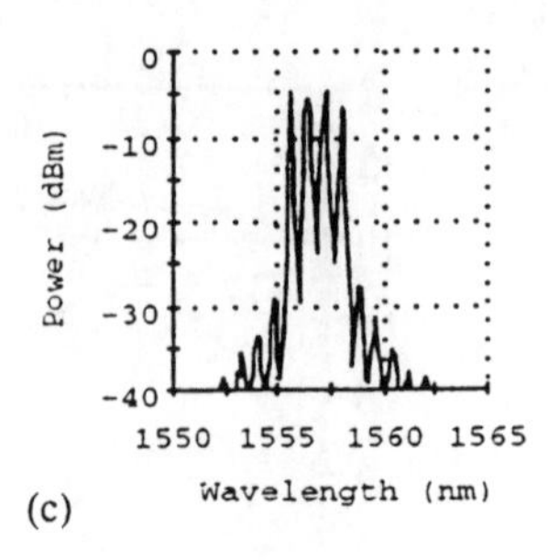

(c)

Figure 4. Numerical worst-channel eye diagram (a) and total output optical spectrum (b) at 9 Mm for ideal exponential-DM, assuming 25 nm-wide EDFA gain bandwidth. Optical spectrum (c) in case of 15 nm-wide EDFA gain bandwidth, showing strong amplitude imbalance between channels. [resolution: 0.1 nm].

3.3. AMPLITUDE STABILISATION BY SLIDING NARROWBAND FILTERS

For illustrative purpose, worst-channel eye diagram and total output optical spectrum are presented in Figure 4, in the case of 40 km amplifier spacing and exponential DM. It is known that the different channel powers follow the fibre chromatic-dispersion slope, in result from the soliton effect [21]. In addition, narrowband filters provide soliton amplitude stabilisation, and are able to partly compensate EDFA gain variation over the wavelength multiplex [21]. This is one of the advantages of this passive control technique. In our numerical tool, EDFA gain shape is simply modelled by a 25 nm-wide Lorentzian-like shape. It results a gain difference of approximately 0.04 dB per amplifier between central and outermost channels. Considering 40 km amplifier-spacing case, this would give a cumulated linear gain difference as high as 9 dB over 9 Mm. Nevertheless, we observe in Figure 4 that the power of each channel remains at its soliton power value, confirming that in-line narrowband filtering provides automatic gain equalisation.

Figure 4 also shows optical spectrum in a very unfavorable case where a 15 nm-wide Lorentzian shape is assumed (this somewhat low value will be artifactly encountered in the loop setup, see below). In that case, the cumulated linear gain difference is as high as 24 dB over 9 Mm and cannot be compensated for by narrowband filtering. Channels are no longer correctly transmitted, and transmission is not feasible over 9 Mm. However, this extreme situation will not be met with practical straigth-line systems.

In conclusion, we have shown that in 4×10 Gbit/s WDM transmission over 9 Mm with sliding filters, N-fold DM with large number of steps yields near-ideal transmission as determined by exponential DDF. With

DM based on fewer steps, the management rule is found to be determinative for optimal performance, that remains acceptable for a 4-step DM 60 km amplifier spacing. In addition, sliding filters provide an automatic feedback compensating gain curvature of realistic EDFA's. On the basis of these numerical results, 4-step DM with 60 km amplifier spacing is experimentally implemented in a single-span recirculating loop [22], detailed in the next section.

4. Experimental Study of the Step-wise Dispersion Management

4.1. EXPERIMENTAL SETUP

The setup is shown in Figure 5. On the transmitter side, 20 ps-wide solitons are generated by four independent Integrated-Laser-Modulators (ILM) driven at 10 GHz. The multiplex central wavelength is tuned near 1557.3 nm (0.8 nm channel spacing). Complementary outputs of a transmission analyser are used to code the soliton streams through $LiNbO_3$ modulators. In order to decrease channel interaction, adjacent channels are orthogonally polarised. On the receiver side, channels are selected by a high-finesse 20 GHz filter. This filter is slided during propagation in order to match the selected-channel wavelength. The selected signal is input to a 10 Gbit/s receiver for BER measurements. Receiver acceptance window is estimated near 55–60 ps.

The recirculating loop contains a single-span, 60 km-long 4-step DDF fibre approaching the rule of reference [16]; the average attenuation is 0.23 dB/km, including splices. The average chromatic dispersion is measured to be 0.54 ps/nm·km, which is somewhat higher than the optimum value of 0.4 ps/nm·
km we had targeted [19]. Three Fabry-Pérot filters (free-spectral-range = 0.8 nm; Finesse = 1) provide narrowband filtering and are made to slide at a 7 GHz/Mm rate. The equivalent filter width is estimated to be near 0.6 nm, value that matched the higher actual fibre average dispersion.

Finally, the loop includes two 1480 nm-pumped EDFA's for loss compensation of fibre and of extra components, namely an acousto-optic switch, an in-and-out 3 dB coupler and an additional flat-top filter to select t he 1558 nm EDFA peak gain. In 4×10 Gbit/s experiments, an additional Mach-Zehnder filter was introduced to flatten the gain profile. Electrical delays control the pulse loading, recirculation and refreshing sequence.

4.2. PERFORMANCE OF 2, 3 AND 4×10 GBIT/S WDM TRANSMISSION

Well-behaved system operation was verified by obtaining error-free transmission (BER $< 10^{-9}$) at 1×10 Gbit/s over 19 Mm, the maximum distance

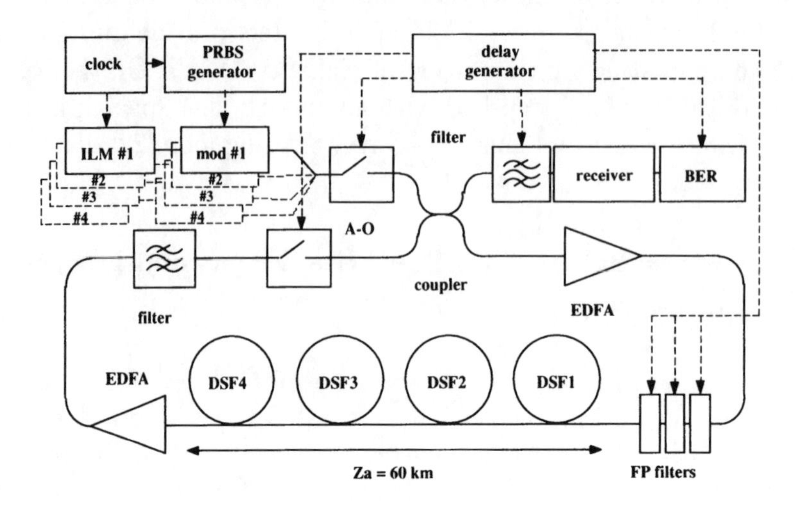

Figure 5. Experimental recirculating loop setup working at 4×10 Gbit/s bit rate and comprising a single-span 60 km-long, 4-step DDF. Average dispersion is 0.54 ps/nm·km and filters are slided at 7 GHz/Mm rate. ILM: integrated laser modulator; AO: acousto-optic; DSF: dispersion shifted fibre; F-P: Fabry-Pérot.

achievable with the electrical delays. When reversing the fibre span (from lower to higher dispersion sides), a strong penalty was observed, showing that, with such large amplifier spacing, correctly designed DDF fibre is already suitable to remove dispersive waves produced by average solitons. This result matches the transmission distance obtained with large effective-area fibre and 63 km amplifier spacing [23].

Wavelength-division multiplexing at 2×10 Gbit/s was performed next. Error-free distance (EFD) is obtained over 15 Mm for the worst channel, as shown in Figure 6. This last result represents a significant improvement of previous single-channel 20 Gbit/s, 10 Mm, 63 km amplifier spacing (but without DM) reported in the literature [22].

We consider then 3×10 Gbit/s transmission. Longer-wavelength channels propagate over 9 Mm while the shortest-wavelength channel is error-free up to 7.0 Mm. This results also represents a record 30 Gbit/s transmission performance with 60 km amplifier spacing.

Two EDFA's are used for a single fibre span. It results an accentuated gain difference of approximately 0.04 dB between channels for each round trip. This would give a cumulated linear gain difference as high as 5 dB over 7 Mm. Nevertheless, we observed that the powers remain nearly equal for the three channels (note: there is no initial pre-emphasis). This feature confirms that guiding filters provide automatic gain equalisation [20], as

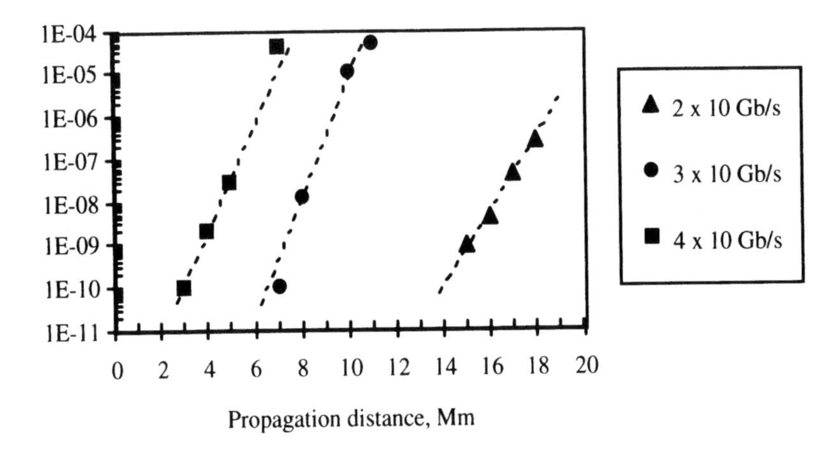

Figure 6. Experimental BER evolution against propagation distance in N $\times 10$ Gbit/s transmission, with N = 2, 3 and 4 channels.

discussed in the previous section.

As concerns transmission of 4 channels at 10 Gbit/s, the maximum (linear) gain difference between central and outermost channels is estimated to be near 16 dB over 7 Mm. This high value, indirectly due to short loop length, cannot be compensated for by guiding-filter stabilisation. For this reason, we inserted an additional filter of opposite curvature, in order to obtain a locally flatter gain. Error-free transmission is then obtained for all channels up to 3.9 Mm (see Figure 6).

4.3. COMPARISON BETWEEN EXPERIMENTAL AND NUMERICAL RESULTS

Experimental performance (i.e. Q = 3 at 9 Mm) for the 40 Gbit/s configuration is found substantially worst than theoretically expected (Q = 8 at 9 Mm, see Section 3). In order to identify the reasons for this discrepancy, numerical simulations taking into account actual loop configuration were carried out. The gain shape of both EDFA was modelled by a 25 nm-wide Lorentzian profile and the actual dispersion profile used in the experiment was included. Worst-channel Q-factor evolution is plotted in Figure 7, taking account for the actual 60 ps receiver acceptance-window. From these simulations, one can first observe that the agreement between theory and experiment is quite good. This validates the reliability of our simulation tool in quantitative predictions. Second, we identify that error-

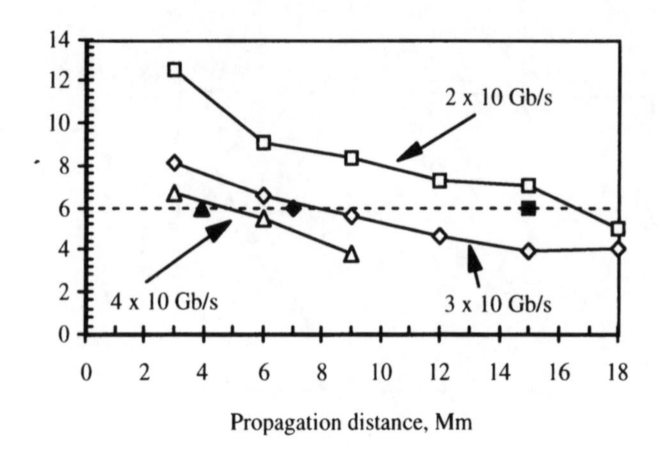

Figure 7. Filled symbols: experimental propagation distance at Q = 6 (BER = 10^{-9}) for 2, 3 and 4×10 Gbit/s WDM transmission respectively. Open symbols: numerical results taking account for actual loop parameters. A very good agreement is found, that demonstrates the validity of the numerical tool.

free distance degradation with respect to the expected one is due first to the discrepancy between the optimal 0.4 ps/nm·km average dispersion and the experimental one of 0.54 ps/nm·km, second to the noise of the additional amplifier and third to the narrower receiver acceptance-window. The situation is summarised in Figure 8 in the case of 4×10 Gbit/s transmission, showing the numerical Q-factor evolution for the actual 4-step loop setup and the ideal 4-step profile.

Both curves are extracted from Figures 2 and 7 and shows that the loop setup decreases the Q-factor by more than 4 points at 9 Mm. Due to the single-span loop, imperfect management effects add up. This would not be the case in straight-line links, where dispersion fluctuations average out. For this type of system, the optimal dispersion predicts an error-free distance well over 10 Mm. This practical aspect is discussed in the next Section 5.

In conclusion of this section, experimental error-free distances over 15, 7 and 3.9 Mm have been obtained in 2, 3 and 4×10 Gbit/s WDM soliton transmission with an amplifier spacing as large as 60 km, while using 4-step dispersion management and sliding filters. These results are found in good agreement with numerical predictions, with a substantial penalty stemming from the actual single-span loop arrangement and non-optimal average dispersion value.

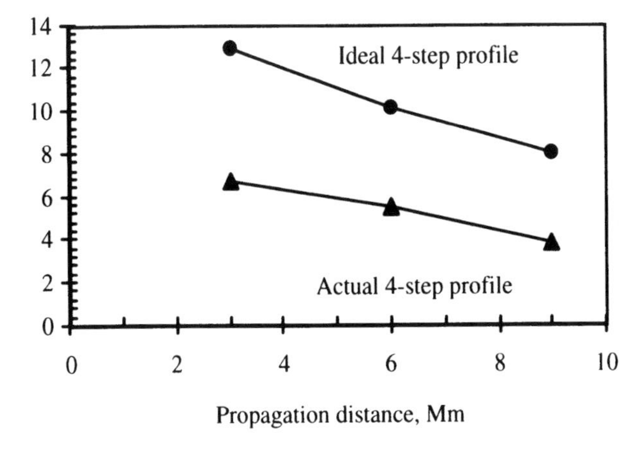

Figure 8. Numerical comparison between worst-channel Q-factor evolution against propagation distance in the case of 4 × 10 Gb/s WDM transmission with actual loop parameters (0.54 ps/nm·km average dispersion, 2 EDFA's per fibre span) and with ideal straight-line parameters (0.40 ps/nm·km average dispersion and 1 EDFA per fibre span).

5. Practical Considerations Concerning Chromatic Dispersion Uncertainty

Due to manufacturing tolerance, zero-dispersion wavelength λ_0 fluctuates from one to another fibre. One may consider that λ_0 follows truncated Gaussian distribution, centred around the manufacturers target value and with a standard deviation within a 5 to 10 nm range. It corresponds to a dispersion standard deviation of 0.4 to 0.8 ps/nm·km. In all practical systems, this uncertainty is too large to be acceptable. The difficulty is generally solved by sorting and selecting fibre dispersion within tighter tolerance. As concern DM case, the dispersion uncertainty is quite large compared to the difference between the dispersion values of the fibre steps (see Table 1 for instance). From a practical point of view, the important question is then the following: as it will be necessary to measure the chromatic dispersion of the fibres and to arrange them into dispersion bins or classes, what is the maximum width of these classes from which fibres can be randomly selected to build the different spans, while avoiding too high Q-factor penalties?

To address this question, new simulations were performed with λ_0 class widths of $\Delta = 0.5$ nm, 1.25 nm and 2.5 nm, corresponding to dispersion class widths of 0.04, 0.1 and 0.2 ps/nm·km (assuming a dispersion slope of 0.08 ps/nm^2·km). As the class widths are relatively small compared to the initial standard deviation, we assumed a uniform chromatic dispersion distribution

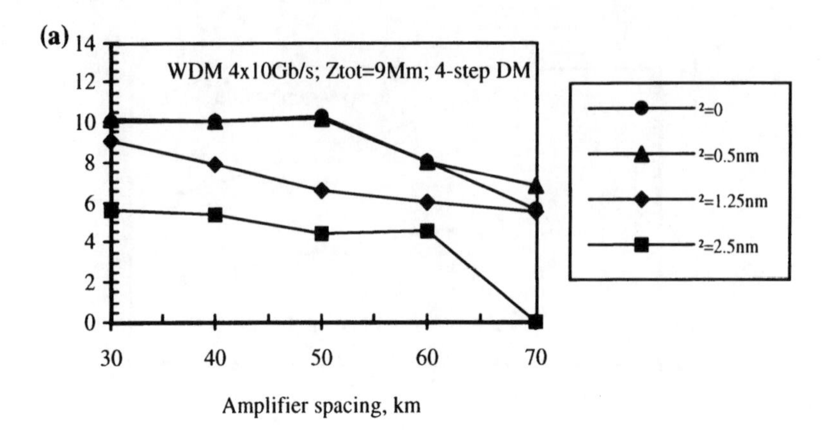

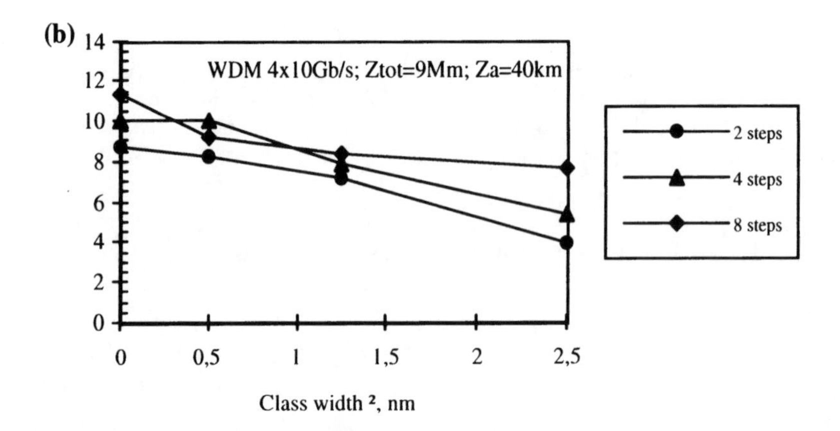

Figure 9. Numerical evaluation of the effect of zero-dispersion-wavelength fluctuation of fibres composing the different steps of the amplifier span. (a) Q-factor evolution of 4-step DM system vs. amplifier spacing, for zero-dispersion wavelength uncertainty ranging from $\Delta = 0$ to $\Delta = 2.5$ nm, (b) Q-factor evolution of 40 km-amplifier spacing vs. zero-dispersion wavelength uncertainty, for 2, 4 and 8-step DM.

within each class. Figure 9(a) shows the evolution of the worst-channel Q-factor against the amplifier spacing, for a 4-step DM, 4×10 Gbit/s system over 9 Mm. The λ_0 class width Δ ranges from 0 nm (ideal management, see Figure 2) up to 2.5 nm. In Figure 9(b), worst-channel Q-factor over 9 Mm is plotted against λ_0 class width assuming a 40 km amplifier spacing.

One first observes from Figure 9(a) that performance is not sensitive to fibre fluctuations when the λ_0 class width is within $\Delta = 0.5$ nm. Above 0.5 nm, we verified that this degradation is not due to actual average dispersion, that remains very close to the 0.4 ps/nm·km target, as this average value has an uncertainty roughly equal to dispersion class width divided by the square root of the number of amplifier spans. In fact, the degradation comes from the fact that the actual N-step dispersion profile in each span is different from the ideal one. In case of λ_0 class width equal to 1.25 nm, Q-factors remain acceptable for 40 km-amplifier spacing ($Q = 8$). This is no more the case for larger amplifier spacing with, for instance, Q-factor approaching 6 for 50 km-amplifier spacing. Second, at given amplifier spacing, one observe from Figure 9(b) that system performance is less sensitive to dispersion fluctuation for larger number of steps.

To summarise, systems with 4-step 60 km-amplifier spacing support 0.5 nm uncertainty in λ_0 value without any penalty. With 4-step 40 km-amplifier spacing however, uncertainty in λ_0 up to 1.25 nm can be accepted, while the Q-factor penalty remains reasonably small.

6. Conclusion

In conclusion, we have numerically and experimentally studied dispersion management issues in long amplifier-spacing WDM soliton systems. We considered 4×10 Gbit/s WDM system over transoceanic distance, exploiting step-wise dispersion-management to limit channel interaction and passive sliding filtering to control jitter and amplitude fluctuation.

First, we numerically determined the impact of DM rule and the relation between amplifier spacing and required number of management steps while maintaining acceptable performance. As concerns 4×10 Gbit/s 60 km amplifier-spacing systems over 9 Mm, 4-step management according to the "FOR"rule gave a Q-factor equal to 8.0, only 1.5 points lower than in the case of ideal exponential dispersion-decreasing management.

Second, experimental error-free distances over 15, 7 and 3.9 Mm have been obtained in 2, 3 and 4×10 Gbit/s WDM transmission with an amplifier spacing as large as 60 km, 4-step dispersion management and sliding filters. These results are found in good agreement with numerical predictions, with a substantial penalty stemming from the actual single-span loop arrangement and non-optimal dispersion, that would not occur in straight-line links. This validates the simulation tool reliability in quantitative predictions and confirms that in straight-line links of same configuration, Q-factors above 8 should be obtained up to 9 Mm for 4×10 Gbit/s WDM transmission and 60 km amplifier spacing.

Third, we studied system performance sensitivity to dispersion fluctuations in practical straight-line systems. We showed that sorting zero-dispersion wavelength λ_0 of the fibre pieces composing the amplifier spans within 0.5 nm eliminates any performance degradation with respect to the theoretical N-step design. Shorter 40 km amplifier-spacing systems admit higher λ_0 uncertainty (i.e. 1.25 nm) with a Q-factor as high as 8.

Finally, our experimental and theoretical results have been obtained with usual 1480 nm-pumped EDFA's and standard DSF. Drastic improvements of these results should be achievable by using 980 nm-pumping and large- effective-area fibre.

Acknowledgements

The Authors would like to thank P. Bousselet for realising the optical amplifiers and J.P. Hebert and D. Lesterlin for providing the ILM components.

References

1. L. F. Mollenauer, S. G. Evangelides and H. A. Haus: Long-distance soliton propagation using lumped amplifiers and dispersion shifted fiber, *J. Lightwave technol.*, **9**, (1991), p.194.
2. L. F. Mollenauer et al: Measurement of timing jitter in filter-guided soliton transmission at 10 Gbit/s and achievement of 375 Gbit/s-Mm error-free at 12.5 & 15 Gbit/s, *Opt.Lett.*, **19**, (1994), p.704.
3. M. Suzuki, I. Morita, N. Edagawa, S. Yamamoto, H. Taga and S. Akiba: Reduction of Gordon-Haus timing jitter by periodic dispersion compensation in soliton transmission, *Electron. Lett.*, **31**, (1995), p.2027.
4. M. Nakazawa and H. Kubota: Optical soliton communication in a positively and negatively dispersion-allocated optical fibre transmission line, *Electron. Lett.*, **31**, (1995), p.216.
5. N. Smith, W. Forysiak and N. J. Doran: Reduced Gordon-Haus jitter due to enhanced power solitons in strongly dispersion managed systems, *Electron. Lett.*, **32**, (1996), p.2085.
6. M. Nakazawa, E. Yamada, H. Kubota and K. Suzuki: 10 Gbit/s soliton data transmission over one million kilometres, *Electron. Lett.*, **27**, (1991), p.1270.
7. E. Maunand *et al.*: Parametric study of chromatic dispersion influence in 20 Gbit/s, 20 Mm regenerated soliton systems with up to 140 km amplifier spacing, *Electron. Lett.*, **32**, (1996), p.1022.
8. L. F. Mollenauer, E. Lichtman, G. T. Harvey, M. J. Neubelt, B. M. Nyman: Demonstration using sliding-frequency guiding filters of error-free transmission over more than 20 Mm at 10 Gbit/s, single-channel, and more than 13 Mm at 20 Gbit/s in a two-channel WDM, *Electron. Lett.*, **29**, (1993), p.910.
9. F. Pitel *et al.*: Experimental assessment of wavelength margin in 20 Gbit/s WDM soliton system over 18 Mm distance, *Electron. Lett.*, **21**, (1996), p.1957.
10. P. V. Mamyshev and L. F. Mollenauer: Pseudo-phase-matched four-wave mixing in soliton wavelength-division multiplexing transmission, *Opt. Lett.*, **21**, (1996), p.396.
11. A. Hasegawa, S. Kumar and Y. Kodama: Reduction of collision-induced time jitters in dispersion-managed soliton transmission systems, *Opt. Lett.*, **21**, (1996), p.39.
12. M. Suzuki, I. Morita, N. Edagawa, S. Yamamoto and S. Akiba: 20 Gbit/s-based soliton WDM transmission over transoceanic distances using periodic compensation of dispersion and its slope, *ECOC'96*, paper **ThB.3.4**, (1996).
13. L. F. Mollenauer, P. V. Mamyshev, M. J. Neubelt: Demonstration of soliton WDM transmission at up to 8 $\times$ 10 Gbit/s, error-free over transoceanic distances, *OFC'96*, **PDP-22**, San Jose, (1996).
14. M. Nakazawa *et al.*: 100 Gbit/s WDM soliton transmission over 10,000 km using in-line synchronous modulation and optical filtering, *Proceedings of OFC'97*, **PD21**, (1997).
15. F. Favre, D. Le Guen and F. Devaux: 4$\times$20 Gbit/s soliton WDM transmission over

2,000 km with 100 km dispersion-compensated spans of standard fibre, *Electron. Lett.*, **33**, (1997), p.1234.
16. A. Hasegawa, Y. Kodama and A. Maruta: Recent progress in dispersion-managed soliton transmission technologies, *Optical Fiber Technology*, **3**, (1997), p.197.
17. W. Forysiak, F. M. Knox and N. J. Doran: Average soliton propagation in periodically amplified systems with stepwise dispersion-profiled fibre, *Opt. Lett.*, **19**, (1994), p.174.
18. E. Kolltveit, J. P. Hamaide and O. Audouin: Optimum two-step dispersion management in WDM soliton systems, *Electron. Lett.*, **32**, (1996), p.1858.
19. S. Cardinal, E. Desurvire, J. P. Hamaide and O. Audouin: Wavelength- multiplexed transoceanic soliton systems:comparison step-wise exponential management techniques, *Electron. Lett.*, **33**, (1997), p.77.
20. L. F. Mollenauer, S. G. Evangelides, J. P. Gordon: Wavelength division multiplexing with solitons in ultra-long distance transmission using lumped amplifiers, *J. Lightwave technol.*, **9**, (1991), p.362.
21. P. V. Mamyshev and L. F. Mollenauer: Wavelength-division-multiplexing channel energy self-equalization in a soliton transmission line by guiding filters, *Opt. Lett.*, **21**, (1996), p.1658.
22. J. P. Hamaide, B. Biotteau, F. Pitel, O. Audouin, E. Desurvire, P. Bousselet, J. P. Hebert and D. Lesterlin: Experimental investigation of N$\times$10 Gbit/s WDM soliton transmission with large 60 km amplifier spacing and 4-step dispersion-decreasing management, *ECOC'97*, **3**, (1997), p.107.
23. J. P. Hamaide, F. Pitel, P. Nouchi, B. Biotteau, J. Wirth, P. Sansonetti and J. Chesnoy: Experimental 10 Gbit/s sliding frequency-guiding soliton transmission up to 19 Mm with 63 km amplifier spacing using large effective- area fiber management, *ECOC'95*, Paper **Th.A.3.7**, (1995).
24. F. Favre, D. Le Guen and M. L. Moulinard: Robustness of 20 Gbit/s 63 km span 6 Mm sliding-filter controlled soliton transmission, *Electron. Lett.*, **31**, (1995), p.1600.

(Received January 16, 1998)

REMARKABLE FEATURES OF DM SOLITONS: IMPLICATIONS FOR HIGH SPEED AND WDM SYSTEMS

N. J. DORAN, W. FORYSIAK, J. H. B. NIJHOF,
A. M. NICULAE AND A BERNTSON
Aston University
Division of Electronic Engineering & Computer Science
Aston Triangle, Birmingham B4 7ET, UK.

Abstract. Dispersion managed solitons are numerically investigated both for single channel and WDM applications. We show the dependence of enhanced power on dispersion map strength and depth for a wide range of parameters, for which we can obtain stable pulses, with no evidence of significant radiation. For strong dispersion maps, WDM interactions are shown to be substantially weaker than predicted by current analytic models.

1. Introduction

In soliton transmission the fibre dispersion and nonlinearity are intimately related. Ideal solitons require both dispersion and nonlinearity and, for a specific pulse duration, the soliton power is simply proportional to the dispersion. In real systems, where fibre loss is present, the relationship between dispersion and nonlinearity is modified according to the average power of the propagating solitons, giving rise to the notion of a so-called *average soliton* [1, 2]. The average soliton, has almost identical properties to an ideal soliton, apart from a rapid variation in pulse power, which must take place over a distance short compared to the soliton period, to ensure pulse stability. The remarkable discovery of very recent years is that by suitable arrangement of the fibre dispersion a new type of soliton, which has been called a *dispersion managed*, or *DM soliton*, may have improved stability over solitons where the dispersion is constant [3, 4, 5]. At first glance, this is surprising. However, on closer examination, it may not be so unexpected, since the perturbations associated with soliton transmission (loss, amplifier noise, inter-soliton interactions) introduce periodic disturbances, so that

A. Hasegawa (ed.), New Trends in Optical Soliton Transmission Systems, 303–316.

some matching behaviour of the dispersion is likely to lead to improved stability.

The most important perturbation in amplified systems (where solitons are likely to be employed) is the periodic gain/loss cycle, and the first example of dispersion management was the deliberate variation of the dispersion to compensate for the effect of fibre loss. The basic idea was first proposed by Tajima in 1987 [6] and is as follows: since the soliton power variation is exponential in distance, the dispersion should also be exponentially decreased to continuously maintain the balance between dispersion and nonlinearity. In principle, ideal soliton propagation is then restored. In practice, such fibre is difficult to manufacture, and therefore, subsequently, there have been several proposed schemes for approximating this exponential dispersion by piecewise constant dispersion fibre [7, 8, 9, 10]. It has been found, for example, that if an appropriate prescription is applied, then maps made up of a few stages (4 is usually sufficient) can effectively eliminate the loss/gain perturbation for single solitons or single wavelengths.

Very recently, it has been found that a new and simple form of dispersion management—the general term we shall use to describe a system with intentionally and designed dispersion variations—can be employed to reduce and control perturbations due to amplifier noise and interactions between channels in WDM systems [11, 12, 13, 14]. This form of dispersion management, where the transmission fibre is made up of fibres of both signs of dispersion, gives rise to a new type of stable propagating pulse, called a DM soliton, whose parameters vary periodically in concert with the dispersion map, and most importantly, whose power can be significantly enhanced. It now appears that DM solitons have properties which make them even better than conventional solitons for applications in high-speed communication systems, and in many cases, obviate the need for more complex control mechanisms such as sliding-guiding filters or active modulation.

This chapter concentrates on a numerical investigation of the effects of dispersion management which are important for the reduction of pulse timing jitter and inter-pulse interaction for high speed future communication systems. The chapter is arranged as follows. Section 2 studies a simple two-stage dispersion management map, and presents results for lossless single pulses illustrating the remarkable stability of DM solitons. The power enhancement as a function of the dispersion map parameters is identified as the key feature of these solitons. A detailed investigation gives useful information for optimum design of DM soliton systems and reveals particularly beneficial feature for WDM sytems. In Section 3 we present results on collisions in WDM channels showing a marked decrease in residual frequency shift. Thus, we show how this type of DM virtually eliminates the effect of loss on WDM interactions.

2. Two-Stage Dispersion Managed Solitons: Enhanced Power and Stable Pulses

The first issue to resolve is whether DM solitons in strong dispersion managed systems are stable with respect to the generation of radiation. If in the lossless case such pulses are non radiative, it gives confidence that if the correct pulse energy and initial form are chosen then they will indeed be stable over very long amplified chains in real systems.

In this section, we will identify parameter ranges where such stable pulses can be found. We first consider the stability of soliton pulses in a dispersion map with idealised lossless fibre composed of two fibres, each of length l and dispersion $\ddot{\beta}_1$ and $\ddot{\beta}_2$, of opposite signs, where $\ddot{\beta}_{1,2} = \ddot{\beta}_{av} \pm \Delta\ddot{\beta}/2$. Extensive numerical investigations show that for a very wide range of dispersion maps, stable pulses can be found. The general features of DM solitons are that their pulse power is enhanced, with respect to conventional solitons in fibre with the same uniform, average dispersion, and secondly, that the pulse undergoes a periodic evolution which has a chirp free point in the centre of each fibre section.

The detailed features depend on the particular dispersion map used, and can be accurately described in terms of two parameters: the map strength $S = l\Delta\ddot{\beta}/T^2$, where T is the pulse width (FWHM), and the map depth $\delta = \Delta\ddot{\beta}/\ddot{\beta}_{av}$. We have found stable pulses for a wide range of these parameters with the key features described above. These stable pulses were obtained using a numerical procedure based on full integration of the nonlinear Schrödinger equation, which starts from an approximate solution, confirms that the pulse dynamics are converging, and averages the pulse shape at the extrema of these oscillations to rapidly converge to the stable pulse shape, which is then confirmed by further integration with the averaging removed.

Figure 1 shows the dependence of power enhancement on the map strength S for three values of map depth δ, to illustrate the wide range of parameters for which stable pulses have been obtained using this numerical procedure. In addition to the numerically obtained points, an empirical fit to these data points is shown, which is discussed in detail below. The DM soliton pulses observed at each of the data points are indeed stable, and show no numerical evidence for significant radiation, and the pulses are periodically well-reproduced. If we, for example, follow the pulse peak power over several 100's of cycles there is no numerically measurable change.

To illustrate this remarkable stability, we show in Figure 2 three pulse evolutions at the mid-points of the fibre (unchirped positions). The centres of these pulses are Gaussian-like (we display the pulse shapes on a log power scale), while the tails are exponential with periodic zeros. A very

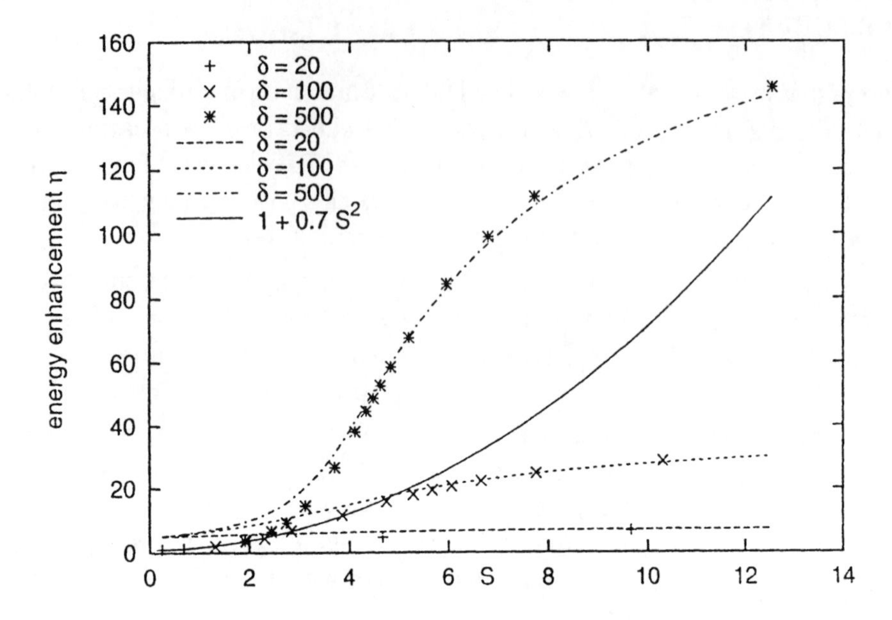

Figure 1. The power enhancement η as a function of map strength S for δ=500, 100 and 20. Dotted lines equation (2), continuous line equation (1).

important point here is that the tails are observed at very low powers so that they carry very little energy. In these examples (which are typical) the tails contain less than 0.01 % of the total pulse energy. Thus effectively the pulse shapes are indeed Gaussian with consequently weak interaction between adjacent pulses.

It is interesting to note the evolution of the pulse shape during one cycle. We have shown the pulses at the fibre mid points, where there is no chirp, but the phase is constant in each region with phase shifts π, at the boundaries of the oscillations (at the points where the field goes through zero). In contrast, at the boundary between the two fibres, the pulse is strongly chirped and much more sech-like in shape. Figure 2 also shows single cycle evolutions for three examples, illustrating the changing pulse shape within each period of the dispersion map. The obvious implication of these calculations is that the required pulse in a communication system depends on where it is launched into the cycle. If the start is at a fibre boundary then a chirped pulse should be launched; if the cycle starts with a half length of either type of fibre, then an unchirped pulse is appropriate. This observation remains broadly true when loss is included but the exact degree of initial chirp will depend on the amplifier spacing relative to the dispersion map. In the absence of DM there are two special points in an amplifier cycle where the pulses have minimum chirp, and thus the number

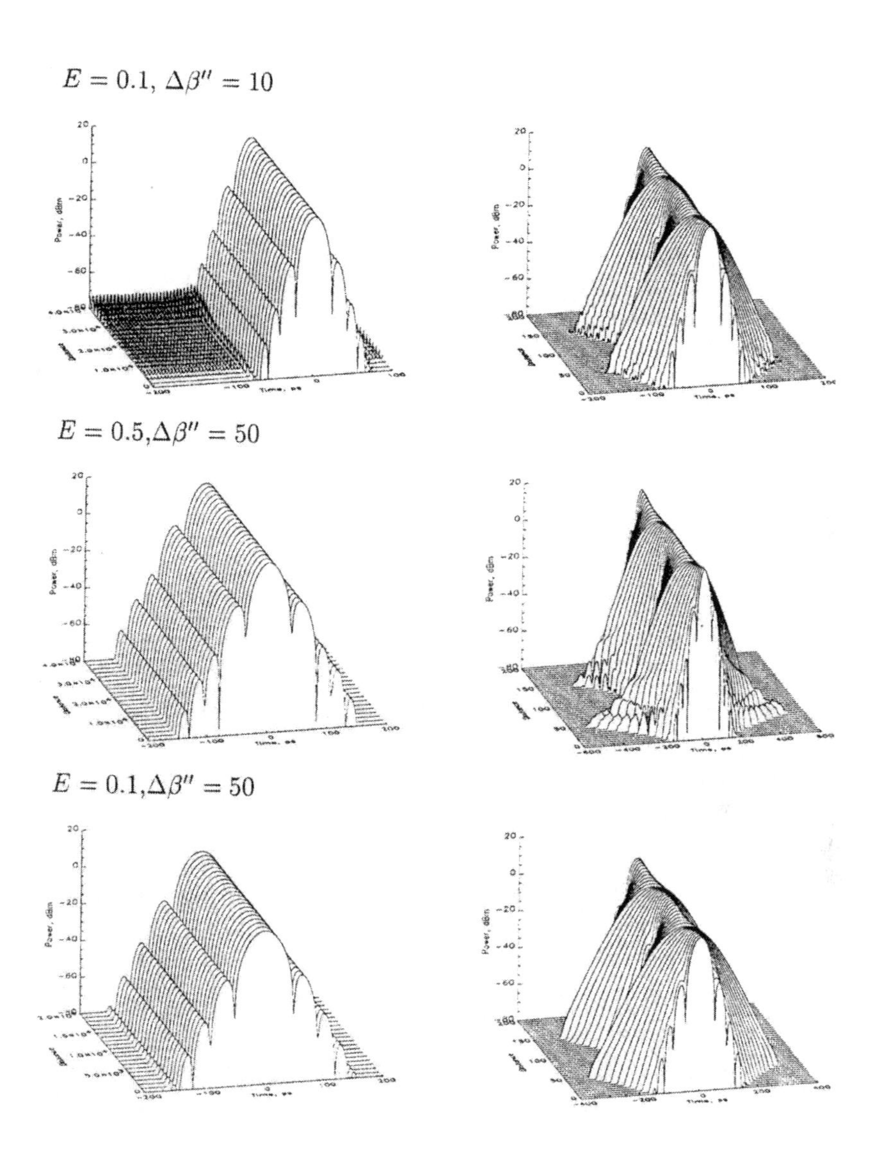

Figure 2. The pulse power at the midpoints of the anomalous fibre for three points from Figure 1. (i) $S = 6.8, \delta = 500, \eta = 98.9$; (ii) $S = 3.7, \delta = 500, \eta = 26.8$; (iii) $S = 3.9, \delta = 100, \eta = 11.7$. Also shown is the single period evolution for each pulse.

and positions of the minimum chirp points will depend on the relationship between the loss and DM cycles.

The energy enhancement is an important and essential feature of dispersion managed solitons. It permits, for example, a very low average dispersion whilst at the same time maintaining a useful pulse energy. In Figure

1, power enhancements of up to 140 are displayed. It is clear that the power enhancement depends not only on S but also on δ, the map depth. In general, for large power enhancements δ should be large, i.e. the local dispersions should be much greater than the average. In reference [3] we derived the empirical relationship

$$\eta = 1 + 0.7S^2 \tag{1}$$

which is a good representation of numerical observations provided δ is large ($\delta > 100$) and $S < 3$. For large S, however, the power enhancement saturates (as illustrated in the figure) so that it approaches, but never exceeds, the soliton power in the local highly anomalous dispersion fibre. In order to extend the region of applicability of our empirical model, we have derived a semi-empirical relation for enhanced power including δ. A simple form of this relation, which is valid for strong maps with $S > 4$, is given by

$$\eta(S,\delta) = \frac{a\delta}{S}\left(S - b + \sqrt{(S-b)^2 + \frac{cS}{\delta}}\right), \tag{2}$$

where $a = 0.2, b = 3.7$ and $c = 180$.

This function is plotted in Figure 1 for comparison with the numerical results for the three values of δ obtained there. Excellent agreement is obtained for large S, while for $S < 2$, equation (1) gives a good fit. The region, $S \sim 4$, is transitional and the fit depends on δ.

An interesting consequence of the results of Figure 1 is that DM gives a natural compensation for higher order dispersion (dispersion slope). In conventional soliton WDM systems the dispersion slope required increasing channels power with wavelength in proportion to the fibre dispersion. In DM soliton systems, however, the variation of power enhancement factor η, with $\ddot{\beta}_{av}$, offers a mechanism to maintain equal powers in each channel, even in the presence of third order dispersion. For simplicity, consider the case where the dispersion slope is the same in each fibre. In this case, the map strength S is equal for all wavelengths, but since $\ddot{\beta}_{av}$ increases with wavelength, the map depth δ decreases with wavelength. Thus, in principle the decrease in the power enhancement factor η with δ provides a means to equalise the channel powers. Equation (2) shows that such a dependence on δ occurs for large S and the results displayed in Figure 1 show that for all values of S the power enhancement increases with δ. This is an important and unexpected additional benefit of DM. This feature is illustrated in Figure 3.

Thus far we have considered only the case with anomalous average dispersion. We have also investigated the case of no net dispersion i.e. $\ddot{\beta}_{av} \to 0$ [15]. In this case, there is only one free parameter S. We can thus obtain

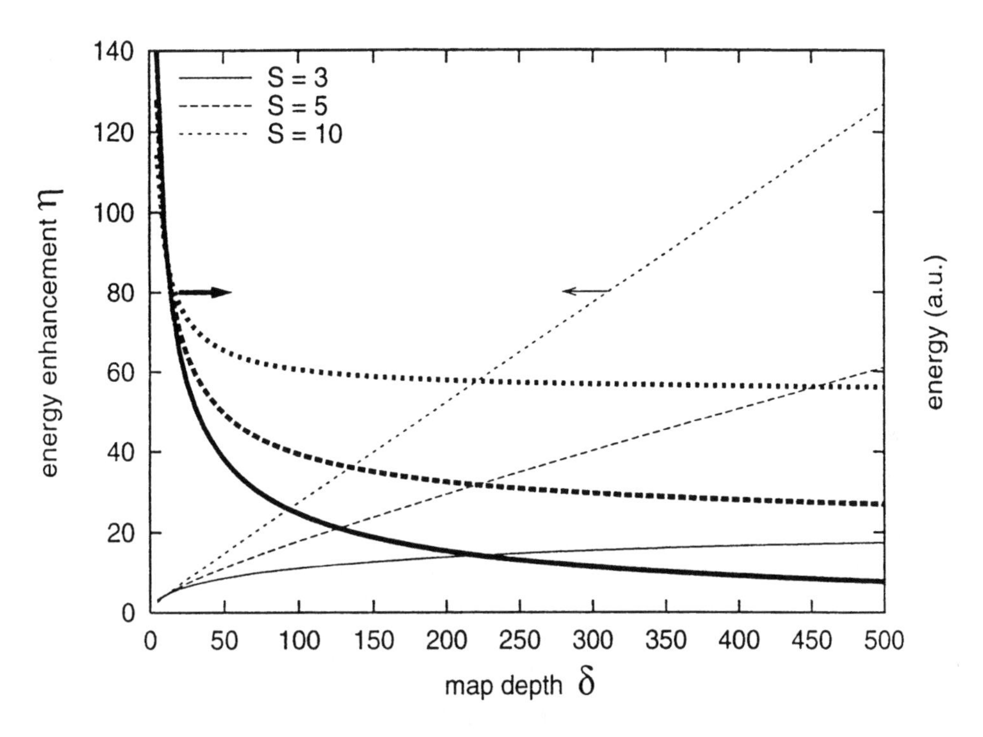

Figure 3. The power enhancement and pulse energy as a function of δ the map depth using equation (2).

all the information from one set of calculations which can be scaled to any arbitrary net zero dispersion map (or at least any symmetric map). Thus all that is needed is a knowledge of how the power varies with pulse duration. Figure 4 shows the results of calculation using the example of $\Delta\ddot{\beta} = 10$ ps^2/km and $l = 80$ km. The figure shows that the stable pulse duration decreases with power in the anomalous fibre section. As the power becomes small, the linear limit is approached and the curves converge to the point where $S = 4$. This is a curious result since the linear system can of course support any pulse duration, but it indicates a stability with respect to nonlinearity of a particular pulse duration. Thus we could consider designing a system around this result using the relationship between compensation i nterval, pulse duration and dispersion ie. $l = 4T^2/\Delta\ddot{\beta}$. For example, at 10 Gbit/s where 20 ps pulses are used, in standard fibre with $\Delta\ddot{\beta} = 40$, the optimum compensation interval is 40 km. This is close to the span used in several recent experiments for very long distance propagation [16, 17]. In contrast, the compensation interval for dispersion shifted fibre,

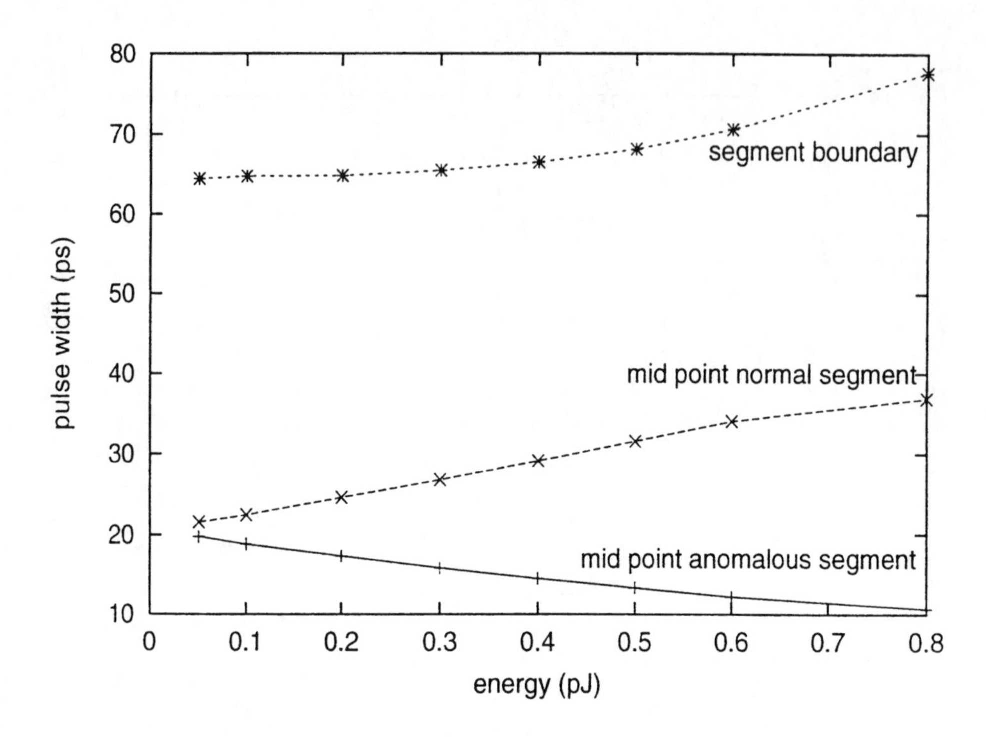

Figure 4. Pulse width as a function of pulse power for net zero dispersion.

for which $\Delta\ddot{\beta} \approx 1$, is excessively long, and would, for example, give poor compensation for amplifier induced jitter.

We have found from further numerical work that it is even possible to obtain non-radiating pulses for a net normal dispersive system. We have found it necessary to keep the net value of dispersion small but this result does imply that soliton-like propagation should be possible for systems with net normal dispersion and this may be important in WDM systems.

A variational approach with a Gaussian ansatz, where we allow the four parameters width, chirp, amplitude and phase, allows us to complete a power dependence with map strength S, for various values of average dispersion. This is shown in Figure 5 where we show contours of constant average dispersion. The strong solid line is for $\ddot{\beta}_{av} = 0$ and the crosses are the numerical points obtained from Figure 4. We see again the importance of the critical map strength S_c which is ~4 from numerics ~4.8 in this approximate approach. We see finite power at zero dispersion for $S > 4.8$. This is a most remarkable conclusion. If S is chosen greater than 4 then soliton-like pulses are stable at zero dispersion for finite power. This means, for example, that WDM is possible around the zero dispersion with both

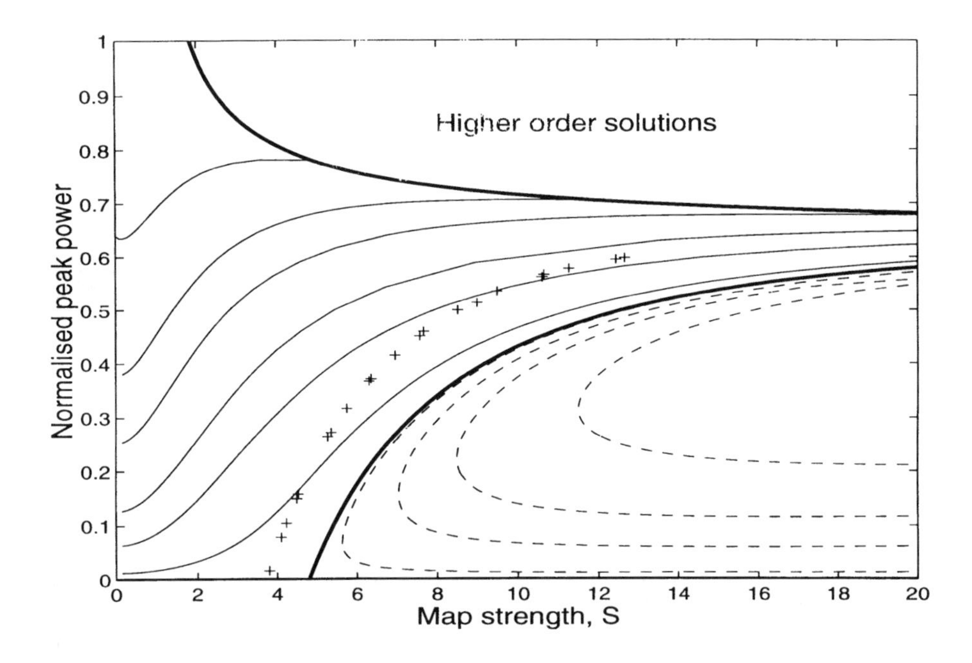

Figure 5. A contour plot of constant $\ddot{\beta}_{av}$ in the plane of the map strength S and peak power P obtained from a variational calculation. The lower bold line is for zero average dispersion. The crosses are numerically obtained results from the full NLSE for $\ddot{\beta}_{av} = 0$.

normal and anomalous net dispersion. The energy variations are reduced around the zero dispersion for larger S. Third order dispersion also does not lead to pulse break-up at zero dispersion. Thus, there is a clear message for system designers. The map should be chosen such that $S > 4$. Interaction problems will be the limit as to how large S should be, but operation in the region around zero dispersion should minimise jitter and permit WDM without the need for dispersion slope compensation.

3. Interactions Between DM Solitons in WDM Systems

In this section, we investigate interactions between DM solitons in different WDM channels, using both analytical and numerical approaches. As in the above, we consider a two stage map and concentrate on the simplest case of a two channel system.

In lossless constant dispersion fibre, it is well known that there is no exchange of energy between solitons from different WDM channels, so that there is only a time shift induced by collision. With dispersion management this may no longer be the case, because in general, a discontinuity

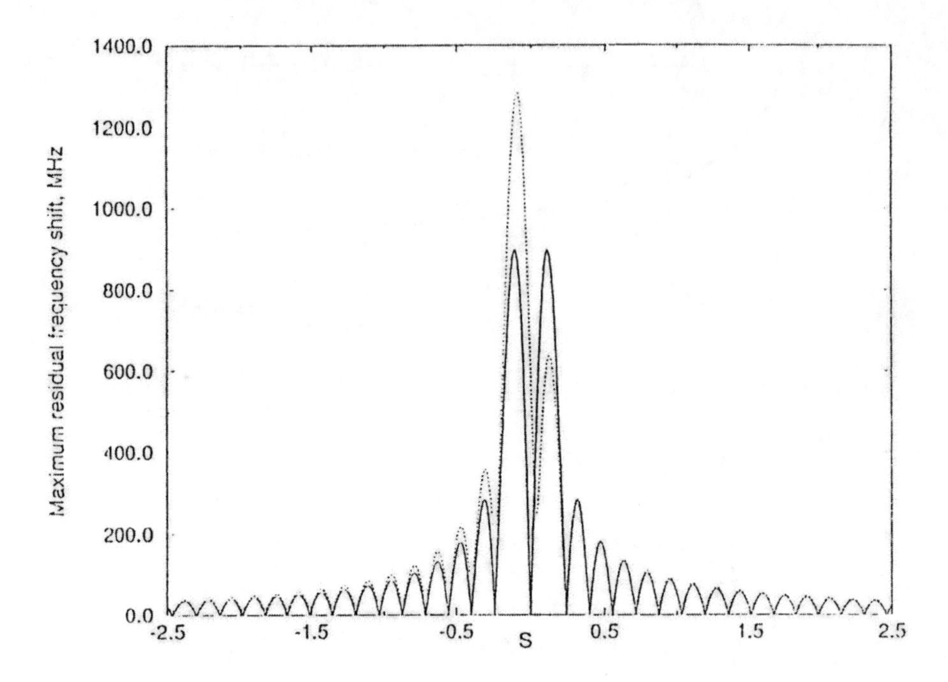

Figure 6. Frequency shift as function of map strength for $L_1 = L_2$ =25 km, L_a =50 km, $\ddot{\beta} = 0.64$ ps^2/km. Continuous line loss=0, dotted line loss=0.2 dB/km.

in dispersion, like the presence of fibre loss, destroys the symmetry of the collision process, and a net frequency shift of each soliton involved in the collision is to be expected.

Recently, it has been noted that the dynamics of DM soliton collisions in DM fibre are strongly modified due to the counter-propagating, zigzag motion induced by the dispersion map [12, 13, 14]. Thus, for a strong map, each overall collision involves a number of relatively fast, local collisions, whose frequency shifting effects can be averaged out. This process can be analysed, and quantitative predictions can be made from the analytic model of Mollenauer et al. [18] who obtained simple expressions for the residual frequency shift arising from soliton collisions in amplified systems comprised of lossy and dispersion-varying fibre. Figure 6 shows the predicted dependence of these frequency shifts on the strength of the map S, for both ideal, lossless and lossy fibres. The other parameters are pulse durations, $T = 20$ ps, $l = 25$ km, a channel spacing of 200 GHz and $\ddot{\beta}_{av} = 0.64$ ps^2/km, giving a collision distance of 50 km.

Figure 6 shows a regular sequence of zero's (for the lossless case) or near zero's (in the lossy case) as a function of the map strength S. In fact, the conditions for which these minima exist, can be predicted by simple analysis of the collision geometry within the dispersion map. The minima occur whenever the overall collision is made up of exactly an integer number of local collisions. Similarly, the maxima occur when the sum is equal to (n + 1/2) local collisions, which is the worst case for residual frequency shift. Thus, for any lossless system it is possible to choose an appropriate map strength S such that the collision induced frequency shift is exactly zero.

Another remarkable feature of Figure 6 is that the inclusion of loss has almost no effect for $S > 0.5$, as the two curves shown become almost indistinguishable. In the lossy case, the minima are close to but not exactly zero, whereas the maxima are essentially the same. Thus DM is effective in eliminating the collision asymmetry induced by loss provided that $S > 0.5$, which would be the case in most practical systems. Indeed, for these parameter values, DM virtually eliminates the effect of loss-induced frequency shifting provided $S > 1$.

In the presence of loss, Figure 6 shows a significant asymmetry for weak dispersion maps, indicating that as far as this effect is concerned, it is preferable to have the anomalous dispersion fibre first in the link. This observation reflects the well-established result that two-step piece-wise approximations to exponential variation of the soliton power, following the principle introduced by Tajima [6], is also effective reducing collision-induced frequency shifts. In this case, the 50 : 50 length split is far from optimum. However, it is arguable that even the optimal piece-wise exponential dispersion profiles are of little practical importance, because strong DM with $S > 1$ obviates the need either for piece-wise exponential dispersion management or even continuous, "ideal"exponential tapering of the dispersion.

The analytic results obtained above are applicable, strictly speaking, only to the adiabatic case where small variations in the pulse width and peak power occur and do not include the power enhancement. Figure 7 shows a comparison of this analysis with numerical calculations for strong dispersion maps using the stable pulses, obtained numerically, as described in Section 2. These numerical results show qualitative agreement with the analysis, but a striking reduction in the overall frequency shift. This surprising result, bearing in mind the significantly enhanced power of the DM solitons, and therefore the increased strength of cross-phase modulation during the interaction, can be explained by considering the dynamics of the soliton collisions. The maxima in the residual frequency shift response occur for the "worst case"when the first full collision of two DM solitons occurs at the boundary between the two fibres of the dispersion map. At this point, however, the nonlinear DM soliton interaction is significantly reduced due to the increased pulse widths at the boundary points. The benefit from the increased pulse widths dominates, in terms of residual frequency shift, over the potentially detrimental increase in the pulse amplitudes.

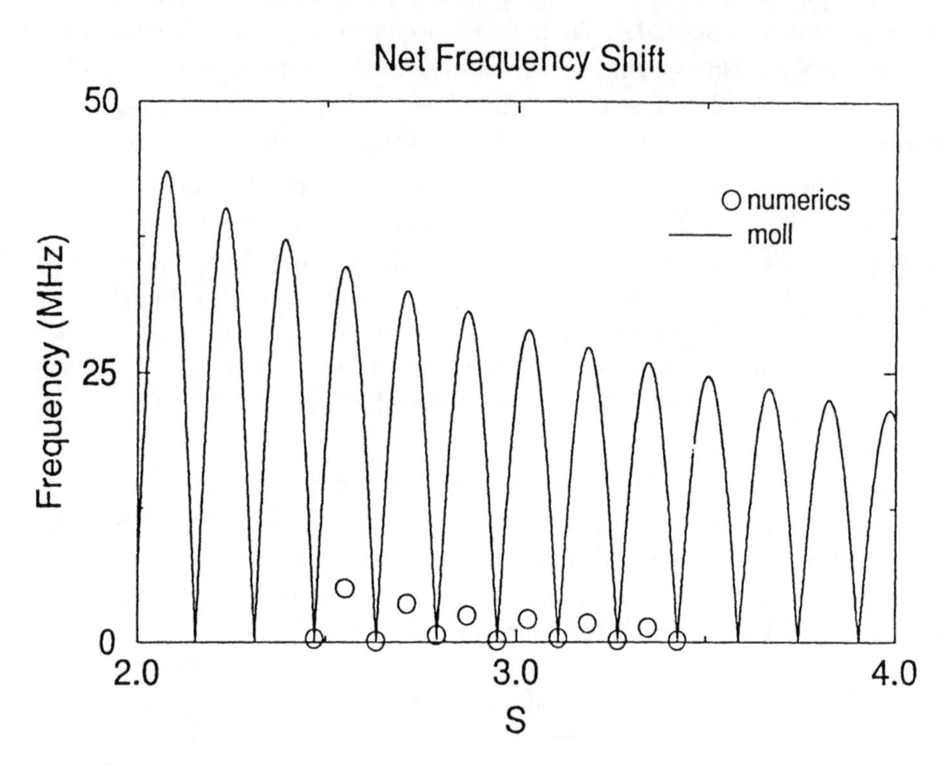

Figure 7. Frequency shift against S. Continuous line analytic [18], circles numerical. Parameters as in Figure 6.

These results show clearly that DM solitons are particularly suitable for WDM applications, and that somewhat surprisingly, DM is at least as effective as ideal exponential dispersion tapering. For a given set of parameters (channel spacing, average dispersion) there is a sequence of "ideal" maps where the residual frequency shift is almost zero. However, for strong dispersion management, such careful map design may not be necessary because the residual frequency shifts are inherently strongly suppressed by the modified DM soliton collision geometry and dynamics.

4. Conclusions

Dispersion managed (DM) solitons hold the key to future high data rate soliton systems. The features of enhanced power and reduced WDM interactions identified and quantified above make DM solitons suitable for simply designed systems.

The theoretical description of DM solitons is emerging but the numerical evidence is still intriguingly at odds with many of the current models. The results on single pulses presented here indicate that in conventional maps pulses with a Gaussian centre and periodic exponential tails are the norm. The tails contain little energy and thus the interaction for small power enhancements may be reduced. Thus very strong power enhancement could have an important role in WDM and jitter limited systems.

Acknowledgement

This work was supported by the UK Engineering and Physical Sciences Research Council.

References

1. Blow, K. J. and Doran, N. J.: Average soliton dynamics and the operation of soliton systems with lumped amplifiers, *IEEE Photon. Technol. Lett.*, **3**, (1991), pp.369-371.
2. Hasegawa, A. and Kodama, Y.: Guiding-center soliton, *Phys. Rev. Lett.*, **3**, (1991), pp.161-164.
3. Smith, N. J., Knox, F. M., Doran, N. J., Blow, K. J. and Bennion, I.: Enhanced power solitons in optical fibres with periodic dispersion management, *Electron. Lett.*, **32**, (1996), pp.54-55.
4. Smith, N. J., Doran, N. J., Knox, F. M. and Forysiak, W.: Energy scaling characteristics of solitons in strongly dispersion-managed fibres, *Opt. Lett.*, **26**, (1996), pp.1981-1983.
5. Kumar, S. and Hasegawa, A.: Quasi-soliton propagation in dispersion-managed optical fibres, *Opt. Lett.*, **22**, (1997), pp.372-374.
6. Tajima, K.: Compensation of soliton broadening in nonlinear optical fibers with loss, *Opt. Lett.*, **12**, (1987), p.54.
7. Forysiak, W., Knox, F. M. and Doran, N. J.: Average soliton propagation in periodically amplified systems with stepwise dispersion-profiled fiber, *Opt. Lett.*, **19**, (1994), p.174.
8. Hasegawa, A., Kumar, S. and Kodama, Y.: Reduction of collision-induced time jitters in dispersion-managed soliton transmission systems, *Opt. Lett.*, **21**, (1996), p.39.
9. Kolltveit, E., Hamaide, J-P. and Andouin, O.: Optimum two-step dispersion management in WDM soliton systems, *Electron. Lett.*, **32**, (1996), p.1858.
10. Forysiak, W., Devaney, J. F. L., Smith, N. J. and Doran, N. J.: Dispersion management for wavelength division multiplexed soliton transmission, *Opt. Lett.*, **22**, (1996), pp.600-602.
11. Smith, N. J., Forysiak, W. and Doran, N. J. : Reduced Gordon-Haus jitter due to enhanced power solitons in strongly dispersion managed systems, *Electron. Lett.*, **32**, (1996), pp.2085-2086.
12. Golovchenko, E. A.: Collision-induced timing jitter reduction by periodic dispersion management in soliton WDM transmission, *Electron. Lett.*, **33**, (1997), pp.735-737.
13. Devaney, J. F. L., Forysiak, W., Niculae, A. M., and Doran, N. J.: Soliton collisions in dispersion managed WDM systems, *Opt. Lett.*, (1997), to be published.
14. Kumar, S., Kodama, Y. and Hasegawa, A.: Optimal dispersion management schemes for WDM soliton systems, *Electron. Lett.*, **33**, (1997), p.459.
15. Nijhof, J. H. B., Doran, N. J., Forysiak, W. and Knox, F. M.: Stable soliton-like propagation in dispersion managed systems with net anomalous, zero and normal dispersion, *Electron. Lett.*, **33**, (1997), p.1726.
16. Harper, P., Knox, F. M., Kean, P. N., Bennion, I. and Doran, N. J.: 10 Gbit/s soliton propagation over 5250km in standard fiber with dispersion compensation,

OFC'97, **ThN1**, (1997), p.304.

17. Yamada, E., Kubota, H., Yamamoto, T., Sahara, A. and Nakazawa, M.: Numerical and experimental comparison between dispersion-allocated soliton, RZ and NRZ pulses at zero group velocity dispersion, *ECOC'97*, **3**, (1997), pp.327-330.
18. Mollenauer, L. F., Evangelides, S. G. and Gordon, J. P.: Wavelength division multiplexing with solitons in ultra-long distance transmission using lumped amplifiers, *J. Lightwave Technol.*, **9**, (1991), p.362.

(Received January 13, 1998)

TRANSMISSION SYSTEMS BASED ON DISPERSION-MANAGED SOLITONS : THEORY AND EXPERIMENT

T. GEORGES AND F. FAVRE

France Télécom
CNET Lannion DTD/RTO,
2, av. P. Marzin, 22307 Lannion Cedex, France

Abstract. The propagation of solitons in a dispersion managed link can be mainly modeled with the evolution of two parameters γ and C, related to the spectral width and the chirp. Steady propagations are shown to be possible if the average dispersion lies in the anomalous domain.With the same conditions, periodical propagations are both theoretically and experimentally demonstrated. With the help of a perturbation theory, the jitter and the signal to noise ratio are theoretically evaluated. The latter is experimentally shown to be the low power limit of terrestrial systems based on non dispersion shifted fiber. In addition, soliton interaction is theoretically investigated and shown to be the upper power limit of terrestrial systems based on non dispersion shifted fiber. Finally, wavelength and power margins of a single channel 20 Gbit/s soliton transmission over 11 amplifier spans of 102 km are presented. With a proper third-order chromatic dispersion management, 1 Tbit/s should be transmitted over 1,000 km.

1. Introduction

The transmission of 20 Gbit/s path-averaged solitons with in-line control is possible over unlimited distances [1]. In such experiments, Kerr nonlinearity is not a limitation but is used to separate the signal from the noise. Unfortunately, this transmission technique does not allow the increase of the bit rate of a single channel to much more than 40 Gbit/s nor the increase of the overall bitrate with help of Wavelength Division Multiplexing because of the Four Wave Mixing [2]. The other drawback is the incompatibility

A. Hasegawa (ed.), New Trends in Optical Soliton Transmission Systems, 317–340.

to the Non Dispersion Shifted Fiber (NDSF) or standard fiber (SF) in the terrestrial networks.

The solution to these limitations may be the dispersion management, as in the case of NRZ transmission. Two kinds of dispersion management can be considered. The first one, for which the dispersion profile follows approximately (by step or not) the exponential decrease of the signal power, should allow the transmission of Tbit/s over unlimited distances (with WDM, [3, 4, 5]). This management is however not suitable for terrestrial links because of the difficulty of its implementation. The second type of dispersion management, for which the local dispersion can be large compared to the average value and not correlated to the signal power, is compatible with terrestrial links. It was recently demonstrated to be the most efficient solution to upgrade the already installed systems of terrestrial networks. A first soliton transmission was demonstrated at 10 Gbit/s on non dispersion shifted fiber [6]. At bit rates of 20 Gbit/s, 10×20 Gbit/s and 16×20 Gbit/s respectively, the longest transmission distances ever reported on non-dispersion shifted fibers (NDSF) were achieved with this technique [7, 8].

The latter technique is also promising on dispersion shifted fibers (DSF) for submarine applications. The first 20 Gbit/s single channel transmission on a straight line of 8,100 km was demonstrated with this technique [9]. In addition, at a bit rate of 10 Gbit/s, this technique was demonstrated to be superior to pure NRZ technique [10]. In each experiment presented in [6]-[10], the transmission line was based on the same components as that for NRZ transmission and the transmission distance was considerably increased by the use of dispersion-managed solitons.

The propagation of dispersion-managed solitons is now widely studied, both numerically and analytically [11]- [22]. The jitter was demonstrated to be strongly reduced with this technique [13, 20, 21]. However, to the author's best knowledge, no rule for the optimization and the design of the systems was presented. This is probably due to the fact that in many cases, the transmission distance is given by a tradeoff between the signal to noise ratio and the soliton interaction as pointed out in reference [8]. And no theoretical analysis of the soliton interaction in a dispersion managed system was presented.

The aim of this paper is to analyse the different limitations of the transmission systems based on a strong dispersion management. The propagation of dispersion-managed solitons is first discussed. Simple rules giving the steady operation conditions (soliton-like propagation) are derived. From this modeling, the effects of amplifier noise is analysed. Both the jitter and the signal to noise ratio are calculated. Finally, an analytical modeling of soliton interaction is presented. Both single and alternate polarizations are studied. These rules are the base of system optimization. As a check,

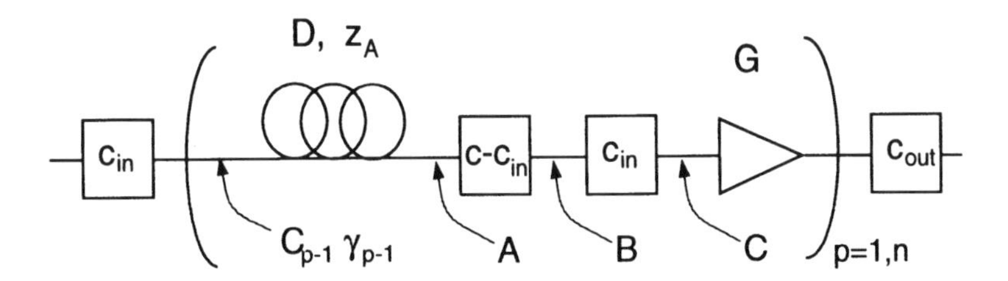

Figure 1. Schematic of the transmission link.

these rules are compared to a 20 Gbit/s transmission experiment. And finally, a comparison between systems based on path-averaged solitons and on dispersion-managed solitons is drawn.

2. Modeling

2.1. SINGLE PULSE PROPAGATION

For sake of simplicity, only periodic lines with a single type of in- line fiber are considered. Dispersive elements are inserted at amplifier location (c) and at emitter ($c_{\rm in}$) and receiver ($c_{\rm out}$) ends. The original idea of reducing the average chromatic dispersion of the line with dispersion compensation was presented in reference [11]]. The generic transmission line is sketched in Figure 1.

The propagation of a pulse $u(z,t) = \sqrt{a(z)}q(z,t)$ in such a line with dispersion $D(z)$, energy evolution $a(z)$ can be approximated by the nonlinear Schrödinger equation

$$iq'_z + \frac{1}{2}D(z)q''_{tt} + a(z)|q|^2q = 0\,. \tag{1}$$

The energy $E = \int_{-\infty}^{+\infty}|q|^2dt$ is a constant of the motion of equation (1) and the pulse energy along the line is equal to $a(z)E$.

The normalized form of the propagation equation (1) is interesting for the comparison of different transmission systems. The normalization requires the choice of a chromatic dispersion reference $\beta_{\rm ref}$ and of a time reference $\tau_{\rm c}$. The latter can be chosen arbitrary. However, $\beta_{\rm ref}$ is often chosen to be the chromatic dispersion of one of the fibers of the link, or to be the average chromatic dispersion of the link or to be in-between these values. The time reference is in general of the order of the pulsewidth. The real time T and chromatic dispersion β are derived from their normalized counterparts t and D by $T = t\tau_{\rm c}$ and $\beta = D\beta_{\rm ref}$. From $\beta_{\rm ref}$ and $\tau_{\rm c}$, the characteristic length $Z_{\rm c} = \tau_{\rm c}^2/|\beta_{\rm ref}|$, the characteristic power $P_{\rm c} = \lambda A_{\rm eff}/(2\pi n_2 Z_{\rm c})$ and the characteristic energy $E_c = P_c \tau_c$ are derived. Real distance and power are thus $Z = zZ_c$ and $P = |u|^2 P_c$. The effective area A_{eff} and nonlinear index n_2 depend of the fiber. When different fibers are used, a single normalization must be retained and the value of $a(z)$ must be modified at the junction of two different fibers. A decrease of the effective area or an increase of the nonlinear index would correspond to an amplification. It must finally be reminded that the propagation equation (1) is suitable for the propagation of a pulse in a weakly birefringent fiber with many random changes of the birefringent axis [23]. This is the case of the real transmission fibers.

Many groups have theoretically [11]-[20] and experimentally [6]-[10], [24]-[28] shown that stable periodic propagations could be obtained in periodical links for with a strong dispersion management. In addition, numerical simulations show that the pulse shape tends to become Gaussian when $Dz_A\overline{\Omega^2} > 1$, where $\overline{\Omega^2}$ is the square of the quadratic bandwidth of the pulse [15, 19]. For this reason, a Lagrangian formalism [17, 18, 29], compatible to any kind of pulse shape is used to analyse the propagation. The ansatz

$$q(z,t) = \sqrt{B} \exp\left[-(1+ib)\frac{(t-\alpha)^2}{2W^2} + i\phi - i\omega\right] \tag{2}$$

is an exact solution of equation (1) in absence of nonlinearity. In the linear regime, the pulsewidth and the chirp parameter b can be related to two parameters γ and C so that

$$\begin{aligned} \gamma W^2 &= 1 + \gamma^2 C^2 \\ b &= -\gamma C\,. \end{aligned} \tag{3}$$

The parameter γ is related to the square of the quadratic bandwith by $\overline{\Omega^2} = \gamma/2$ and is constant in the linear regime. The square of the quadratic pulsewidth is $\overline{\tau^2} = W^2/2$. The parameter C corresponds to the cumulative dispersion ($C\prime_Z = D$). $E = BW\sqrt{\pi}$ and the pulsation ω are constant. The temporal position α and the pulsation are related by the fiber dispersion D by

$$\alpha'_z + D_\omega = 0 \tag{4}$$

Equation (1) is equivalent to $\frac{\delta \mathbf{L}}{\delta \mathbf{q}^*} = 0$, where the Lagrangian $\mathcal{L}$ is defined by reference [17, 18, 29]

$$2\mathbf{L} = i(q'_z q^* - q q'^*_z) - D|q'_t|^2 + a|q|^4\,. \tag{5}$$

With the ansatz (2), the average Lagrangian $L = \int_{-\infty}^{+\infty} \mathbf{L} dt$ becomes

$$L = \frac{E}{4}\left[-4\phi'_z + b'_z 2b\frac{W'_z}{W} - D\left(\frac{1+b^2}{W^2} + 2\omega^2\right) + 4\alpha\omega'_z + aB\sqrt{2}\right] . \qquad (6)$$

The nonlinearity modifies the evolution of W, b and Φ. These evolutions are derived from the average Lagrangian and are written as references [17, 18]

$$\begin{aligned} W'_z + \frac{Db}{W} &= 0 && (7a) \\ b'_z + D\frac{(1+b^2)}{W^2} - \frac{\alpha E}{W\sqrt{2\pi}} &= 0 && (7b) \\ \phi'_z + \frac{D}{2}\left(\omega^2 + \frac{1}{W^2}\right) - \frac{5aE}{4W\sqrt{2\pi}} - \alpha\omega'_z &= 0 && (7c) \end{aligned}$$

and equation (4) is still valid. Equations (7a-b) can be transformed in

$$\begin{aligned} C'_z &= D + \left(C^2 - \frac{1}{\gamma^2}\right)\frac{aE}{W^3\sqrt{2\pi}} \\ \gamma'_z &= -2\gamma C\frac{aE}{W^3\sqrt{2\pi}} . \end{aligned} \qquad (8)$$

It comes that the spectral width (related to γ) can be modified by the nonlinearity, increased when $C < 0$ and decreased when $C > 0$. In the normal dispersion domain $D < 0$, the spectrum broadens, which is a well known result.The second term in the derivative of C can be considered as a nonlinear dispersion. It can counterbalance the linear chromatic dispersion in the anomalous dispersion domain $D > 0$, when $\gamma|C| < 1$ (i.e. when the temporal broadening is less than a factor of $\sqrt{2}$). This is the base of the soliton propagation. Equations (8) apply of course to more complicated transmission lines as that descibed in Figure 1.

We now focus on the particular line sketched in Figure 1 with a constant fiber chromatic dispersion $D = 100$ and length $z_A = 0.2$, a lumped dispersion compensation $c = -\eta D z_A$ (η is the compensation ratio) and an amplification $G = 100$. Since $Dz_A \gg 1$, parameter C evolves much along the span. However, due to the compensation, when $|(1-\eta)Dz_A| < 1$ or $0.95 < \eta < 1.05$ in this particular case, parameter C is weakly modified from one input span to the next one. The series (C_p, γ_p) of the parameters at the input of each span, obtained by successive integrations of equations (8) over a span, is now considered.

Of particular interest are the fixed points of the series (C_p, γ_p) because they correspond to a steady propagation. With an energy $E = 75$ and $\eta =$

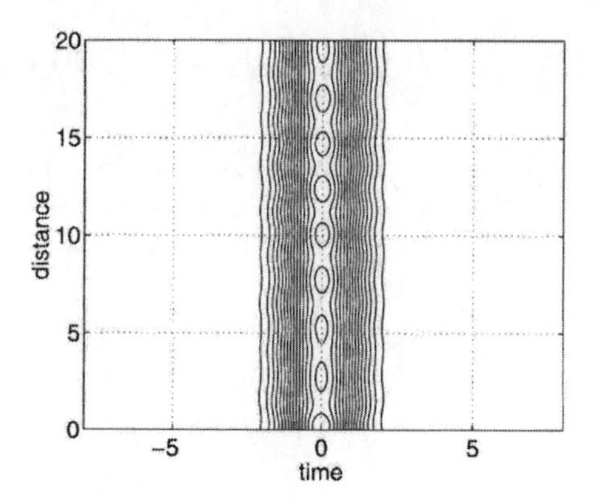

Figure 2. Steady propagation for $Dz_{\mathrm{A}} = 20, G = 100, z_{\mathrm{A}} = 0.2, \eta = 0.99$ and $E = 75$.

0.99 (the average dispersion is), a single fixed point $(C, \gamma) = (-1.72, 0.675)$ is found. As a check of the theory, equation (1) is directly integrated with input data corresponding to the fixed point and the field at point B of the span is plotted in Figure 2. The pulse is Fourier transform limited $(C = 0)$ at point B in each span. The conditions of this steady propagation were empirically found in reference [16]. As pointed out in references [16, 22], the pre-chirp is a key element for the control of the nonlinearity. If different input parameters are used, the series $(C_{\mathrm{p}}, \gamma_{\mathrm{p}})$ lie on closed trajectories, indicating stable periodic propagation. These trajectories are plotted in Figure 3 for $\gamma = 2/3$ and four different values of pre-chirp, -1.7 (close to the fixed point), -1, 0 and 1. The periods of the propagation on each trajectory are 4.2, 4.39, 5.64 and 8.52 respectively.

Let us come back to the real world with a 100 km non dispersion shifted fiber with 16.5 ps/nm·km ($\beta = -21$ ps^2/km) chromatic dispersion, 80 μm^2 effective area, 0.2 dB/km loss and $n_2 = 2.5{\times}10^{-20}$m^2/W. The normalization leads to $\beta_{\mathrm{ref}} = \beta/\mathrm{D} = -0.21$ ps^2/km, $Z_{\mathrm{c}} = Z_{\mathrm{A}}/z_{\mathrm{A}} = 500$ km, $\tau_{\mathrm{c}} = \sqrt{|\beta_{\mathrm{ref}}|Z_{\mathrm{c}}} = 10.25$ ps and $E_{\mathrm{c}} = 16.2$ fJ. The steady propagation requires a pre-chirp of $-1.72\beta_{\mathrm{ref}}Z_{\mathrm{c}} = 181$ ps^2 or -142 ps/nm, a pulsewidth $\tau = 2\tau_{\mathrm{c}}\sqrt{\log(2)/\gamma} = 20.8$ ps and a pulse energy of 1.2 pJ. At 20 Gbit/s, the corresponding amplifier output energy is 12 mW (or 10.8 dBm). These data are not too far away from the experimental data of references [7, 8].

With the same parameters of the line, the energy E was varied from 0.1 to 1000. In each case, the trajectories found are topologically equivalent to those of Figure 3 and a single fixed point was observed. As a new check of the validity of the theory, the results were compared to a direct integration of equation (1) for the fixed points. A good agreement was observed up to

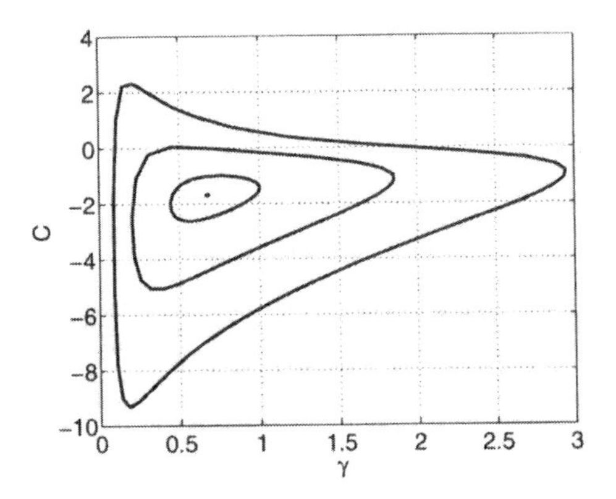

Figure 3. Phase diagram of the propagation for $Dz_A = 20, G = 100, z_A = 0.2, \eta = 0.99$ and $E = 75$.

$E = 400$. The value of the parameter $\gamma(E)$ of the fixed points is plotted in Figure 4. For small values of the energy $E, \gamma D z_A \ll 1$ which implies small pulsewidth variations. The propagation is thus within to the path-averaged soliton regime, for which $\overline{\Omega^2}$ or γ are proportional to E^2. More explicitely,

$$\gamma_{\text{PAS}} = \frac{1}{6}\left[\frac{E(G-1)}{\overline{D}G\log(G)}\right]^2 .$$

The latter relation is also plotted in Figure 4. It comes that for $E < 2$, γ is given by the relation of the PAS. Beyond this value of energy, γ varies little with E (especially in the range $40 < E < 400$). This means that the pulse characteristics leading to a steady propagation does not depend much on the energy and could explain the robustness of the pulse with respect to the in-line power [7]. In addition, for a given γ (or spectral width), the steady propagation requires an enhanced energy [19, 20]. This provides a better resistance against the amplifier noise (jitter and signal to noise ratio). Finally, it appears that there is a maximum value of γ (approximately 0.78 for $E = 170$) for a steady propagation. This implies that there is a maximum intra-span broadening acceptable for a stable propagation, indicating that an increase of the bit rate of a single channel or of the amplifier span may require a reduction of the in-line fiber chromatic dispersion. An important point must be noticed here. In opposition to the path-averaged soliton regime, a pulse for which $\gamma \neq \gamma(E)$ can propagate on a closed trajectory similar to those plotted in Figure 3, without an emission of a dispersive wave. In consequence, a change in energy will change the parameters of the propagation but not increase the dispersive wave.

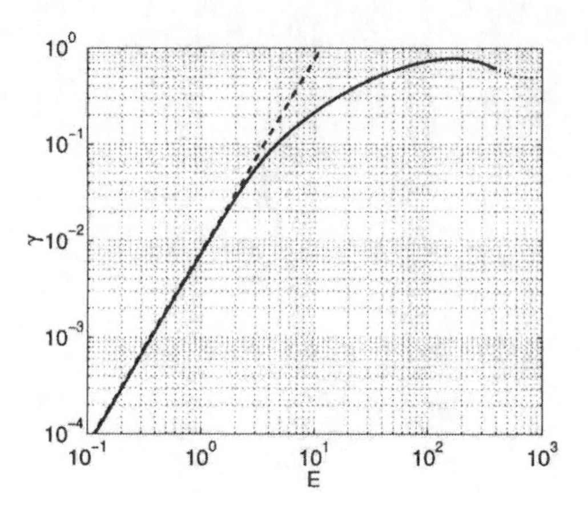

Figure 4. γ at the fixed point (solid line) and γ_{PAS} (dashed line) versus E, with the parameters of Figure 3.

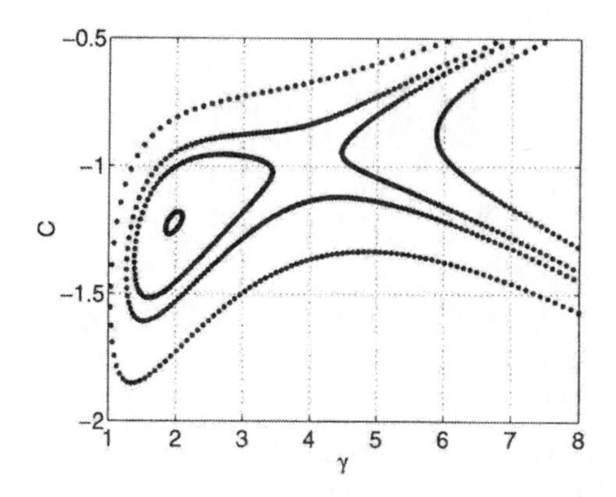

Figure 5. Phase diagram of the propagation for $Dz_{\mathrm{A}} = 20, G = 100, z_{\mathrm{A}} = 0.2$, $\eta = 1.00225$ and $E = 75$.

The value of the compensation ratio η was varied from 0.95 to 1.05 with $E = 75$. The same topology as that of Figure 3 with a single fixed point was obtained up to $\eta \approx 1.0025$. Just below this critical value, the trajectories far from the fixed point open (Figure 5). The area with closed trajectories shrinks when η becomes closer and closer from the critical value. Beyond this critical value, all the trajectories are open and no steady propagation is possible (Figure 6).

For the steady propagation of Figure 2, the nonlinear phase shift per span is approximately $\phi = 0.65$ radians. After 10 spans, a nonlinear phase

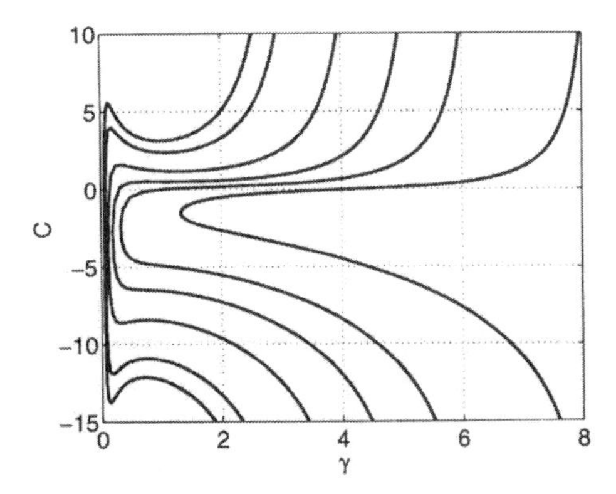

Figure 6. Phase diagram of the propagation for $Dz_A = 20, G = 100, z_A = 0.2, \eta = 1.01$ and $E = 75$.

shift greater than 2π is accumulated without pulse distorsion. The nonlinearity is controlled by the dispersion management. In comparison, the maximum power in NRZ transmission systems is limited by the Kerr effect. The waveform of the pulses is distorted by a combination of the nonlinear phase shift and the local chromatic dispersion, which induces inter- symbol interferences and a reduction of the system eye opening. On the other hand, the amplifier noise increases linearly with the transmission distance in an amplified system. For a given signal to noise ratio at receiver end, an increase of the transmission distance requires an increase of the pulse energy (and thus the pulse power). It comes that the maximum transmission distance of a NRZ transmission system is limited by the maximum nonlinear phase shift. In transmissions on NDSF at 1.55 μm signal wavelength (large chromatic dispersion), the latter was experimentally found to be 0.6π in 10 Gbit/s transmission systems [30] and 0.25π in 20 Gbit/s systems [31].

The RZ format was often considered as less performant than the NRZ format, since the peak power must be increased in order to keep the same pulse energy. For given transmission distance and signal to noise ratio, the nonlinear phase shift is larger for the RZ format. Soliton transmission in strongly dispersion managed systems demonstrates that RZ pulses with a proper management are not directly limited by the total nonlinear phase shift and can thus provide better system performance. As a proof, the 20 Gbit/s soliton transmission over 2,400 km [8] would require a total phase shift of 1.8π with NRZ format, to be compared to 0.25π, the maximum value reported today [31]. In reference [8], it was pointed out that the maximum in-line power and transmission distance are of course not directly limited by the accumulation of the nonlinear phase shift but by the soliton interaction.

2.2. PERTURBATION

A weak perturbation is now added so that the propagation equation can be written as

$$iq'_z + \frac{1}{2}D(z)q''_{tt} + a(z)|q|^2 q = \varepsilon P[q] \,, \tag{9}$$

where $\varepsilon \ll 1$. Following the methodology of reference [29] and neglecting the dispersive wave, the propagation equations in presence of the perturbation are now obtained by

$$\frac{\partial L}{\partial X} - \frac{\partial}{\partial z}\left(\frac{\partial L}{\partial X'_z}\right) = \left[\int_{-\infty}^{+\infty} \varepsilon P[q] \frac{\partial q^*}{\partial X} dt + c.c.\right] \,, \tag{10}$$

where X is one of the 6 parameters B, W, b, ω, α and ϕ of the ansatz (2) and *c.c.* is for complex conjugate. The right hand side of equation (10) induces small variation $\delta X'_z$ of the parameters X written as

$$\delta X'_z = \varepsilon \int_{-\infty}^{+\infty} f(x) P[q] dt + c.c. \tag{11}$$

where

$$\begin{aligned} f_E &= -iq^* \\ ih_\alpha &= W^2 f_\omega/(1-ib) = (t-\alpha)q^*/E \\ if_W/W &= f_b/2(1-ib) = \left[\frac{(t-\alpha)^2}{W^2} - \frac{1}{2}\right] q^*/E \\ f_\Phi &= \left[2\frac{(t-\alpha)^2}{W^2} - 3\right] q^*/4E \,. \end{aligned} \tag{12}$$

3. Noise

Let us now consider the noise $n(t)$ emitted by the lumped amplifier. It is assumed to be white so that it satifies the equations

$$\begin{aligned} \langle n(t) n^*(t') \rangle &= n_{sp} \frac{G-1}{N_0} \delta(t-t') \\ \langle n(t) n(t') \rangle &= 0 \,, \end{aligned} \tag{13}$$

where n_{sp} is the excess noise factor of the amplifier, N_0 is the number of photons in an energy unit and the brackets indicate an averaging of the random variable. The noise of an amplifier located at $z = z_{\rm amp}$ is a perturbation $P = in(t)\delta(z - z_{\rm amp})/\sqrt{a(z_{\rm amp}}$, where the value of a is taken at amplifier output. This perturbation modifies the soliton parmeters X. Applying equations (11), the modification δX of parameter X induced by the noise $n(t)$ can formally be written as

$$\delta X = \frac{i}{\sqrt{a}} \int_{-\infty}^{+\infty} f_X(t) n(t) dt + c.c. \,. \tag{14}$$

From equations (13) and (14), the correlation between two perturbations δX and δY becomes

$$\langle \delta X \delta Y \rangle = n_{\rm sp} \frac{G-1}{a N_0} \int_{-\infty}^{+\infty} f_X(t) f_Y^*(t) dt + c.c. \,. \tag{15}$$

Focusing on the three parameters E, α and ω that play an important role in the system performance, it comes

$$\frac{\langle \delta E^2 \rangle}{2E^2} = \frac{\langle \delta \omega^2 \rangle}{\gamma} = \frac{\langle \delta \alpha^2 \rangle}{W^2} = \frac{\langle \delta \alpha \delta \omega \rangle}{b} = n_{\rm sp} \frac{G-1}{a E N_0} \tag{16}$$

Fluctuations in energy are related to the signal-spontaneous beat noise at receiver end and timing and frequency fluctuations are related to jitter. With help of equation (16), the jitter variance σ^2 at the end of line sketched in Figure 1 becomes after some algebra

$$\sigma^2 = n_{\rm sp} \frac{(G-1)}{E N_0} \sum_{p=1}^{n} \left\{ \gamma_{\rm p}^{-1} + \gamma_{\rm p} \left[(n-p) \overline{D} z_{\rm A} + C_{\rm p} + c_{\rm out} \right]^2 \right\} , \tag{17}$$

where $\overline{D} z_{\rm A}$ is the cumulative dispersion of the whole span. In addition, it is found that the jitter variance does not depend on the relative location of the amplifcation and of the dispersion compensation. Of course, the location of the compensation influences the jitter if the dispersion compensation is lossy (it would increase the noise factor if located in front of the amplifier) or nonlinear (it would modify the propagation if located after the amplifier).

At the fixed point, $C_{\rm p} = c_{\rm i}$. In addition, if the pulse is Fourier transform limited ($C = 0$) at receiver end, $c_{\rm out} = -c_{\rm in}$. With the latter assumptions and after replacing a summation by an integration, it comes

$$\langle \delta \alpha^2 \rangle = \frac{n_{\rm sp}(G-1)}{E N_0} \frac{z}{z_{\rm A}} \left(\frac{1}{\gamma} + \frac{\gamma z^2}{3} \right) , \tag{18}$$

where $z = n z_{\rm A}$ is the total system length. This formula is very similar to that of the PAS regime. Simply the soliton energy is larger for a given average chromatic dispersion, which reduces the value of the jitter. These analytical results on jitter are in good agreement with numerical ones [13, 21].

The jitter can also be evaluated everywhere within a span. Let us consiter the $(n+1)^{th}$ span. The jitter variance is given by equation (17) where c_{out} is replaced by $c(z)$, the cumulative dispersion within the $(n+1)^{th}$ span. In the line with $Dz_A = 20$ and typically 10 or 20 spans, the value of $c(z)$ increases from 0 to 20 and then decreases down to 0.2. In comparison, the values of $(n-p)\overline{D}z_A$ and C_p are of the order of 1. In consequence, the jitter variance evolves more between two consecutive spans than from the end of a span to the next one. This is in agreement with the results of reference [21].

The other important source of system degradation of the noise is the signal-spontaneous beat noise at receiver end. Neglecting the spontaneous-spontaneous beat noise, the Q factor, the standard deviation of energy fluctuations and the signal to noise ratio (SNR) are related at receiver end by

$$Q = \frac{E}{\sigma_E} = \sqrt{\frac{SNR}{2}} \,. \tag{19}$$

Using equation (16), it comes that after a transmission length z,

$$SNR = \frac{EN_0}{n_{sp}(G-1)} \frac{z_A}{z} \,. \tag{20}$$

Again, the value of the SNR is formally equivalent to that of the PAS regime, but is enhanced by the larger energy.

3.1. INTERACTION OF TWO ORTHOGONALLY POLARIZED SOLITONS

The propagation of two orthogonally polarized pulses $u_j(z,t) = \sqrt{a(z)}q_j(z,t)$ $(j = 1,2)$ in a periodic transmission line with arbitrary dispersion $D(z)$ and energy evolution $a(z)$ can be approximated by the system of nonlinear Schrödinger equations

$$\begin{aligned} iq'_{1z} + \frac{1}{2}D(z)q''_{1tt} + a(z)\left[|q_1|^2 + \kappa|q_2|^2\right]q_1 &= 0 \\ iq'_{2z} + \frac{1}{2}D(z)q''_{2tt} + a(z)\left[|q_2|^2 + \kappa|q_1|^2\right]q_2 &= 0\,, \end{aligned} \tag{21}$$

where $\kappa = 1$ for orthogonally polarized solitons. This set of equations is valid in fibers with randomly varying birefringence as discussed in reference

[32]. It must be noticed that the collision of two solitons of two different channels is described by the same system of equations with $\kappa = 2$.

The Lagrangian associated to system (21) is $\mathbf{L} = \mathbf{L}_1 + \mathbf{L}_2 + \mathbf{L}_{12}$, where

$$\begin{aligned} 2\mathbf{L}_j &= i(q'_{jz}q^*_z - q_j q'^*_{jz}) D|q'_{jt}|^2 + a|q_j|^4 \quad \text{for } j = 1, 2 \\ \mathbf{L}_{12} &= a\kappa |q_1|^2 |q_2|^2 \end{aligned} \tag{22}$$

and system (21) is equivalent to $\frac{\delta \mathbf{L}}{\delta q^*_j} = 0$ for j = 1, 2. The ansatz must take into account the frequencies ω_j and the times α_j of the pulses,

$$q_j(z,t) = \sqrt{B_j} \exp\left[-(1 + ib_j)\frac{(t-\alpha_j)^2}{2W_j^2} - i\omega_j t + i\phi_j\right] \quad \text{for } j = 1, 2\,. \tag{23}$$

As in the case of the single pulse propagation, the parameters W_j and b_j are related to γ_j and C_j by $\gamma_j = (1 + b_j^2)/W_j^2$ and $b_j = -\gamma_j C_j$. Following the methodology of reference [33] and using the average Lagrangian $L = \int_{-\infty}^{+\infty} \mathbf{L}dt$, the propagation equations are obtained by

$$\frac{\partial}{\partial z}\left(\frac{\partial L}{\partial X'_z}\right) = \frac{\partial L}{\partial X}\,, \tag{24}$$

where X is one of the 12 parameters $B_j, W_j, b_j, \omega_j, \alpha_j$ and ϕ_j $(j = 1, 2)$ of the ansatz (23). Applying equation (24), it comes

$$\begin{aligned} \alpha'_{jz} &= -D\omega_j && (25a) \\ E_1\alpha'_{1z} &= -E_2\omega'_{2z} = \frac{2a\kappa E_1 E_2}{\sqrt{2\pi}}\frac{\alpha}{W^3}\exp\left(-2\frac{\alpha^2}{W^2}\right) = af && (25b) \\ \gamma'_{jz} &= -2\gamma_j C_j F_j && (25c) \\ C'_{jz} &= D + (C_j^2 - \gamma_j^{-2})F_j\,, && (25d) \end{aligned}$$

where $W^2 = W_1^2 + W_2^2$, $2\alpha = (\alpha_1 - \alpha_2)$ is the soliton separation and

$$E_j F_j = \frac{\alpha}{\sqrt{2\pi}}\left[\frac{E_j^2}{W_j^3} + \kappa\frac{E_1E_2}{W^3}\left(1 - 4\frac{\alpha^2}{W^4}\right)\exp\left(-2\frac{\alpha^2}{W^2}\right)\right]\,. \tag{26}$$

The evolution of phases ϕ_j are not taken into account since they do not play any role in the interaction. It turns out that soliton interaction mainly affects the frequency of the pulses so that the solitons attract each other. The soliton parameters γ_j and C_j are slightly affected as well. It must be noticed that soliton collision is modelized by equations (25) and (26) as well with $\kappa = 2$.

Now consider a couple of orthogonally polarized solitons, otherwise identical with $\omega_1(0) = \omega_2(0) = 0$. Their initial separation is $2\alpha = 8(\alpha_1(0) = -\alpha_2(0) = 4)$. The parameters of the line and of the initial pulses correspond to the steady propagation described in the previous section: $D = 100, z_A = 0.2, c = -19.8, E = 75, G = 100, c_{in} = -1.72$ and $\gamma = 0.675$. Following equations (25) and (26), the series $\{\alpha_1(p)\}$ of the first soliton temporal position in point B of section p is plotted versus transmission distance in Figure 7. Also plotted are the values obtained by numerical integration of equations (21). The accuracy of the analytical tool is demonstrated by this result.

The initial repelling of the solitons, in opposition with the classical behavior in the path-averaged regime, is explained as follows : the total frequency shift $\delta\omega$ accumumated all along the fiber in an amplification section (equation 25(b)) sees in average, only a fraction $\rho < 1$ of the total fiber chromatic dispersion but sees the whole dispersion compensation. At point B, the time shift related to $\delta\omega(> 0)$ in the whole amplification section is thus equal to $\delta\omega[Dz_A(1-\rho) - z_A - c_{in}]$ and is positive when ρ is small enough. At point C (Figure 1), it is even larger whereas in point A it is always negative. The frequency shifts $\delta\omega(p)$ induced by soliton interaction in the previous p sections are responsible for a negative time shift $-\delta\omega(p)z_A$. The sum of the two terms is positive at point B in the first sections (repelling) and becomes negative (attraction) when $\delta\omega(p)$ is large enough (Figure 7). The value of α_1 before (point A) and after (point C) the lumped compensation is also plotted in Figure 7. The same process is responsible for soliton interaction in a link with 100 % dispersion compensation. In this case, if the in-line chromatic dispersion is $D > 0$, the solitons repel each other all along the link and if $D < 0$, they attract each other. The collision distance at point B is about 18. At half the collision distance, $z = 9$, the solitons already overlap at compensator input. They overlap at compensator output at $z = 19$.

The collision distance has thus not as clear as a meaning as in soliton propagation with constant dispersion. However, when the initial separation is $2\alpha = 8$, the collision distance of orthogonally polarized path-averaged solitons with a constant chromatic dispersion is about 500, 25 times larger than in figure 3. Two factors contribute to the reduction of the collision distance of dispersion-managed solitons : The first factor is the enhanced soliton energy since the interaction strength is proportional to the pulse energy (equations (25) and (26). The second (and main) factor is the pulse broadening. According to equation (25b), the attraction force is equal to $a(z)f/E$. Factor $f$depends on C and is the smallest when the pulse is

Fourier transform limited ($C = 0$). It increases first with increasing value of $|C|$ due to an increased overlapping between the pulses. It reaches a maximum for $W = 2\alpha/\sqrt{3}$ or $C = \pm 5.45$ when $\alpha = 4$. The value of the maximum is

$$f_{\mathrm{max}}/E = \frac{3\sqrt{3}}{4e^{3/2}\sqrt{2\pi}} \frac{\kappa E}{\alpha^2} . \tag{27}$$

The value of the maximum interaction force decreases slowly with soliton separation : it scales as α^{-2} to be compared to $\exp(-4\alpha)$ with a path-averaged soliton.

There are different sources of system degradation due to soliton interaction. The first one is related to the frequency shift $\delta\omega$ induced by the interaction and to the use of an optical filter at receiver end. The interacting solitons with nonzero frequency shift are more attenuated by the filter than single pulses. This is responsible for a reduction of the eye opening of the system. The penalty increases in dense WDM systems for which narrow filters are used for channel demultiplexing. The second one is still related to the frequency shift and a non optimum post-compensation of the dispersion. This induces a jitter proportionnal to the frequency shift and to the error ΔC of the the post-compensation. The third one corresponds to a jitter related to the time shift at point B (that is with optimum post-compensation of the dispersion). This time shift is also related to $\delta\omega$. The relative influence of the three limits depends on the particular design of each system. In consequence, since the main sources of system performance degradation are related to $\delta\omega$, the comparison of the value of $\delta\omega$ seems to be a good criterion for the evluation of soliton interaction.

3.2. INTERACTION OF TWO PARALLEL POLARIZED SOLITONS

Let us focus now on the single polarization interaction of solitons corresponding to the same line and the same steady operating conditions as for Figure 2. The soliton separation 2α is first computed by a direct integration of the nonlinear Schrodinger equation. The data are presented in Figure 8. The two solitons are initially separated by $2\alpha = 8$, they are either in phase (circles) or in opposition phase (crosses) and otherwise identical. It appears from Figure 7 that the interaction is almost phase difference independent in opposition to what is known of standard path- averaged solitons.

This means that the first-order terms of the interaction, which are phase difference dependent, may be neglected. The only phase difference independent terms of the interaction are those of the right hand side of equations (21) with $\kappa = 2$. To confirm this idea, equations (25) and (.26) are integrated with $\kappa = 2$. The results are plotted in Figure 8 with a thick solid

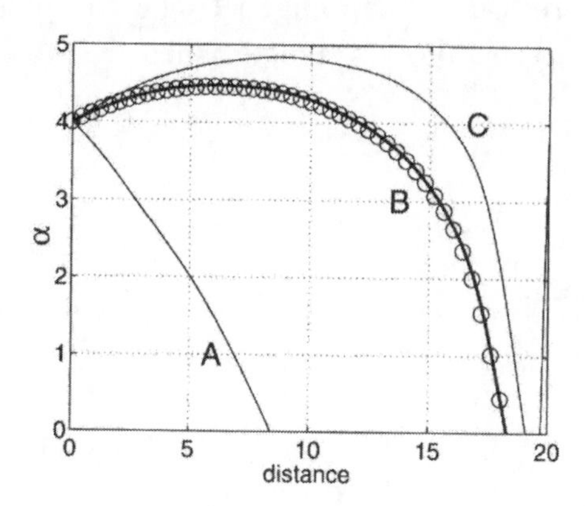

Figure 7. Interaction of two orthogonally polarized solitons: Evolution of α_1, analytic (point B, thick solid line, point A and C, lower and upper thin solid line) and numeric (circles) results. $D = 100, z_A = 0.2, c = -19.8, G = 100, E = 75, C(0) = -1.72$ and $\gamma(0) = 0.675$.

line and a good agreement is obtained with the numerical simulations. The frequency shift and the timing positions solitons before (point A) and after (point C) the lumped compensation are also plotted in Figure 8. With a perturbation theory, out of the scope of this paper, it can be shown that the first-order term of the interaction is strongly reduced by the pulse broadening and the related chirp of the pulses.

The collision distance is about 15. This value is to be compared to 43, the collision distance in the frame of the path-average soliton. There is a factor of 3 between these two values. With a smaller soliton separation, $2\alpha = 5$, the collision distances are 6.5 and 9.5 respectively for the dispersion-managed soliton and the path-averaged soliton. For a soliton separation of $2\alpha = 8$, the ratio of the collision distance of path-averaged solitons to that of dispersion-managed solitons is much smaller (43/15) with single polarization than with alternate polarizations (500/19). This is explained as follows : with orthogonally polarized solitons, soliton interaction is a second-order effect and is increased by the soliton broadening induced by the dispersion management. With paralelly polarized solitons, the soliton interaction is still a second-order effect (increased by the pulse broadening) to be compared to a (phase difference dependent) first-order effect in the case of the path-averaged soliton regime.

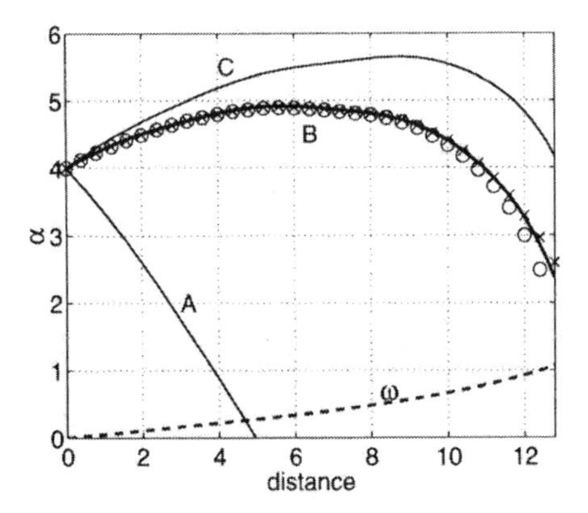

Figure 8. Interaction of two parallelly polarized solitons: Analytical (point B, thick solid line, point A and C, lower and upper thin solid line) and numerical (in-phase: circles, opposition-phase: crosses) values of α and of the frequency shift (dashed line).

4. Experiment

4.1. SET-UP

The experimental set-up is shown on Figure 9. The outputs of a CW laser tuned around 1553 launched in an electroabsorption modulator (EAM) in order to generate 25 ps pulses at 10 GHz. The EAM output was encoded with a lithium niobate modulator (EOM) fed with a $2^{15}-1$ pseudorandom binary sequence at 10 Gbit/s. The EOM output was optically time and polarization multiplexed. Pre-chirping was achieved by using 2.27 km of dispersion compensating fiber (DCF) with $D = -115$ ps/nm·km. The 20 Gbit/s stream was gated by an acoustooptic switch and partially launched (20 %) in a recirculation loop through a 80 : 20 coupler. The loop contained 102 km-span of standard fiber (SF) with $D = 16.4$ ps/nm·km, $dD/d\gamma = 0.05$ ps/nm^2· km and $\alpha = 0.21$ dB/km followed by 17.3 km of DCF with $D = -95$ ps/nm·km and $dD/d\gamma = -0.2$ ps/nm^2·km. The resulting zero-dispersion wavelength for the loop was 1549 nm with a dispersion slope reduced to 0.015 ps/nm^2·km.

Two 980 nm pumped EDFA with a noise factor of 6 dB were used in the loop. The loop gain was equalized by a tunable optical filter (OF). The pump of the second amplifier was modulated for gating the circulating stream. At the output, pulse fitting was achieved by using 11.5 km of standard fiber [7]. The signals were filtered by a 0.25 nm-FWHM tunable optical bandpass filter (OBF). The 20 towards 10 Gbit/s demultiplexing was achieved by a polarization insensitive EAM. The time demultiplexed

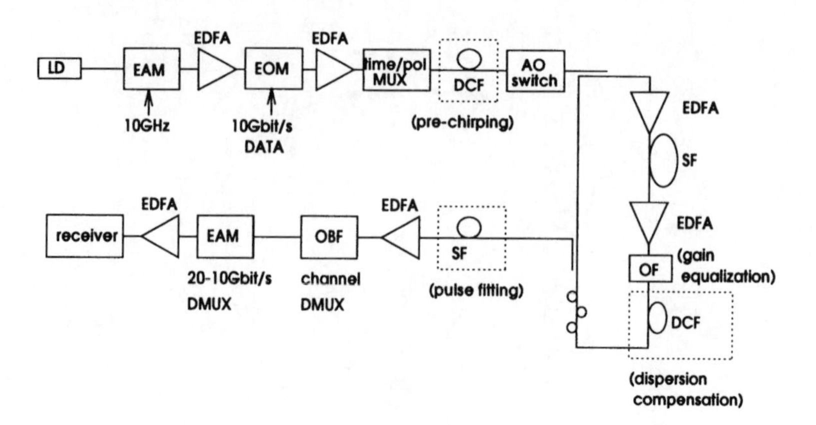

Figure 9. Experimental set-up.

signal at 10 Gbit/s fed an optical receiver. BER was measured using an error performance analyser at 10 Gbit/s.

4.2. SPECTRUM EVOLUTION

The parameter γ of Figure 3 is of practical interest for the experimental point of view since it is directly related to the bandwidth of the signal. According to Figure 3, the evolution of the bandwidth versus the propagation distance should be periodical. This is illustrated by Figure 10. The period is about 2,700 km for the large power and increases when the power is reduced. A good agreement is obtained with the analytical modeling. It was also confirmed experimentally that the evolution of the spectral width strongly depends on the value of the pre-chirp. In addition, with smaller compensation ratios, the period is reduced and propagations over many periods could be observed.

4.3. SYSTEM MARGIN

Single-wavelength transmission was studied in order to evaluate in-line power and wavelength margins. Figure 11 shows the BER limits measured after 1,100 km of standard fiber in the power-versus-wavelength axis. It can be seen that a BER $< 10^{-9}$ transmission was achieved within a 7 dB wide in-line power range over more than 10 nm. The maximum distance with BER $< 10^{-9}$ was 2,400 km.

The lower in-line power limit at 10^{-9} error rate, 2 dBm is wavelength independent over 9 nm. The limitation is attributed to the signal to noise ratio, which is effectively wavelength independent.

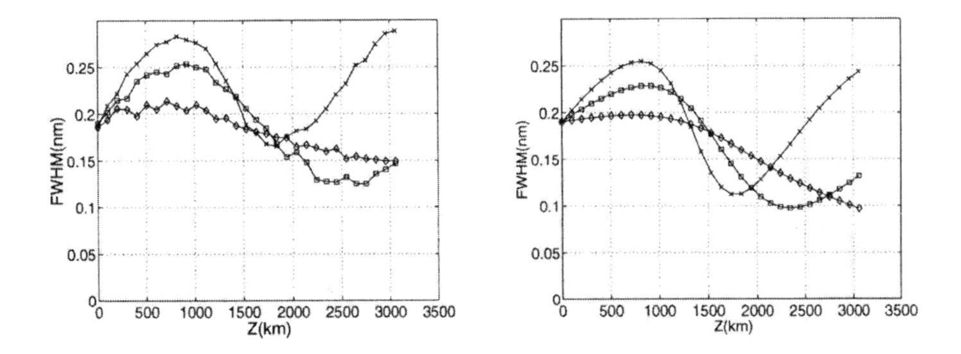

Figure 10. Evolution of the spectral width at 1558 nm for three values of the in-line power , 11 dBm (crosses), 9 dBm (squares) and 4 dBm (diamonds). Top: experimental, bottom: analytical.

The signal to noise ratio can be measured ($\mathrm{SNR_m}$) as the ratio of the spectral density of the signal to the spectral density of the noise in a bandwidth much smaller than the signal bandwidth. For a Gaussian pulse, the relation between $\mathrm{SNR_m}$ and SNR of equation (20) is

$$\mathrm{SNR} = T\Delta\nu\sqrt{\pi/\log(2)}\mathrm{SNR_m} \, , \tag{28}$$

where $\Delta\nu$ is the FWHM bandwidth of the signal. The value of the calculated $\mathrm{SNR_m}$ is about 22 dB and its measured value is only 16 dB. The difference may become from a noisy signal at loop input, from the noise accumulated by the other amplifier in the loop and from the initial fluctuations of the gain in the loop due to the loop closure. In principle 12.5 dB should be enough to ensure 10^{-9} error rate. The difference comes from the imperfection of the receiver and from the fact that the spontaneous-spontaneous beat noise should not be neglected. The increase of the power limit on the side of the window is due to the non flatness of the gain bandwidth.

The jitter variance can be calculated with equation (17). It increases with the signal wavelength (from 1549 nm to 1560 nm) and decreases with the in-line power (from 2 to 11 dBm). The minimum jitter, 0.2 ps, is at 1549 nm with 11 dBm. The maximum jitter, 1 ps, is at 1560 nm for 2 dBm in-line power. Jitter is therefore not a limitation for such a terrestrial system.

The frequency shift induced by soliton interaction is now investigated with equations (25) and (26). It is found to increase with in-line power and to be almost wavelength independent. This is the reason why it is considered to be responsible for the upper power limit of the system.

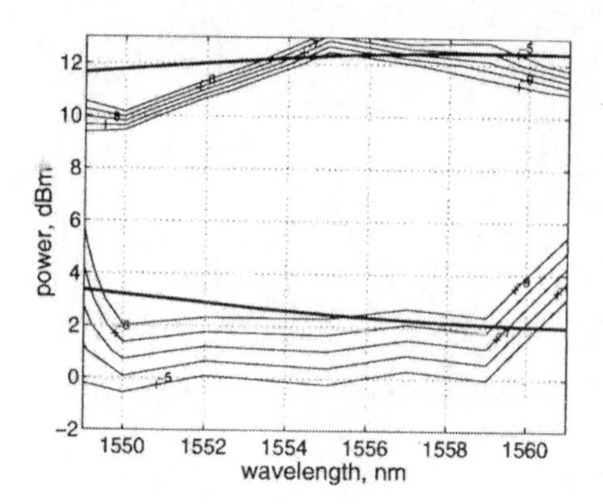

Figure 11. Experimental $\log_{10}$ (BER) of a 20 Gbit/s soliton transmission over 1,100 km (thin solid lines). Theoretical limits (thick solid lines).

In Figure 11, theoretical limits are drawn in thick solid lines. The lower thick solid line corresponds to SNR = 250. The evaluation of the SNR takes into account the 0.3 nm Gaussian filtering (wavelength demultiplexing) and the 35 ps gating (time demultiplexing) at the receiver end. The increase of the power for SNR = 250 at the lower signal wavelengths is due to a lower energy gated by the temporal demultiplexer, because of broader pulses. The upper thick solid line corresponds to a 5 GHz frequency shift (about 1/4 of the initial spectral bandwidth) induced by the soliton interaction after 11 spans. The exact value of the frequency shift was fitted, however this limit of the interaction is shown to be almost independent of the signal wavelength (or of the average chromatic dispersion) within 10 nm, in agreement with the experiment. This is a special feature of the dispersion-managed soliton. The decrease of the upper power limit around the zero dispersion wavelength may be due to inter-symbol interference, because the pulses are temporally broadened. For the theoretical curves, the nonlinear effect on the dispersion compensating fiber (DCF) and the noise from the amplifier following the DCF were neglected. These assumptions are valid since the input power of the DCF is much smaller than the input power of the standard fiber and the output power of the DCF is much larger than that of the in-line fiber. The lower power limit is due to signal to noise ratio and the upper power limit to the soliton interaction. The experimental increase of the limit due to SNR for wavelengths smaller than 1550 nm and greater than 1559 nm is partly due to the limited bandwidth of the amplifiers. In addition, in the lower wavelength side, the signal spectral width strongly increases and may be another source of penalty for the system.

4.4. COMPARISON TO A SOLITON TRANSMISSION SYSTEM WITH CONSTANT DISPERSION

In terrestrial systems based on solitons, the maximum transmission distance is given by a tradeoff between signal to noise ratio and soliton interaction. It is thus proposed in this section to compare the interaction in systems based on the same pulse energy.

When increasing the transmission distance, the SNR is reduced and the frequency shift induced by the soliton interaction increases. The lower power limit is thus increased and the upper power limit is decreased. The maximum transmission distance is given by a tradeoff between SNR and soliton interaction. At 1553 nm signal wavelength, the experimental maximum transmission distance is 2,400 km with a signal power of 6.6 dBm [8, 26]. With these parameters, the theoretical frequency shift and SNR are calculated to be about 4 GHz and 300 respectively, close to the limit values.

With twice as short initial pulses (12.5 ps), the theoretical upper power limit related to the soliton interaction is found to be almost the same as for 25 ps pulses. This is in agreement with the experiment presented in reference [26]. The same transmission distance can be expected. This proves that such a transmission system based on dispersion-managed solitons is robust with respect to initial pulsewidth.

An important question now must be answered. Does the increased soliton interaction reduce the performance of soliton transmission system with strong dispersion management compared to systems without dispersion management. As a first comparison, a transmission over the single mode fiber without compensation can be considered. The condition for the path-averaged soliton regime, $Z_A < Z_c$ requires the use of solitons with pulsewidths greater than 80 ps. This is of course not compatible with a 20 Gbit/s transmission.

As a second comparison, a soliton transmission on a fiber with constant dispersion 0.066 ps/nm·km (equal to the average chromatic dispersion of the dispersion managed system aforementionned) is considered. If the same pulsewidth is used (25 ps), soliton interaction limits the transmission distance to about 80 spans. However, the soliton energy is 6.8 times smaller than in the case of dispersion management and the transmission distance is limited by the SNR to 3 spans.

As a third comparison, with the same chromatic dispersion, shorter pulses are used in order to increase the soliton energy. Because of the limit of the path-averaged soliton regime, the pulse width cannot be decreased by a factor much greater than 2.5. With 10 ps pulses, the ratio Z_A/Z_c is equal to 1.25. The soliton energy is still 2.7 times smaller than in the case of dispersion management and the transmission distance is limited to about 9 spans by the signal to noise ratio.

In all cases, the dispersion management increases the maximum transmission distance.

5. Conclusion

In conclusion, the transmission of solitons in dispersion managed links was both theoretically and experimentally analysed. A good agreement was found between theory and experiment. The analytical tools should be useful for system optimization. The advantages of such propagation is that in opposition to the NRZ transmission system, the system performance is not directly affected by the accumulation of the nonlinear phase, and in opposition to the standard soliton transmission system, there is no critical relation between the pulsewidth and the in-line power. In consequence, the transmission distance at 20 Gbit/s on NDSF with large amplifier spacing is increased compared to that of NRZ systems and the power margins are large at 1,100 km and 100 km amplifier spacing. The maximum transmission power is indirectly limited by the nonlinear phase shift through soliton interaction. More work is needed in order to explicitely relate the soliton interaction and the bit error rates. As for the other transmission systems, the lower power limit is due to amplitude fluctuations. The use of soliton controls should be investigated in order to improve the transmission distance. The jitter is found to be negligible for such terrestrial distances. Finally, this transmission technique was experimentally shown to be compatible to Wavelength Division Multiplexing. A global rate of 320 Gbit/s over 1,100 km was already demonstrated [8]. Improvement up to 1 Tbit/s should be possible by increasing the flat bandwidth of the amplifier and by improving the third-order dispersion management. The analytical tools developped in this paper could of course be used for optimization of submarine long-haul transmission systems.

Acknowledgments

We would like to thank P. Vandamme for helpful discussions and M. Monerie for his constant support. This work has been partially carried out under the ESTHER contract of European Economic Community/ACTS program.

References

1. Aubin, G., Montalant, T., Moulu, J., Nortier, B., Pirio, F., Thomine, J. B. and Devaux, F.: 20 Gbit/s soliton transmission over transoceanic distances with a 105 km amplifier span, *Electron. Lett.*, **31, n°13**, (1995).
2. Mamyshev, P. V. and Mollenauer, L. F.: Pseudo phase matched four-wave mixing in soliton WDM transmission, *Opt. Lett.*, **21**, (1996).
3. Mollenauer, L. F., Mamyshev, P. V. and Neubelt, M. J.: Demonstration of soliton

WDM transmission at up to 8x10 Gbit/s error-free, over transoceanic distances, *OFC'96*, **PD22**.

4. Nakazawa, M., Suzuki, K., Kubota, H., Sahara, A. and Yamada, E.: 100 Gbit/s WDM soliton transmission over 10,000 km using in-line synchronous modulation and optical filtering, *OFC'97*, paper **PD21**.
5. Hasegawa, A., Kumar, S. and Kodama, Y.: Reduction of collision-induced time jitters in dispersion-managed soliton transmission, *Opt. Lett.*, **21**, (1996), pp.39-41.
6. Knox, F. M., Harper, P., Kean, P. N., Bennion, I. and Doran, N. J.: Soliton transmission at 10 Gbit/s over 2022 km of standard fibre with dispersion compensation, *ECOC'96*, **WeC.3.2**, (1996), pp.101-104.
7. Le Guen, D., Favre, F., Moulinard, M. L., Henry, M., Michaud, G., Mace, L., Devaux, F., Charbonnier, B. and Georges, T.: 200Gbit/s 100 km-span soliton WDM transmission over 1000 km of standard fibre with dispersion compensation and pre-chirping, *OFC'97*, Dallas, paper **PD17**.
8. Le Guen, D., Favre, F., Moulinard, M. L., Henry, M., Michaud, G., Devaux, F., Charbonnier, B. and Georges, T.: 320 Gbit/s soliton WDM transmission over 1100 km with 100 km dispersion- compensated spans of standard, *ECOC'97*, Post-Deadline Session, pp.25-28.
9. Edagawa, N., Morita, I., Suzuki, M., Yamamoto, S., Taga, H. and Akiba, S.: 20 Gbit/s, 8100 km straight-line single-channel soliton- based RZ transmission experiment using periodic dispersion compensation, *Proc.of 21st ECOC'95*, Brussels, Post-deadline paper **ThA3.5**, (1995), pp.983-986.
10. Jacob, J., Golovchenko, E. A., Philipetski, A. N., Carter, G. M. and Menyuk, C. R.: Long-haul, 10 Gbit/s error-free transmission of NRZ data and dispersion-managerd solitons using the same transmission system, *ECOC'97*, Edinburgh, pp.1.11-1.14.
11. Kubota, H. and Nakazawa, M.: Partial soliton communication system, *Opt. Comm.*, **87**, (1992), p.15.
12. Knox, F. M., Forysiak, W. and Doran, N. J.: 10 Gbit/s soliton communication systems over standard fiber at 1.55 μm and the use of dispersion compensation, *IEEE J. of Lightwave Technol.*, bf 13, (1995), pp.1955-1963.
13. Suzuki, M. et al.: Reduction of Gordon-Haus timing jitter by periodic dispersion compensation in soliton transmission, *Electron. Lett.*, **31**, (1995), p.992.
14. Gabitov, I. and Turitsyn, S. K.: Average pulse dynamics in a cascaded transmission system with passive dispersion compensation, *Opt. Lett.*, **21**, (1997), pp.327-329.
15. Grudinin, A. B. and Goncharenko, I. A.: Increased amplifier spacing in soliton system with partial dispersion compensation, *Electron. Lett.*, **33**, (1997), pp.1602-1603.
16. Georges, T.: Extended path-averaged Soliton regime in highly dispersive fibers, *Optics Letters*, **22**, (1997), p.679.
17. Gabitov, I., Shapiro, E. G. and Turitsyn, S. K.: Optical pulse dynamics in fiber links with dispersion compensation, *Opt. Comm.*, **134**, (1997), pp.317-329.
18. Malomed:, B. A.: Pulse propagation in a nonlinear optical fiber with periodically modulated dispersion: variational approach, *Optics Comm.*, **136**, (1997), pp.313-319.
19. Smith, N. J., Knox, F. M., Doran, N. J., Blow, K. J. and Bennion, I.: Enhanced power solitons in optical fibres with periodic dispersion-management, *Electron. Lett.*, **32**, (1996), pp.54-55.
20. Nakazawa, M., Kubota, H., Sahara, A. and Tamura, K.: Marked increase in the power margin through the use of a dispersion-allocated soliton, *IEEE Photon. Techn. Lett.*, **8**, (1996), pp.1088-1090.
21. Smith, N. J., Forysiak, W. and Doran, N. J.: Gordon-Haus jitter reduction in enhanced power soliton systems, *OFC'97*, Dallas, paper **ThN5**.
22. Georges, T. and Charbonnier, B.: Pre-chirping and dispersion compensation for long-haul 20-Gbit/s soliton transmission at 1.55 μm on non dispersion shifted fibers, *OFC'97*, Dallas, paper **WH2**.
23. Wai, P. K. A., Menyuk, C. R. and Chen, H. H.: Stability of solitons in randomly

varying birefringent fibers, *Opt. Lett.*, **16**, (1991), pp.1231-1233.

24. Grudinin, A. B., Goncharenko, I. A., Gray, S. and Payne, D. N.: *Proc. of 21st ECOC'95*, Brussels, paper **TuL1.3**, (1995), pp.295-298.
25. Favre, F., Le Guen, D., Moulinard, M. L., Henry, M., Michaud, G., Devaux, F., Legros, E., Charbonnier, B. and Georges, T.: Demonstration of soliton transmission at 20 Gbit/s over 2200 km of standard fibre with dispersion compensation and pre-chirping, *Electron. Lett.*, **33**, (1997), pp.511-512.
26. Favre, F., Le Guen, D. and Devaux, F.: 4×20Gbit/s soliton WDM transmission over 2000 km with 100 km dispersion-compensated spans of standard fibre, *Electron. Lett.*, **33**, (1997), pp.1234-1235.
27. Yamada, E., Kubota, H., Yamamoto, T., Sahara, A. and Nakazawa, M.: Numerical and experimental comparison between the dispersion-allocated soliton, RZ and NRZ pulses at zero group velocity dispersion, *ECOC'97*, Edinburgh, (1997), pp.3.327-3.330.
28. Grudinin, A. B., Durkin, M., Ibsen, M., Laming, R. I., Schiffini, A., Franco, P., Grandi, E. and Romagnoli, M.: Straight-line 10 Gbit/s soliton transmission over 100 km of standard fibre with in-line chirped fibre grating for partial dispersion compensation, *Electron. Lett.*, **33**, (1997), pp.1572-1573.
29. Hasegawa, A. and Kodama, Y.: *Oxford series in optical and imaging science*, **7**, (1995).
30. Kikuchi, N., Sasaki, S. and Sekine, K.: 10 Gbit/s dispersion- compensated transmission over 2245 km conventional fibers in a recirculating loop, *Electron. Lett.*, **31**, (1995), pp.375-377.
31. Tkach, R. W., Derosier, R. M., Forghieri, F., Gnauk, A. H., Vengsarkar, A. M., Peckham, D. W., Zyskind, J. L., Sulhoff, J. W. and Chraplyvy, A. R.: Transmission of 8×20 Gbit/s channels over 232 km of conventional fiber, *IEEE Photon. Techn. Lett.*, **7**, (Nov. 1995).
32. Wai, P. K. A., Menyuk, C. R. and Chen, H. H.: Effects of randomly varying birefringence on soliton interactions in optical fibers, *Opt. Lett.*, **16**, (1991), pp.1735-1737.
33. Anderson, D.: *Phys. Rev.*, **A6**, (1983), pp.3135-3137; T. Ueda and Kath, W. L.: *Phys. Rev.*, **A42**, (1990), pp.563-570.

(Received November 17, 1997)

TRANSMISSION AND COLLISION OF QUASI-SOLITONS IN DISPERSION MANAGED OPTICAL FIBERS

S. KUMAR
Institut für Theoretische Optik
University of Jena, Max-Wien-Platz-1
07743, Jena, Germany

AND

A. HASEGAWA
Faculty of Engineering, Osaka University
2-1 Yamada-oka, Suita, Osaka 565, Japan

Abstract. Combination of pre-chirp and varying dispersion profile produces soliton like nonlinear pulse having a stationary structure (quasi-soliton). In quasi-soliton system, the interaction with the neighbouring pulse is small and therefore, allows $\sim$ 100 Gbits/sec transmission over ultra-long distance with reasonable power. The varying dispersion profile provides an additional pulse confinement and quasi-soliton systems have better transmission characteristics than those with two-step alternating dispersion profile.

1. Introduction

The soliton transmission in the periodically dispersion compensated fibers has attracted considerable attention recently [1]-[7]. Gabitov and Turitsyn [4] have obtained an asymptotic form of the oscillating localised pulse with chirp in a dispersion managed fiber system. Georges and Charbonnier [6] have shown that the pre-chirping of the input solitons can reduce the dispersive wave generation and soliton interaction. In this paper, we employ essentially the same principle as in reference [6]. We show that the transmission capacity can be further improved by tailoring the dispersion profile of the fiber with anomalous dispersion in a specific way. With this dispersion profile, the equation of motion is described by the quadratic potential in

A. Hasegawa (ed.), New Trends in Optical Soliton Transmission Systems, 341–353.

addition to the usual self-trapping potential. The fundamental stationary mode of this equation varies its functional shape from secant-hyperbolic to Gaussian depending on the input power. Consequently, we call this stationary localised mode as quasi-soliton.

The basic mechanism behind the quasi-soliton transmission is as follows. Initially prechirped pulse propagates in the transmission fiber with anomalous dispersion. The pulse gets compressed due to pre-chirping and fiber nonlinearity. After some distance, the chirp reverses the sign and dispersion dominates. As a result pulse begins to expand. At the output end of the anomalous dispersion fiber we recover the initial pulse, but with exactly opposite chirp. This is possible only if we tailor the dispersion of the fiber. With dispersion tai loring, the nonlinear chirp and dispersion induced chirp act in such a way that the pulse maintains its functional shape. To restore the initial chirp, a normal dispersive element such as fiber grating or dispersion compensating fiber may be used. Using the stationary phase approximation, we have found that the pulse compresses and expands within a unit cell, and yet the functional form is maintained throughout.

In quasi-soliton system, the interaction between neighbouring pulses is reduced as compared to the soliton systems. This is due to the Gaussian tail and chirp of the pulse. We have also compared the quasi-soliton system and the soliton in two-step alternating dispersion profile, and found that for the given strength of dispersion management and average dispersion of the unit cell, quasi-soliton systems have reduced interaction and better power margins. This is because quasi-soliton has a large chirp as compared to the two-step soliton. The combination of pre-chirp and varying profile gives rise to a constant quadratic potential term which is responsible for the pulse confinement and Gaussian tail. Since the Gordon-Haus jitter is inversely proportional to energy, it may be expected that the quasi-soliton systems have less Gordon-Haus jitter than that in the two-step soliton systems.

2. Propagation in Fiber with Anomalous Dispersion

We first consider the propagation in the anomalous dispersion fiber of length Z_1. The "renormalised" amplitude u of the light wave envelope $q(= a(Z)u)$ in a fiber with loss Γ and variable dispersion $d(Z)$, and with $a(Z) = \exp(-\Gamma Z)$, satisfies

$$i\frac{\partial u}{\partial Z'} + \frac{d_e(Z')}{2}\frac{\partial^2 u}{\partial T^2} + \alpha|u|^2 u = 0, \quad 0 \le Z \le Z_1 , \tag{1}$$

where

$$Z' = \int_0^Z a^2(Z)dZ, \quad d_e(Z') = d(Z)/a^2(Z), \tag{2}$$

and α is the normalised nonlinear coefficient. Here Z is normalised by the initial dispersion. Let us now construct a stationary localised solution by introducing a chirped pulse through reference [7]

$$u = \sqrt{p(Z')}\nu(p(Z')T, Z')\exp[iC(Z')T^2/2], \tag{3}$$

where p and C are real functions of Z'. Substituting equation (3) in equation (1), we obtain

$$i\frac{\partial \nu}{\partial Z'} + \frac{d_e p^2}{2}\frac{\partial^2 \nu}{\partial \tau^2} + \alpha p|\nu|^2\nu = \frac{K_1 \tau^2 p}{2}\nu, \tag{4}$$

where $K_1(Z) = (\dot{C} + C^2 d_e)/p^3$ and $\tau = pT$. We choose the dispersion profile such that K_1 is constant. If $d_e p^2 \propto p$ and $K_1 = 0$, equation (4) gives the ideal NLS equation for ν, therefore, we assume the constraint $d_e p = \text{const}(= 1)$ in equation (4) to obtain

$$i\frac{\partial \nu}{\partial Z''} + \frac{1}{2}\frac{\partial^2 \nu}{\partial \tau^2} + \left[\alpha|\nu|^2 - \frac{K_1\tau^2}{2}\right]\nu = 0, \tag{5}$$

where $Z'' = \int_0^{Z'} p(s)ds$. ν can be interpreted as the wave function of a quasi particle in a combined potential of self-trapping and the chirp $K_1\tau^2$. If $\alpha = 0$, equation (5) represents the wave equation for the linear harmonic oscillator, while if $K_1 = 0$, equation (5) gives the ideal NLS equation. The stationary solutions of equation (5) can be obtained by setting

$$\nu = f(\tau)\exp[iK_2 Z''] \tag{6}$$

to get

$$\frac{1}{2}\frac{d^2 f}{d\tau^2} + \alpha f^3 - K_2 f - (K_1/2)\tau^2 f = 0. \tag{7}$$

Equation (7) is solved numerically. Figure 1 shows $f(\tau;\alpha,K_1)$ (solid line) with $\alpha = 0.5$ and $K_1 = 1$, and the broken line shows the soliton solution, $(\eta/\sqrt{\alpha})\text{sech}[\eta T]$ of the same pulse width and corresponds to the local dispersion. The soliton requires about 2.9 times the power of the quasi-soliton represented by $f(\tau;\alpha,K_1)$ for $K_1 = 1$. As K_1 increases, the function f approaches the Gaussian shape. When $K_1 = 0$, $f = (\eta/\sqrt{\alpha})\text{sech}(\eta T)$ (complete nonlinear regime) and when $\alpha = 0$, the fundamental eigenfunction of the harmonic oscillator is given by $f = \exp(-\sqrt{K_1}\tau^2/2)$ (linear regime). In general, the function $f(\tau)$ interpolates between secant hyperbolic and Gaussian shapes, but always has a Gaussian tail for non-zero K_1.

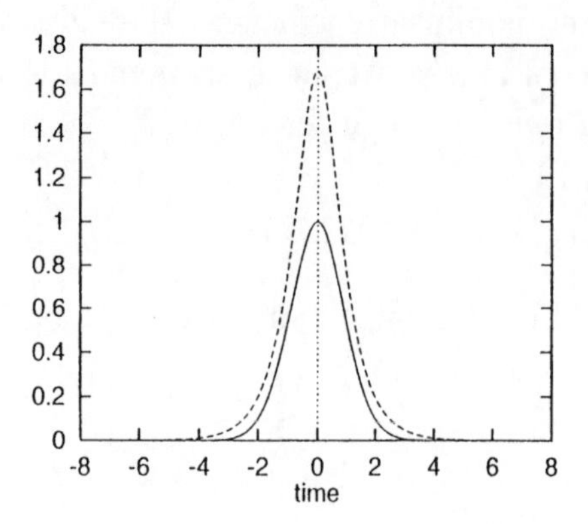

Figure 1. Plot of $f(\tau;\alpha,K_1)$ versus time τ (solid line) with $K_1 = 1$, $K_2 = -0.15915$ and $\alpha = 0.5$. The broken line shows the soliton solution of same pulse width.

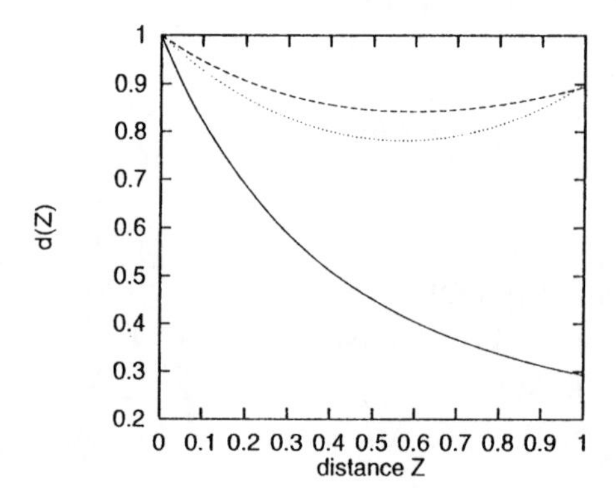

Figure 2. The dispersion profile for various $C(0)$ and Γ with $K_1 = 1$. Solid line shows the profile with $C(0) = -1.34$, $\Gamma = 0.36$, broken line with $C(0) = -0.4$, $\Gamma = 0.1$, and dotted line with $C(0) = -0.8$, $\Gamma = 0$.

The corresponding dispersion profile is given by [7]

$$d_e(Z') = \cosh(\delta Z') + \frac{C(0)}{\delta}\sinh(\delta Z'), \quad \delta = \sqrt{K_1 + C(0)^2}, \tag{8}$$

and

$$d(Z) = d_e(Z')\exp(-2\Gamma Z). \tag{9}$$

Figure 2 shows the dispersion profile for different values of $C(0)$ and Γ. For the large value of $C(0)$ or Γ, the dispersion profile has large variations and the profile given by equation (9) has to be approximated with two or more number of steps for a practical design. When the chirp $C(0)$ and K_1 become zero, we obtain the soliton solution from equation (7) and the dispersion profile becomes $\exp(-2\Gamma Z)$ as expected.

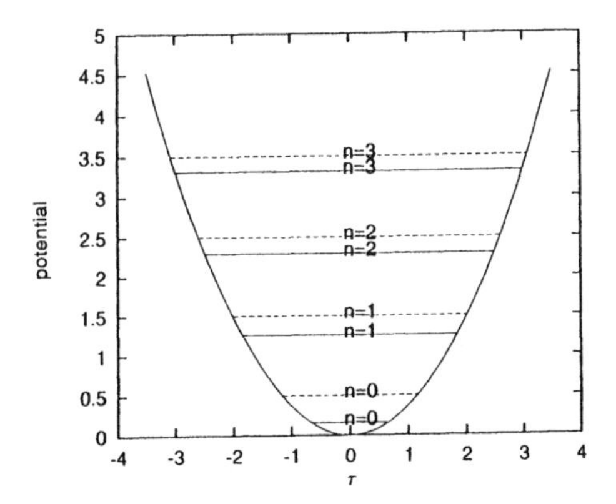

Figure 3. Discrete levels of the system (7) (solid line). For comparison, discrete levels of the harmonic oscillator is also shown (broken line). $K_1 = 1$.

Using equations (6) and (3), the envelope q may be written as

$$q(T, Z) = \frac{\exp(-\Gamma Z)}{\sqrt{d_e(Z')}} f(T/d_e) \exp[iC(Z')T^2/2 + i\beta(Z')], \tag{10}$$

where

$$C(Z') = \dot{d}_e/d_e^2, \; \beta(Z') = K_2 \int_0^{Z'} p(s)ds, \tag{11}$$

with $d_e(Z')$ given by equation (8).

When $K_1 = 0$, the solution given by equation (10) reduces to the chirped soliton solution derived by Moores [8], with $d_e(Z') = \exp[C(0)Z']$. In this limit, very good soliton compression can be achieved since the pulse maintains the secant hyperbolic shape throughout. Equation (10) describes the fundamental mode of the NLS equation with the dispersion profile given by equation (9). In the absence of nonlinearity ($\alpha = 0$), the higher order eigenfunctions of the system (7) are given by the Hermite polynomials, i.e.

$$\nu_n(\tau, Z'') = \frac{1}{\sqrt{2^n n!}} \left(\frac{\sqrt{K_1}}{\pi} \right)^{1/4} H_n(K_1^{1/4}\tau) \exp(-\sqrt{K_1}\tau^2/2 - i\lambda_n Z'') \tag{12}$$

with eigenvalues $\lambda_n = (n + 1/2)\sqrt{K_1}$. In the presence of the nonlinearity, there exist nonlinear eigenfunctions similar to the modes given by equation (12). The solid line in Figure 3 shows the eigenvalues of the system (7) and the broken line shows the eigenvalues of the linear harmonic oscillator. As can be seen, as n increases the eigenvalues of the system (7) approach that of the linear harmonic oscillator, indicating that the effect of nonlinearity becomes smaller for the higher order modes.

In this paper, we study only the fundamental mode for the long-distance communiation applications. From equations (11) and (8), we see that the sign of chirp is reversed at $Z = Z_c$ given by $Z'(Z_c) = (1/\delta)\tanh^{-1}[-C(0)/\delta]$. The next step is to reverse the chirp so that we recover the launched pulse having the same initial chirp. This can be accomplished by the use of normal dispersive element such as grating.

3. Propagation in the Fiber Grating

The effect of grating with dispersion $d_N(Z)$, placed after the transmission fiber is to multiply the signal by the transfer function $h(\Omega)$ such that

$$q(T, Z_a) = \int h(\Omega)\tilde{q}(\Omega, Z_1)\exp[i\Omega T]d\Omega, \tag{13}$$

where $h(\Omega) = \exp(i\Omega^2 g/2)$, $g = \int_{Z_1}^{Z_a} d_N(Z)\mathrm{d}Z$, and

$$\tilde{q}(\Omega, Z_1) = \int q(T, Z_1)\exp[-i\Omega T]dT. \tag{14}$$

Substituting equation (10) in equation (14) and using the stationary phase approximation, we obtain

$$\tilde{q}(\Omega, Z_1) \approx \exp(-\Gamma Z_1)\sqrt{\frac{2i\pi}{C_1}}f\left(\frac{-\Omega}{C_1 d_1}\right)\exp[-i\Omega^2/(2C_1) + i\beta_1] \tag{15}$$

with $d_e(Z'(Z_1)) \equiv d_1$, $C(Z'(Z_1)) \equiv C_1$, etc. Using equation (15) in equation (13) and again using the stationary phase approximation, we obtain

$$q(T, Z_a) \approx \exp[-\Gamma Z_1]\sqrt{w}f(wT)\exp[iC_2T^2/2 + i\beta_1], \tag{16}$$

where $w = 1/d_1[1 - gC_1]$, $C_2 = C_1/[1 - gC_1]$. From equation (16), we see that the initial pulsewidth and chirp can be recovered if

$$d_1[1 - gC_1] = \pm 1 \quad \text{and} \quad C(0) = \pm C_1 d_1. \tag{17}$$

The solution of equations (17) give the approximate value of g. The exact value of g for chirp cancellation can be obtained from evaluating the integral in equation (13) numerically. The dispersion compensating fiber may also be used instead of grating, since the pulse is essentially linear after the long propagation in transmission fiber.

4. Numerical Simulations

We carried out the numerical simulation of equation (1) with the following parameters: fiber dispersion (at $Z = 0$) is -0.5 ps^2/km, wavelength $= 1.55$

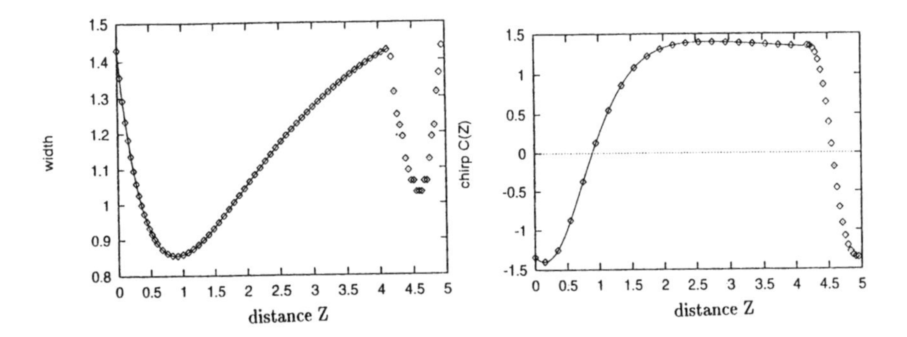

Figure 4. Plot of pulse width and chirp as a function of distance. Solid line shows the theoretical results, squares show the simulation results. The length of the grating is exaggerated to show the chirp changes clearly. Parameters: $C(0) = -1.34$, $\Gamma = 0.36$, $K_1 = 1$, $Z_1 = 4.15$, $\alpha = 0.5$ and $g = 0.835$.

μm, $A_{eff} = 50\ \mu\text{m}^2$, nonlinear coefficient $n_2 = 3.18 \times 10^{-16}\text{cm}^2/\text{W}$, fiber loss = 0.2 dB/km, amplifier spacing = 66.4 km, integrated dispersion g of the grating = 6.687 ps^2. We have used a suitable time scale t_0 and length scale z_0 such that $t/t_0 = T$ and $z/z_0 = Z$, where t and z are real time and distance respectively. By assuming $t_0 = 2.83$ ps, z_0 becomes 16 km. The pulse width of the quasi-soliton (Figure 1) is found to be 1.47 t_0 and the peak power is 12 mW. The Gaussian pulse of the proper pulse width could be a good approximation to the pulse $f(\tau)$ shown in Figure 1. Due to the scaling freedom in the NLS equation, the normalised parameters can also correspond to different pulsewidths and dispersions. For example, the normalised parameters shown in the figures could as well correspond to initial dispersion -2 ps^2/km and pulse width 8.32 ps.

Figure 4 shows the theoretical (solid line) and simulation results (squares) for a unit cell. The chirp is computed numerically using the formula given in reference [9]. Initially the pulse is pre-chirped such that it undergoes compression. The non-linearity induced chirp also enhances the compression. However, after some distance, the fiber dispersion dominates and the chirp reverses its sign at $Z = Z_c$. Thereafter, the pulse begins to expand. There exists a length Z_1 such that we recover the initial pulse width but with exactly opposite chirp. The grating reverses the chirp so that we restore the initial chirp. We carried out the simulation over a length of 8,000 km and found that the stationary mode propagates with almost no distortion. The chirp C is computed at each amplifier position and it is found that even after a length of 8,000 km, the chirp is almost same as the initial chirp indicating that maximum transmission distance is limited only by amplifier noise and higher order effects, which are not included in the simulations.

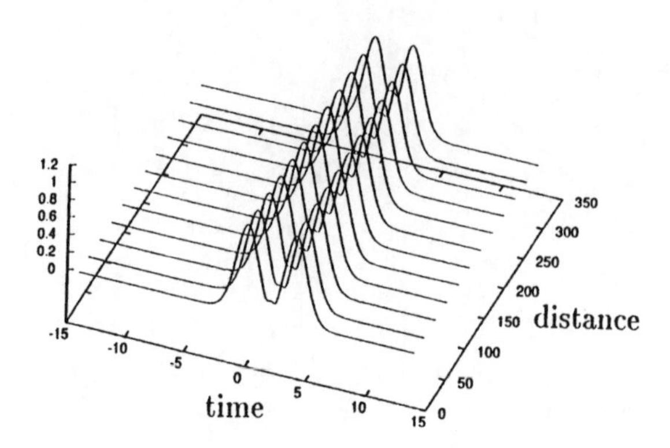

Figure 5. Quasi-soliton interaction over a distance of 5,000 km. Parameters: $C(0) = -1.34$, $Z_1 \approx Z_a = 4.15$, $K_1 = 1$, $\alpha = 0.5$, $g = 0.835$ and $\Gamma = 0.36$.

4.1. QUASI-SOLITON INTERACTION

Since the quasi-soliton has a Gaussian tail and large chirp, it may be expected that the interaction between neighbouring channels is reduced. Figure 5 shows the simulation results of two quasi-soliton interaction. Two 4.16-ps quasi-solitons separated by 11 ps, propagate over a distance of 2,000 km with almost no interaction. They collide after propagating a distance of 8000 km. But two solitons of the same peak power (12 mW) separated by 11 ps would collide after 80 km. The comparision can also be made on the basis of equal average dispersion. In the present example, the average dispersion of the unit cell ($\equiv (\int_0^{Z_1} d(Z)dZ - g)/Z_a$) is -0.0149 ps^2/km. Comparing with the collision distance in the case of solitons propagating in the same constant dispersion, it is found that there is an improvement by a factor of $\sim$2.6 for quasi-solitons. In this example, quasi-soliton has an energy enhancement factor $\sim$5.7 as compared to the case of ideal soliton. For a practical implementation, the dispersion profile given by equation (9) may be stepwise approximated. The varying profile is approximated by a single step by keeping the the same average dispersion of the anomalous fiber section to obtain a two-step alternating dispersion profile. The nonlinear pulse propagation in the 2-step alternating dispersion profile is quite similar to the case of enhanced power solitons [5]. The amplitude and chirp are optimised to obtain the quasi-stable propagation [10]. It is found that the soliton in the two-step profile requires less energy than the quasi-soliton for the same average dispersion of the unit cell. In our parameter region, the accumulated dispersion of the anomalous fiber section is 7.67 ps^2, and the average dispersion of the unit cell is -0.0149 ps^2/km. The quasi-soliton requires the energy 2.8 times that being required by the soliton in two-step profile. Figure 6 shows the interaction of two 4.16 ps pulses separated by

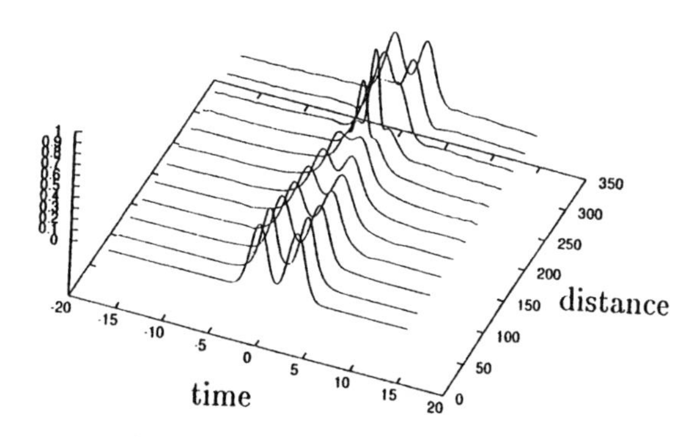

Figure 6. Soliton interaction in a two-step profile. Parameters: $C(0) = -0.32$, $Z_1 \approx Z_a = 4.15$, dispersion of the first step $= 0.231$, $\alpha = 0.5$, $g = 0.835$ and $\Gamma = 0.36$. The input excitation is a Gaussian pulse with peak amplitude 0.6. Pulse width, separation and average dispersion are same as that of Figure 5.

11 ps in the two-step profile. As can be seen, the pulses collide after propagating a distance of 3,600 km, showing an improvement factor ~2.2 for quasi-solitons. This is because for the given strength of dispersion management ($\propto$ accumulated dispersion of the transmission fiber + g) and average dispersion, the maximum chirp of quasi-soliton is much more than that of two-step soliton. The transmission of highly chirped and yet well-confined pulse with Gaussian tail is possible if we tailor the dispersion in accordance with equation (9). For a practical design, it may be necessary to have two or more step approximation to the profile (9).

Since the soliton in the two-step profile requires much less energy than the quasi-soliton, it may be expected that for the given strength of dispersion anagement and average dispersion, the reduction in the Gordon-Haus jitter is better in the case of quasi-soliton as compared to the soliton in two-step profile.

5. Quasi-soliton Collisions

Let us consider the soliton collision between two channels separated by the angular frequency ΔB. equation (1) admits a Galilean invariant solution, i.e.,

$$u_i = \sqrt{p} f(p[T - T_i]) e^{-i(\kappa_i T - C[T-T_i]^2/2 + \lambda_i)}, \quad i = 1, 2 \tag{18}$$

$$dT_i/dZ = -\kappa_i d(Z). \tag{19}$$

Here $\{u_i, T_i, \kappa_i, \lambda_i\}$ represent the renormalised envelope, time position, frequency and phase factor of the ith channel, respectively. In the absence of perturbation, the frequency κ_i is constant. However, due to collision with the other channel, the frequency and amplitude become functions of Z. Let the initial separation be ΔB, i.e. $\kappa_1(0) = -\kappa_2(0) = \Delta B/2$ and let the initial temporal separation between two channels be T_0, i.e. $T_1(0) = -T_2(0) = T_0/2$. Integrating equation (19), we obtain

$$T_1(Z) = -T_2(Z) \approx \frac{T_0 - \Delta B \int_0^Z d(s)ds}{2}. \tag{20}$$

The perturbation seen by the channel 1 due to collision with the other channel is given by $2i\alpha a^2(Z)|u_2|^2 u_1$. Following the adiabatic perturbation technique [11], we obtain

$$\begin{aligned} \frac{d\kappa_1}{dZ} &= \frac{2\alpha(pa)^2}{g_0} \int f^2(\tau)\frac{d}{d\tau}[f^2(\tau - s_0(Z))]d\tau \\ &= \frac{-2\alpha(pa)^2}{g_0}\frac{d}{ds_0} \int f^2(\tau + s_0/2) f^2(\tau - s_0/2)d\tau, \end{aligned} \tag{21}$$

where $s_0(Z) = 2pT_1(Z)$. To simplify Eq.(21), we approximate $f(\tau)$ by $\exp(-\gamma\tau^2)$ for some constant γ, to obtain the frequency shift at the end of the anomalous dispersion fiber as

$$\kappa_1(Z_1) = \kappa_1(0) + X_0 \int_0^{Z_1} (pa)^2 s_0(Z) \exp(-\gamma s_0^2)dZ, \tag{22}$$

where $X_0 = 2\alpha\sqrt{\pi\gamma}/g_0$.

In the normal dispersion fiber, the pulse is linear and therefore, do not undergo any frequency shift, i.e., $\kappa_1(Z_a) = \kappa_1(Z_1)$. However, due to dispersion, the pulse position undergoes a shift by $\kappa_1 g$, i.e.

$$\begin{aligned} T_1(Z_a) &= T_1(Z_1) + \kappa_1 g \\ &= (T_0 - \Delta B \bar{d} Z_a)/2, \end{aligned} \tag{23}$$

where $\bar{d}$ is the average dispersion of a unit cell.

After computing the frequency shift in each unit cell and adding, we obtain the frequency shift at $Z = \infty$, $\Delta K (\equiv \kappa_1(\infty) - \kappa_1(0))$ as

$$\Delta K = X_0 \sum_{m=0}^{\infty} \int_0^{Z_1} (pa)^2 s_m \exp(-\gamma s_m^2)dz \tag{24}$$

with $s_m(Z) = p\left(T_0 - \Delta B\left[m\bar{d}Z_a + \int_0^Z d(s)ds\right]\right)$.

Figure 7 shows the frequency shift $|\Delta K|$ as a function of the initial separation T_0, obtained from equation (24), which is in agreement with the direct numerical simulation of equation (1) (shown in '+'). In contrast

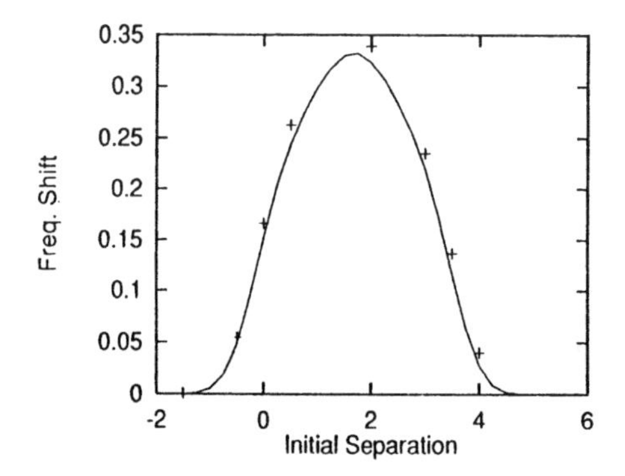

Figure 7. Plot of the permanent frequency shift ΔK versus initial separation between two pulses. Solid line shows the theoretical results and '+'shows the numerical results. $\Delta B = 8$, other parameters same as Figure 4.

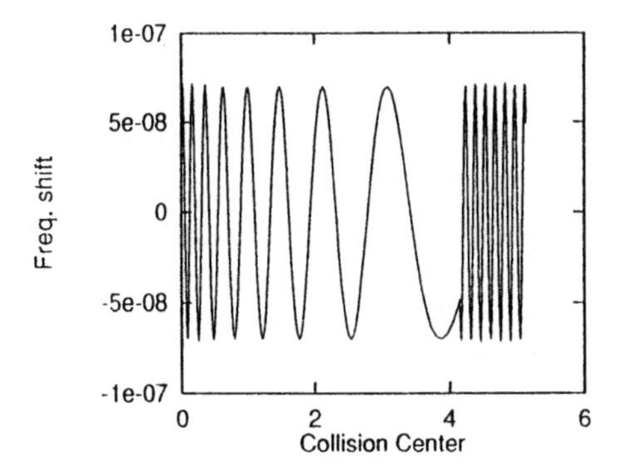

Figure 8. Plot of the permanent frequency shift ΔK versus collision center. Parameters same as Fig. 7. Length of the grating is exaggerated.

to the sech solitons, the completely overlapped pulses at the input do not produce the maximum frequency shift. In our parameter region, the maximum frequency shift occurs for $T_0 \approx 1.75$. This is because the quasi-solitons undergo compression in the begining and if $T_0 \approx 1.75$, maximum overlap between pulses occurs when they are maximally compressed.

As the initial separation becomes larger, the collision becomes nearly complete and the frequency shift becomes negligibly small. But the frequency shift ΔK oscillates depending on where the collision occurs within a unit cell. Figure 8 shows the frequency shift ΔK due to a complete collision as a function of center of collision within a unit cell $[0, Z_a]$. Although the frequency shift ΔK is a periodic function of the collision center in the co-ordinate $\hat{Z} = \int_0^Z d(s)ds$ with constant period $\bar{d}Z_a$, in the real co-ordinate Z, the period seems to increase in the range $[0, Z_1]$ due to varying dispersion profile of equation (9).

It is found that even for the channel separation that is twice the spectral width of quasi-soliton ($\Delta B = 3$), the frequency shift is only of the order of 10^{-10}. Therefore, within the given bandwith, more number of channels can be accommodated without increasing the collision-induced jitters.

6. Conclusions

In conclusion, we have shown that the short pulse (~4 ps) transmission in dispersion managed optical fibers over ultra-long distance is possible at moderate powers. The chirped stationary solution of the NLS equation is derived and a novel dispersion profile is proposed to support the stationary mode. The functional form of the stationary mode varies from secant-hyperbolic to Gaussian depending on the input power. We have shown that the interaction between two neighbouring quasi-solitons is considerably reduced due to Gaussian tail and high chirp. After comparing with the soliton in a two-step alternating dispersion profile, it is found that quasi-soliton has reduced interaction and better power margins.

We thank Prof. Y. Kodama for helpful discussions and Mr. M. Wald for help in computer calculations.

References

1. Suzuki, M., Morita, I., Edagawa, N., Yamamoto, S., Taga, H. and Akiba, S.: Reduction of Gordon-Haus jitter by periodic dispersion compensation in soliton transmission, *Electron. Lett.*, **31**, (1995), pp.2027-28.
2. Haus, H. A., Tamura, K., Nelson, L. E. and Ippen, E. P.: Stretched-pulse additive mode-locking in fiber ring lasers: theory and experiment, *IEEE J. Quantum Electron.*, **QE-31**, (1995), pp.591-598.
3. Nakazawa, M., Kubota, H., Sahara, A. and Tamura, K.: Marked increase in the power margin through the use of a dispersion-allocated soliton, *IEEE Photon. Technol. Lett.*, **8**, (1996), pp.1088-90.
4. Gabitov, I. and Turitsyn, S. K.: Averaged pulse dynamics in a cascaded transmission system with passive dispersion compensation, *Opt. Lett.*, **21**, (1996), pp.327-329.
5. Smith, N. J., Knox, F. M., Doran, N. J., Blow, K. J. and Bennion, I.: Enhanced power solitons in optical fibers with periodic dispersion management,*Electron. Lett.*, **32**, (1996), pp.54-55.
6. Georges, T. and Charbonnier, B.: Reduction of the dispersive wave in periodically amplified links with initially chirped solitons, *IEEE Photon. Techn. Lett.*, **9**, (1997) , pp.127-129.
7. Kumar, S. and Hasegawa, A.: Quasi-soliton propagation in dispersion-managed optical fibers, *Opt. Lett.*, **22**, (1997) , pp.372-374.
8. Moores, J. D.: Nonlinear compression of chirped solitary waves with and without phase modulation, *Opt. Lett.*, **21**, (1996), pp.555-557.

9. Belanger, P. A. and Belanger, N.: RMS characteristics of pulses in nonlinear dispersive lossy fibers, *Opt. Comm.*, **117**, (1995), pp.56-60.
10. Wald, M., Uzunov, I. M., Lederer, F. and Wabnitz, S.: Optimization of periodically dispersion compensated breathing soliton transmission, to be published in *IEEE Photon. Technol. Lett.*.
11. Hasegawa, A. and Kodama, Y.: *Solitons in Optical Communications*, Oxford U. Press, Oxford, (1995), p.181.

(Received November 17, 1997)

LONG-HAUL SOLITON WDM TRANSMISSION WITH PERIODIC DISPERSION COMPENSATION AND DISPERSION SLOPE COMPENSATION

I. MORITA, M. SUZUKI, N. EDAGAWA, K. TANAKA, AND S. YAMAMOTO
KDD R&D Laboratories
2-1-15 Ohara Kamifukuoka, Saitama 356, Japan

Abstract. 20 Gbit/s-based soliton WDM transmission experiments using periodic dispersion compensation and dispersion slope compensation were demonstrated. Accumulated dispersion slope was compensated with two methods. One was the method using periodical individual dispersion compensation. By using this technique, 60 Gbit/s (20 Gbit/s× 3WDM) transmission over 8,000 km was demonstrated. The other was the method using a dispersion-flattened transmission fiber. 160 Gbit/s (20 Gbit/s×8WDM) transmission over 4,000 km using periodically dispersion compensated dispersion-flattened fiber was also demonstrated.

1. Introduction

Aggregate system capacity has been dramatically increasing by using the wavelength-division-multiplexing (WDM) transmission technologies and has exceeded 100 Gbit/s even for transoceanic applications [1, 2, 3, 4]. In such systems, soliton WDM transmission is quite attractive, because soliton-based systems have potential to carry higher channel bit rate signals than NRZ systems thus they are expected to need smaller number of channels for the same system capacity. The key technological issues in soliton WDM transmission are the reduction of collision induced timing jitter and four wave mixing. To mitigate these effects, several techniques, such as dispersion tapering fiber spans [4, 5, 6], sliding frequency guiding filters [5] and inline synchronous modulation [4], have been proposed. So far, 80 Gbit/s (10 Gbit/s × 8WDM) transmission over 9,000 km using sliding-frequency guiding filters [5] and 160 Gbit/s (20 Gbit/s × 8WDM) transmission over

A. Hasegawa (ed.), New Trends in Optical Soliton Transmission Systems, 355–365.

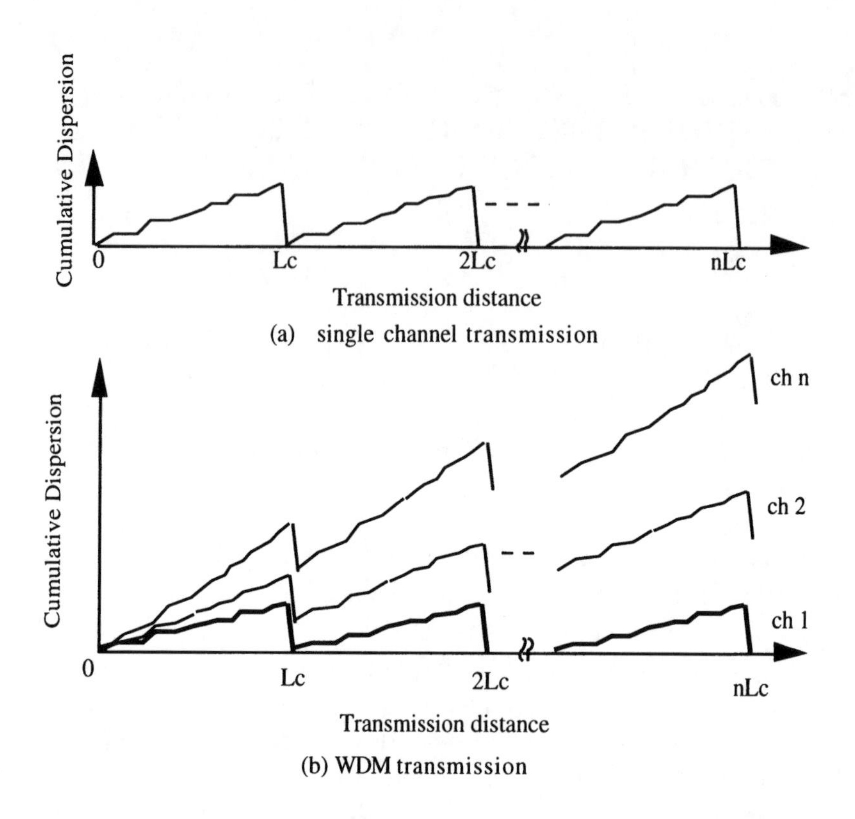

Figure 1. Dispersion map of periodic dispersion compensation scheme.

10,000 km using inline synchronous modulation [4] have been demonstrated. In both demonstrations, dispersion-tapered fiber span was also used.

For single-channel transmission, we have proposed a transmission scheme using periodic dispersion compensation [7, 8] to reduce the accumulation of Gordon-Haus timing jitter [9], which is one of the major constraints in soliton transmission systems. In this transmission scheme, the cumulative chromatic dispersion is offset periodically along the system by the dispersion compensation fiber (DCF) with negative (normal) dispersion and the total system dispersion is close to zero as shown in Figure 1(a). Since the Gordon-Haus timing jitter is due to the translation of random variation of the soliton center frequency, caused by ASE noise of optical amplifiers and fiber nonlinearity, through the fiber dispersion [9], the accumulated timing jitter does not increase significantly in quasi-zero-dispersion systems.

In soliton-WDM systems, however, the difference in chromatic dispersion for each channel (i.e. dispersion slope) affects the system performance significantly, because the dispersion compensation optimized for one channel can not provide enough dispersion compensation for other channels due

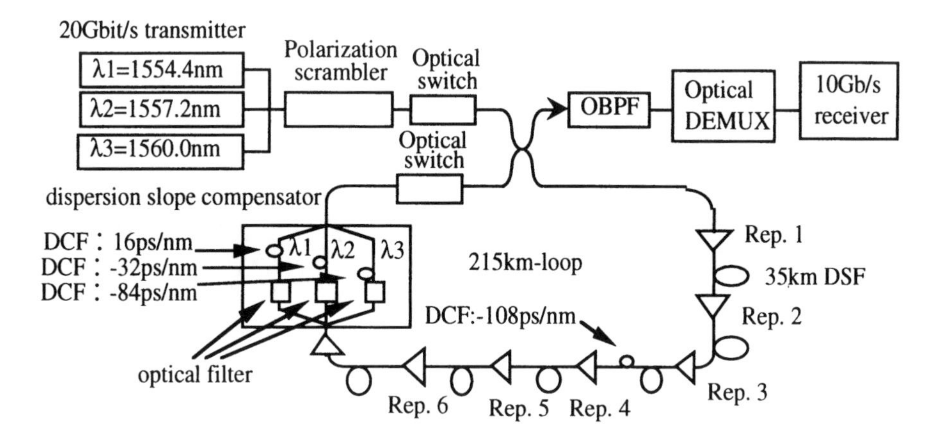

Figure 2. Experimental setup for 20 Gbit/s × 3WDM transmission using periodical individual dispersion compensation.

to the dispersion slope of the transmission line as shown in Figure 1(b).

In this paper, we conducted experimental investigation of two methods in order to overcome this problem. One was the method using periodical individual dispersion compensation [10], and the other was the method using a dispersion-flattened fiber (DFF) [13, 14]. The experimental results on 20 Gbit/s-based soliton WDM transmission using these methods were shown.

2. Soliton WDM Experiments Using Periodical Individual Accumulated Dispersion Compensation

2.1. EXPERIMENTAL SETUP

We conducted soliton WDM experiments using periodical individual dispersion compensation [10]. In this transmission scheme, the accumulated dispersion for each channel were compensated individually in the dispersion slope compensators which were inserted periodically in the transmission line.

Figure 2 shows a schematic diagram of the experimental setup. The experiments were undertaken in a 215 km fiber loop. We used three 20 Gbit/s optical time-division-multiplexing (OTDM) transmitters. Signal wavelengths were 1554.4 nm (ch1), 1557.2 nm (ch2) and 1560.0 nm (ch3). In each 20 Gbit/s transmitter, a 20 Gbit/s optical soliton data stream was produced by time-division-multiplexing 10 Gbit/s RZ data pulses which were generated with a DFB-LD, a sinusoidally-driven electroabsorption (EA) modulator

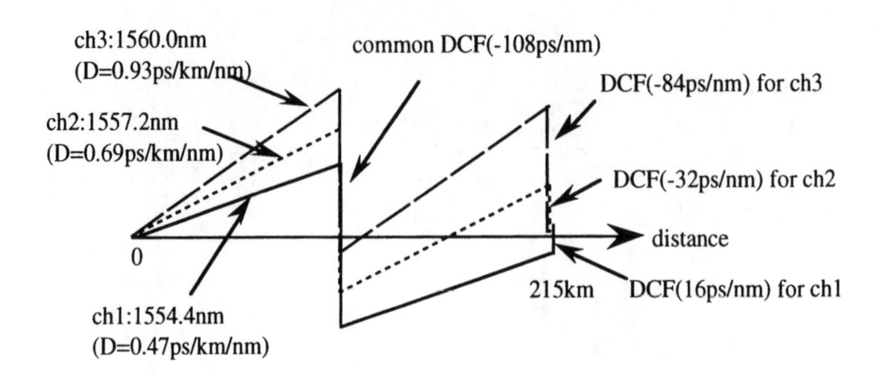

Figure 3. Dispersion map of 215 km loop for each channel.

and two $LiNbO_3$ intensity modulators operated at 10 Gbit/s by a $2^{15}-1$ pseudorandom binary sequence. The pulse width obtained was 10 ps to 15 ps. In addition, each 20 Gbit/s OTDM signal was modulated at 20 GHz by a $LiNbO_3$ phase modulator in order to improve transmission characteristics [11, 12]. After combining three 20 Gbit/s signals, a polarisation scrambler was used to suppress the polarization hole burning effect of the erbium-doped-fiber-amplifier (EDFA) repeaters.

The transmission line comprises six spans of 35 km-long dispersion-shifted fiber (DSF) and DCF. The average dispersion slope of the DSF was 0.084 ps/nm^2·km and the average chromatic dispersion for ch1, ch2 and ch3 were 0.43 ps/nm·km, 0.69 ps/nm·km and 0.93 ps/nm·km, respectively. To compensate for the accumulated dispersion, the common-DCF with -108 ps/nm was placed after the third span and the residual accumulated dispersion for each wavelength after 215 km fiber transmission was compensated with the individual DCFs for each channel (DCF(+16 ps/nm) , DCF(-32 ps/nm) and DCF(-84 ps/nm) for ch1, ch2 and ch3, respectively) in the dispersion slope compensator. In the dispersion slope compensator , the transmitted signals were divided into three paths with an optical fiber coupler and each channel was selected with optical filters placed on each path and then was recombined with an optical fiber coupler following the individual DCFs. Figure 3 shows the dispersion map of the loop for each channel.

In the receiver, the desired channel was selected by an optical bandpass filter and the transmitted 20 Gbit/s signals were optically time division-demultiplexed to 10 Gbit/s data stream with an optical gate using a sinusoidally-driven EA modulator. The clock recovery was done with a 20 GHz phase-locked-loop.

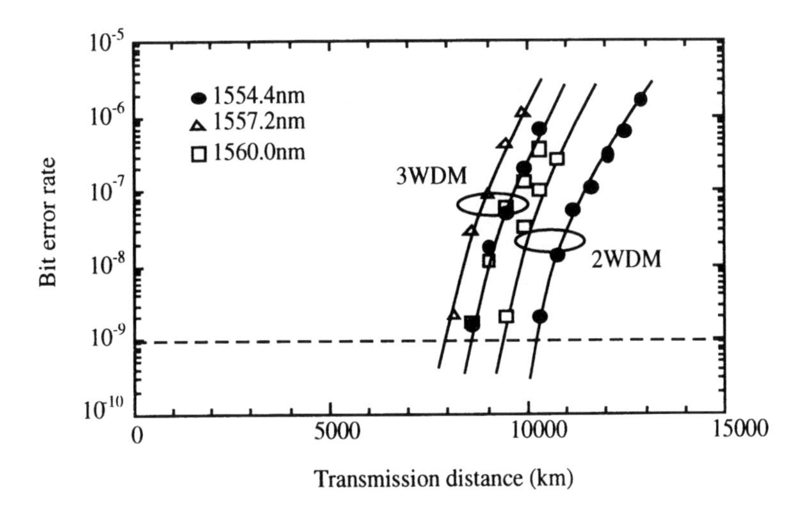

Figure 4. BER vs. transmission distance for 20 Gbit/s-based WDM experiment using periodical individual dispersion compensation.

2.2. 20 GBIT/S × 2WDM AND 20 GBIT/S × 3WDM EXPERIMENTS

We have conducted two-channel WDM and three-channel WDM transmission experiments. The nominal repeater output power was about 3 dBm for two-channel transmission and 5 dBm for three-channel transmission, respectively. No significant waveform distortion and timing jitter were observed after 9,000 km transmission in 2WDM even through the estimated number of collisions during the transmission was about 2,000. This indicates the effectiveness of the periodic dispersion and its slope compensation to reduce the soliton-collision timing jitter in WDM soliton transmission, since the frequency shift due to the soliton-soliton collision is not translated into timing jitter in quasi-zero-dispersion systems. To evaluate the transmission performance, the bit error rate (BER) of the received 10 Gbit/s signals were measured. Figure 4 shows the BER obtained as a function of the transmission distance.

A bit rate of about 10^{-9} was achieved over 9,500 km for two-channel transmission and over 8,000 km for three-channel transmission. In these experiments, the effects of FWM seemed to be quite small. This may be attributed partly to large local dispersion and partly to large channel spacing, since FWM product during the collision can be reabsorbed after rapid collision, which can be regarded as quasi-lossless collision [5]. From these results, we confirmed the applicability of the periodic dispersion and its

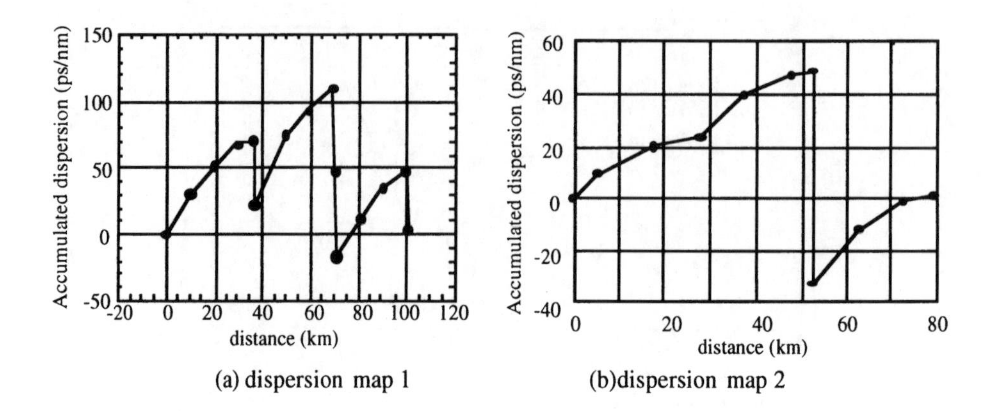

(a) dispersion map 1 (b)dispersion map 2

Figure 5. Dispersion map of DFF loop.

slope compensation to soliton-WDM transmission systems.

3. Soliton WDM Experiment Using a Dispersion-flattened Fiber

In the preceding section, we described the method using periodical individual dispersion compensation. For a practical viewpoint, however, this approach is rather complicated, unsuitable for dense-WDM and not so easy to implement. So we investigated the transmission characteristics of soliton WDM using a dispersion flattened fiber (DFF) [13, 14], which is simpler and more attractive for practical use.

3.1. EFFECT OF FIBER DISPERSION OF DFF

First, to see the fundamental transmission characteristics of DFF, we conducted the signal-channel and two-channel soliton transmission experiments using fiber loops with two types of dispersion map: dispersion map 1 and dispersion map 2. Figure 5 shows these dispersion maps schematically.

In dispersion map 1, the transmission line comprises three spans of DFF and DCF. The average span length was 33 km. The DCF was used at the end of each span to compensate for the rather large dispersion of the transmission fiber (about 4.4–1.5 ps/nm·km). The span average dispersion was 0.7 ps/nm·km. To reduce the jitter effect, another DCF was placed at the end of the second span to compensate for the accumulated dispersion along the loop. The system average dispersion and dispersion slope at 1555 nm were 0.03 ps/nm·km and -0.0007 ps/nm^2·km, respectively.

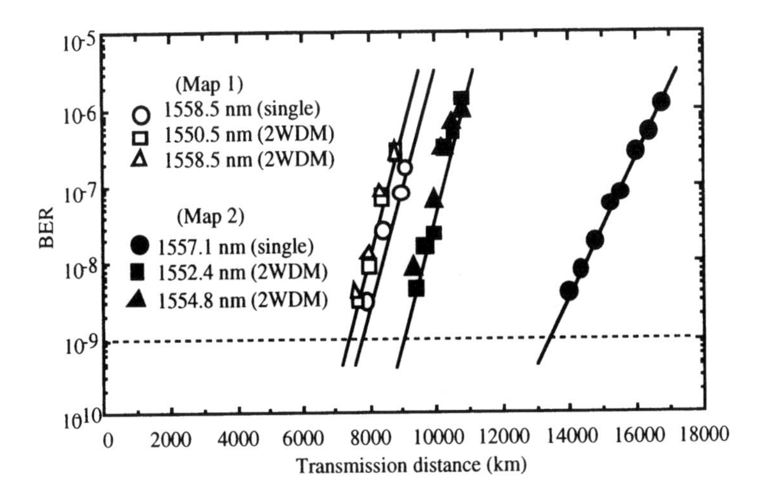

Figure 6. BER vs. transmission distance for single-channel and two-channel soliton transmission using two types of DFF loop.

In dispersion map 2, the transmission line comprises three spans of about 27 km-long DFF and DCF. The average dispersion and dispersion slope of the DFF at 1555 nm were 1.1 ps/nm·km and 0.0005 ps/nm^2·km, respectively. The accumulated chromatic dispersion of the loop was compensated for by the DCF with −82 ps/nm and the system average dispersion was 0.02 ps/nm·km.

Figure 6 shows the bit error rate measured as a function of transmission distance for single-channel and two-channel WDM transmission experiments in both dispersion maps. In these experiments, we used the same 20 Gbit/s OTDM transmitter and receiver as described in the previous section. As shown in Figure 6, in the dispersion map 1, the maximum transmission distance was limited to only about 8,000 km even in the case of single-channel transmission. By changing dispersion map so as to decrease the chromatic dispersion of DFF, from dispersion map 1 to dispersion map 2, the transmission performance was significantly improved. A bit rate of about 10^{-9} was achieved over 13,000 km for single-channel transmission and over 9,000 km for two-channel transmission.

3.2. 160 GBIT/S (20 GBIT/S × 8WDM) TRANSMISSION EXPERIMENT

We have conducted eight-channel 20 Gbit/s soliton WDM transmission using dispersion map 2 [14]. Figure 7 shows a schematic diagram of the experimental setup. We used four 20 Gbit/s × 2WDM transmitters as a 160

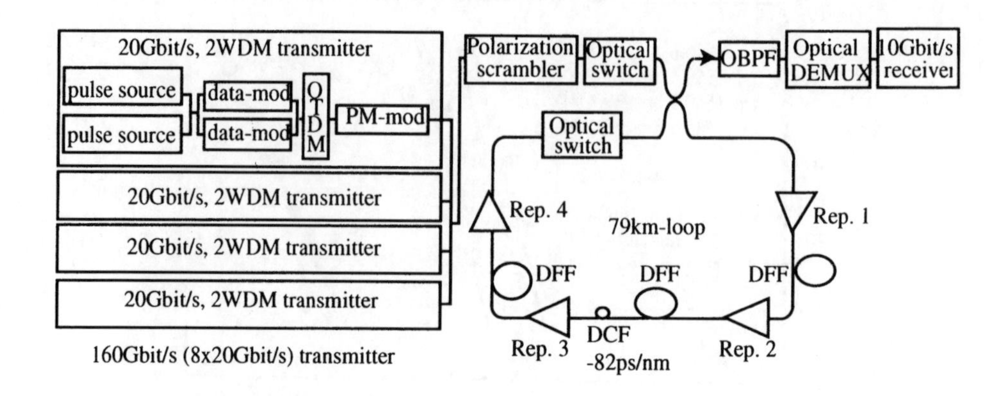

Figure 7. Experimental setup for 160 Gbit/s (20 Gbit/s × 8WDM) transmission.

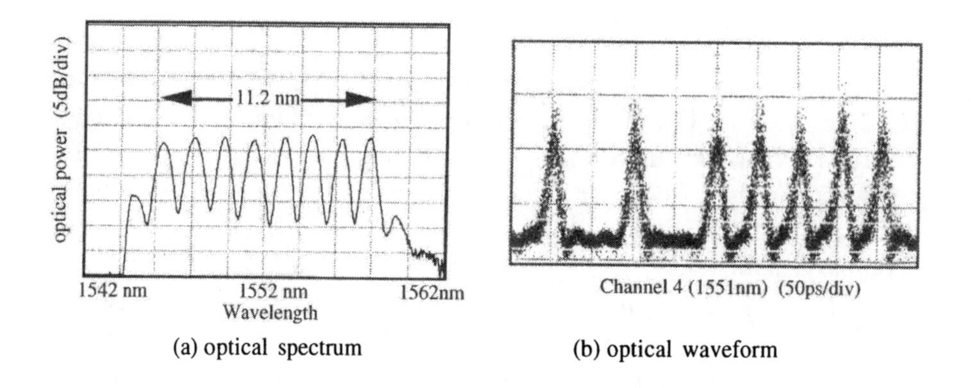

Figure 8. Optical spectrum and waveform after 4,400 km transmission.

Gbit/s transmitter. The wavelength of eight channels ranging from 1546 nm to 1558 nm was equally spaced by 1.6 nm. Each 2WDM transmitter consists of two EA-modulator based soliton pulse generators with different wavelengths. These two 10 GHz optical pulse sequences were combined with an optical fiber coupler and then data-modulated with a $2^{11}-1$ pseudorandom binary pattern and multiplexed to 20 Gbit/s by using OTDM. The average repeater output power was about 7.8 dBm.

Figure 8(a) shows the optical spectrum after 4,400 km transmission. Quite flat gain response was obtained. Note that the small peaks outside the signal wavelength band (1546 nm–1558 nm) in Figure 8(a) represent

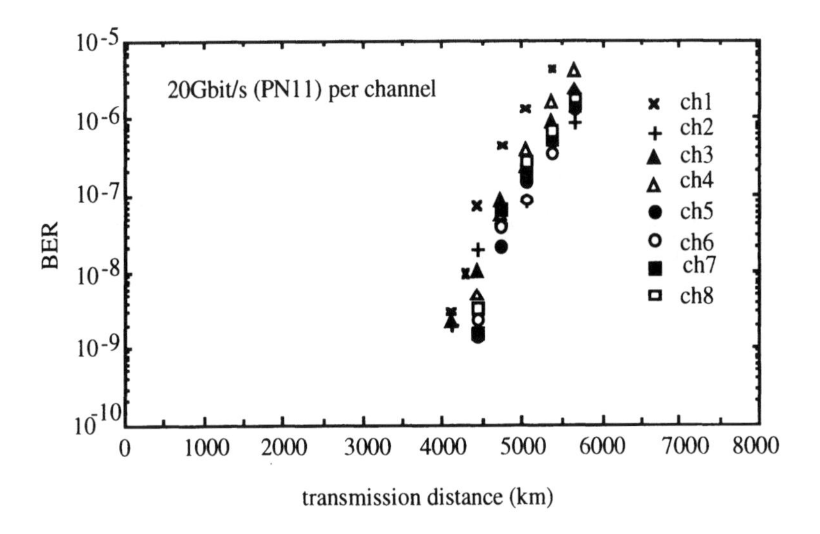

Figure 9. BER vs. transmission distance for 160 Gbit/s (20 Gbit/s × 8WDM) experiment.

the ASE accumulation at the transmissible peaks of the FP guiding filter. Figure 8(b) shows the typical waveform after 4,400 km transmission for channel 4. Significant waveform distortion and timing jitter were not observed after transmission, since the effect of collision-induced timing jitter can be reduced due to small residual dispersion in periodic dispersion compensation scheme. Figure 9 shows BER as a function of transmission distance for eight channels. A BER of less than 10^{-9} was obtained for all the channels after over 4,000 km transmission. The differences in peak power and chromatic dispersion for eight channels were within $+/-0.5$ dB and $+/-0.003$ ps/nm·km, respectively, and quite flat transmission performance has been confirmed over 11 nm wavelength window. To increase the number of channels with keeping higher channel bit rate, this transmission scheme using periodically dispersion compensated dispersion flattened fiber has been confirmed useful through this experiment.

In these experiments, however, the DFF has not optimized yet. The average loss of the DFF, for example, was 0.33 dB/km, which was larger than conventional transmission fiber's value. Therefore, if the parameters of DFF are refined, further progress in transmission performance can be expected.

4. Conclusion

We have conducted 20 Gbit/s-based soliton WDM transmission experiments using periodic dispersion compensation and dispersion slope compensation. Dispersion slope was compensated with two methods. One was the method using periodical individual dispersion compensation and the other was the method using a DFF. 60 Gbit/s (20 Gbit/s × 3WDM) transmission over 8,000 km using periodical individual dispersion and 160 Gbit/s (20 Gbit/s × 8WDM) transmission over 4,000 km using periodically dispersion compensated DFF were demonstrated. These results indicate the transmission scheme using periodic dispersion compensation and dispersion slope compensation is a quite useful and attractive way to increase the number of channels keeping higher channel bit rate in ultra large capacity transoceanic transmission systems.

Acknowledgment

The authors would like to express their thanks to Drs. H. Murakami and S. Akiba of KDD R&D laboratories for their continued encouragement.

References

1. Taga H., Imai K., Takeda N., Suzuki M, Yamamoto S. and Akiba S.: 10WDM × 10 Gbit/s Long-distance transmission experiment using a dispersion slope compensator and non-soliton RZ pulse, *OAA '97*, Victoria, **TuB2**, (1997).
2. Bergano N., Davidson C. R., Mills M. A., Corbett P. C., Menges R., Zysking J. L., Sulhoff J. W., Srivastava A. K. and Wolf C.: Long-haul WDM transmission using 10 Gb/s channels: A 160 Gb/s (16 ××10 Gb/s) 6,000 km demonstration, *OAA '97*, Victoria, post-deadline paper, **PDP-9**, (1997).
3. Nissov M., Bergano N. S., Davidson C. R., Rottwitt K., Menges R., Corbett P. C. and Innis D.: 100 Gb/s (10 ×10 Gb/s) WDM transmission over 7,200 km using distribution Raman amplification, *IOOC-ECOC'97*, Edinburgh, post-deadline paper, **TH3A**, (1997).
4. Nakazawa M., Suzuki K., Kubota H., Sahara A. and Yamada E.: 160 Gbit/s WDM (20 Gbit/s × 8channels) soliton transmission over 10,000 km using in-line synchronous modulation and optical filtering, *OAA '97*, Victoria, post-deadline paper, **PDP-10**, (1997).
5. Mamyshev P.V. and Mollenauer L.F.: Pseudo-phase matched four-wave mixing in soliton wavelength-division multiplexing transmission, *Optics Lett.*, **21**, (1996), pp.396-398.
6. Mollenauer L.F., Mamyshev P.V., Neubelt M.J.: Demonstration of soliton WDM transmission at up to 8 ×10 Gbit/s, error-free over transoceanic distances, *OFC'96*, San Jose, post-deadline paper, **PD. 22**, (1996).
7. Suzuki M., Morita I., Edagawa N., Yamamoto S., Taga H., and Akiba S.: Reduction of Gordon-Haus timing jitter by periodic dispersion compensation in soliton transmission, *Electron. Lett.*, **31**, (1995), pp.2027-2029.
8. Edagawa N., Morita I., Suzuki M., Yamamoto S., Taga H. and Akiba S.: 20 Gbit/s, 8,100 km straight-line single-channel soliton-based RZ transmission experiment using periodic dispersion compensation, *ECOC'95*, Brussels, post-deadline paper,

Th.A.3.5, (1995), pp.983-986.

9. Gordon J. P. and Haus H. A.: Random walk of coherently amplified solitons in optical fiber transmission, *Opt. Lett.*, **11**, (1986), pp.665-667.
10. Suzuki M., Morita I., Edagawa N., Yamamoto S. and Akiba S.: 20 Gbit/s-based soliton WDM transmission over transoceanic distances using periodic compensation of dispersion and its slope, *Electron. Lett.*, **33**, (1997), pp.691-692.
11. Morita I., Suzuki M., Edagawa N., Tanaka K., Yamamoto S. and Akiba S.: Performance improvement by initial phase modulation in 20 Gbit/s soliton-based RZ transmission with periodic dispersion compensation, *Electron. Lett.*, **33**, (1997), pp.1021-1022.
12. Georges T. and Charbonnier B.:Reduction of the dispersive wave in periodically amplified links with initially chirped solitons, *IEEE Photonics Technol. Lett.*, **9**, (1997), pp.127-129.
13. Edagawa N., Morita I., Suzuki M., Yamamoto S., Tanaka K., and Akiba S.: Long distance soliton WDM transmission using a dispersion-flattened fiber, *OFC'97*, Dallas, post-deadline paper, **PD19**, (1997).
14. Suzuki M., Morita I., Edagawa N., Yamamoto S. and Akiba S.: 160 Gbit/s (8 × 20 Gbit/s) soliton WDM transmission experiment using dispersion flattened fiber and periodic dispersion compensation, *IOOC-ECOC'97*, Edinburgh, **WE2B**, (1997).

(Received November 17, 1997)

ULTRAFAST NETWORKS USING HIGH-SPEED RZ OPTICAL PULSES FOR TRANSMISSION, ROUTING AND PROCESSING

D. COTTER, J. K. LUCEK, P. GUNNING, D. G. MOODIE,
A. J. POUSTIE, K. J. BLOW AND R. J. MANNING
BT Laboratories
Martlesham Heath, Ipswich IP5 3RE, UK.

1. Introduction

The processing power of computers is expected to increase apace for at least the next 10–15 years. Industry experts predict that by the year 2011 microprocessor chips will run at clock rates around 10 GHz and achieve processing speeds of 100,000 million instructions per second [1]. Increasingly these powerful individual machines will be connected in networks to search and retrieve remote information, recognise patterns within the information, make intelligent inferences, process and present the information to the user in easily accessible formats and with almost imperceptible delay. The crucial component of this vision is a network capable of providing bursty high-bandwidth data transfer on instantaneous demand.

There is already a large mismatch between the services that can be offered by first-generation optical networks (those based on WDM transmission and routing) and the requirements of users. Currently this mismatch is overcome by overlaying the optical network with electronic service layers (such as SDH, ATM, Frame Relay and Internet protocols). However, there is a widening gap between the future communications needs of powerful computing systems and the capabilities of first-generation optical networks and existing electronic overlays. Increasingly users will desire network services supporting bursty data traffic with high instantaneous bandwidth ($\geq$1 Gbit/s) and ultra-low latency. Figure 1 is an historical overview of the development of transmission techniques in telecommunications. The 1950s saw the development of frequency-division multiplexing (FDM) in the electrical radio-frequency domain to increase the capacity of telephone transmission cables. The wide-spread use of electronic digital transmission and switching did not occur until the 1970s because it was harder to achieve—it required

A. Hasegawa (ed.), New Trends in Optical Soliton Transmission Systems, 367–380.

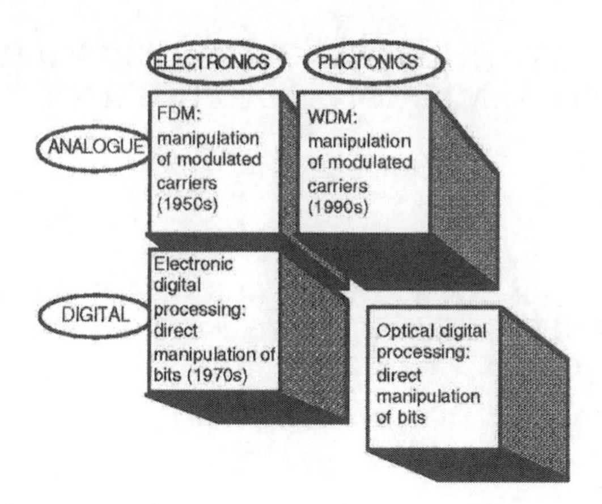

Figure 1. Development of transmission and processing in telecommunications.

more complex component technologies—but once established, it became the basis of all our modern-day telecommunications and computing. In the late 1990s, we are entering the photonics era with the widespread deployment of WDM point-to-point transmission and also the first experiments on WDM networking in the field. WDM is inherently an analogue transmission technique, and in many ways is the direct optical equivalent of FDM. Like FDM, it is being used to increase the transmission capacity of cables and networks. However, with the introduction of WDM we are also witnessing the rediscovery of many of the drawbacks of analogue transmission that long ago proved to be the fatal limitation of FDM: spectral non-uniformity due to the concatenation of amplifiers; inter-channel crosstalk arising from non-ideal filters and imperfect switches; limits imposed by the nonlinearity of the transmission channel; the accumulation of noise in amplifier chains; and the need for complex control schemes for frequency/wavelength stabilisation. Not only do these factors impose limits in WDM transmission, they also restrict the size, scalability and topologies for practical WDM networks. In contrast, OTDM, which uses return-to-zero (RZ) pulses, lends itself to optical digital processing techniques—such as all-optical 3R regeneration and bit-serial processing—which are under development in some research laboratories. Optical digital processing—the 'fourth quadrant'in Figure 1—can serve to overcome the limitations of analogue optical transmission and will also allow greater freedom in future advanced networks.

The technologies for OTDM transmission and networking are currently undergoing a revolution. Fibre-based devices, such as the nonlinear-loop mirror [2], which were prevalent in research laboratories until recently, are giving way to more compact semiconductor versions [3] for channel selection, demultiplexing, clock recovery, dispersion compensation, and soon also for all-optical 3R regeneration. For example, it is possible to fabricate an optical add-drop multiplexer as a photonic integrated circuit on a single semiconductor chip [4]. The recent advances in technology are not restricted to all-optical devices; electro-absorption modulators can now provide switching windows as short as 6 ps and modulation depths as great as 40 dB [5]. They have been used to demultiplex OTDM at 80 Gbit/s [6], and operation as high as 160 Gbit/s is now believed to be feasible [7].

These recent advances in ultrafast optical devices could open OTDM techniques to a much wider range of future applications. Until recently OTDM has been seriously considered only for future use at the highest levels of large national and international telecommunications networks, where traditionally the capacity demand has been greatest. However, some device technologies are beginning to reach such a level of development that, for the first time, we can seriously consider using OTDM in future networks much closer to the end user-such as ultra-high-speed local area networks (LANs) and multi-processor interconnection networks.

2. Ultra-High-Speed Local Area Networks

An example of this new use of OTDM is a prototype local area network (LAN), recently constructed at BT Labs., called SynchroLan [8, 9]. This network is based on optical time-division multiple access (TDMA) with a 40 Gbit/s aggregate capacity, and is being used to interconnect high-end computers. The LAN design uses ultrafast optical technology only very sparingly to minimise costs. Only one optical pulse source is required for the entire LAN, and very few optical components are required to read and write data at the computer interfaces.

The ever-increasing power of computers will place increasingly stringent demands on interconnection networks, and optical TDMA technology can offer an attractive solution for the future. In current computer networks, bandwidth is typically offered on a best-available basis according to demand. In networks based on ATM and Internet protocols, the latency is dominated by delays within the switches and routers, and can be orders of magnitude higher than the physical time of flight. With increasing processor speeds, the latency measured in clock cycles becomes greater. In addition to being very high, the latency varies from packet to packet (or cell to cell), depending on the volume of other traffic. For real-time applications and distributed computing the time and processing overhead of smoothing out delay and bandwidth variations may be unacceptable. It may be advantageous to provide connection-oriented inter-connection with very high guaranteed bandwidth, zero delay variation, speed-of-light latency and

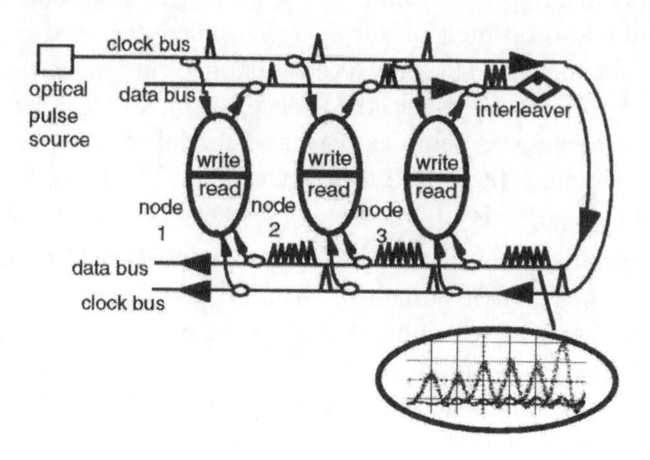

Figure 2. Schematic of the SynchroLan prototype. Inset shows a sampling oscilloscope trace of six bit-interleaved channels (scale = 25 ps/division).

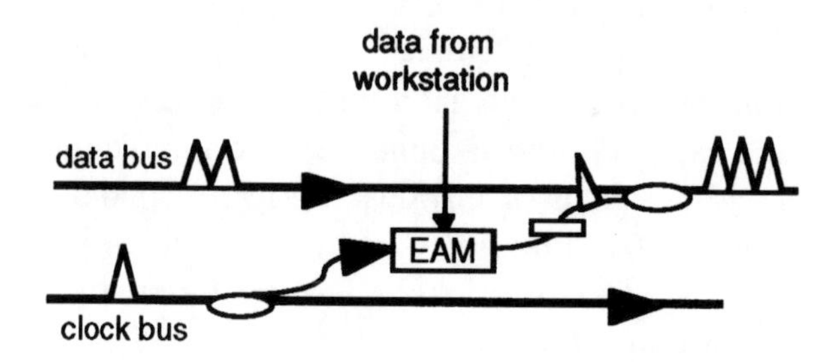

Figure 3. WRITE section of a node interface (EAM = electro-absorption modulator).

broadcast capability. Bit-interleaved optical TDMA networks [10] can fulfil these criteria. Access to the shared medium is by allocation of time-slots. The bandwidth of the individual time-channels is one that can be handled by the electronic computer interfaces, but the aggregate rate can be very high.

The prototype ultra-high speed LAN built at BT Labs. (SynchroLan) uses the re-entrant dual bus topology shown in Figure 2. In this scheme, a train of 4 ps optical pulses (at the data channel base rate of 2.5 GHz) is transmitted along the clock bus from a single head end, and these pulses are used at each computer interface in the processes of both reading and writing data on the LAN. In the first version of SynchroLan [8], the clock and data signals travelled along the same fibre to simplify synchronisation. The fibre was polarisation-maintaining (pm), and the clock and data were distinguished using polarisation by sending them along the slow and fast axis of the pm fibre, respectively. However, we have recently developed a more practical solution using standard (non-pm) fibre. This means that existing fibre infrastructure can be used (such as the small-diameter fibre cables that have been installed in many buildings using the 'blown fibre'technique [11, 12]). We have established that when optical pulses are sent along separate fibre strands within the same blown fibre cable, the relative jitter at the output due to environmental effects is negligible (less than 600 fs rms measured over a 300 m path and several hour time period). This means that rather than using the two polarisation axes of pm fibre to distribute clock and data, separate fibres within the same cable can be used instead.

To allow a computer to write data, its network interface taps off a copy of the clock pulse train, which is then modulated with data using an electro-absorption modulator (EAM). At BT Labs. EAMs have been developed that have the desired combination of high modulation depth (> 30 dB) and low polarisation sensitivity (1 dB) [5, 7]. As shown in Figure 3, the newly created data channel is inserted onto the data bus in an appropriate time-slot, via a controllable optical delay device and a tap coupler. To read data (Figure 4), copies of the clock and data channels are tapped from the bus. The clock pulses are used to drive a suitable demultiplexing device in the computer interface, which selects the required time-channel from the aggregated TDMA data stream. In contrast to the optical add-drop multiplexers used in conventional OTDM, no optical pulse sources are required within the computer interfaces.

For very high aggregate data rates (greater than 200 Gbit/s) the demultiplexer used in the read section of the network interface would be an all-optical device, such as an integrated interferometric semiconductor switch [13]. The SynchroLan prototype has an aggregate capacity of 40 Gbit/s and so it is sufficient to use EAMs to perform the channel selection in the read section. These devices exhibit a highly non-linear characteristic of absorption versus applied voltage, and state-of-the-art components can provide switching windows as short as 6 ps [7]. These properties make the devices very suitable for selecting the required channel with high rejection of the unwanted channels. The optical clock is converted to an electrical signal and applied to the electro-absorption modulator via an impulse generator. The non-linear characteristic of the modulator results in the opening of a 15 ps optical transmission window every clock period when driven by the 30 ps electrical pulses from the impulse generator. This allows a single 2.5 Gbit/s channel to be selected from the aggregated 40 Gbit/s data. Channel

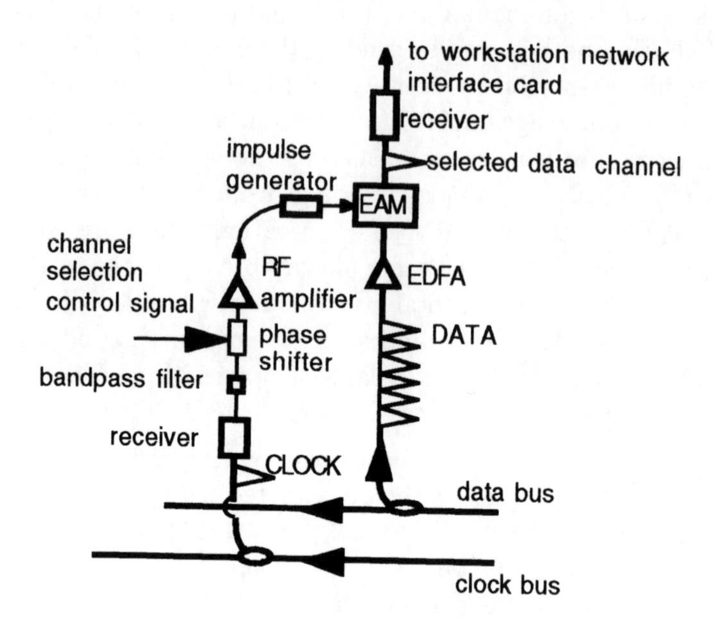

Figure 4. READ section of a network interface.

selection in the read section is performed using an addressable phase shifter in the electrical path under the control of the attached computer interface.

Currently the SynchroLan prototype includes three computer interfaces and is approximately 300 m in length using standard installed blown-fibre cable. Measurements have shown that data communication over the LAN at single-channel data rates of 2.5 Gbit/s is error-free, and high-end work stations connected to the LAN have successfully communicated during sessions of many hours without any loss of data. The bandwidth is constant and there is no delay variation. Because each node has access to all the channels on the bus, broadcasting and multicasting are easily implemented. The LAN is logically equivalent to an N×N cross-bar switch, where N is the number of nodes. However, in contrast to the cross-bar switch, the LAN scales as N instead of N^2, and merges the transmission and switching into one.

A very important advantage of the optical-TDMA approach is that it is amenable to optical digital processing; for example, to extend the reach, the entire contents of the LAN backbone could be digitally regenerated using a *single* 40 Gbit/s all-optical regenerator [14, 15]. Moreover recent measurements on state-of-the-art EAMs indicate the feasibility of upgrading the aggregate capacity of the LAN to 160 Gbit/s [7].

3. Ultrafast Self-Routing Packet Networks

Looking further into the future, photonic technology is also being used at BT Labs. in the development of an ultrafast self-routing network (i.e. a network in which packets find their own optimum path from source to destination). Such a network is intended to provide multi-Gbit/s interconnections for both connection-oriented and connectionless traffic in future distributed computing environments, to support applications running on processors some 10–1000 times more powerful than those of today. Fixed-length data packets composed of picosecond optical pulses at peak rates around 100 Gbit/s will be routed through the network, with header data processing 'on the fly'using ultrafast photonic logic, and with no or very minimal buffering in the optical domain.

A basic stability criterion for such a network is that each pipelined routing decision made 'on the fly' within the network must be completed within a single packet time slot duration (e.g. a few nanoseconds for a ~500-bit packet length at 100 Gbit/s), and therefore must be as simple as possible. A routing decision involves processes at two distinct levels of granularity—the bit level and the packet level. For ultrafast networks, processes at the bit level require photonic devices with response times at least as fast as the bit period (on the picosecond time scale), whereas high-level processing of packets can be performed using high-speed electronics at the packet rate (on the nanosecond scale).

Despite the recent dramatic advances in photonic technology, when compared with VLSI electronics, photonic logic devices are still at a very primitive stage of development—they are relatively bulky, poorly integrated, have limited cascadability and are power-hungry—and are likely to remain so for the foreseeable future. For example, an optical add-drop multiplexer integrated on a single semiconductor chip [4] has the functionality of a single AND/NAND gate, and although it represents a superb state-of-the-art achievement in photonic integrated circuit technology, a silicon chip of the same size (4.0 × 1.5 mm) can provide around 0.3 million electronic logic gates using today's technology. Therefore, as with the ultrafast LAN, the design approach that is being adopted for self-routing networks is to use simple and sparse processing in the ultrafast domain, and to rely on only a small number of photonic devices with relatively primitive functionality. Indeed, the self-routing network being developed by BT uses only a few non-cascadable photonic AND gates at each node.

The basic building block in the network design is the method of binary self-routing shown in Figure 5. The optical packet contains a payload, with

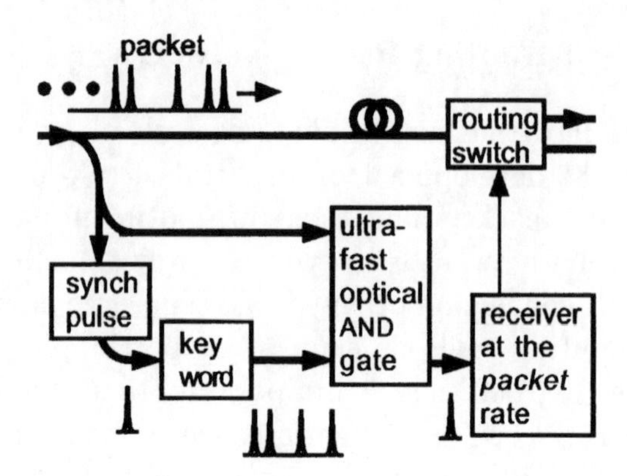

Figure 5. Self-routing binary switch using keyword address recognition.

its own higher-level header and data, preceded by around 1–2 bytes which carry the destination address needed for routing in the photonic network. The entire packet, both payload and optical routing header, is transmitted at 100 Gbit/s in RZ format. In the self-routing switch, a copy of the packet is passed to a photonic word-recognition unit, which makes a simple binary decision on whether or not the packet address-header matches a specific 'keyword'that represents the location address of the switch. The result of that decision determines the configuration of an opto-electronic routing switch. This switch is timed to operate during a brief guard band (~1 ns duration) between successive packets. A short length of fibre acts as a buffer, to compensate for the combined time delay of the word-recognition and routing-switch circuits.

Figure 5 shows the arrangement used in an experimental demonstration of binary self-routing [16, 17]. Packets at 100 Gbit/s peak line rate were first generated using a soliton-compressed laser source and a reconfigurable silica-on-silicon planar lightwave circuit (PLC). At the routing node an ultrafast synchronisation pulse, derived individually from each incoming packet, was converted into a fixed 6-bit keyword (representing a network address) using a further PLC. The packet address-headers and keyword were selected from a special code set [18] that ensures that if (and only if) the address-header of a packet entering the node did not match the keyword, the optical AND gate produced an output which was used to trigger the electro-optic cross-bar routing switch. The ultrafast AND gate used in the experiments was based on four-wave mixing in a semiconductor optical amplifier. This method of word recognition requires just one optical AND gate to recognise a full multi-bit word, regardless of its length. It provides

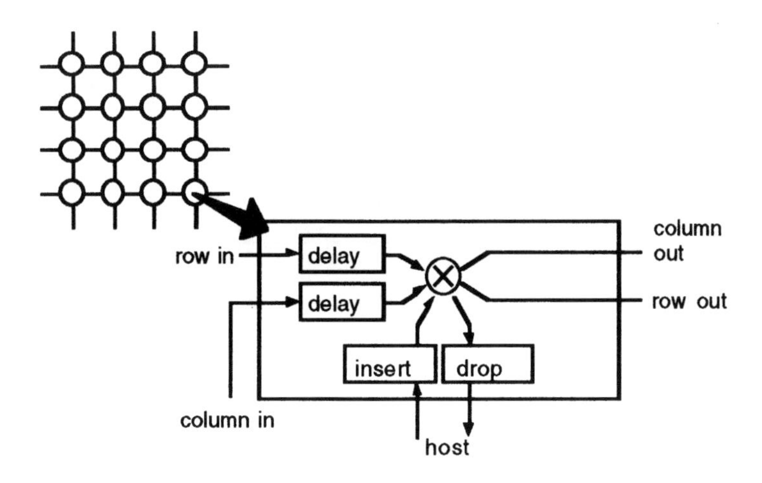

Figure 6. Structure of a 2-connected node in a mesh network. The optical delays on the inputs bring the packet time slots into alignment at the switch. Contention is resolved using deflection routing without buffers ('hot-potato 'routing).

a simple and practical approach to address-recognition and routing in networks with large address sets. Moreover, the packets can be transmitted and received in an asynchronous fashion, avoiding completely the difficult problem of distributing a global bit-level (picosecond precision) clock signal. We have thus successfully demonstrated for the first time the generation of 100 Gbit/s packet headers, ultrafast optical address recognition and binary self-routing.

A key research challenge today is to devise a self-routing network that can provide efficient routing and high throughput, yet requires minimal bit-level and packet-level processing. Researchers at BT Labs. are therefore developing primitive self-routing techniques for use in scalable mesh topologies (Figure 6), with ultrafast keyword recognition and binary routing as the basic building blocks [19]. The abundant bandwidth provided by ultra-high-speed networking permits counter-traditional engineering tradeoffs to be made—bandwidth can be traded for simplified processing and photonic hardware. For example, it is advantageous to use address-header coding techniques that require more header bits, but can be implemented using simpler photonic logic. Similarly, routing schemes that provide submaximal network throughput, but need much less bit-level processing, are attractive.

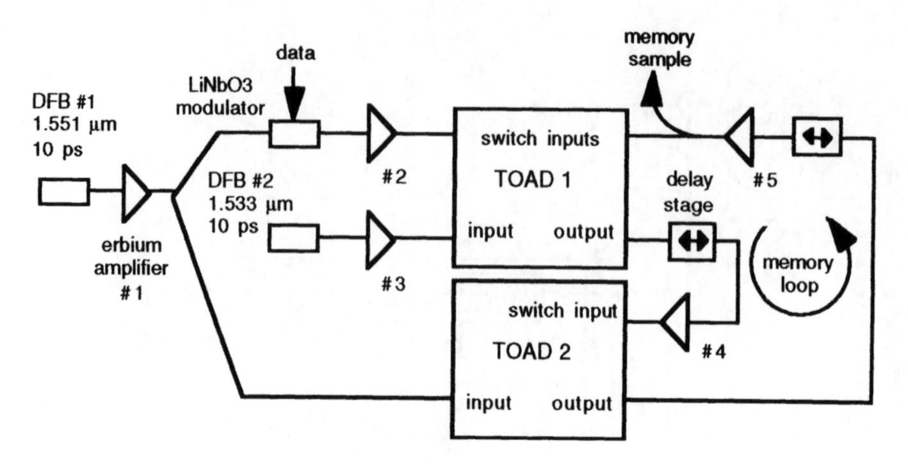

Figure 7. Experimental arrangement of an all-optical regenerative memory.

4. Optical Digital Processing

Although our network designs are based on the principle of minimising the amount of ultrafast processing required (such as the use of a single AND gate for header address recognition), a practical implementation will be greatly helped by using a modest additional amount of ultrafast processing of somewhat higher complexity. One of the key building blocks to implement more advanced information processing is the function of 'memory'. For some applications, it may be desirable not only to provide stable storage of optical pulses but also to process the stored data within the memory. For example, the memory system might provide amplitude equalisation such that any pulse amplitude modulation in the original input data is removed. In addition, if the memory has a variable threshold level for stable storage then the system can discriminate between input pulses of differing amplitudes and self select which pulses to store. Here we demonstrate how both these functions can be achieved in an all-optical regenerative memory architecture as shown in Figure 7 [20]. The memory architecture is a combination of two nonlinear optical switching elements (TOAD/SLALOM [21, 22]). Stable pulse storage for *several hours* is obtained by full all-optical regeneration of the optical pulse data pattern after *each* circulation of an optical fibre delay line [20]. Nonlinear optical switches based on semiconductor optical amplifiers have some advantages over all-fibre switching devices in that they require a low switching energy (<1 pJ per pulse) and can be relatively compact to reduce latency in the optical processing system [23].

The key to understanding how the memory circuit can achieve pulse amplitude restoration and thresholding is to examine the sequential shaping

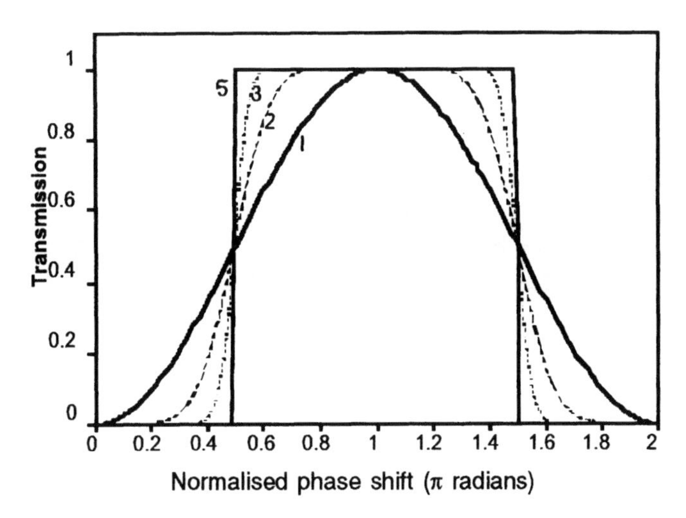

Figure 8. Calculated transmission of the memory circuit after 1, 2, 3 and 5 circulations.

function of concatenated nonlinear-optical switching gates. In a single interferometric gate, the transmission function is proportional to $1 - \cos(\Delta\Phi)$ where $\Delta\Phi$ is the relative phase shift between the counter propagating pulses in the loop [21, 22, 24]. However, on each circulation of the memory this transmission function is reapplied to the stored and regenerated pulses and the overall transmission function rapidly evolves to become a square window as shown in Figure 8 [25]. It can be seen from Figure 8 that initial input pulses which induce a phase shift between $\pi/2$ and $3\pi/2$ are stored and their amplitudes are equalised due to the flat top of the transmission function. Similarly, pulses with initial phase shifts outside this final square window have amplitudes which are reduced to zero and hence are not stored in the memory. In this example where the pulse energies are chosen to give exactly a π phase shift, the minimum threshold for pulse storage is a phase shift of $\pi/2$ but the width of the square window, and hence the storage threshold, can be varied by adjusting the pulse energies to give more or less than π phase shift.

The details of the regenerative memory circuit shown in Figure 7 can be found in reference [20]. We experimentally demonstrated amplitude equalisation and the storage threshold of the optical memory by inputting a combination of a fixed and a variable amplitude binary data sequence into the memory. As shown in Figure 9, the overall binary sequence applied once to the $LiNbO_3$ modulator was 11101101, comprised of an electronically combined fixed amplitude 01001000 sequence and a variable amplitude 10100101 sequence. The upper oscilloscope traces show the initial data se-

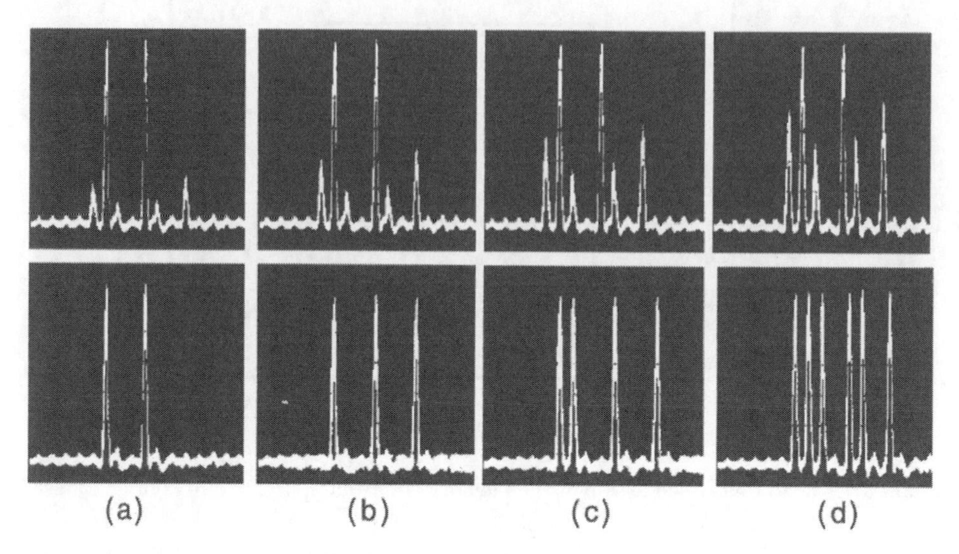

Figure 9. Experimental oscilloscope traces of the input pulse binary sequence (upper) and the stored sequence (lower) for increasing amplitudes of the variable pattern (a → d) — see text.

quence which is input to the memory only *once* in each case. The lower trace shows the actual data sequence that was stored in the optical memory. Note that the stored pulses are equalised in amplitude compared to the input pulses by virtue of the square transmission function. The storage threshold is approximately 0.4 of the amplitude of the fixed binary sequence and initial pulses with amplitudes below this level are not stored in the memory. This threshold level could be varied by changing the gain of erbium amplifier #5 (Figure 7) to induce more than phase shift. As the threshold level was progressively lowered for a fixed input pulse pattern as in Figure 9(b-upper), the stored pulse pattern followed the sequence as in Figures 9(b-lower) to 9(d-lower).

We have thus demonstrated an all-optical regenerative memory with variable storage threshold and amplitude restoration. This functionality should be scaleable to high data rates approaching 100 Gbit/s [26] and to low latency memory stores using integrated TOAD/SLALOM devices [27].

5. Conclusions

Ultra-high bit-rate systems, although less mature than WDM, will provide the key to future photonic networks with high connection bandwidth, massive capacity, low latency and simplified access and control. Research programs are underway in various laboratories to develop the components and systems that will be needed to establish OTDM as a future commercial technology. Advances in photonic devices are bringing ultra-high bit rate networking—at speeds towards 100 Gbit/s and beyond—much closer to practical reality. These advances make it possible now to envisage the use of OTDM techniques, not just in the highest layers of national and international networks, but also much closer to the user—such as the recent world-first demonstrations at BT Labs. of a prototype LAN (SynchroLan) using 40 Gbit/s optical-TDMA and a 100 Gbit/s packet self-routing switch for multi-processor interconnection.

The key to these developments is the use of RZ pulses and the introduction of digital techniques in the optical domain. In the future we expect to see further major advances in optical digital processing and applications of these techniques in ultrafast networks.

References

1. Hyde, J.: The Future of Computing, *ECS50 Symposium*, (1997) Department of Electr-onics and Computer Science, University of Southampton, UK.
2. Doran, N. J. and Wood, D.: Non-linear optical loop mirror, *Optics Letters*, **13**, (1988), pp.56-59.
3. Ellis, A. D., Patrick, D. M., Flannery, D., Manning, R. J., Davies, D. A. O. and Spirit, D. M.: Ultra-high-speed OTDM networks using semiconductor amplifier-based processing nodes, *Journal of Lightwave Technology*, **13**, (1995), pp.761-770.
4. Jahn, E., Agrawal, N., Pieper, W., Ehrke, H. J., Francke, D., Furst, W. and Weinert, C. M.: Monolithically integrated nonlinear Sagnac interferometer and its application as a 20 Gbit/s all-optical demultiplexer, *Electronics Letters*, **32**, (1996), pp.782-783.
5. Moodie, D. G., Harlow, M. J., Guy, M. J., Perrin, S. D., Ford, C. W. and Robertson, M. J.: Discrete electroabsorption modulators with enhanced modulator depth, *Journal of Lightwave Technology*, **14**, (1996), pp.2035-2043.
6. Marcenac, D. D., Ellis, A. D. and Moodie, D. G.: 80 Gbit/s OTDM using electro-absorption modulators, *Proceedings of IOOC/ECOC'97, Edinburgh, UK*, **3**, (1997), pp.23-26, Institution of Electrical Engineers, London.
7. Moodie, D. G., Cannard, P. J., Dann, A. J., Marcenac, D. D., Ford, C. W., Reed, J., Moore, R. T., Lucek, J. K. and Ellis, A. D.: Low polarisation sensitivity buried hetero-structure electroabsorption modulators for ultra high speed networks, *Proceedings of IOOC/ECOC'97, Edinburgh, UK*, **1**, (1997), pp.171-174, Institution of Electrical Engineers, London.
8. Lucek, J. K., Gunning, P., Moodie, D. G., Smith, K., Ellis, A. D. and Pitcher, D.: SynchroLan: A 40 Gbit/s optical-TDMA LAN, *Electronics Letters*, **33**, (1997), pp.887-888.
9. Gunning, P., Lucek, J. K., Moodie, D. G., Smith, K., Pitcher, D., Badat, Q. and Siddiqui, A. S.: 40 Gbit/s optical-TDMA LAN over 300 metres installed blown fibre, *Proceedings of IOOC/ECOC'97*, Edinburgh, UK, **4**, (1997), pp.61-64, Institution of Electrical Engineers, London.
10. Perrier, A. P. and Prucnal, P. R.: High-dimensionality shared-medium photonic switch, *IEEE Transactions on Communications*, **41**, (1993), p.224.
11. Cassidy, S. A. and Reeve, M. H.: A radically new approach to the installation of optical fibre using the viscous flow of air, *BT Technology Journal*, **2**, (1984), pp.56-

60.
12. Cobb, K. W., Jenkins, P. D., MacKenzie, I. and Stockton, D. J.: Fibre campus networks, *British Telecommunications Engineering*, **11**, (1992), pp.112-117.
13. Gunning, P., Lucek, J. K., Nesset, D., Collins, J. V., Ford, C. W., Pitcher, D., Smith, K., Cotter, D., Jahn, E., Agrawal, N. and Siddiqui, A. S.: Optical-TDMA LAN incorporating packaged integrated Mach-Zehnder interferometer channel selector, *Electronics Letters*, **33**, (1997), pp.1404-1405.
14. Lucek, J. K. and Smith, K.: All-optical signal regenerator, *Optics Letters*, **18**, (1993), pp.1226-1228.
15. Pender, W. A., Widdowson, T. and Ellis, A. D.: Error free operation of a 40 Gbit/s all-optical regenerator, *Electronics Letters*, **32**, (1996), pp.567-568.
16. Cotter, D., Lucek, J. K., Shabeer, M., Smith, K., Rogers, D. C., Nesset, D. and Gunning, P.: Self-routing of 100 Gbit/s packets using 6-bit 'keyword'address recognition, *Electronics Letters*, **31**, (1995), pp.2201-2202.
17. Cotter, D., Tatham, M. C., Lucek, J.K., Shabeer, M., Smith, K., Nesset, D., Rogers, D. C. and Gunning, P.: *Ultrafast all-optical signal processing for packet switching, in Photonic Networks*, pp.401-413, ed. Prati, G., (1996) Springer-Verlag, Berlin.
18. Cotter, D. and Cotter, S. C.: Algorithm for binary word recognition suited to ultrafast nonlinear optics, *Electronics Letters*,**29**, (1993), pp.945-946.
19. Cotter, D. and Tatham, M. C.: 'Dead reckoning'-a primitive and efficient self-routing protocol for ultrafast mesh networks, *IEE Proceedings Communications*, **144**, (1997), pp.135-142.
20. Poustie, A. J., Blow, K. J. and Manning, R. J.: All-optical regenerative memory for long term data storage, *Optics Communications*, **140**, (1997), pp.184-186.
21. Sokoloff, J. P., Prucnal, P. R., Glesk, I. and Kane, M.: A terahertz optical asymmetric demultiplexer (TOAD), *IEEE Photonics Technology Letters*, **5**, (1993), pp.787-790.
22. Eiselt, M., Pieper W. and Weber, H. G.: SLALOM; Semiconductor laser amplifier in a loop mirror, *IEEE Journal of Lightwave Technology*, **13**, (1995), pp.2099-2112.
23. Poustie, A. J., Manning, R. J. and Blow, K. J.: All-optical circulating shift register using a semiconductor optical amplifier in a fibre loop mirror, *Electronics Letters*, **32**, (1996), pp.1215-1216.
24. Blow, K. J., Manning, R. J. and Poustie, A. J.: Nonlinear optical loop mirror with feedback and a slow nonlinearity, *Optics Communications*, **134**, (1997), pp.43-48
25. Nayar, B. K., Finlayson, N. and Doran, N. J.: Concatenated all-optical loop mirror switches, *Journal of Modern Optics*, **40**, (1993), pp.2372-2377.
26. Manning, R. J. and Sherlock, G.: Recovery of a π phase shift in $\sim$12.5 ps in a semiconductor laser amplifier, *Electronics Letters*, **31**, (1995), pp.307-308.
27. Jahn, E. et al.: Monolithically integrated asymmetric Mach-Zehnder interfer-ometer as a robust all-optical add/drop multiplexer for OTDM systems, *Electronics Letters*, **32**, (1996), pp.216-217.

(Received December 25, 1997)

40 GBIT/S TRANSMISSION USING OPTICAL TIME-DIVISION MULTIPLEXING AND DEMULTIPLEXING

G. ISHIKAWA
Fujitsu Laboratories Ltd.
4-1-1 Kamikodanaka, Nakahara-ku, Kawasaki 211-88, Japan

Abstract. In preparation for the next step up from today's 10 Gbit/s optical transmission systems, we are now studying about 40 Gbit/s optical transmission systems. We employed optical time-division-multiplexing (OTDM) and demultiplexing techniques. As part of this work, we developed a $LiNbO_3$ integrated OTDM modulator and a $LiNbO_3$ polarization-independent demultiplexer. Using these devices, we achieved two-WDM-channel 40 Gbit/s transmissions (totaling 80 Gbit/s) over a 100 km conventional single-mode fiber and a 667 km dispersion-shifted fiber.

1. Introduction

As the next generation up from 10 Gbit/s transmission systems, 40 Gbit/s optical [1]-[5] and electrical devices [6]-[9] and related transmission experiments [10]-[18] are being actively studied and discussed. There are two key issues to be resolved in order to achieve 40 Gbit/s long-haul transmissions. One issue is the techniques to use for high-speed optical modulation and demodulation. At present, data rates are restricted more by the speed of the electronic devices used in the transmitter and receiver than that of the optical devices. We expect that optical time-division multiplexing and demultiplexing techniques can compensate for this limitation [19, 20]. Utilizing the high speed and various functions of $LiNbO_3$ optical devices, we developed a 40 Gbit/s $LiNbO_3$ integrated optical time-division multiplexing (OTDM) modulator and a $LiNbO_3$ polarization-independent demultiplexer, both of which can be driven by 20 GHz electrical devices [1, 10, 11]. The other issue is the transmission distance limitation due to chromatic dispersion, self-phase modulation (SPM), and polarization-mode dispersion. Because the dispersion tolerance at 40 Gbit/s is only a few tens of ps/nm, dispersion

A. Hasegawa (ed.), New Trends in Optical Soliton Transmission Systems, 381–390.

compensation techniques are indispensable for not only transmission over a conventional single-mode fiber (SMF) with large dispersion but also over a dispersion-shifted fiber (DSF). Maintaining the required optical signal-to-noise ratio requires a higher optical power level. At a higher optical power level, the waveform distortion due to fiber nonlinearities, such as the interaction between the SPM and the chromatic dispersion, is more severe. Therefore, optimization of the optical power level is very important.

This paper introduces our 40 Gbit/s $LiNbO_3$ integrated OTDM modulator and $LiNbO_3$ polarization-independent demultiplexer. It also describes our two-WDM-channel 40 Gbit/s (totaling 80 Gbit/s) OTDM transmission experiments over a 100 km conventional SMF [11] and a 667 km DSF [10] with optimum dispersion compensation and optical power level.

2. 40 Gbit/s $LiNbO_3$ Integrated OTDM Modulator

Figure 1 shows the structure and operation of our OTDM modulator, which integrates a 1×2 balanced bridge-type switch and two Mach-Zehnder modulators with a bandwidth of over 20 GHz. This modulator produces 40 Gbit/s OTDM signals driven by complementary 20 GHz sinusoidal clocks and independent 20 Gbit/s NRZ data signals. The chirping parameter α of this OTDM modulator is dominant in the clock modulation. The push-pull configuration achieved a zero-chirp. In the optical multiplexing, optical interference occurs between the skirts of the two 20 Gbit/s RZ signals. This interference is varied by slight asymmetric waveguide parameters and stress, both of which are unavoidable in the fabrication process. We then added two optical phase controllers before an intersecting coupler in the final multiplexing stage. By varying the applied voltage, we obtained an adequate 40 Gbit/s waveform for transmission using anti-phase interference.

3. $LiNbO_3$ Polarization-independent Demultiplexer

Figure 2 shows the structure and operation of our Ti:$LiNbO_3$ polarization-independent demultiplexer. We introduced an intersecting splitter to serve as a polarization-beam-splitter, whose intersecting length was optimized to produce a polarization extinction ratio of more than 20 dB. In addition, to achieve a high modulation efficiency independent of the polarization state of the input signal, the TE-mode element was converted to TM-mode by inserting a $\lambda/2$ plate into a sawed groove. The following 1×2 switches demultiplexed a 40 Gbit/s TM-mode signal into complimentary 20 Gbit/s RZ signals. The optical bandwidth of the 1×2 switches was more than 20 GHz and the driving voltage at 20 GHz was 6.7 V. At the final stage, each polarization-beam-combiner multiplexed the 20 Gbit/s RZ signals having same bit-sequence, resulting in complimentary 20 Gbit/s RZ signals.

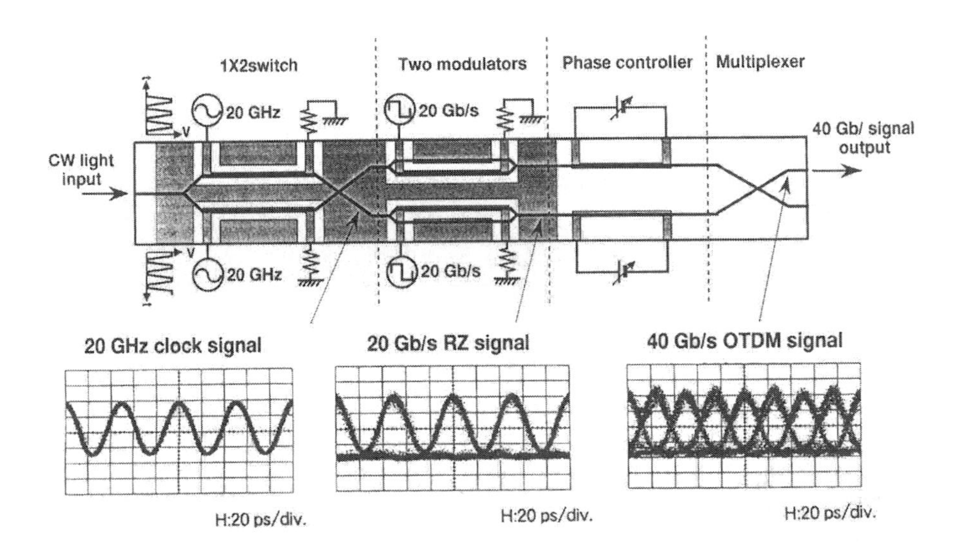

Figure 1. Structure and operation of 40 Gbit/s Ti:$LiNbO_3$ integrated OTDM modulator.

To avoid optical interference during multiplexing, another $\lambda/2$ plate was inserted as a TM/TE converter behind one 1×2 switch and orthogonal-polarized signals were multiplexed. The polarization-dependent loss was 0.5 dB.

4. 2 ch×40 Gbit/s Transmission Experiments over a 100 km SMF

4.1. CONFIGURATION OF TRANSMITTER AND RECEIVER

Figure 3 shows our experimental setup. The two-channel CW lasers of 1552 nm and 1555 nm were simultaneously modulated into 40 Gbit/s signals by a $LiNbO_3$ integrated OTDM modulator. The transmission line consists of two 50 km SMF sections (totaling 100 km and having an average chromatic dispersion of +18.6 ps/nm·km) and one in-line optical amplifier repeater. Dispersion compensating fibers (DCFs) with a dispersion of −930 ps/nm were set up before the in-line amplifier and the preamplifier. At the receiver end, one 40 Gbit/s signal was selected with cascaded optical band-pass filters (3 and 1 nm), demultiplexed into 20 Gbit/s RZ signals by a $LiNbO_3$ polarization-independent demultiplexer, and detected by a commercial PIN photodiode with an optical bandwidth of more than 40 GHz (NEL KEPD1510VPG). The equalized 20 Gbit/s signal was then electrically demultiplexed into a 10 Gbit/s signal using HBT D-FF IC [21], and the bit-error-rate (BER) was measured. By using a phase-locked loop, we

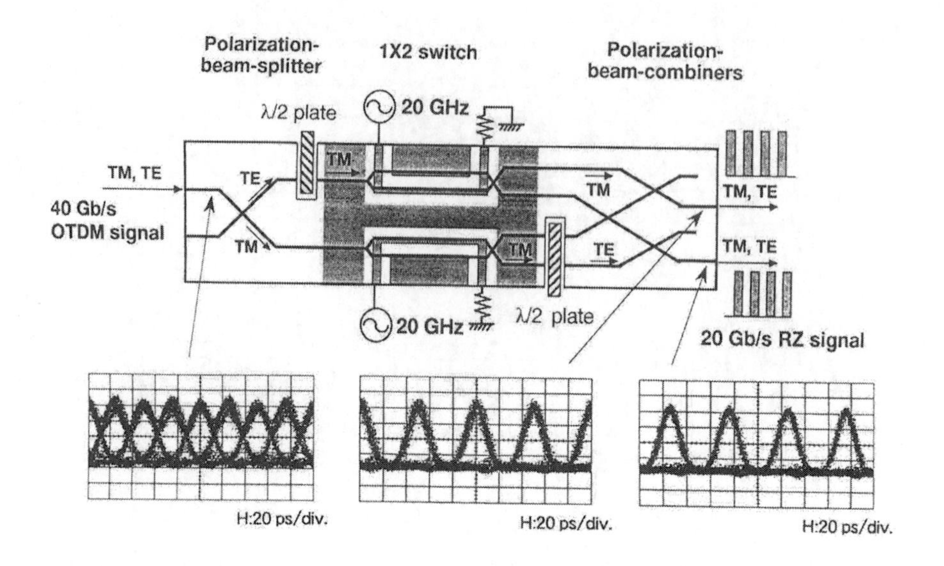

Figure 2. Structure and operation of Ti:LiNbO$_3$ polarization-independent demultiplexer.

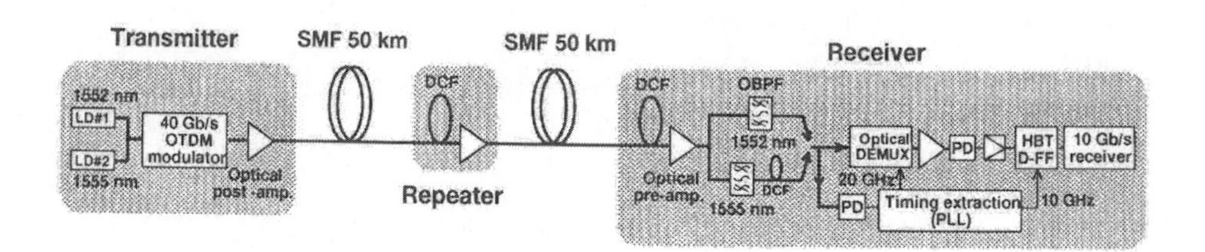

Figure 3. Experimental setup for 2 ch×40 Gbit/s transmission over a 100 km SMF.

extracted 20 GHz and 10 GHz clock signals for optical and electrical demultiplexing from another 40 Gbit/s OTDM signal divided just behind the optical preamplifier. We defined the receiver sensitivity as the input power of the measured channel into the optical preamplifier, and the power penalty at BER of 10^{-9}. The noise figures of all optical amplifiers were 6 dB.

4.2. TOLERANCE OF DISPERSION COMPENSATION AND OPTICAL POWER

First, to optimize the dispersion compensation for each channel, we experimentally evaluated the dispersion compensation tolerance. Figure 4(a) shows the experimental results of total dispersion (SMF + DCF) dependence of the power penalty in the repeaterless 50 km transmission for a single 1552 nm channel. Assuming that the allowable power penalty is less than 1 dB, the total dispersion tolerance was 30 ps/nm. This shows that the strictly complete dispersion compensation of the transmission line is required. Next, we experimentally evaluated the optimum repeater output power. Figure 4(b) shows the optical power tolerance with complete dispersion compensation. In the repeaterless 50 km transmission, no penalty transmission was achieved over the wide power range of +1 to +7 dBm. In the 100 km transmission with one in-line amplifier, waveform distortion caused by SPM was significant in the high-power region, and optical SNR degradation became serious in the low-power region. The minimum penalty was 1.0 dB at the fiber-launched power of +3 dBm.

4.3. 2 CH×40 GBIT/S TRANSMISSION EXPERIMENT

Based on the above results, we performed two WDM-channel 100 km SMF transmission experiments. The fiber-launched power of each channel was set to its optimum of +3 dBm. DCFs were set to completely compensate for dispersion at 1552 nm for each repeater span (total dispersion = +930 ps/nm). For 1555 nm, an additional DCF of −22 ps/nm was inserted in the receiver to compensate for the dispersion slope in the transmission line. Figure 5(a) shows the measured optical spectrum. Figure 5(b) shows the eye-patterns detected by a commercial 32 GHz O/E converter (HP83440D) after transmission with complete dispersion compensation. Figure 5(c) depicts the BER characteristics. We obtained power penalties of less than 1 dB for both channels.

5. 2×40 Gbit/s Transmission Experiments over a 667 km DSF

Figure 6 shows the experimental setup. The transmission line consists of ten DSF sections and nine optical amplifier repeaters, each with a noise figure of 6 dB. The length of each fiber section was about 66 km and the average loss 0.2 dB/km. The total transmission distance was 667 km. The average zero-dispersion wavelength was 1562.5 nm, and the dispersion slope was 0.0735 ps/nm^2·km. A dispersion compensating fiber (DCF), a 1.3 mm zero-dispersion single-mode fiber with +19 ps/nm·km, was inserted in only the receiver. The configurations of the transmitter and the receiver were

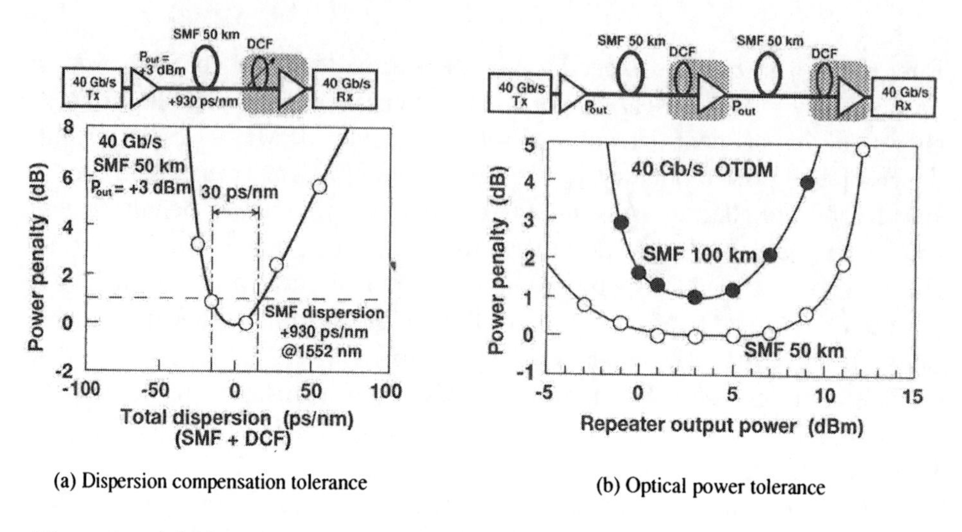

Figure 4. (a) Dispersion compensation tolerance and (b) Optical power tolerance in single channel transmission at 1552 nm.

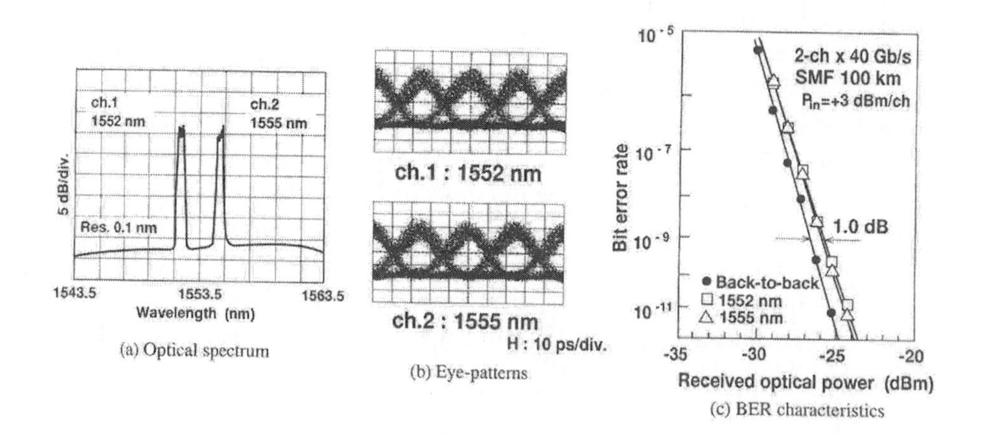

Figure 5. (a) Measured optical spectrum, (b) Eye-patterns and (c) BER characteristics for 2 ch×40 Gbit/s transmission over a 100 km SMF.

the same as described in Section 4.

At first, in the single-channel transmission experiments, spectra broadening due to modulation instability occurred at an average zero-dispersion wavelength of 1562.5 nm. Also, we had to set the signal wavelengths shorter than 1562 nm with dispersion compensation. Figure 7 shows the repeater

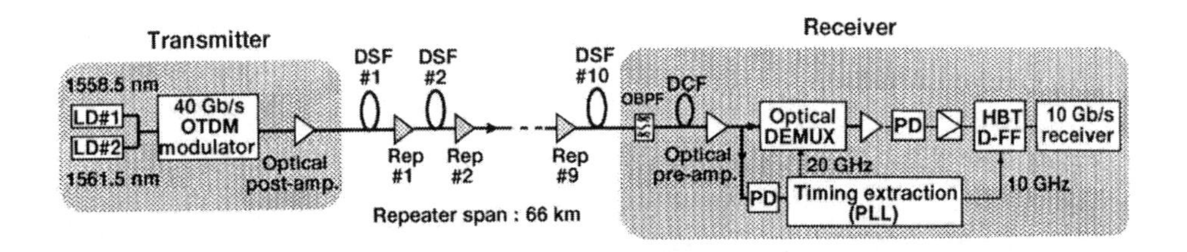

Figure 6. Experimental setup for 2 ch×40 Gbit/s transmission over a 667 km DSF.

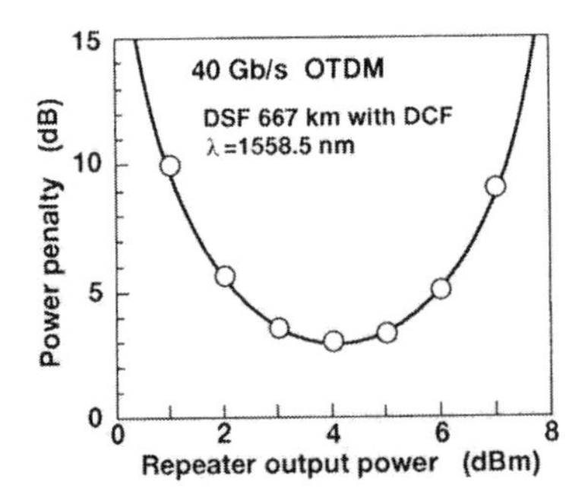

Figure 7. Repeater output power tolerance in single-channel transmission at 1558.5 nm.

output power tolerance for single-channel transmission at 1558.5 nm, which corresponds to the gain peak of the cascaded optical amplifier repeaters. We completely compensated for the chromatic dispersion of -196 ps/nm with 10.3 km DCF. Above +5 dBm, the waveform distortion due to SPM is significant and the penalty increases rapidly. On the other hand, as the repeater output power decreases, the power penalty becomes larger because the optical SNR is degraded after transmission. We found the optimum power to be +4 dBm.

We then performed two-WDM-channel 667 km transmission experiments. We set the wavelengths to 1558.5 nm (ch.1) and 1561.5 nm (ch.2). A received channel was selected through cascaded optical bandpass filters of 1 nm and 3 nm. We used the DCFs of 10.3 km (+196 ps/nm) and 2.6 km (+49 ps/nm) for each channel, respectively. Figure 8(a) shows the optical spectra and (b) received eye-patterns. The difference in signal peak levels was caused by the gain slope of the cascaded optical amplifier repeaters, which results in the difference in waveform distortion due to the SPM effect. For this wavelength allocation, we set the repeater output power at

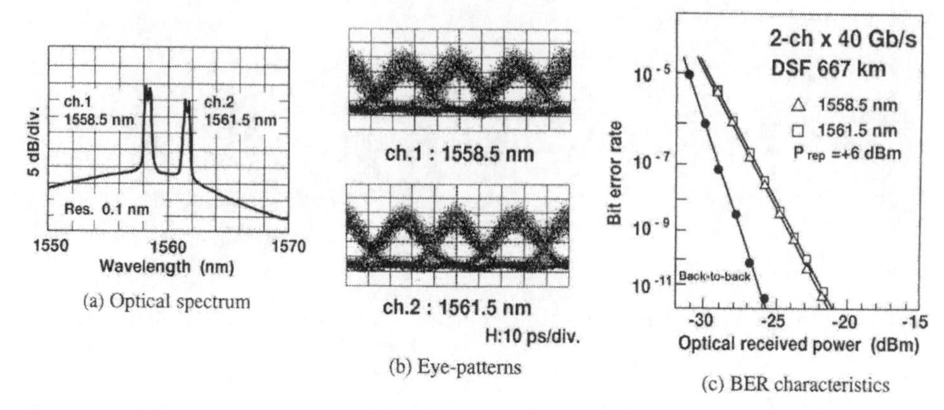

Figure 8. (a) Measured optical spectrum, (b) Eye-patterns and (c) BER characteristics for 2 ch×40 Gbit/s transmission over a 667 km DSF.

+6 dBm (~+3 dBm/ch.) to suppress the SPM effect for 1558.5 nm. Figure 8(c) shows the BER characteristics. The back-to-back sensitivities at a BER of 10^{-9} for both wavelengths were the same as −27.4 dBm. After a 667 km transmission, the power penalties were 3.2 dB (ch.1) and 3.6 dB (ch.2), which were primarily caused by optical SNR degradation. If we can improve the noise figure of optical amplifier repeaters, these penalties will decrease.

6. Conclusion

We developed a 40 Gbit/s $LiNbO_3$ integrated OTDM modulator and a $LiNbO_3$ polarization-independent demultiplexer driven by 20 GHz electrical devices. Using these devices, we achieved two-WDM-channel 40 Gbit/s transmissions (totaling 80 Gbit/s) over a 100 km conventional SMF and a 667 km DSF. Our results confirmed the feasibility of the OTDM scheme for higher bit-rate transmission systems, and established guidelines for 40 Gbit/s system design with optimum dispersion compensation and optical power against chromatic dispersion and SPM.

Acknowledgments

The author would like to thank H. Ooi and Y. Akiyama for their contributions to the work on this entire paper, and to M. Doi and M. Seino for providing a Ti:$LiNbO_3$ integrated OTDM modulator and a polarization-independent demultiplexer. We also thank T. Ihara for providing an HBT D-FF IC, and H. Nishimoto, H. Hamano, Drs. H. Kuwahara and T. Chikama for their encouragement.

References

1. Doi, M., Taniguchi, S., Seino, M., Ishikawa, G., Ooi, H. and Nishimoto, H.: 40 Gb/s Integrated OTDM Ti:LiNbO$_3$ Modulator, *PS'96 Tech. Digest*, **PThB1**, (1996), pp.172-173.
2. Noguchi, K., Mitomi, O., Noguchi, K. and Miyazawa, H.: Low-voltage and broad-band Ti:LiNbO$_3$ modulators operating in the millimeter wavelength region, *OFC'96 Tech. Digest*, **ThB2**, (1996), pp.205-206.
3. Madabhushi, R., Uematsu, Y. and Kitamura, M.: Wide-band Ti:LiNbO3 optical modulators with reduced microwave attenuation, *IOOC/ECOC'97 Tech. Digest*, **vol.2**, (1997), pp.29-32.
4. Takeuchi, H., Tsuzuki, K., Sato, K., Yamamoto, M., Itaya, Y., Sano, A., Yoneyama, M. and Otsuji, T.: NRZ operation at 40 Gb/s of a compact module containing an MQW electroabsorption modulator integrated with a DBR laser, *IEEE Photon. Technol. Lett.*, **9**, (1997), pp.572-574.
5. Ido, T., Tanaka, S. and Inoue, H.: MQW electroabsorption modulators for 40-Gb/s TDM systems, *OFC'97 Tech. Digest*, **WG5**, (1997), pp.140-141.
6. Yoneyama, M., Sano, A., Hagimoto, K., Otsuji, T., Murata, K., Imai, Y., Yamaguchi, S., Enoki, T. and Sano, E.: A 40-Gbit/s optical repeater circuits using InAlAs/InGaAs HEMT digital IC chip set, *IEEE MTT-S Digest*, **WE1D-2**, (1997), pp.461-464.
7. Tsuda, H., Miyamoto, Y., Sano, A., Kato, K., Imai, Y. and Hagimoto, K.: 40 Gb/s baseband-type optical receiver module using a waveguide photodetector and a GaAs MESFET distributed amplifier IC, *OECC'96 Tech. Digest*, **19A1-2**, (1996), pp.506-507.
8. Muramoto, Y., Takahata, K., Fukano, H., Kato, K., Kozen, A., Nakajima, O. and Matsuoka, Y.: 46.5-GHz-bandwidth monolithic receiver OEIC consisting of a waveguide PIN photodiode and a distributed amplifier, *IOOC/ECOC'97*, Post-deadline paper, (1997), pp.37-40.
9. Bach. H.-G., Bertenburg, R.M., Bulow, H., Jacumeit, G., Mekonnen, G.G., Umbach, A., Unterborsch, G., Veith, G. and van Waasen, S.: Ultrafast monolithic InP-based photoreceiver module detecting a 40 Gbit/s optical TDM RZ modulated pulse sequence, *IOOC/ECOC'97 Tech. Digest*, **Vol. 4**, (1997), pp.101-104.
10. Ishikawa, G., Ooi, H., Akiyama, Y., Taniguchi, S. and Nishimoto, H.: 80-Gb/s (2×40-Gb/s) transmission experiments over 667-km dispersion-shifted fiber using Ti:LiNbO$_3$ OTDM modulator and demultiplexer, *ECOC'96*, Post-deadline, **TuC.3.3, 5.**, (1996), pp.37-40.
11. Ooi, H., Ishikawa, G. and Akiyama, Y.: 2-ch×40-Gb/s OTDM transmission over 100 km of 1.3 mm zero-dispersion single-mode fiber, *OECC'97 Tech. Digest*, **9B3-3**, (1997), pp.134-135.
12. Kuwano, S., Takachio, N., Iwashita, K., Otsuji, T., Imai, Y., Enoki, T., Yoshino, K. and Wakita, K.: 160-Gbit/s (4-ch 40-Gbit/s electrically multiplexed data) WDM transmission over 320-km dispersion-shifted fiber, *OFC'97*, Post-deadline paper, **PD25**, (1996).
13. Hagimoto, K., Yoneyama, M., Sano, A., Hirano, A., Kataoka, T., Otsuji, T., Sato, K. and Noguchi, K.: Limitations and challenging of single-carrier full 40-Gbit/s repeater system based on optical equalization and new circuit design, *OFC'97 Tech. Digest*, **ThC1**, (1997), pp.242-243.
14. Sano, A., Kataoka, T., Tomizawa, M., Hagimoto, K., Sato, K., Wakita, K. and Kato, K.: Automatic dispersion equalization by monitoring extracted-clock power level in a 40-Gbit/s, 200-km transmission line, *ECOC'96 Tech. Digest* , **Vol. 2**, (1996), pp.207-210.

15. Suzuki, K., Ohkawa, N., Murakami, M. and Aida, K.: Unrepeatered 40 Gb/s RZ signal transmission over 240 km conventional single mode fiber, *IOOC/ECOC'97 Tech. Digest*, **Vol. 3**, (1997), pp.27-30.
16. Lee, W.S., Garthe, R.A., and Hadjifotiou, A., Modeling and experimental comparison of a 40 Gbit/s OTDM system over a transmission distance of 560 km, *OFC'97 Tech. Digest*, **TuD1**, (1996), pp.14-15.
17. Garthe, D., Saunders, R.A., Lee, W.S. and Hadjifotiou, A.: Simultaneous transmission of eight 40 Gbit/s channels over standard single mode fiber, *OFC'97*, Post-deadline paper, **PD20**, (1997).
18. Dong, L., Cole, M.J., Ellis, A.D., Durkin, M., Ibsen, M., Gusmeroli, V. and Laming, R.I.: 40 Gbit/s 1.55 mm transmission over 109 km of non-dispersion shifted fibre with long continuously chirped fibre grating, *OFC'97*, Post-deadline paper, **PD6**, (1997).
19. Kawanishi, S., Takara, H., Morioka, T., Kamatani, O., Takiguchi, K., Kitoh, T. and Saruwatari, M.: 400 Gbit/s TDM transmission of 0.98 ps pulses over 40 km employing dispersion slope compensation, *OFC'96*, Post-deadline paper, **PD-24**, (1996).
20. Marcenac, D.D., Ellis, A.D. and Moodie, D.G.: 80 Gbit/s OTDM using electroabsorption modulators, *IOOC/ECOC'97 Tech. Digest*, **Vol. 3**, (1997), pp.23-26.
21. Ihara, T., Oikawa, Y., Yamamoto, T., Tomofuji, H., Hamano, H., Ohnishi, H. and Watanabe, Y.: InGaP/GaAs HBT-IC chipset for 10-Gb/s optical receiver, *IEEE GaAs IC Symposium Tech. Digest*, **J.3**, (1996), pp.262-265.

(Received November 17, 1997)

BROADBAND AND GAIN-FLATTENED OPTICAL FIBER AMPLIFIERS

M. NISHIMURA
Sumitomo Electric Industries, Ltd.
1, taya-cho, Sakae-ku, Yokohama 244, Japan

Abstract. Broadband and gain-flattened erbium-doped fiber amplifiers are being widely developed to meet the requirements posed by wavelength-division-multiplexed transmission systems/networks pursuing the ultimate capacity. One approach is the improvement of erbium-doped fiber itself for broader gain bandwidth or better gain flatness. In addition to the well-known aluminum-codoped silica fibers and the fluoride glass fibers, rather new materials, such as tellurite glass, or hybrid configurations which utilize two or more erbium- doped fibers with different glass compositions are being examined. Another approach is the use of gain-equalizing optical filters. One of the most promising gain equalizing techniques is the long-period fiber grating. In order to maintain the gain flatness even under variable operational conditions, control schemes of the optical fiber amplifiers are also a critical issue.

1. Introduction

Although erbium-doped fiber amplifiers (EDFAs) are considered to be highly matured, further improvement of the performance is still a challenge toward realization of the ultimately large capacity optical fiber transmission based on wavelength-division-multiplexing (WDM). It is obvious that the gain bandwidths of EDFAs should be as large as possible in order for many WDM signal channels to be accommodated. It should be noted, however, that one of the important issues here is the wavelength dependence of the gain, which generally causes an imbalance in the optical signal power and the signal-to-noise ratio among the WDM channels in cascaded EDFA chains and limits the usable bandwidth.

A. Hasegawa (ed.), New Trends in Optical Soliton Transmission Systems, 391–402.

In order to quantitatively examine the gain flatness, a factor of merit has been defined as the ratio of the maximum peak-to-peak gain variation (ΔG) in a specified wavelength range to the average gain (G) [1]. This relative gain variation ($\Delta G/G$) can be used to roughly estimate the gain excursion in EDFA chains. To minimize the relative gain variation in a wavelength range as wide as possible is, of course, the primary target, but it is also essential not to deteriorate other performances of the EDFA, such as the noise figure (NF) or the energy conversion efficiency.

There are two major approaches to broadband and/or gain-flattened EDFAs. One is the improvement of erbium-doped fiber (EDF) itself, and the other is the use of gain-equalizing optical filters. Since gain equalizers function primarily as optical attenuators exhibiting certain wavelength-dependent loss spectra, use of any gain equalizer tends to deteriorate the pump efficiency or the maximum signal output power of the EDFA, especially when it is placed at the output of the amplifier. Therefore, it would be more preferable in general, if the performance of the EDFA can be improved without using gain equalizers. In this paper, various kinds of EDFs being examined for improving the gain bandwidth or the gain flatness are reviewed first, and then gain equalizing techniques will be discussed. Finally, the control schemes of EDFAs which are employed in order to maintain the gain flatness under variable operational conditions, such as change of the input signal power level or the number of WDM channels, will also be discussed.

2. Erbium-Doped Fibers for Broadband and Gain-Flattened EDFAs

2.1. ALUMINUM-CODOPED SILICA FIBERS

It is well known that codoping aluminum broadens the gain bandwidth of the erbium-doped silica-based fiber [2]. In general, the spectral gain bandwidth depends upon the concentration of aluminum, and the higher the aluminum concentration is, the larger the gain-flattened bandwidth is [3]. Figure 1 shows typical gain spectra measured for erbium-doped silica-based fibers with different aluminum concentrations. It can be seen from Figure 1 that by increasing the aluminum concentration, the gain peak near 1535 nm is suppressed and the flat-gain band around 1550 nm is broadened.

It has been demonstrated [4] that, by carefully optimizing the pumping conditions and the length of such relatively high Al concentration EDF, excellent gain-flatness can be achieved in the wavelength range from 1544 to 1557 nm. The relative gain variation in this 13 nm wavelength band is estimated to be as small as 1.4 %. In order to achieve the best gain flatness, however, it is necessary, in general, for the population inversion level in the

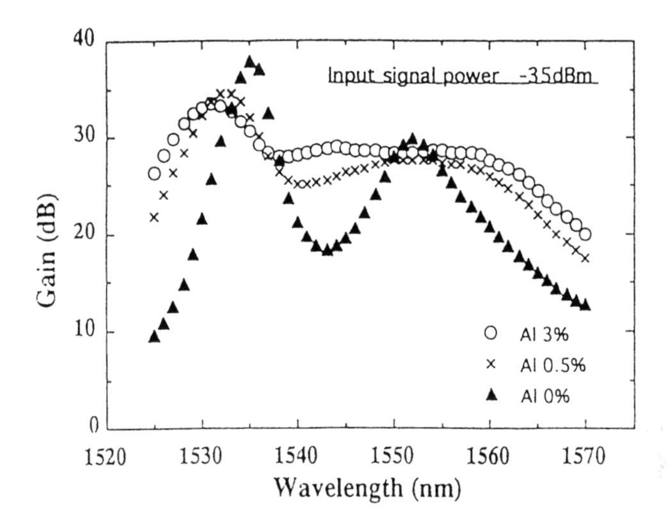

Figure 1. Gain spectra of erbium-doped silica-based fibers with different aluminum concentrations.

EDF to be relatively high. This tends to limit the pumping efficiency of the gain-flattened EDFAs utilizing only high Al concentration EDF.

2.2. FLUORIDE GLASS FIBERS

The fluoride-based EDF is one of the good candidates for broader bandwidths [5], although some concern over productivity and/or reliability of the fluoride-based fiber still remains. The relative gain variation is typically 5 % in a wavelength range of 25 nm or more. Although the relative gain variation is only moderately small, the bandwidth more than 25 nm without gain equalizers is very attractive especially for the terrestrial networks with relatively short transmission distances employing several inline EDFAs.

One apparent drawback has been said to be the limited noise figure performance, since it cannot be pumped at 980 nm, while the silica-based EDF exhibits the quantum-limited NF when pumped at 980 nm. Recently, it has been demonstrated [6] that the noise performance of the fluoride-based EDF can be improved by pumping the fiber at 970 nm through the excited state absorption of the pump light. It appears, however, that the amplification efficiency is sacrificed to a considerable degree in this scheme.

2.3. HYBRID CONFIGURATIONS

2.3.1. *Hybrid EDFAs*

It might be difficult for one type of EDF having a certain glass composition to satisfy all the requirements. The "hybrid" EDF configuration, which consists of two (or more) EDFs with different glass compositions exhibiting different gain spectra, is an attractive approach. If two EDFs have specific gain profiles with different features opposite to each other (for example, positive and negative gain slopes in a certain wavelength range), it should be possible to optimize the gain flatness just by cascading those EDFs with appropriate lengths. In addition, by carefully making use of the dependence of the gain profiles upon the population inversion levels in the EDFs, you may have a considerable degree of freedom of design in optimizing the gain flatness without compromising the efficiency under arbitrary operational conditions.

It has been found that a combination of Al-codoped silica-based EDF and P/Al-codoped silica-based EDF is one of the most promising hybrid configurations for gain-flattened EDFAs [7, 8]. Figure 2 shows the gain spectra calculated from measured absorption and emission cross sections for Al-codoped EDF and P/Al-codoped EDF assuming various levels of the population inversion in each EDF. While, around 1550 nm, the Al-codoped EDF has a positive gain slope, which decreases as the inversion level is increased, in the same wavelength range, the P/Al-codoped EDF exhibits a negative gain slope, which becomes steeper when the inversion level is higher. By adjusting the lengths of those two EDFs with the inversion level in each fiber taken into account, you can design the hybrid EDFA which exhibits a specified gain and a nearly zero gain slope around 1550 nm even under heavily saturated conditions. Similar characteristics in a combination of Al-codoped EDF and P/Yb-codoped EDF have also been utilized to construct gain-flattened EDFAs [9].

The hybrid configuration has also been adopted in order to primarily improve the noise performance of the fluoride-based EDFAs [10]. This can be realized by placing a 980 nm-pumped silica-based EDF in front of a 1480 nm-pumped fluoride-based EDF. Because of a low NF of the first-stage silica-based EDF, the overall NF of the hybrid EDFA is improved to the level comparable to those of the standard silica-based EDFAs. In addition, the gain flatness can be improved to some extent by balancing the gain profiles of the two EDFs which are different. Figure 3 shows an example of the gain spectrum measured for the hybrid silica-based/fluoride-based EDFA by the locked-inversion technique [11]. In this example, the relative gain variation is reduced to 3.6 % in a 27 nm bandwidth.

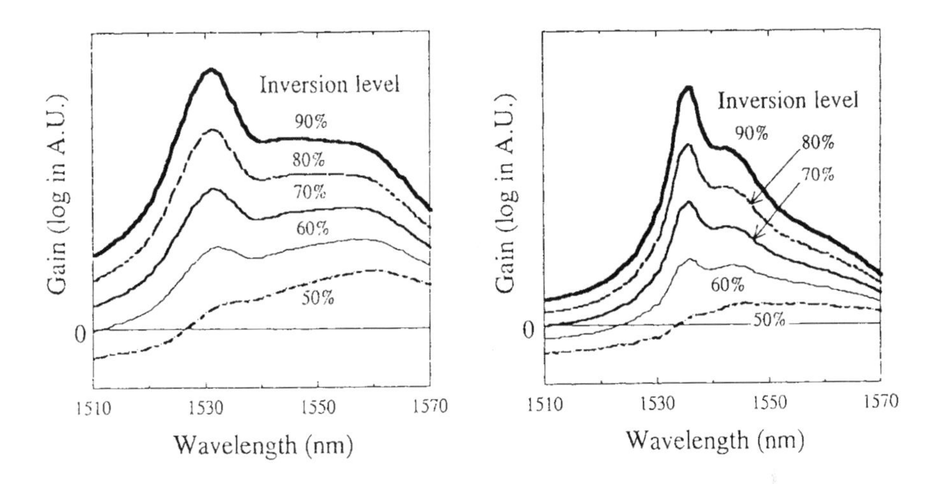

Figure 2. Gain spectra calculated from measured absorption and emission cross sections for Al-codoped EDF (left) and P/Al-codoped EDF (right) assuming various levels of the population inversion in each EDF.

2.3.2. *Performance of Hybrid Silica-Based EDFAs*

Since silica-based fibers are generally preferable, compared with fluoride glass fibers or other elaborate glass materials, considering general performance, ease of splicing, productivity and reliability, the hybrid silica-based EDFAs consisting of the Al-codoped and the P/Al-codoped EDFs is an attractive choice. It should be worthwhile to pursue the ultimate performance of this type of EDFAs.

In order to achieve a low NF, a high pump efficiency and gain flatness simultaneously, the optimum configuration of the hybrid silica-based EDFA has been thoroughly examined [8]. Figure 4 illustrates one of the optimized configurations, which consists of the first stage amplifier with the Al-codoped EDF pumped at 980 nm and the second stage amplifier with the hybrid EDF pumped at 1480 nm. By optimizing the lengths of the three EDF sections used, it is possible to achieve excellent gain flatness with the relative gain variation as small as 1 % in a wavelength range of 1545 nm to 1557 nm and yet realize a low NF and a high pumping efficiency in the EDF. Figure 5 shows an example of the optical spectrum observed at the output of a cascaded chain of five similar type EDFAs with a total gain of 150 dB for 8 channel WDM signals. The maximum signal level variation is as small as 1.5 dB. Expansion of the flat-gain bandwidth of the hybrid silica-based EDFA to 25 nm or more has also been examined by optimizing the glass compositions of the Al-codoped and the P/Al-codoped

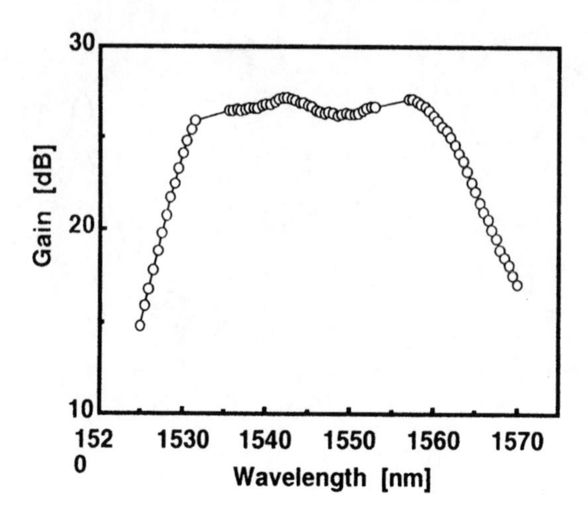

Figure 3. An example of the gain spectrum of a hybrid silica-based/fluoride-based EDFA.

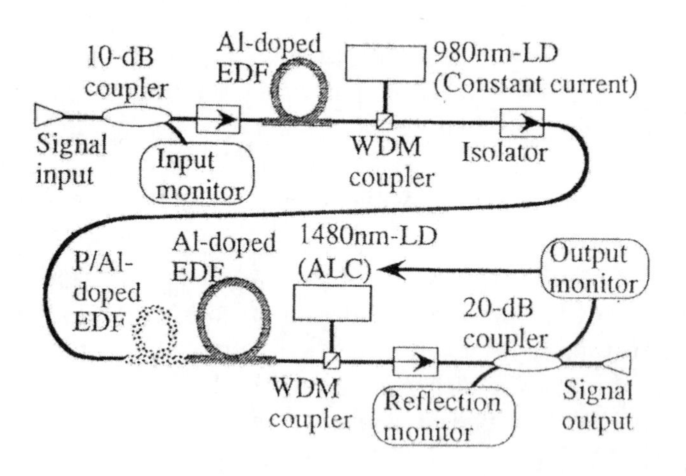

Figure 4. A configuration of the hybrid silica-based EDFA designed for a low NF, a high pump efficiency and gain flatness.

EDFs and the lengths of the fibers [12]. Figure 6 shows the gain spectrum of such an EDFA measured by utilizing the locked-inversion technique [11]. For comparison, a typical gain spectrum of a fluoride-based EDFA is also plotted in the same figure. The bandwidth-maximized hybrid silica-based EDFA exhibits the relative gain variation of 5.8 % in a 25 nm bandwidth

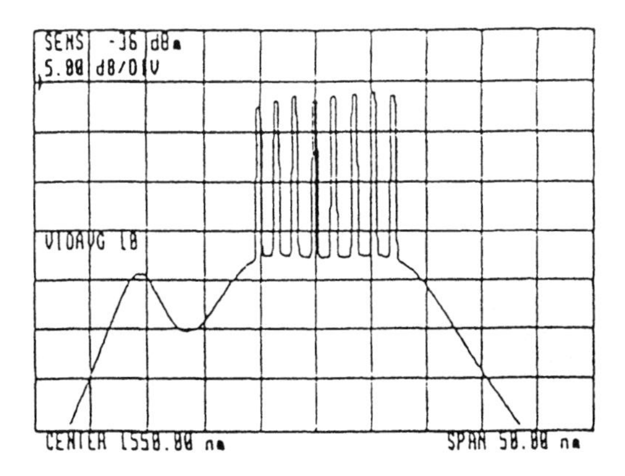

Figure 5. An example of the optical spectrum observed after five cascaded hybrid silica-based EDFAs with 8 channel WDM signals (total gain = 150 dB).

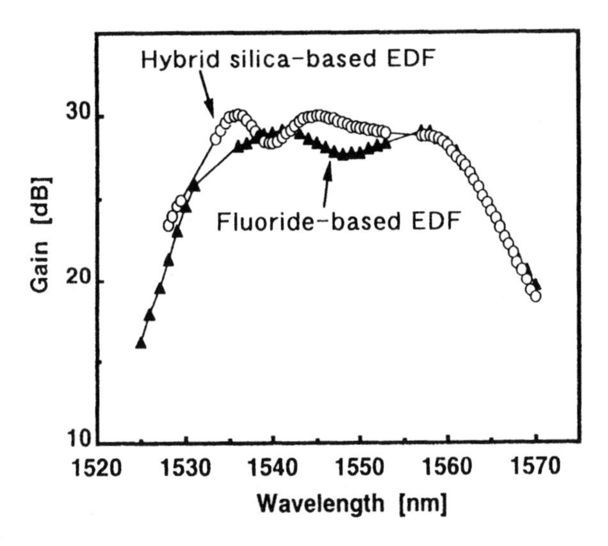

Figure 6. Gain spectra of the bandwidth-maximized hybrid silica-based EDFA and a fluoride-based EDFA for comparison.

from 1535 nm to 1560 nm, which appears to be comparable to that of the fluoride-based EDFA (5.3 % in this specific example).

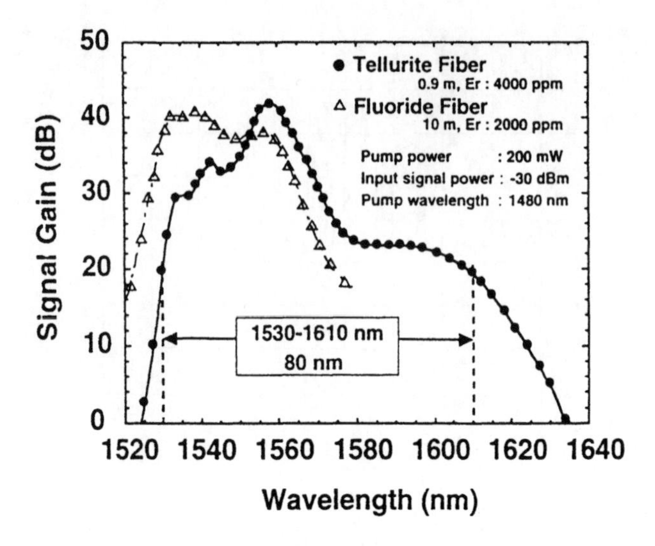

Figure 7. The gain spectrum of the tellurite-based EDFA reported in reference [18].

2.4. BANDWIDTH EXPANSION TOWARD 1.6 μM

In pursuing the ultimate capacity of transmission by increasing the number of WDM signal channels, there is a strong demand for further gain bandwidths far more than those of the ordinary EDFAs. One approach is the EDFAs designed to operate in the longer wavelength range [13, 14]. By using relatively long EDFs and pumping them properly, it is possible to achieve the signal gain in a wavelength range from 1570 nm to 1600 nm (with no gain in the 1530–1560 nm band). It seems, however, that the pumping efficiency of such amplification is considerably lower than those of the standard EDFAs operating around 1550 nm. Ultra-broadband EDFAs employing this 1580 nm-band EDFA and the standard 1550 nm-band EDFA in a parallel configuration have been demonstrated [15, 16].

An alternative approach to the ultra-wide gain bandwidth is the erbium-doped tellurite glass fiber [17]. It has recently been demonstrated [18] that the fiber exhibits an extremely broad gain spectrum nearly 80 nm from 1530 nm to 1610 nm, although it is not necessarily a flat-gain bandwidth. Figure 7 shows the small signal gain spectrum of the tellurite fiber reported in reference [18].

3. Gain Equalizers for Gain-Flattened EDFAs

3.1. GAIN EQUALIZING TECHNIQUES

Another approach to the gain-flattened EDFA is the use of gain-equalizing optical filters [19]. Since they function primarily as a wavelength-dependent attenuator, use of any optical filter tends to lower the efficiency or the maximum output power, especially when it is placed at the output of the EDFA. Elaborately designed optical filters, however, have good potential for achieving the ultimate gain flatness.

Various techniques have been proposed and developed to realize gain equalizers having the loss profiles which precisely match the gain spectra of the EDFAs. Among those are planar lightwave circuits (PLCs) [20], long-period fiber gratings [21], Fabry-Perot etalon filters [22], periodic tapered fiber filters [23] and so on.

3.2. FIBER-GRATING-BASED GAIN EQUALIZERS

The long-period fiber grating is believed to be one of the most promising techniques because of its low insertion loss, structural simplicity, compactness and gain-equalizing performance. The technique is based on the coupling between the fundamental mode and the cladding modes caused by the grating written inside the fiber with a relatively long period (typically several hundred microns). By properly designing the grating period, the length of the grating and the depth of the index modulation, it is possible to obtain a nearly Gaussian-shaped loss spectrum which has almost any arbitrary peak wavelength, spectral bandwidth and peak loss. Even asymmetric gain profiles of the EDFAs can be equalized by a combination of two, three or more fiber gratings properly designed.

Figure 8 shows an example of the output spectrum of the gain-flattened EDFA employing a long-period-fiber-grating-based gain equalizer. It can be seen that the gain is sufficiently flattened in an approximately 30 nm band. Gain-flattening by the long-period fiber gratings in a nearly 40 nm band has already been demonstrated [24].

4. Control of Gain-Flattened EDFAs

Finally, it should be noted that, in any type of EDFA, the best gain flatness is obtained only under designed operational conditions. When the input signal power level is changed and the signal gain becomes off the designed value, this would normally cause a change of the population inversion in the EDF, resulting in a change of the spectral gain profile or the gain flatness of the EDFA. The automatic gain control (AGC) is known to be effective in maintaining the optimum gain flatness even under varying operational

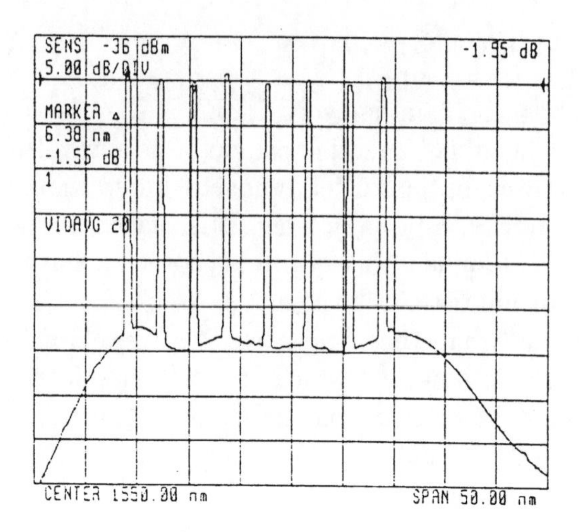

Figure 8. An example of the output spectrum of a gain-flattened EDFA employing a fiber-grating-based gain equalizer.

conditions. Various types of the AGC schemes have been proposed and demonstrated. Those AGC schemes have been proved to be highly effective especially for changes in the number of WDM signal channels inputted to the EDFA, because, in principle, they maintain the output signal power of each channel, so long as the input signal power of each channel is unchanged.

The AGC might, however, pose another problem in other cases, from the viewpoint of the system operation. For example, the output signal power of each EDFA becomes variable depending upon the input signal power, which may be varied by attenuation change of transmission lines or reconfiguration of optical path networks. Use of a electrically-controllable variable optical attenuator in the EDFA [25] responds to the problem to some degree. It has been demonstrated that, even under variable input signal power conditions, both the gain flatness and the output power level of each channel can be maintained by the proposed scheme [25]. Another possible approach to the dynamic gain equalization is use of variable optical filters whose gain spectra are electrically controllable. Lattice-type optical circuits [26] may be used as such variable gain equalizers.

It should be noted here that, in any control schemes of the EDFA, one of the most critical issues is the speed of the response. Recently, it has been reported that, in order for operating channels to survive abrupt changes in channel loading, the response time of the EDFA should be as short as 1 μs in a chain of ten cascaded EDFAs, for example [27]. Further development is necessary to realize (economically) the broadband and gain-flattened EDFAs which satisfy all those requirements.

5. Summary

Significant efforts are currently being made to expand the gain bandwidth of the EDFA which is really "usable"for ultra-high capacity WDM transmission. The spectral gain profile of the EDFA should be not only broadband but also as flat as possible. The ultimate performance will possibly be achieved through a combination of the improvement of the erbium-doped fiber itself for a broader gain spectrum and the development of gain equalizers with higher precision. Further development is also required to realize the EDFA which has proper control over the gain flatness and the output signal power levels with sufficiently short response time even under dynamically changing operational conditions.

References

1. Bayart, D., et al.: Impact of fluoride-based EDFAs' gain flatness on the design of a WDM amplifier cascade, *Technical Digest of Optical Fiber Communication Conference (OFC'95)*, **TuP2**, (1995), pp.78-79.
2. Poole, S. B.: Fabrication of Al_2O^3 codoped optical fibres by a solution-doping technique, *Proceedings of the Fourteenth European Conference on Optical Communication (ECOC'88)*, **1**, (1988), pp.433-436.
3. Kashiwada, T., et al.: Spectral gain behavior of Er-doped fiber with extremely high aluminum concentration, *Technical Digest of Optical Amplifiers and their Applications Topical Meeting (OAA'93)*, **MA6**, (1993), pp.104-107.
4. Yoshida, S., et al.: Gain-flattened EDFA with high Al concentration for multistage repeatered WDM transmission systems, *Electron. Lett.*, **31**, (1995), pp.1765-1767.
5. Clesca, B., et al.: 1.5 μm fluoride-based fiber amplifiers for wideband multichannel transport networks, *Optical Fiber Technology*, **1**, (1995), pp.135-157.
6. Yamada, M., et al.: Low-noise and gain-flattened fluoride-based Er^{3+} -doped fibre amplifier pumped by 0.97 μm laser diode, *Electron. Lett.*, **33**, (1997), pp.809-810.
7. Kashiwada, T., et al.: Gain-flattened optical-fiber amplifiers with a hybrid Er-doped-fiber configuration for WDM transmission, *Technical Digest of Optical Fiber Communication Conference (OFC'95)*, **TuP1**, (1995), pp.77-78.
8. Kakui, M., et al.: Design optimization of hybrid erbium-doped fiber amplifiers for WDM transmission systems, *Optical Fiber Technology*, **3**, (1997), pp.123-133.
9. Wysocki, P., et al.: Dual-stage erbium-doped, erbium/ytterbium codoped fiber amplifier with up to +26 dBm output and 17 nm flat spectrum, *Technical Digest of Optical Amplifiers and their Applications Topical Meeting (OAA'96)*, **SaA2**, (1996), pp.199-202.
10. Yamada, M., et al.: A low-noise and gain-flattened amplifier composed of a silica-based and a fluoride-based Er^{3+}-doped fiber amplifier in a cascade configuration, *IEEE Photon. Technol. Lett.*, **8**, (1996), pp.620-622.
11. Hansen, S. L., et al.: Gain tilt of erbium-doped fiber amplifiers due to signal-induced inversion locking, *IEEE Photon. Technol. Lett.*, **4**, (1993), pp.409-411.
12. Kakui, M., et al.: Gain-flattened hybrid silica-based Er-doped fiber amplifiers designed for more than 25 nm-bandwidth without gain equalizers, *Technical Digest of Optical Amplifiers and their Applications Topical Meeting (OAA'97)*, **TuD4**, (1997), pp.142-145.

13. Massicott, J. F., et al.: High gain, broadband, 1.6 μm Er^{3+} doped silica fibre amplifier, *Electron. Lett.*, **26**, (1990), pp.1645-1646.
14. Yamashita, I., et al.: Er^{3+}-doped fibre amplifier operating at wavelengths of 1.55 and 1.60 μm, *Electron. Lett.*, **32**, (1996), pp.1102-1103.
15. Yamada, M., et al.: Broadband and gain-flattened amplifier composed of a 1.55 μm-band and a 1.58 μm-band Er^{3+}-doped fibre amplifier in a parallel configuration, *Electron. Lett.*, **33**, (1997), pp.710-711.
16. Sun, Y., et al.: Ultra wide band erbium-doped silica fiber amplifier with 80 nm of bandwidth, *Technical Digest of Optical Amplifiers and their Applications Topical Meeting (OAA'97)*, **PD2**, (1997).
17. Wang, J. S., et al.: Tellurite glass: a new candidate for fiber devices, *Optical Materials*, **3**, (1994), pp.187-203.
18. Mori, A., et al.: 1.5 μm broadband amplification by tellurite-based EDFAs, *Technical Digest of Optical Fiber Communication Conference (OFC'97)*, **PD1**, (1997).
19. Tachibana, M., et al.: Erbium-doped fiber amplifier with flattened gain spectrum, *IEEE Photon. Technol. Lett.*, **3**, (1991), pp.118-120.
20. Taba, H., et al.: Demonstration of optical FDM based self-healing ring network employing arrayed-waveguide-grating ADM filters and EDFAs, *Proceedings of the 20th European Conference on Optical Communication (ECOC'94)*, **1**, (1994), pp.263-266.
21. Vengsarkar, M., et al.: Long-period fiber grating as gain-flattening and laser stabilizing devices, *Technical Digest of the tenth International Conference on Integrated Optics and Optical Fibre Communication (IOOC'95)*, **5**, (1995), pp.3-4.
22. Taga, H., et al.: 110 Gbit/s (22 × 5 Gbit/s), 9,500 km transmission experiment using 980 nm pump EDFA 1R repeater without forward error correction, *Technical Digest of Optical Amplifiers and their Applications Topical Meeting (OAA'96)*, **PD5**, (1996).
23. Cullen, T. J., et al.: EDFA gain flattening using periodic tapered fibre filters, *Technical Digest of Optical Amplifiers and their Applications Topical Meeting (OAA'97)*, **WC2**, (1997), pp.231-234.
24. Wysocki, P. F., et al.: Erbium-doped fiber amplifier flattened beyond 40 nm using long-period grating, *Technical Digest of Optical Fiber Communication Conference (OFC'97)*, **PD2**, (1997).
25. Kinoshita, S., et al.: Low-noise and wide-dynamic-range erbium-doped fiber amplifiers with automatic level control for WDM transmission systems, *Technical Digest of Optical Amplifiers and their Applications Topical Meeting (OAA'96)*, **SaA5**, (1996), pp.211-214.
26. Fukutoku, M., et al.: 25 nm bandwidth optical gain equalization for 32- channel WDM transmission with a lattice type optical circuit, *Technical Digest of Optical Amplifiers and their Applications Topical Meeting (OAA'96)*, **FA4**, (1996), pp.66-69.
27. Sun, Y. and Srivastava, A. K.: Dynamic effects in optically amplified networks, *Technical Digest of Optical Amplifiers and their Applications Topical Meeting (OAA'97)*, **MC4**, (1997), pp.44-47.

(Received November 17, 1997)

PASSIVELY MODE-LOCKED SEMICONDUCTOR LASERS STABILIZED BY SUBHARMONIC ELECTRICAL INJECTION AND THEIR APPLICATION TO MILLIMETER WAVE PHOTONICS

M. TSUCHIYA, T. HOSHIDA AND T. KAMIYA

Department of Electronic Engineering, the University of Tokyo
Hongo 7-3-1, Tokyo 113-8656, Japan

Abstract. A review is presented, based on our experimental and theoretical works, of mode-locked monolithic semiconductor lasers with optical pulse train output stabilized by the subharmonic electrical injection and their performance in the millimeter wave photonic application. The following issues are emphasized: (1) the novel method of pulse train stabilization and its mechanism, (2) the application of a mode-locked semiconductor laser as an optical mm-wave oscillator and (3) the unique functionality in the compact optical mm-wave transmitter applications.

1. Introduction

Electromagnetic waves of millimeter wavelength (mm-wave) are expected to play indispensable roles in the future broadband wireless systems. Currently, however, there exists an obstacle preventing this frequency resource from being cost-effective and realistic; huge transmission loss. The concept of mm-wave links via optical fiber networks is attractive [1, 2] to overcome it by enabling not only low-loss transmission but also advanced mm-wave signal distribution. For its practical implementation, considerable attention has been attracted recently to the quest for high quality optical signal generation methods at repetition rates of millimeter-wave frequencies or beyond [1, 3, 4]. Desired features for those are (I) high performance at mm-wave frequencies or higher, (II) minimized phase noise, (III) minimized requirement on driving electronics and (IV) compactness (reduced number of independent devices in other words).

A. Hasegawa (ed.), New Trends in Optical Soliton Transmission Systems, 403–418.

Among the high frequency optical signal sources reported so far [5]-[14], mode-locked monolithic semiconductor lasers (ML-MSLs, [15]) have advantageous features as follows. First of all, the monolithic structure is beneficial in (IV). As for (I), passively ML-MSLs can generate pulses at repetition rates beyond the mm-wave frequencies [16]- [20]. Although its drawbacks in (II) had been the large phase noise and the difficulty in synchronization with external circuits, those have been settled by the recent progress of the hybrid mode-locking (H-ML) technique [21]. It will be further attractive with respect to (III) if the H-ML scheme is provided with external signals of lower frequency. Although optical short pulse injection at subharmonics of mode-locking frequency is considerably attractive from such a viewpoint [22, 23], it needs an additional laser and mismatches (IV).

The subharmonic hybrid mode-locking (SH-ML) [24]- [27] is a solution to the requirement, where the phase noise of passively ML-MSLs is reduced significantly by *subharmonic electrical injection.* This scheme seems, thus, attractive in all the aspects (I – IV). It is, therefore, interesting to examine the potential of SH-ML scheme in the field of mm-wave photonics. First of all, we briefly look over the typical performance of SH-ML-MSLs in the next section. For the details of the performance, our previous publication [27] should be referred. In the third section, the physical mechanism behind the SH-ML operation is discussed with a numerical simulation method. In Section 4, we mention the optical mm-wave transmitter performance of SH-ML-MSLs and its potential for mm-wave photonics is investigated.

2. Stabilization of Passively ML-MSLs by Subharmonic Electrical Injection

2.1. MSL DEVICE

Figures 1 shows the laser we have mainly investigated: a 1.55 μm distributed Bragg reflector (DBR) semiconductor laser [28] consisting of a saturable absorber (SA, 75 mm), a gain (750 mm), a phase control (PC, 150 mm) and a DBR (200 mm) sections monolithically. The former two sections include three compressively strained InGaAs quantum wells and a graded-index separate-confinement heterostructure. The latter two sections are made of InGaAsP with a 1.3 mm wavelength bandgap and have Pt heaters on top with which the optical path lengths can be varied by the thermo-optic effect. The laser was passively mode-locked around 33 GHz with a gain section current of 100 mA and the SA reverse bias voltage V_{SA} of 0.5 V to 0.7 V. Typical temporal and spectral shapes of pulses are shown in Figures 2(a) and 2(b), respectively. The time-bandwidth product of pulses stays around 0.35, which is close to the transform limit.

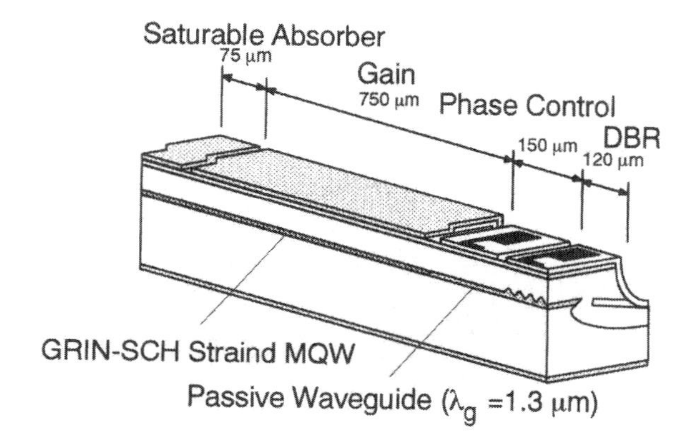

Figure 1. Schematic of a monolithic semiconductor laser.

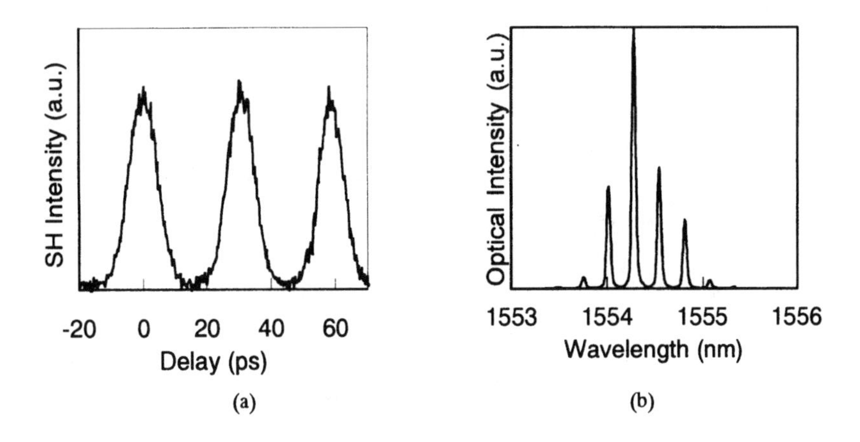

Figure 2. A correlation trace (a) and optical spectra (b) of the monolithic semiconductor laser under a passive mode-locking condition.

2.2. PHASE NOISE REDUCTION

Figures 3 show a phase noise reduction experiment setup and RF spectra around 33 GHz under (a) passive mode-locking and (b) third-order SH-ML ($n = 3$) conditions. It is clearly seen that the sideband noise is significantly suppressed by the electrical injection of $f_{\mathrm{rep}}/3$ into the SA section, i.e. under a (3rd order) SH-ML condition. In Figure 4, a single sideband noise (SSBN) under passive, H-ML, and SH-ML conditions are shown. One can see that SSBN under the hybrid mode-locking conditions is dominated by the driving circuit below 30 kHz and by the laser itself above 100 kHz.

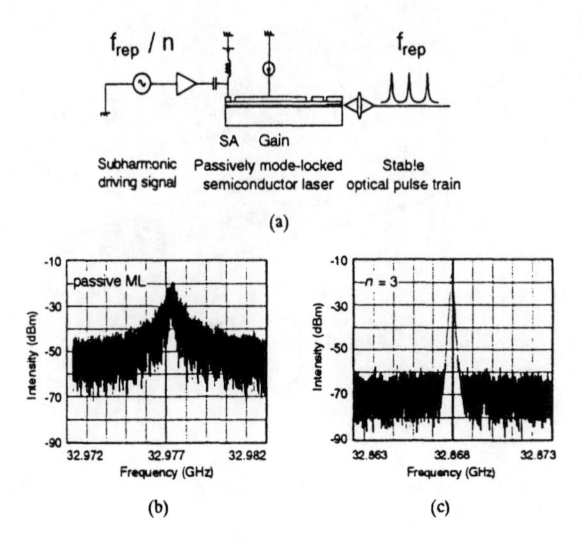

Figure 3. (a) Experiment setup of phase noise reduction. Millimeter-wave spectra around 33 GHz under (b) passive mode-locking and (c) third-order SH-ML conditions. Resolution bandwidth = 100 kHz.

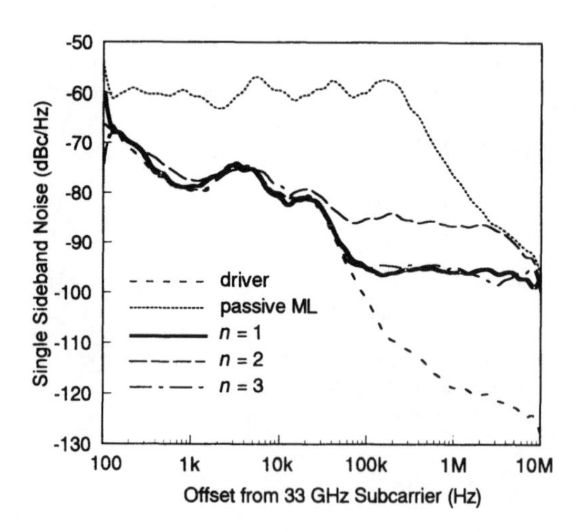

Figure 4. SSBN is plotted as functions of frequency offset.

We applied the SH-ML scheme to another MSL and extended the frequency up to the 60 GHz range which is considered to be important for various applications such as optical fiber-linked wireless communication, broadcasting and LAN. The 55 GHz MSL is shown in Figure 5(a). It is similar to the one in Figure 1 but is shorter and has a gain (675 mm) and

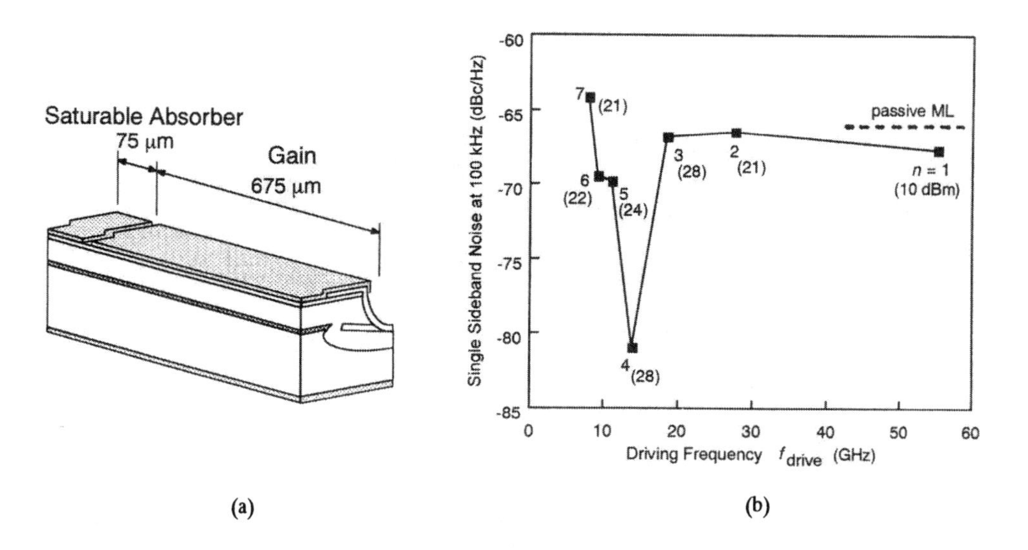

Figure 5. (a) A 55GHz ML-MSL and (b) Subharmonic order dependence of the phase noise reduction.

an SA (75 mm) sections only. Figure 5(b) shows SSBN at 100 kHz offset obtained for each subharmonic order. The largest noise reduction (−15 dB) was observed for $n = 4$ indicating the SH-ML scheme is quite effective also for a 55 GHz ML-MSL pulse train. The SSBN dependence on n can be explained as follows : the electrical modulation becomes less efficient with increasing driving frequency ($n < 4$), and the applicable injected power is limited to avoid falling into the self-pulsation mode for the lower frequency ($n > 4$) where the driving frequency approaches the relaxation oscillation frequency.

2.3. AMPLITUDE MODULATION

One potential problem in the SH-ML scheme is that the subharmonic driving signals might induce unwanted amplitude modulation (AM) at their frequencies [25]. Figures 6(a)-(c) show synchronously- scanned streak camera traces of the optical pulse trains generated from the 33 GHz ML-MSL under n = 1, 2 and 3 conditions, respectively. In the trace (c), the pulse train shows periodic amplitude variation indicating existence of a 11 GHz component. It should be noted here that no such AM symptom is seen for $n = 2$ as well as for $n = 1$. This implies that AM be negligible under the second-order SH-ML case.

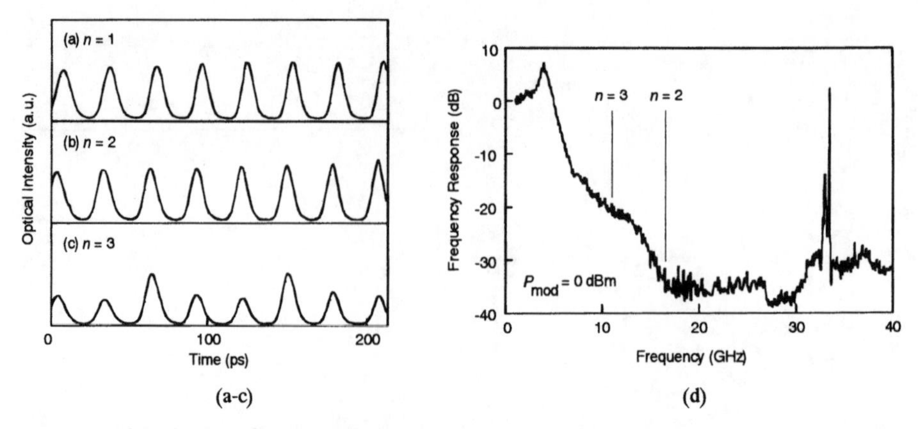

Figure 6. Synchronously-scanned streak camera traces of the optical output generated under (a) $n = 1$, (b) $n = 2$ and (c) $n = 3$ conditions. (d) Small signal modulation response of the laser under a passively mode-locked condition. 0 dBm modulation signal was applied to the saturable absorber section. The vertical axis is not calibrated.

The origin for the significant difference between the $n = 2$ and 3 cases was clarified by investigating the small signal response of the device (Figure 6(d)). The response at $f_{\mathrm{mod}} = 11$ GHz ($n = 3$) was approximately 15 dB higher than that at $f_{\mathrm{mod}} = 16.5$ GHz ($n = 2$), which corresponds fairly well to the difference in the AM level in Figures 6(b) and 6(c). It is thus implied that AM can be diminished if a laser is driven at a subharmonic frequency where the modulation response of the device is negligible.

2.4. LOCKING BANDWIDTH

The ML-MSL locking bandwidth is an important parameter because the frequency trimming to a strictly fixed frequency-window is indispensable for any system application in mm-wave photonics. If we define the locking bandwidth as the range of SSBN < -90 dBc/Hz, the locking bandwidths for 23 dBm RF input were 24 MHz, 10 MHz and 56 MHz in the $n = 1$, 2 and 3 SH-ML cases, respectively (Figure 7). It should be noted that the locking bandwidth for $n = 3$ increases drastically with the RF input power and is larger rather than for $n = 1$ when the RF input power is 23 dBm.

3. Mechanism for the Stabilization by Subharmonic Electrical Injection

It is quite interesting to investigate the stabilization mechanism in the SH-ML scheme from a viewpoint of device physics. It is also a technologically important issue in order to make the potential of ML-MSL clear and establish a design principle. For this purpose, we have performed a numerical simulation of pulse generation and stabilization actions of a ML-MSL

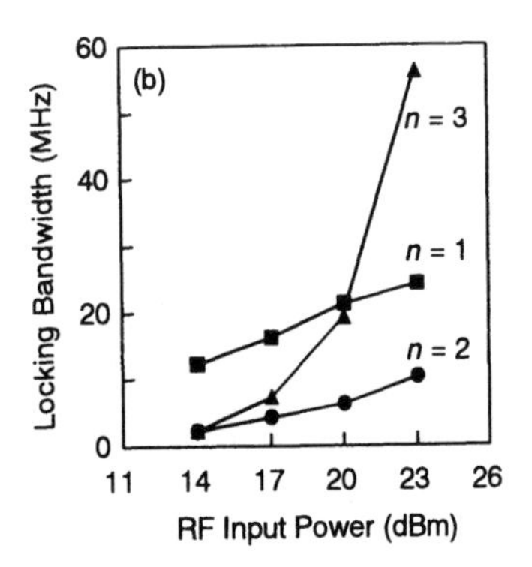

Figure 7. Locking bandwidth measured as functions of $P_{\rm mod}$.

device. In the simulation, we applied the traveling wave equation to the multi-section laser diode;

$$\frac{\partial E^{\pm}}{\partial t} \pm \frac{1}{\nu_g}\frac{\partial E^{\pm}}{\partial z} = \frac{1}{2}\{\nu_g \Gamma(n,S)\}E^{\pm}\,.$$

Here, $E^{\pm}$ is the optical field traveling toward $\pm z$ directions, ν_g the group velocity of light, Γ the confinement factor and $g(n,S)$ the gain coefficient. The dynamics of carrier density n in the gain and SA sections is dealt with by using the following rate equation,

$$\frac{\partial n}{\partial t} = \frac{J(t)}{ed} - (A_x n + B_x n^2 + C_x n^3) - \nu_g \Re(g(n,S))S\,,$$

where $S = | E^+ |^2 + | E^- |^2$ is the photon density, $J(t)$ is the injection current density, q the elemental charge, d the thickness of active layer, n the carrier density and Ax, Bx, Cx the coefficients for non-radiative, radiative and Auger recombination processes, respectively. Especially in the SA region, current, Bx and Cx terms are ignored, and the following expression is derived:

$$\frac{\partial n}{\partial t} = -\frac{n}{\tau_a} - \nu_g \Re(g(n,S))S\,.$$

Here, τ_a is the carrier lifetime. The gain coefficients for the gain and SA sections are given by

$$g(n,S) = (1 + i\alpha_g)\frac{dg}{db}(n - N_0)$$

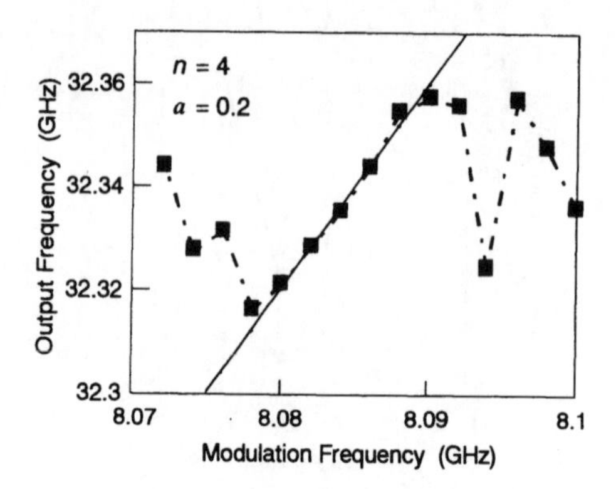

Figure 8. Optical pulse repetition frequency f_{rep} vs. subharmonic electrical injection frequency f for the modulation. The solid line indicates the relationship $4f = f_{\mathrm{rep}}$.

and

$$g(n,S) = (1 + i\alpha_a)a_0 \frac{n - N_0}{N_0} \frac{1}{1 + \varepsilon_a S},$$

respectively, where ε_g and ε_a are the gain saturation coefficients, α_g and α_a are the chirp parameters. The DBR section was modeled with a temporal filter having the first order Gauss response.

The most interesting aspect of the SH-ML mechanism is how the electrical input of subharmonic frequency stabilizes the optical pulse train. There could be two possible stories; (1) the nonlinear I–V characteristics of the SA section diode gives rise to the electrical harmonic component generation which stabilizes the pulse train, and (2) the electrical injection modulates the physical parameters of SA section, which interacts with the passive mode-locking dynamics. In the latter, the nonlinearity needed for the frequency multiplication is brought by the mode-locking action of the laser. In order to investigate the latter interesting mechanism, we introduce the modulation of subharmonic frequency to the linear absorption coefficient a_0 or to the carrier lifetime τ_a, which can be expressed by

$$a_0 = a_0(1 + a\sin(2\pi f)), \quad \tau_a = \tau_a(1 + a\sin(2\pi f t)).$$

Here, a is the modulation depth. Note that the former corresponds to the absorption change induced by the quantum confined Stark effect and the latter to the carrier escape rate modulation. We found out that the optical pulse repetition frequency follows the harmonic frequency of the electrical injection in both cases (see the case of a_0 modulation ($n = 4$) in Figure 8). The results obtained in the simulation suggest thus that the inherent nonlinear nature of ML-MSL is a possible origin of the SH-ML mechanism through either modulation of a_0 or τ_a.

4. Optical mm-wave Transmitter Application of ML-MSL

4.1. ML-MSL AS AN OPTICAL MM-WAVE OSCILLATOR

As described in the previous sections, the phase noise of optical mm-wave signal generated from passively ML-MSL devices can be reduced down to those of electrical oscillators. Therefore, if a stabilized ML-MSL is combined with an optical modulator of IF frequencies, such a sub-system can be supposed to be an optical mm-wave transmitter and generate an IF-modulated optical mm-wave signal. Here, the ML-MSL acts as an optical mm-wave oscillator. The optical mm-wave transmitter scheme consisting of an ML-MSL and an IF optical modulator has been reported [29].

On the other hands, among several possible architectures of optical mm-wave links, there is a configuration where a control station consists of two optical sources for an IF signal and a mm-wave carrier, and outputs of the two are mixed at each base station (Figure 9(a)) [30]. This is advantageous in some aspects including the transmission bandwidth of the system [31, 32]. In such a scheme, it is a key issue to realize efficient remote mm-wave up-conversion in a compact manner. In the following, a novel and significantly simplified technique which we have proposed and demonstrated is described.

4.1.1. *Optical IF Signal Up-conversion by Nonlinear Photo-detection*

The experimental setup is schematically shown in Figure 9(b). The second harmonic component contained in ML-MSL output was used as a mm-wave signal (65.622 GHz). The optical IF signal was generated by intensity modulation of a 1543 nm distributed feedback (DFB) semiconductor laser. Each of the two optical signals was amplified, combined by a 3 dB fiber coupler, and lead to the nonlinear photodetection stage.

We used a wideband waveguide p-i-n photodiode (NEL, KEPD1510VPG) as a nonlinear photodetector. This device has an optical saturation power of approximately 7 mW and a bandwidth well beyond 50 GHz when it is reverse biased at 3 V [33]. The response was degraded as the input optical power increased especially under low reverse bias conditions. Such sub-linear characteristics enable the mixing of the optical signals (Figure 9(c)). The inset of Figure 9(c) shows a typical mm-wave spectrum of the photodiode output with an IF signal at 400 MHz. Upper and lower sidebands were clearly observed 400 MHz off the ML-MSL signal (65.622 GHz).

The dynamic range of the present configuration was investigated by two-tone measurements. The power levels of up-converted signal, inter-

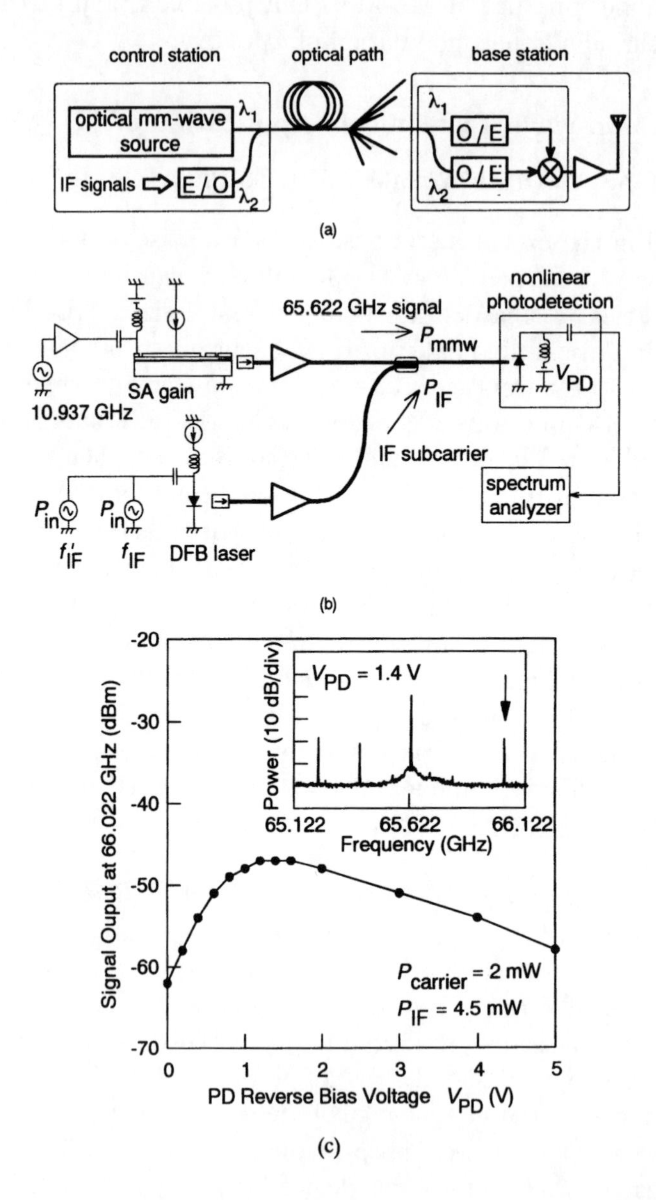

Figure 9. (a) A fiber-optic link architecture for mm-wave distribution using a base station up-converter fed optical mm-wave signal from a control station. (b) Experimental setup for mm-wave up-conversion using nonlinear photodetection technique. The optical mm-wave source is a subharmonically hybrid mode-locked semiconductor laser. (c) Up-converted signal level as a function of photodiode bias voltage. The inset shows a typical output spectrum of the nonlinear photodetector. $f_{IF} = 400$ MHz, $P_{in} = -5$ dBm.

modulation, and noise were measured as functions of injected IF power to the DFB laser. A spurious-free dynamic range of 93.3 dB$\cdot$Hz$^{2/3}$ was ob-

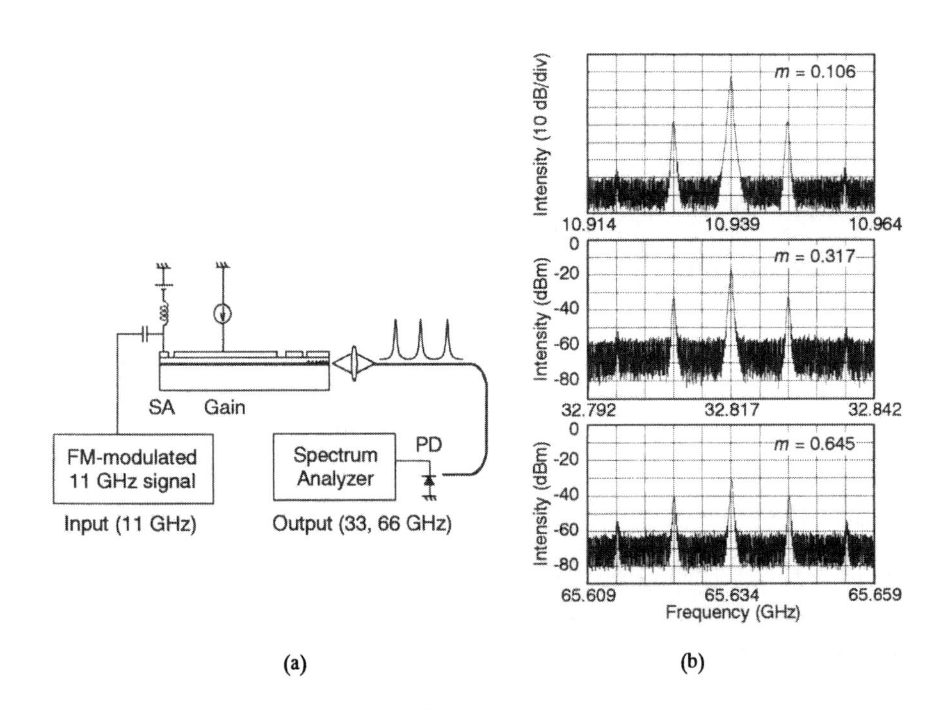

Figure 10. (a) Experimental setup for frequency modulated subcarrier signal generation. (b) Spectra of input signal and generated subcarriers at 33 GHz and 66 GHz frequency modulated by a 10 MHz rectangular wave (from the top). Resolution bandwidth = 300 kHz.

tained, which is large enough for some applications such as wireless local area networks and cellular phone systems employing code-division multiple access.

This technique, enabling flexible optical mm-wave transmission with simplified base station configuration, would provide a promising way to construct realistic broadband mm-wave wireless systems.

4.2. ML-MSL AS A FUNCTIONAL OPTICAL MM-WAVE TRANSMITTER

Properties of an electrical oscillator can be controlled by an external electrical signal to some extent. It is also the case for an ML-MSL as an optoelectronic millimeter wave oscillator. Furthermore, ML-MSLs have more than one electrodes with which the optical properties of waveguide can be trimmed, leading to a chance to modulate optical mm-wave signals of ML-MSL without using external modulators. In the following, novel methods are proposed and demonstrated for the generation of angle-modulated millimeter-wave optical signals: direct modulation of frequency and phase of millimeter-wave subcarriers generated from an ML-MSL under SH-ML conditions.

4.2.1. *Frequency Modulation*

The pulse repetition frequency in the SH-ML operation can be deviated according to the change in the driving signal frequency. Within the locking bandwidth, there exists no significant increase of phase noise [26]. Such locking property of an SH-ML laser makes it possible to generate frequency-modulated subcarrier signals by applying a frequency-modulated electrical signal with frequency modulation (Figure 10). In this scheme the frequency multiplication from 11 GHz to 66 GHz, brought by the SH-ML operation and the short pulse nature, provides advantages such as frequency reduction in driving electronics and enhancement of modulation index. Figures 10 show a result of a preliminary experiment. The signal injected to the SA section was a 10.939 GHz carrier that was frequency-modulated by 10 MHz rectangular wave with a modulation index of 0.106 (Figure 10(b) (top)). Figures 10(b) (middle and bottom) show the corresponding spectra at 32.817 GHz and 65.634 GHz, respectively. It should be noted here that the relative sideband intensities were enlarged with (sub)carrier frequencies as expected. The modulation indices deduced from the spectra, 0.317 and 0.645 for 33 GHz and 66 GHz subcarriers respectively, agree very well with a theoretical prediction.

4.2.2. *Phase Modulation*

Another approach to modulating the subcarrier is provided by the PC section modulation. Since the current injection to the PC section varies the optical cavity length and induce the detuning, it is expected that the timing of outgoing pulses be shifted. This effect can be utilized to phase-modulate the generated subcarrier signals. Recently Georges et al. [34] reported such an RF phase-control scheme in a resonantly modulated semiconductor laser. The use of an SH-ML-MSL provides additional attractive features; higher harmonic components that are generated due to the short pulse nature extend the subcarrier frequency and reduction of electrical driving frequency.

Figure 11(a) shows the experimental setup to demonstrate the phase modulation capability. Resultant range of phase shift was more than 2.0 radian and 4.1 radian for 33 GHz and 66 GHz, respectively. In Figure 11(b), the DC phase shift is plotted as a function of I_{PC}. In a range from 30 mA to 60 mA, the phase increased linearly with a modulation sensitivity of -0.33 ps/mA, which corresponds to 0.068 radian/mA and 0.137 radian/mA for 33 GHz and 66 GHz components, respectively. The invariance of pulse shape

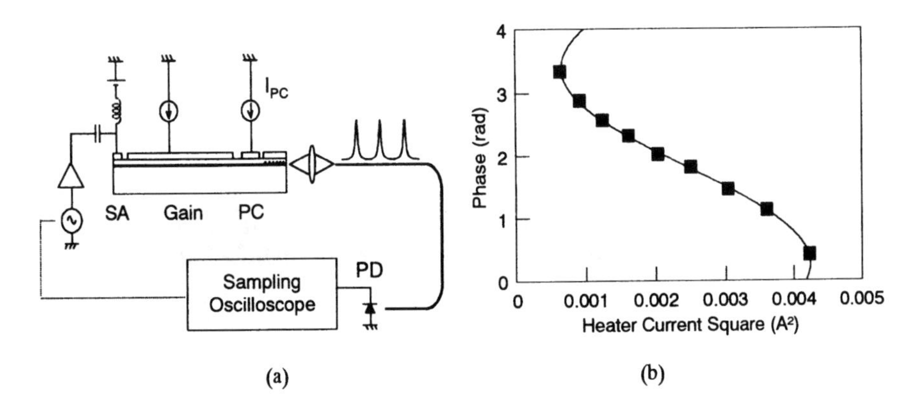

Figure 11. (a) Experimental setup for phase modulation of subcarrier signal. (b) The RF phase-shift of the generated subcarrier signals as a function of I_{PC}.

was confirmed by autocorrelation measurements. The results of the RF phase-shift can be explained by analogy with injection-locked electrical oscillators [35]; the phase angle θ between the injected signal and the oscillator is given by $sin\theta = (f_{\text{free}} - nf)/D$, where D is the locking bandwidth. For the device used in the above experiment, the phase-modulation bandwidth is limited to several Hz by the slow response of the thermo-optic effect. The use of the carrier injection effect to modify the laser cavity length [36] is expected to greatly increase the phase-modulation bandwidth. Indeed, 100 MHz phase modulation was demonstrated in a separate experiment by using a device with a PC section having a carrier injection electrode (Figure 12).

5. Conclusion

We have reviewed, based on our experimental and theoretical works, mode-locked monolithic semiconductor lasers with optical pulse train output stabilized by the subharmonic electrical injection and their performance in the millimeter wave photonic application. We have emphasized (1) the novel method of pulse train stabilization and its mechanism, (2) the application of a mode-locked semiconductor laser as an optical mm-wave oscillator and (3) the unique functionality in the compact optical mm-wave transmitter applications. Those features seem to the authors to be interesting from the mode-locked diode laser physics point of view and to be attractive for the implementation of the future cost-effective millimeter wave signal distribution systems.

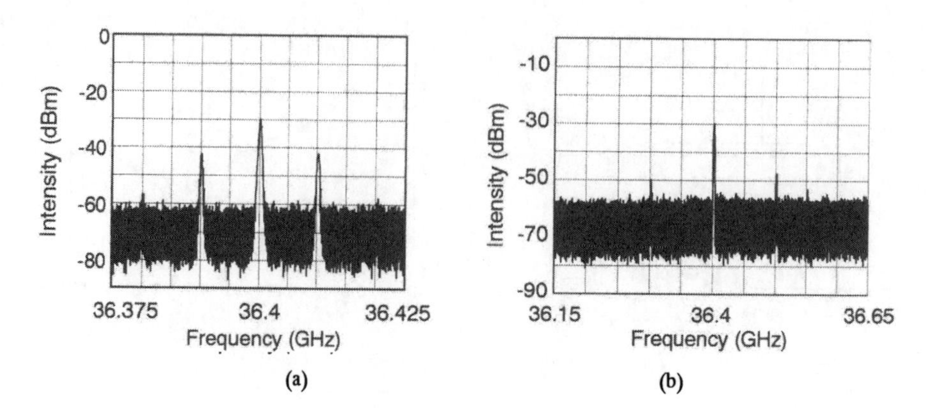

Figure 12. RF spectra of phase-modulated optical mm-wave signal generated directly from the ML-MSL. (a) 10 MHz phase modulation and (b) 100 MHz phase modulation.

Acknowledgments

The authors thank Dr. Y. Ogawa of the Oki Electric Industry Co., Ltd., Prof. H. F. Liu of the Melbourne University, Mr. T. Jujo, Mr. Y. Itaka and Mr. K. Nishikawa of the University of Tokyo for their support and discussion.

References

1. Ogawa, H., Polifko, D., and Banba, S.: Millimeter-wave fiber optic systems for personal radio communication, *IEEE Trans. Microwave Theory Tech.*, **40**, (1992), pp.2285-2292.
2. Komaki, S., Tsukamoto, K., Hara, S., and Morinaga, N.: Proposal of fiber and radio extension link for future personal communications, *Microwave and Optical Technol. Lett.*, **6**, (1993), pp.55-59.
3. O'Reilly, J., and Lane, P.: Remote delivery of video services using mm-wave and optics, *IEEE J. Lightwave Technol.*, **12**, (1994), pp.369-375.
4. Lau, K. Y.: Short pulse and high frequency signal generation in semiconductor lasers, *IEEE J. Lightwave Technol.*, **7**, (1989), pp.400-419.
5. Weisser, S., Larkins, E. C., Czotscher, K., Benz, W., Daleiden, J., Esquivias, I., Fleissner, J., Maier, M., Ralston, J. D., Romero, B., Sah, R. E., Schfelder, A., and Rosenzweig, J.: CW direct modulation bandwidths up to 40 GHz in short-cavity $In_{0.35}Ga_{0.65}As/GaAs$ MQW lasers with undoped active regions, *Proc. 21st Eur. Conf. on Opt. Comm. (ECOC '95)*, **Th.B.3.3**, pp.1015-1018, Brussels, 1995.
6. Nagarajan, R., Levy, S., Mar, A., and Bowers, J. E.: Resonantly enhanced semiconductor lasers for efficient transmission of millimeter wave modulated l ight, *IEEE Photon. Technol. Lett.*, **5**, (1993), pp.4-6.
7. Georges, J. B., Kiang, M.-H., Heppell, K., Sayed, M., and Lau, K. Y.: Optical transmission of narrow-band millimeter-wave signals by resonant modulation of monolithic semiconductor lasers, *IEEE Photon. Technol. Lett.*, **6**, (1994), pp.568-570.
8. Tucker, R. S., Koren, U., Raybon, G., Burrus, C.A., Miller, B. I., Koch, T. L., G.,

Eisenstein, and Shahar, A.: 40 GHz active mode-locking in a 1.5 μm monolithic extended-cavity laser, *Electron. Lett.*, **25**, (1989), pp.621-622.

9. Sato, K., Kotaka, I., Kondo, Y., and Yamamoto, M.: Active mode locking at 50 GHz repetition frequency by half-frequency modulation of monolithic semiconductor lasers integrated with electroabsorption modulators, *Appl. Phys. Lett.*, **69**, (1996), pp.2626-2628.
10. Noguchi, K., Miyazawa, H., and Mitomi, O.: 75 GHz broadband Ti:$LiNbO_3$ optical modulator with ridge structure, *Electron. Lett.*, **30**, (1994), pp.949-951.
11. Chernikov, S. V., Taylor, J. R., Mamyshev P. V., and Dianov, E. M.: Generation of soliton pulse train in optical fibre using two CW singlemode diode lasers, *Electron. Lett.*, **28**, (1992), pp.931-932.
12. Chernikov, S. V., Taylor, J. R., and Kashyap, R.: Integrated all optical fiber source of multigigahertz soliton pulse train, *Electron. Lett.*, **29**, (1993), pp.1788-1789.
13. Swanson, E. A., Chinn, S. R., Hall, K., Rauschenbach, K. A., Bondurant, R. S., and Miller, J.W.: 100-GHz soliton pulse train generation using soliton compression of two phase side bands from a single DFB laser, *IEEE Photon. Technol. Lett.*, **6**, (1994), pp.1194-1196.
14. Wake, D., Lima, C. R., and Davies, P. A.: Optical generation of millimeter-wave signals for fiber-radio systems using a dual-mode DFB semiconductor laser, *IEEE Trans. Microwave Theory Tech.*, **43**, (1995), pp.2270-2276.
15. Lau, K. Y.: Narrow-band modulation of semiconductor lasers at millimeter wave frequencies ($>$ 100 GHz) by mode-locking, *IEEE J. Quantum Electron.*, **26**, (1990), pp.251-261.
16. Sanders, S., Eng, L., Paslaski, J., and Yariv, A.: 108 GHz passive mode locking of a multiple quantum well semiconductor laser with an intracavity absorber, *Appl. Phys. Lett.*, **56**, (1990), pp.310-311.
17. Martins-Filho, J. F., Avrutin, E. A., Ironside, C. N., and Roberts, J. S.: Monolithic multiple colliding pulse mode-locked quantum-well lasers: experiment and theory, *IEEE J. Sel. Topics Quantum Electron.*, **1**, (1995), pp.539-551.
18. Chen, Y.-K., and Wu, M.C.: Monolithic colliding pulse mode-locked quantum-well lasers, *IEEE J. Quantum Electron.*, **28**, (1992), pp.2176-2185.
19. Arahira, S., Oshiba, S., Matsui, Y., Kunii, T., and Ogawa, Y.: Terahertz-rate optical pulse generation from a passively mode-locked semiconductor laser diode, *Opt. Lett.*, **19**, (1994), pp.834-836.
20. Arahira, S., Matsui, Y., and Ogawa, Y.: Mode-locking at very high repetition rates more than terahertz in passively mode-locked distributed-Bragg-reflector laser diodes, *IEEE J. Quantum Electron.*, **32**, (1996), pp.1211-1224.
21. Kim, D. Y., Pelusi, M. D., Ahmed, Z., Novak, D., Liu, H. F., and Ogawa, Y.: Ultra-stable millimeter-wave signal generation using hybrid mode-locking of a monolithic DBR laser, *Electron. Lett.*, **31**, (1995), pp.733-734.
22. Arahira, S., and Ogawa, Y.: Synchronous mode-locking of passively mode-locked semiconductor laser diodes by using optical short pulses repeated at subharmonics of the cavity roundtrip frequencies, *IEEE Photon. Technol. Lett.*, **8**, (1996), pp.191-193.
23. Wang, X., Yokoyama, H., and Shimizu, T.: Synchronized harmonic frequency mode-locking with laser diodes through optical pulse train injection, *IEEE Photon. Technol. Lett.*, **8**, (1996), pp.617-619.
24. Hoshida, T., Liu, H. F., Daza, M.R.H., Tsuchiya, M., Kamiya, T., and Ogawa, Y.: Generation of 33 GHz stable pulse trains by subharmonic electrical modulation of a monolithic passively mode-locked semiconductor laser, *Electron. Lett.*, **32**, (1996), pp.572-573.
25. Hoshida, T., Liu, H. F., Tsuchiya, M., Ogawa, Y., and Kamiya, T.: Extremely low-amplitude modulation in a subharmonically hybrid mode-locked monolithic semiconductor laser, *IEEE Photon. Technol. Lett.*, **8**, (1996), pp.1160-1163.
26. Hoshida, T., Liu, H. F., Tsuchiya, M., Kamiya, T., and Ogawa, Y.: Locking Char-

acteristics of a Subharmonically Hybrid Mode-locked Multisection Semiconductor Laser, *IEEE Photon. Technol. Lett.*, **8**, (1996), pp.1600-1602.

27. Hoshida, T., Liu, H. F., Tsuchiya, M., Ogawa, Y., and Kamiya, T.: Subharmonic Hybrid Mode-Locking of a Monolithic Semiconductor Laser, *IEEE J. of Selected Topics in Quantum Electron.*, **2**, (1996), pp.514-522.
28. Liu, H. F., Arahira, S., Kunii, T., and Ogawa, Y.: Generation of wavelength tunable transform-limited pulses from a monolithic passively mode-locked distributed Bragg reflector semiconductor laser, *IEEE Photon. Technol. Lett.*, **7**, (1995), pp.1139-1141.
29. Ahmed, Z., Novak, D., Waterhouse, R. B., and Liu, H. F.: Optically fed millimetre-wave (37 GHz) transmission system incorporating a hybrid mode-locked semiconductor laser,: *Electron. Lett.*, **32**, (1996), pp.1790-1792.
30. Noël, L., Marcenac, D., and Wake, D.: 120 Mbit/s QPSK radio-fibre transmission over 100 km of standard fibre at 60 GHz using a master/slave i njection-locked DFB laser source, *Electron. Lett.*, **32**, (1996), pp.1895-1897.
31. Schmuck, H.: Comparison of optical millimetre-wave system concepts with regard to chromatic dispersion, *Electron. Lett.*, **31**, (1995), pp.1848-1849.
32. Park, J., Elrefaie, A. F., and Lau, K. Y.: 1550-nm transmission of digitally modulated 28-GHz subcarriers over 77 km of nondispersion shifted fiber, *IEEE Photon. Technol. Lett.*, **9**, (1997), pp.256-258.
33. Kato, K., Kozen, A., Muramoto, Y., Itaya, Y., Nagatsuma, T., and Yaita, M.: 110-GHz, 50 %-efficiency mushroom-mesa waveguide p-i-n photodiode for a 1.55-μm wavelength, *IEEE Photon. Technol. Lett.*, **6**, (1994), pp.719-721.
34. Georges, J. B., Lux, R. A., Yeung, S. P., Lau, K. Y., and Chang, W.: Simultaneous fiber-optic transport and RF phase control of narrow-band millimeter-wave signals using multicontact monolithic semiconductor lasers, *IEEE Photon. Technol. Lett.*, **8**, (1996), pp.953-955.
35. Adler, R.: A study of locking phenomena in oscillators, *Proc. of the I.R.E. and Waves and Electronics*, **34**, (1946), pp.351-357.
36. Liu, H. F., Arahira, S., Kunii, T., and Ogawa, Y.: Frequency-tunable millimetre-wave signal generation using a monolithic passively mode-locked semiconductor laser, *Electron. Lett.*, **32**, (1996), pp.740-741.

(Received December 25, 1997)

ULTRAFAST SOLITON MULTIPLE-ACCESS NETWORKS

K.L. HALL, K.A. RAUSCHENBACH, S.G. FINN, R.A. BARRY,
J.D. MOORES, E.A. SWANSON AND S. R. CHINN

MIT Lincoln Laboratory
C-237, 244 Wood Street, Lexington, MA 02173, USA.

Abstract. Potential advantages of ultra-high speed optical time-division-multiplexed (TDM) systems, operating at single stream data rates of 100 Gbit/s, include increased intelligence within the network, provision of packet service to high-end users, and truly flexible bandwidth-on-demand. A number of key technologies must come together to realize high performance TDM networks. High-speed optical sources, all-optical switches, buffers and rate converters are all required. In this talk, we will give an overview of an appropriate architecture for ultra-high-speed TDM networks, as well as describe recent key technology developments.

1. Introduction

Ultrafast optical time-division-multiplexed (TDM) single channel networks support low access time, high throughput multiple access communications with simplified end equipment compared to multiple-lower-rate channel approaches. These networks have the potential to provide truly flexible bandwidth on demand at burst rates of 100 Gbit/s. Bit-interleaved TDM subsystems have been demonstrated [1]-[9]. However, for local area networks (LANs) and metropolitan area networks (MANs) servicing a combination of high-speed and low-speed users, a 100 Gbit/s slotted TDM network has several important operating advantages over bit-interleaved multiplexing approaches [10, 11]. We describe here architecture and technology for single-stream, 100 Gbit/s local/metropolitan area shared media networks with slotted TDM.

A. Hasegawa (ed.), New Trends in Optical Soliton Transmission Systems, 419–424.

2. Architecture

The network we envision runs at an aggregate rate of 100 Gbit/s and serves a heterogeneous population of users; high-end single users, high speed video servers, terabyte media banks and networks of supercomputers. These users operate from 1–100 Gbit/s. Such a network may serve as a backbone for high-speed network interconnection, support rapid transfer of large data blocks, and provide flexible, bandwidth on demand access for lower-rate users. The architecture is scaleable rates higher than 100 Gbit/s with concomitant scaling of device operating rates.

In these networks, because of the large speed, there are many data packets in flight at one time, a so-called high-latency system. Processing is limited at the 100 Gbit/s rate, and there is no high-speed variable delay for ultrafast pulses. This makes simultaneous guaranteed bandwidth and bandwidth on demand services difficult to implement. Also, it is difficult to insure both efficient and "fair"bandwidth sharing in the presence of moderately loaded or overloaded network traffic conditions. There have been some articles addressing these various problems in the literature [12, 13], but they do not appear to satisfy simultaneously, all of the above criteria. We have developed a frame-based slotted architecture called HLAN (helical local area network) [11, 14] which satisfies the above criteria. In early work, the HLAN architecture was implemented on a helical ring physical structure, but most recently has been explained using a unidirectional bus implementation.

HLAN is a frame-based architecture, which is implemented on a unidirectional bus. Only one bus is required. Traffic flows from guaranteed bandwidth transmitters (GBW), bandwidth on demand transmitters (BOD) and all receivers are downstream of the transmitters.

GBW services are provided to users who request them by the headend. The headend allocates reserved slots to the node on the GBW segment. Users can access the HLAN at their guaranteed rate using only a counter, a flip-flop, a few gates and slot marker detection logic [14]. Using existing GaAs logic, HLAN slot rates of 10–100 Mslots/s can be supported.

Fair and efficient BOD service is provided via the BOD segment. The headend creates credits and distributes them at a given interval. Users who have data to send and a credit can burst their data onto the network when the first empty slot comes by. Users with data to send but no credits, must wait for the headend to distribute more credits. Users who accumulate credits but have no data to send will be limited to a finite number of reserve credits. To prevent lockouts from high-traffic upstream nodes, the headend monitors free slots at the end of the BOD segment. If no free slots are observed, the number of allocated credits is decreased, decreasing the band-

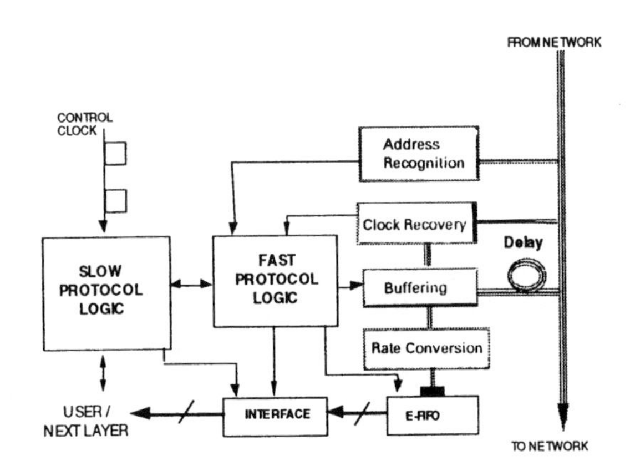

Figure 1. Block diagram of components in a user receiver node.

width available to the individual nodes. If, on the other hand, multiple free slots are observed, the number of allocated credits is increased, returning more bandwidth to the individual nodes. The system is "fair" because all users will get equal throughput when they have data to send. The algorithm is also efficient because no slot that could be used is left empty [11].

3. Technology

Figure 1 shows a block diagram of the receiver node. The components shown in the shaded boxes, at the interface between the optical bus, and the node electronics, are high performance optical devices that must operate at the optical data rate.

After data packets enter the receiver node, via a passive tap, the first function that must be performed is address/header recognition. In the receiver node, the incoming address must be compared with a local address to determine whether or not the packet requires further processing. Address recognition must be performed on-the-fly, at 100 Gbit/s. In the transmitter nodes, header recognition includes determining if there is a credit marker present and if the slot is empty or full.

If the packet address matches the local address, the data packet is stored for rate conversion down to a rate that the node electronics can handle. The related process at the transmitter node is up conversion of lower rate data and storage to support asynchronous network access.

Active fiber loop buffers allow long term storage of ultrafast packets. These buffers contain an amplitude modulator, polarization controller and polarizing elements. These active loop buffers have a number of built in stabilizing mechanisms and have stored for many thousands of circulations [15]-[18]. The built-in stabilizing mechanisms include timing stability due to the loss modulation by the amplitude modulator and pulse energy stability due to additive pulse limiting caused by nonlinear polarization rotation in the fiber loop. Loading of externally generated data packets has been demonstrated at packet rates up to 40 Gbit/s [19].

Rate conversion from the buffer to the electronics rate may be achieved via optical sampling. This scheme has the advantage over demultiplexing that the rate converted streams are time dilated and require no re-ordering of the bits. In addition, careful phase alignment of the sampling stream is not needed. The essence of this scheme is that aliasing brought about by sampling a high speed optical stream at a rate lower than the Nyquist rate gives rise to a time dilated version of the packet. The sampling has been achieved in an optical switch, and rate conversion of data packets from 80 Gbit/s to 100 Mbit/s has been realized. The cross-correlated data are detected and low-pass filtered so that the cross-correlation signal envelope, and not the individual sampling pulse envelopes are measured.

In summary, we have described a candidate architecture, HLAN, for management and control of single-stream 100 Gbit/s local and metropolitan area networks, providing packet switched services (slotted TDM) in a unidirectional bus topology. Key technologies, including 100 GHz optical sources [20], all-optical switches, active fiber loop buffers and rate conversion techniques have been demonstrated.

Acknowledgments

We are grateful to Erich P. Ippen and Hermann A. Haus for stimulating discussions and for important insights into device operation and performance. We acknowledge the experimental work of David Jones, Bryan Robinson, Naimish Patel, Victor Lum and William Wong. Also we thank Claudia I. Fennelly for her assistance. This work was supported by the consortium on Wideband All-Optical Networks and the Defense Advanced Research Projects Agency (DARPA). Opinions, interpretations, conclusions and recommendations are those of the authors and are not necessarily endorsed by the United States Air Force.

New results for this ultrafast TDM network will be posted at www.ll.mit.edu (under Group 65, AON, TDM references).

References

1. Ellis, A. D., Widdowson, T., Shan, X. and Moody, D. G.: Three-node 40 Gb/s OTDM network experiment using electro-optic switches, *Electron. Lett.*, **30**, (1994), p.1333.

2. Suzuki, K., Iwatsuki, K., Nishi, S. and Saruwatari, M.: Error-free demultiplexing of 160 Gbit/s pulse signal using optical loop mirror including semiconductor laser amplifier, *Electron. Lett.*, **30**, (1994), p.1504.
3. Cotter, D., Smith, K., Shabeer, M., Rogers, D. C. and Nesset, D.: Ultrafast self-routing packet networks, *Technical Digest of Optical Fiber Communications Conference*, Optical Society of America, Washington, DC, paper **WJ1**, (1995).
4. Morioka, T., Kawanishi, S., Takara, H. and Saruwatari, M.: Multiple-output 100 Gb/s all-optical demultiplexer based on multichannel four-wave mixing pumped by a linearly chirped square pulse, *Electron. Lett.*, **30**, (1994), p.1959.
5. Cotter, D., Lucek, J. K., Shabeer, M., Smith, K., Rogers, D. C., Nesset, D. and Gunning, P.: *Electron. Lett.*, **31**, (1995), p.1475.
6. Yamamoto, T., Imai, T., Komukai, T., Miyajima, Y. and Nakazawa, M.: Optical demultiplexing and routing of a TDM signal by using four-wave mixing and a novel wavelength router with optical circulators and fiber gratings, *Electron. Lett.*, **31, No.9**, (1995), p.744.
7. Kawanishi, S., Takara, H., Morioka, T., Kamatani, O. and Saruwatari, M.: 200 Gb/s, 100 km time division multiplexed optical transmission using supercontinuum pulses with prescaled PLL timing extraction and all-optical demultiplexing, *Electron. Lett.*, **31, No.10**, (1995), p.816.
8. Ellis, A. D., Widdowson, T., Shan, X., Wickens, G. E. and Spirit, D. M.: Transmission of a true single polarization 40 Gb/s soliton data signal over 205 km using a stabilized erbium fiber ring laser and 40 GHz electronic timing recovery, *Electron. Lett.*, **29, No.11**, (1993), p.990.
9. Ellis, A. D., Davies, D. A. O., Kelly, A. and Pender, W. A.: Data driven operation of semiconductor amplifier loop mirror at 40 Gb/s, *Electron. Lett.*, **31, No.15**, (1995), p.1245.
10. Finn, S. G. and Barry, R. A.: Optical Services in Future Broadband Networks, *IEEE Network Magazine*, (November/December, 1996), p.7.
11. Barry, R. A., Chan, V. W. S., Hall, K. L., Kintzer, E. S., Moores, J. D., Rauschenbach, K. A., Swanson, E. A., Adams, L. E., Doerr, C. R., Finn, S. G., Haus, H. A., Ippen, E. P., Wong, W. S. and Haner, M.: All-Optical Network Consortium-Ultrafast TDM Networks, *J. Lightwave Technol./IEEE J. Select. Areas Commun.*, **14, No.5**, (1996), p.999.
12. Yeh, C., Lin, M., Gerla, M. and Rodrigues, P.: RATO-net: A random-access protocol for unidirectional ultra-high-speed optical fiber networks, *J. Lightwave Technol.*, **8, No.1**, (1990).
13. Watson, G. C. and Tohme, S.: S++-A new MAC protocol for Gb/s local area networks, *IEEE J. Select. Areas Commun.*, **11, No.4**, (1993).
14. Finn, S. G.: HLAN-an architecture for optical multi-access networks, *Technical Digest for the LEOS Summer Top. Mtg.*, (1995), p.45.
15. Doerr, C. R., Wong, W. S., Haus, H. A. and Ippen, E. P.: Additive pulse mode-locking/limiting storage ring, *Opt. Lett.*, **19, No.21**, (1994), p.1747.
16. Moores, J. D., Hall, K. L., LePage, S. M., Rauschenbach, K. A., Wong, W. S., Haus, H. A. and Ippen, E. P.: 20 GHz optical storage loop/laser using amplitude modulation, filtering and fast saturable absorption, *IEEE Photon. Technol. Lett.*, **7, No.9**, (1995), p.1096.
17. Hall, K. L., Moores, J. D., Rauschenbach, K. A., Wong, W. S., Ippen, E. P. and Haus, H. A.: All-optical storage of a 1.25 kbit packet at 10 Gb/s, *IEEE Photon. Technol. Lett.*, **7, No.9**, (1995), p.1093.
18. Moores, J. D., Wong, W. S. and Hall, K. L.: 50 Gb/s optical pulse storage ring using novel rational-harmonic modulation, *Opt. Lett.*, **20, No. 24**, (1995), p.2547.

19. Hall, K. L.: 40-Gbit/s optical packet buffering, *Technical Digest of Optical Fiber Communications Conference*, Optical Society of America, Washington, DC, (1997), p.251.
20. Swanson, E. A., Chinn, S. R., Hall, K. L., Rauschenbach, K. A., Bondurant, R. S. and Miller, J. W.: 100 GHz soliton pulse train generation using soliton compression of two phase side bands from a single DFB laser, *IEEE Photon. Technol. Lett.*, **6, No.10** (1994), p.1194.

(Received November 17, 1997)

STAR PROJECT

A. HASEGAWA
Graduate School of Engineering, Osaka University
Suita, Osaka 565, Japan

1. Introduction

Japanese government established in July 1996 the Science and Technology Basic Action Programs through which government can spend its money more effectively to prepare Japan as a leading science and technology nation of the 21st century.

The main areas of implementation include

1) Expansion of government's R&D investment
2) Construction of innovative R&D systems
3) Promotion of collaborative work among industry, academia and government.

For these purposes, the national budget for technology will increase yearly up to $3.8 billion, or 1 % of GNP, in the year 2000.

The STAR project has been selected as one of the strategically important national R&D projects sponsored by Japanese Ministry of Posts and Telecommunications. It is expected to contribute greatly in upgrading the current one to the next generation network infrastructure using ultra high-speed optical network technologies.

2. Main Objectives

The STAR project aims at constructing terabit per second all-optical networks over global distances based on the optical soliton concept, in order to meet the subscriber's acute demand for multimegabit multimedia services which are expected in the early 21st century. The project has its basis in matching the expected social demand to the technical feasibility required to respond to the demand.

A. Hasegawa (ed.), New Trends in Optical Soliton Transmission Systems, 425–430.

3. Sponsor and Participants

The STAR project is financially supported by the Japanese Ministry of Posts and Telecommunications as one of the national projects in the promotion of science and technology. It was started in 1996 and is expected to continue for ten years. The project is administered by a planning committee established within the Support Center for Advanced Telecommunications Technology Research, Foundation (SCAT). The members of the committee consist of Professor Akira Hasegawa, the Chair, Professors Kazuro Kikuchi and Tetsuya Miki, Drs. Takao Matsumoto and Kazuo Sakai, and Mr. Takayuki Wada, Secretariat. The participating organizations as of 1997 are, in alphabetic order, Fujikura, Fujitsu, Mitsubishi Electric, NEC, Oki, Sumitomo Electric , Toshiba, Osaka University and the University of Electro-Communications.

4. Technical Approach

A terabit per second global network requires a factor of a hundred enhancement in the project of transmission speed and distance (bit-distance project) over the presently available systems. Such an enhancement may be achieved by a combination of wavelength division multiplexing (WDM) and optical time division multiplexing (OTDM) within the gain bandwidth of an Erbium Doped Fiber Amplifier (EDFA), assisted by optical signal processing and switching as well as properly programmed fiber dispersion. A feasible combination of OTDM and WDM will be 100 Gbit/s transmission rate per channel with 10 wavelength channels.

All optical processing is most easily achieved by the use of Return to Zero (RZ) pulses having distance-independent pulse shapes with a reasonable energy per pulse. Soliton satisfies these requirements in addition to their ability to render the maximal bit-distance product per wavelength channel. However, in order to achieve the 100 Gbit/s transmission rate over a global distance, solutions to intrinsic problems of solitons, such as the requirement of peak power proportional to the bit rate squared, inefficient use of bandwidth due to the requirement of large separation between adjacent pulses, and time jitters induced by amplifier noise, should be solved. We believe that these problems can be solved with proper dispersion management and careful preparation of the initial pulse shape.

In addition to the transmission issue, we need to develop proper technologies for the light source, signal processing and switching, adaptable to a terabit per second WDM-TDM combined systems.

The consortium is responsible for the development of the compact light source, mux and demux devices, EDFA with flat gain characteristic, low loss fiber with desired dispersion profiles and all optical add/drop switching devices. It is also expected to provide leadership in theoretical and computational support for the entire project.

5. Key Issues

- Development of new solitons which requires less peak power and weaker interactions.
- Development of WDM/OTDM terabit light source and all-optical signal processing.
- Development of all-optical add/drop switching.
- Development of low noise EDFA with a flat gain profile.
- Development of low loss fiber with a programmed dispersion profile.

6. Research Activities of the Participants

Followings are the brief summary of research activities of participating corporations and universities.

Fujikura

Goal: Development of technologies for manufacturing and diagnosis for optical fibers having stable dispersion and nonlinear characteristics suitable for soliton transmission.

Contents and current status: Cross phase modulation method is being used to study the fiber nonlinear coefficient. To measure the dispersion variation by means of the pulse walk off, a mode locked fiber ring laser is constructed which generates 160 ps with repetition rate of 107 MHz. Studies are being made on the effect of the dispersion properties of base material on the variation of the dispersion characteristic along the direction of propagation. It was found that at the stage of the base material, variation of plus or minus 0.65 ps/km·nm.

Fujitsu

Goal: Development of all optical signal processing element based on four wave mixing (FWM).

Contents: Use of optical fiber for FWM is studied in order to apply for dispersion compensation using phase conjugacy, wavelength conversion, ultra fast switching and wavelength multiplexing/demultiplexing.

Current Status: It is found that there exists an optimum fiber length for FWM due to variation of zero dispersion fluctuation of a fiber and precise control of excitation light source.

Mitsubishi Electric
Contents and Current Status: Construction of ADM devices based on quartz type waveguide grating is being made and its characteristics are studied on simulations. New production schemes for such waveguides by means of CVD and other methods are tried.

NEC
Goal: Demonstration of ultra-fast soliton based transmission with proper control and development of ultra-fast all optical switch.
Contents and current status: To achieve ultra-fast intercontinental optical soliton transmission, proper dispersion management are studied to reduce soliton-soliton interactions in computer simulation. 20 Gbit/s stable soliton transmission is demonstrated in a fiber with a proper combination of DSF and DCF having 60 km amplifier spacings. For the construction of ultra-fast all optical switch, a new device called transient cross-phase modulation is proposed and has been demonstrated at 910 nm wavelength with a GaAs/GaAlAs waveguide. The device is based on a signal light frequency shift induced by transient nonlinear refraction. Currently InGaAsP waveguide is being studied for a switch of the communication wavelength.

Oki Electric
Contents and current status: Research is being made to develop a reflective type EA modulator module to produce 5 ps soliton pulse generation having 20 GHz repetition cycle in order to achieve 20 Gbit $\times$ 2 OTDM soliton. In addition, optical delay elements, which are needed in multiplexing optical signals, are being studied using thermal properties of a fiber.

Sumitomo Electric
Goal: Development of flat gain EDFA amplifiers with gain variation less than 1 % over the wavelength range of 30 nm which will enable the gain variance less than 20 dB for a integrated gain of 2,000 dB over 10,000 km soliton transmission.
Contents: Search is being made for a new glass elements as well as a mean of cascading glass fibers with two different elements to achieve the flattening. Gain equalizers having precisely controllable characteristics which will ultimately be needed to achieve the required level of gain flattening are also studied. In addition methods of controlling the gain is also studied.
Current status: Absorption and florescence spectra are being measured when additional doping of elements such as Ta, Ga, Nb, Mg and B is made to Er. It was confirmed that Ga and Mg can produce expansion of florescence spectra similar to Al. For equalizers, fiber grating and Fabry-Perot etalon are being studied. With respect the gain control, simulations are

being made for various monitoring methods.

Toshiba

Contents and current status: Development of quartz based waveguide optical filter having thermal stability. Based on computer simulation feasibility of thermally stable waveguide type optical filter is demonstrated.

Osaka University

Goal: Theoretical and experimental studies on all optical ultra-fast soliton based transmission and network construction for intercontinental distance.

Contents: Dispersion managed soliton having periodically varying chirp is studied to construct ultra-high speed soliton based transmission line having little interactions between adjacent solitons, reduced effects of amplifier noise, WDM feasibility and requiring no transmission control with elongated amplifier spacing. In addition experiments to test transmission of solitons in such dispersion managed fibers are carried out.

Current status: General theory of soliton transmission in periodically varying dispersion has been derived for weak as well as strong dispersion variations. It is shown that in the weakly dispersion managed system, soliton becomes a dressed soliton having the ideal soliton core dressed by linear wave having chirp. For the strongly dispersion managed system, soliton propagation is described by the nonlinear Schrodinger equation having additional linear parabolic potential. By means of a proper dispersion control, a new soliton called the quasi-soliton has been obtained having a Gaussian tail that satisfies all the requirements stated above. Experimental demonstration is made such that a dispersion managed soliton can propagate without distortion even for amplifier spacing of 80 km.

University of Electro-Communications

Goal: Theoretical and experimental studies on photonic networks mainly based on dense Wavelength Division Multiplexing.

Contents: Basic design parameters for WDM photonic networks are studied to construct very large capacity and flexible transport networks indispensable for future multimedia network environments. These parameters include optimum optical frequency allocations, transport routing control, basic methods for optical frequency control, and so on.

Current Status: Design methods for optimum frequency allocation taking into account routing control flexibility are under study. Some of fundamental optimizing concepts and algrorisms for these design are being studied and compared.

7. Expected Impact

Successful development of global terabit systems will allow world wide subscribers global multimedia access to multimegabits per second networks. This will bring about a truly new era for multimedia networks in which telephones are replaced by personal computers.

Optical components for signal processing and switching to be developed simultaneously in this project also opens a new era in optoelectronics in which a transition from electronics to optics takes place in analogy to the transition from electromechanical systems to electronics in the 1950's. This will produce a significant impact on ultra high speed data processing in the next generation of computers.

(Received December 25, 1997)

ESTHER PROJECT: PRACTICAL IMPLEMENTATION OF SOLITONS

P. FRANCO
ESTHER Project coordinator
Pirelli Cavi S.p.A.,
viale Sarca 222, 20146, Milano, Italy

Abstract. We describe the structure and the activities of the ACTS Project named ESTHER (Exploitation of Soliton Transmission Highways in the European Ring), reporting the experimental results already obtained and the planning for the future activity.

1. Introduction

The ESTHER Consortium is an Advanced Communications Technologies and Services (ACTS) Project funded by the EEC and including the following partners:

- Pirelli Cavi S.p.A. (Italy)
- France Télécom - CNET (France)
- Telefonica Investigación y Desarrollo (Spain)
- Istituto Superiore delle Poste e delle Telecomunicazioni (Italy)
- Câbles Pirelli S.A. (France)
- Fondazione Ugo Bordoni (Italy)
- University of Southampton (Great Britain)
- University of Aveiro (Portugal)
- University of Paderborn (Germany)
- University of Padova (Italy)
- University of Ljubljana (Slovenia)
- Consorzio per la Ricerca e la Educazione Permanente (Italy)

The aim of the ACTS Projects is to promote and to fund the research and the field application of advanced techniques in telecommunications.

In this framework, the ESTHER Project addresses a very advanced field in the domain of optical communications: the upgrading of the existing

A. Hasegawa (ed.), New Trends in Optical Soliton Transmission Systems, 431–443.

telecom networks to higher bit rates, in order to increase the capacity of the installed fiber facilities and giving an answer to the increasing needs caused by the progressive diffusion of broadband services (digital video TV, computer internetworking, etc.).

The transport networks presently installed in the European countries have been designed for use at relatively low bit-rates and no provision has been made to allow immediate upgrade of the network toward higher transmission speeds. This heritage from the past is therefore becoming a critical bottle-neck in the development of pan-European interconnections, where extremely high volumes of traffic will have to be transferred over spans exceeding 2,000 km.

While the current technologies appear unable to address the above mentioned problem, new approaches stemming from advanced R&D will probably solve these difficulties in the long run. Nevertheless, all the solutions demonstrated in high-bit rate optical transmission research are unsuitable to allow network upgrading without requiring an extensive modification of the existing hardware. The ESTHER Project is mainly aimed to provide the network operator with a synergic ensemble of new systems, design tools and recommendations that will make possible to use the already installed network facility with a minimum impact on the cables already present in the link. The main experimental demonstrations will address the very important problem of showing the feasibility of a direct upgrade, in an operating fiber link: on step index fibers to 10 Gbit/s, and on dispersion shifted ones to 40 Gbit/s.

2. Project Objectives

The ESTHER Consortium aims to achieve the following objectives:

- to develop and to test in the transport network robust, cost-effective soliton transmission systems to be used in the upgrade of step-index fiber links to 10 Gbit/s bit rate, up to connection lengths of 500 km.
- to demonstrate by an actual field-trial the viability of soliton techniques for 40 Gbit/s telecommunications over 500–700 km dispersion-shifted fiber links.
- to bring to industrial maturity the preferred soliton sources and any other special device (in line reshaping units, optimized fibers) needed in this application, with particular reference to the full-integration of optical devices and control electronics.
- to define strategies and procedures for the integration of the high bit-rate soliton links into the European network, with particular reference to standardization issues.

The first goal of the ESTHER program will be the demonstration, at 1550 nm, of high bit rate soliton transmission over step-index fiber, which constitute a significant part of the telecommunication network in most of the industrialized countries. Soliton propagation in dispersive fiber has been generally viewed as a not competitive approach to dispersion management, because the short soliton period will cause impairments which limit the total span of the link to 100 km. However, the expertise accumulated by the Partners and the availability of some advanced soliton control devices will allow to demonstrate the extension of nonlinear transmission to distances of about 500 km.

The second goal of the Project will be linked to the upgrading of standard dispersion shifted lines at a bit rate of 40 Gbit/s. The Project should then participate to the build-up of the theoretical and experimental know-how necessary for the assessment of soliton transmission technology for future terrestrial networks based upon very high bit rate OTDM links, i.e. typically 40 Gbit/s. The ESTHER Project plans to implement a preliminary field trial at very high bit-rate, exploiting the same infrastructure used for the SI soliton transmission tests ("compound" cables containing SI and DS fibers).

The activity will then include the design and test of 40 Gbit/s soliton sources and decoders, and the laboratory development of several 40 Gbit/s transmission systems with different architectures. The final objective is to make these systems compatible with the requirements of megametric pan-European OTDM links, and more specifically with long and irregular amplifier (or repeater) spacing resulting from the existing civil engineering infrastructures. To obtain these aims it is necessary the final development of opto-electronic devices, such as transmitters and receivers, and the proper interfaces to standard low bit rate tributaries. In this case the aim is to obtain devices and interfaces reliable, stable and rugged for long field trial application. The goal will be met by addressing with the outmost attention the optimization of all the electronic circuit needed to operate the soliton lasers in reproducible and reliable way in view of mass production of the innovative sources developed in the Project.

3. Project Structure

The ESTHER Project is divided into four technical workpackages.

WORKPACKAGE 1: theoretical background and network analysis.
The main activities are of WP1 are the following:

- synthesis on soliton transmission techniques and individuation of the limits;

- comparison of the other potential transmission techniques;
- network analysis and relationship with other "pan-European network" oriented Projects.

The main objectives of these activities can be summarized as follows:

- identify the attractive applications of soliton technology in the terrestrial European network;
- bring to the network designers the relevant information on soliton technology;
- fill the gap between the WDM-oriented and TDM-oriented Projects aiming at defining the future pan-European network.

To fulfill these aims the Project will use the following technical approach. Various network constraints will be accounted for in comparing the suitability of different transmission techniques. The aim will be to determine the range of situations where soliton techniques offer the most advantageous solution.

The optimization study described in the previous point depends on the bit rate under consideration. Therefore the needs of the pan-European network in terms of traffic capacity will be assessed in order to locate the points where soliton links are required. Both traffic data and the consequent network planning will be projected also analyzing and using scenarios currently under study in other European bodies, independently of the present Project.

A better comparison between the soliton approach and those based on WDM techniques will be seeked by organizing joint meetings with coordinators of WDM oriented Projects. The results obtained from this WP will be integrated with other research activities of WP03 and WP04 related more directly with the in field demonstrators.

WORKPACKAGE 2: specific devices and components.
Partners will develop and build, starting from their complementary bases, appropriate and specific components required for soliton transmission demonstrators and field trials.
In particular the main activities are:

- integrated soliton sources;
- OTDM soliton receiver.

The main objectives of these activities can be illustrated as follows:

- fabrication of two different type of transmitters: mode-locked soliton sources and electro-absorption soliton sources;
- modeling activity of AM and FM mode-locked $LiNbO_3$ waveguide lasers;

- design and fabrication of the electronic control circuit needed to stabilize the emission of the soliton pulse-train from the modelocked solid-state laser;
- realization of a 40–10 Gbit/s soliton OTDM receiver, based on 40 GHz clock extraction and optical gate control to allow the use of 10 Gbit/s TDM electronics.

All the proposed subcomponents will be incorporated into demonstrators of WP03 and WP04. The required key functionalities are soliton sources at 10–20 and 40 Gbit/s emitting at a precise wavelength, transform-limited very-short pulses with very reduced jitter and 40–10 Gbit/s OTDM receiver subcomponents. The development of these subcomponents will then allow the implementation of packaging techniques being developed in parallel within the program for building rugged modules with high performance.

WORKPACKAGE 3: field demonstration on step index fiber links.
The main activities are:

- characterization of the field trial line;
- characterization and installation of the network interfaces and line terminal equipment;
- advanced transmission protocols for long haul soliton transmission in step index fibers;
- 1.5 μm field trial demonstration in step index link (10 Gbit/s over 500 km).

This part of the Project investigates and demonstrate soliton transmission over long step-index fiber links.

The workpackage activities will start from characterizing the field line and in parallel, theoretical investigation and simulation will obtain design data and parameters needed to identify the most convenient modulation format to overcome the well known limits to long-haul transmission in step-index fibers due to soliton interaction.

The experimental test-bed will be installed in the laboratory to allow testing of soliton transmission over 500 km, collecting crucial information in view of the field step-index soliton demonstration.

The successful completion of the aforementioned activities will allow the implementation of a field trial, where the transmission equipment previously developed will be used to transmit real data traffic over the targeted span.

A draft of a standardization proposal for unconventional networks exploiting soliton transmission at 10 Gbit/s at 1.5 μm will be prepared for submission to the relevant international bodies.

WORKPACKAGE 4: field demonstration on dispersion shifted fiber links.
The main activities are:

- theoretical evaluation of 40 Gbit/s soliton transmission;
- laboratory demonstration of 40 Gbit/s soliton transmission;
- field demonstration of 10, 20 and 40 Gbit/s soliton transmission on dispersion shifted fiber.

The main objectives can be summarized as follows:

- to demonstrate by experimental and theoretical means the capabilities of soliton technique for single-wavelength transmission on dispersion-shifted-fiber links of national and pan-European networks;
- in that context, to determine the limits of 40 Gbit/s soliton transmission both without and with in-line or off-line processing;
- to evaluate at such bit rates the system robustness to the dispersion of key in-line fiber parameters;
- to validate the results by field demonstrations at 10, 20 and 40 Gbit/s.

Work will be based upon the use of analytical tools, numerical simulations and laboratory experiment. Both straight line and recirculating loop based (for long distance links) demonstrations will be carried out. Performance will be estimated in terms of jitter levels, Q-factors and bit error rates. Special emphasis will be put on the comparison of 40 Gbit/s compatible in-line and off-line processing techniques. Acceptable levels of polarization mode dispersion for the in-line fiber will be estimated.

Key components for soliton generation, photodetection and optical processing will be provided by WP02. Results obtained at 10, 20 and 40 Gbit/s will allow the definition of field experiments run on a special cable installed in Rome.

4. Field Experiments Setup

The three different experimental setup of the field experiments are reported in Figures 1, 2 and 3.

In the first scheme, a 10 Gbit/s soliton source with orthogonal polarization coding (described in Subsection 4.1) is used to feed a 500 km SI fiber link that is installed in a cable between Roma and Pomezia. The length of the cable is 25 km and the structure of the link is the following: there are 10×50 km fiber spans with 10 commercially available in-line optical amplifiers. The output power of the amplifiers is 14 dBm.

In the second scheme, a 10 Gbit/s soliton source with parallel polarization coding (described in Subsection 4.2) is used to feed a 594 km Telecom Italia DS fiber link. This link is installed between Roma, Pomezia, Latina, Formia, Napoli and Nola and there are 8 commercially available in-line optical amplifiers. The output power of the amplifiers is 14 dBm. The span length is the following: Roma–Pomezia–Latina 75 km; Latina–Formia 84

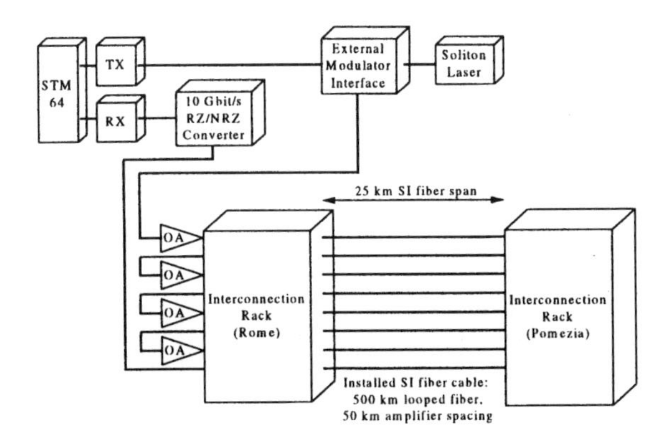

Figure 1. 10 Gbit/s soliton transmission over 500 km SI fiber link.

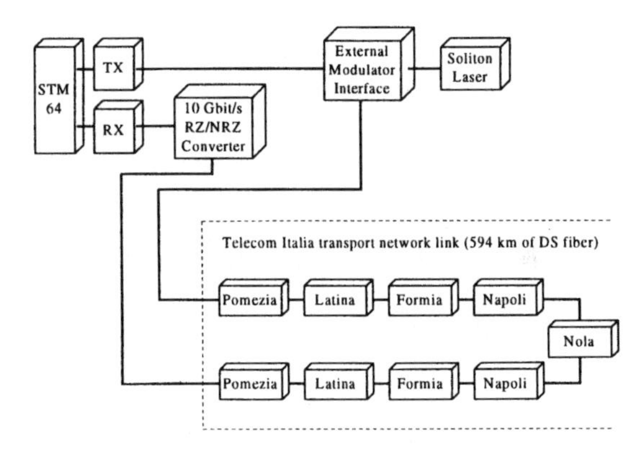

Figure 2. 10 Gbit/s soliton transmission over 594 km "real world" DS fiber link.

km; Formia–Napoli 88 km; Napoli–Nola–Napoli 100 km; Napoli–Formia 88 km; Formia–Latina 84 km; Latina–Pomezia–Roma 75 km.

In the last scheme, a 40 Gbit/s soliton source with orthogonal polarization coding (described in Subsection 4.3) is used to feed a 700 km DS fiber link that is installed in the cable between Roma and Pomezia. The structure of the link is the following: there are 7×100 km fiber spans with 7 commercially available in-line optical amplifiers. The output power of the amplifiers is 14 dBm.

In the three schemes, the 10 Gbit/s line terminal equipment have been modified in order to give two electrical signals (10 GHz clock and 10 Gbit/s data) at the transmitter side. At the receiver side, an optimized RZ to NRZ

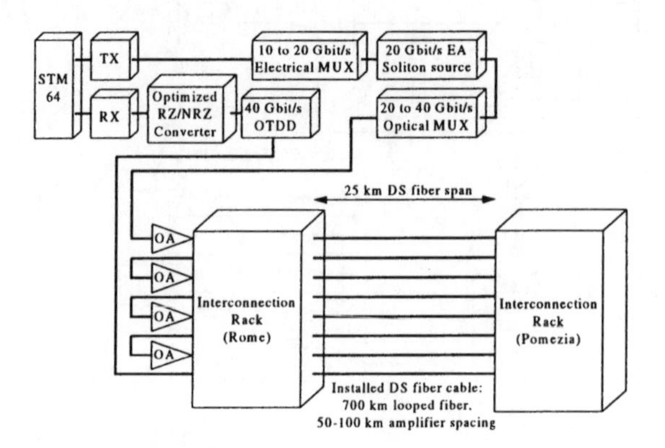

Figure 3. 40 Gbit/s soliton transmission over 700 km DS fiber link.

converter will be included. This converter will be optimized depending on the pulse duration used in the experiments. Moreover, while in both the 10 Gbit/s transmission experiments no optical demultiplexing is required, in the 40 Gbit/s experiment a 40 to 10 Gbit/s optical gating is used before the receiver.

4.1. ORTHOGONAL POLARIZATION 10 GBIT/S SOLITON GENERATOR

The pulsed laser source used for the 10 Gbit/s soliton transmission on SI fiber is depicted in Figure 4: a 5 GHz soliton stream is generated by means of a $LiNbO_3$ waveguide modelocked laser. The output power of the device is 0 dBm and the pulse duration is 50 ps. The stabilization of the laser is obtained by using the PLL-based electronic circuit. The single polarization output of the $LiNbO_3$ laser is amplified by means of a polarization maintaining (PM) optical amplifier (output power 11 dBm) and then sent to the polarization encoder represented in Figure 5.

The 5 GHz pulse stream is split with a 50/50 polarization maintaining coupler and sent to the two amplitude modulators. Two 90/10 polarization maintaining couplers allow to split part of the signal for the bias control circuits. Then, by properly adjusting the fiber length of the two arms, the two modulated pulse streams are combined by the polarization beam splitter with orthogonal polarizations and with a 100 ps (one bit) delay.

The polarization-encoding soliton source requires several electrical support circuits. In particular, the input has to be interfaced to the single 10 Gbit/s serial data stream given by the line terminal equipment (LTE). The latter has to be split into two 5 Gbit/s serial streams using a 10 GHz

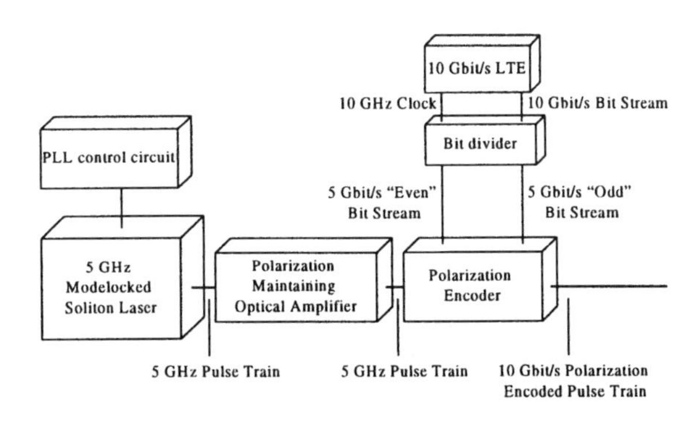

Figure 4. Orthogonal polarization 10 Gbit/s soliton generator.

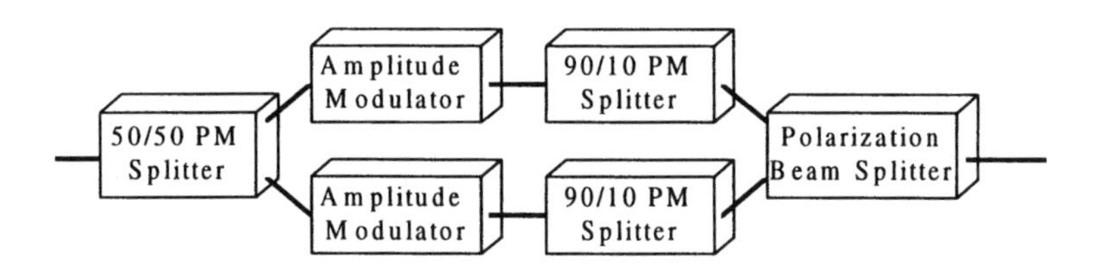

Figure 5. Optical scheme of the polarization encoding device.

to 5 GHz frequency divider and two D-flip-flops to sample and latch the incoming data.

4.2. PARALLEL POLARIZATION 10 GBIT/S SOLITON GENERATOR

The scheme of the laser source is very similar to the previous one and is reported in Figure 6: we used the same $LiNbO_3$ modelocked laser driven and controlled with a 10 GHz PLL circuit, obtaining a pulse duration of 30 ps. At the output we encoded the 10 Gbit/s bit stream given by the LTE by using an amplitude modulator with 10 GHz bandwidth.

4.3. ORTHOGONAL POLARIZATION 40 GBIT/S SOLITON GENERATOR

The scheme of the electro-absorption 40 Gbit/s orthogonal polarization soliton source is depicted in Figure 7. A DFB laser at 1551.3 nm is coupled to a 20 GHz bandwidth MQW electro-absorption modulator (EAM) giving a 20 GHz, 9 ps pulse stream. The EAM output is modulated at 20 Gbit/s by a $LiNbO_3$ electro-optic modulator (EOM) driven by a 20 Gbit/s data

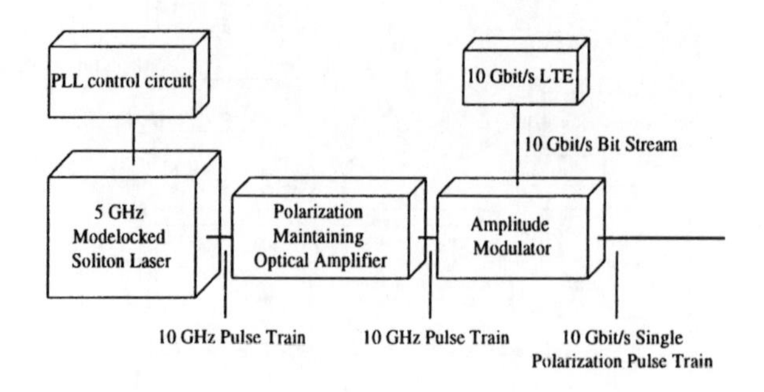

Figure 6. Parallel polarization 10 Gbit/s soliton generator.

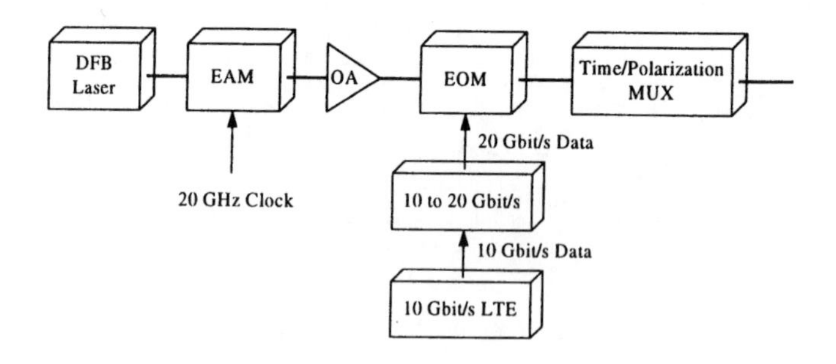

Figure 7. Orthogonal polarization 40 Gbit/s soliton generator.

stream obtained by a 10 to 20 Gbit/s multiplexer. The resolution 20 Gbit/s soliton stream is time and polarization multiplexed at 40 Gbit/s.

At the output of the fiber link an optical gating is necessary in order to demultiplex the signal from 40 Gbit/s to 10 Gbit/s. This is obtained by using the 10 GHz polarization independent EA optical gate represented in Figure 8.

5. Laboratory Experiments Results

5.1. 10 GBIT/S SOLITON TRANSMISSION ON STEP INDEX FIBER

The experimental setup is a 10×50 km SI fiber line fed by the 10 Gbit/s soliton source represented in Figure 4. The average fiber attenuation is 0.21 dB/km, the average group velocity dispersion is 16.5 ps/nm·km and the PMD is 0.05 ps/nm$^{1/2}$.

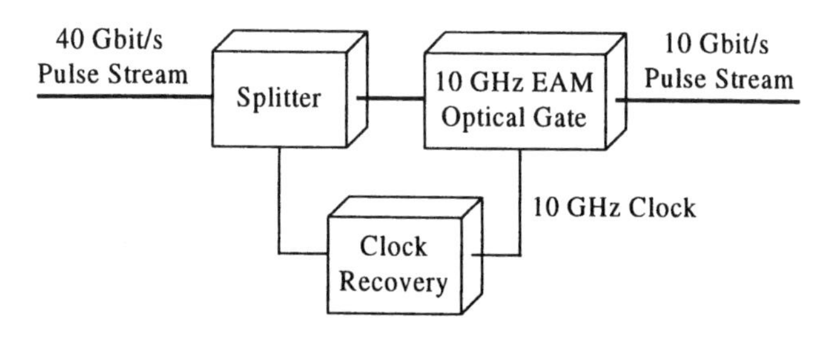

Figure 8. Polarization independent 10 GHz optical gate.

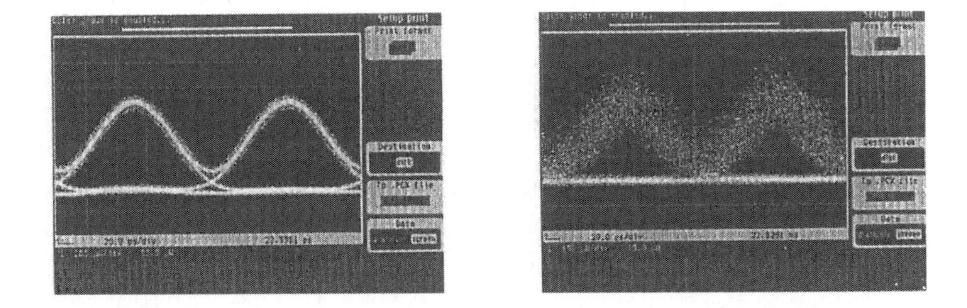

Figure 9. Eye diagram at 0 km and at 300 km.

The experimental results show that good BER performances can be obtained for propagation distances up to 300 km. The eye diagram and the BER curves at the beginning of the line and at 300 km are reported in Figure 9 and Figure 10, showing a 3 dB penalty.

Even if the initial target of 500 km can not be reached, the 300 km propagation still represents the higher distance obtained at 10 Gbit/s on SI fiber with no in-line propagation control.

5.2. 10 GBIT/S SOLITON TRANSMISSION ON DISPERSION SHIFTED FIBER

The experimental setup has the same characteristics of the Roma–Nola–Roma link (see Figure 2). There are 3×80 km, 1×120 km, 3×80 km DS fiber spans with average loss of 0.23 dB/km, average zero dispersion wavelength at 1549 nm and PMD of 0.3 ps/km$^{1/2}$. The soliton source is depicted in

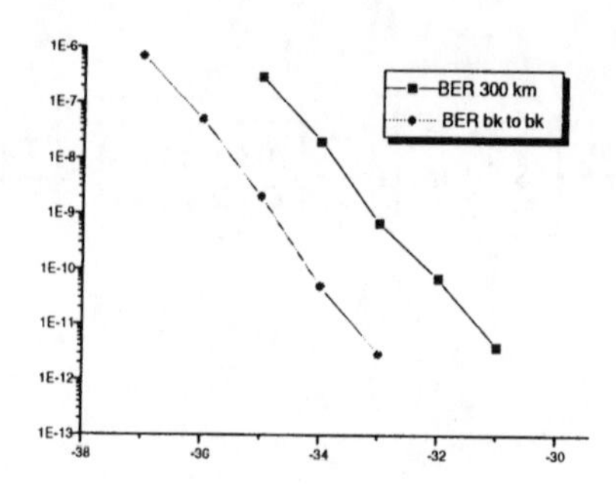

Figure 10. BER curves at 0 km and at 300 km.

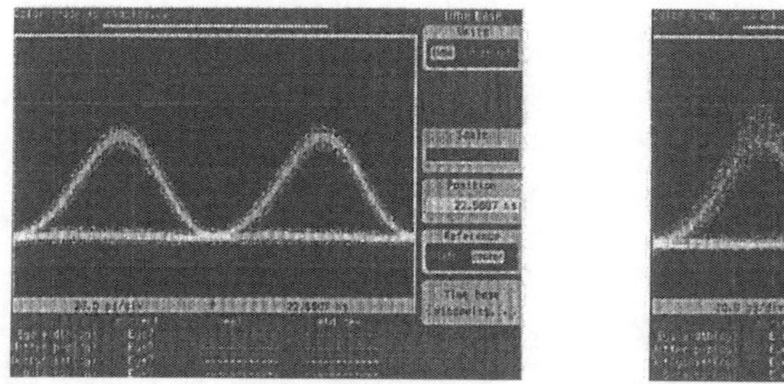
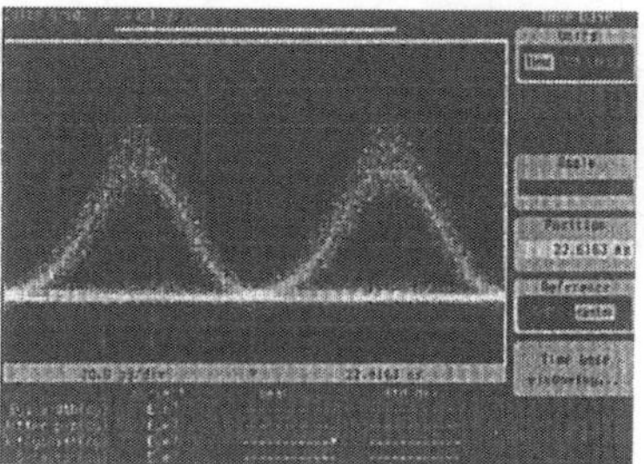

Figure 11. Eye diagrams at 0 km and at 600 km.

Figure 6 and the transmission wavelength is 1561 nm.

The eye diagram and the BER curves at the beginning and at the end of the line are reported in Figure 11 and Figure 12, showing a 2 dB penalty.

5.3. 40 GBIT/S SOLITON TRANSMISSION ON DISPERSION SHIFTED FIBER

The experimental setup is a 8×100 km DS fiber line fed by the 40 Gbit/s soliton source represented in Figure 7. The average fiber attenuation is 0.21 dB/km and the average group velocity dispersion slowly decreases with distance from 0.19 ps/nm·km to 0.12 ps/nm·km. At the receiver a 40 Gbit/s to 10 Gbit/s optical demultiplexer is used.

The Q-factor measurements versus distance for two different dispersion maps are repaorted in Figure 13.

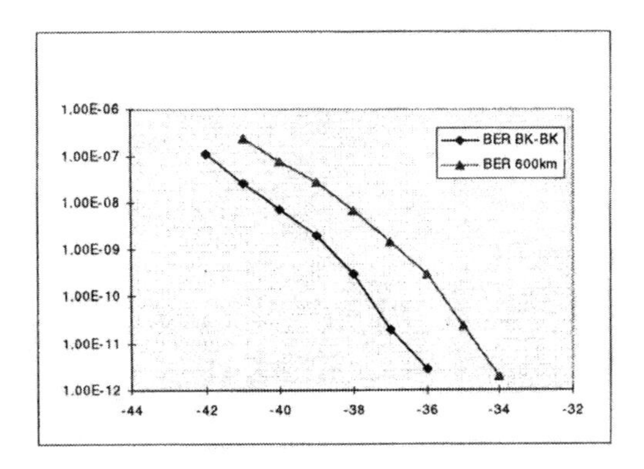

Figure 12. BER curves at 0 km and at 600 km.

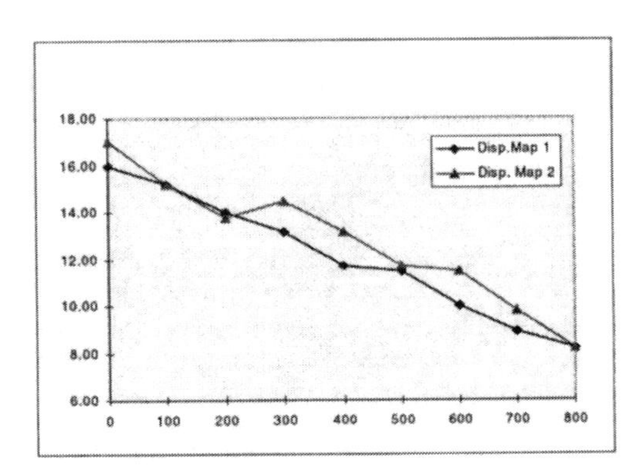

Figure 13. Q-factor measurements for two different dispersion maps.

6. Future Plans and Conclusions

In conclusion, we presented the structure and the activities of the ACTS Project named ESTHER and we showed the main results already obtained.

From the laboratory trials we obtained 10 Gbit/s soliton propagation on SI fiber with no in-line control for distances up to 300 km, and 40 Gbit/s soliton propagation on DS fiber with uncontrolled dispersion maps for distances up to 800 km,.

During the next year (the last year of the Project) all the experiments will be carried out in the field, in order to validate the results already obtained in the laboratory.

(Received November 17, 1997)

SOLITON TRANSMISSION IN THE EUROPEAN NETWORK

J.J.E. REID
Philips Optoelectronics Research
Prof. Holstlaan 4, 5656 AA Eindhoven, The Netherlands

Abstract. The UPGRADE project, supported by the European Union, seeks to provide a solution for the capacity enhancement of the standard single mode optical fibre network. The technological approach, experiments on soliton transmission, system design and the progress of the project is described along with recent results obtained from a field trial in the German national network.

1. Introduction

Within the ACTS (Advanced Communications, Technology and Services) programme of the European Union the principal goal of the UPGRADE project is the provision of a solution for the capacity enhancement of the European backbone trunk-links without the necessity of having to install new types of optical fibre in the network. New optically transparent backbones would be used to link major centres of population and commerce, implying transmission distances of up to 1,000 km at data rates of 10 Gbit/s and higher. Europe is, for the greater part, already cabled with Standard Single Mode, step-index, Fibres (SSMF), while world-wide more than 90 million kilometres of this type of fibre exists in either operational systems or dark fibre links. SSMF comprises some 90% of the total installed fibre base.

For services such as ATM or video-on-demand, a very high capacity transport layer is obligatory. Single channel SDH, or SONET, at a bitrate of 2.5 Gbit/s on SSMF has been the state-of-the-art for several years in the majority of the optical core networks and, in general, most network providers still rely on single-channel backbones of this type. This can be ascribed to a number of factors varying from commercial motive, to network infrastructure. The capacity is however not considered suffi-

A. Hasegawa (ed.), New Trends in Optical Soliton Transmission Systems, 445–458.

cient for use as the backbone of the trunk network, either on a national or pan-European scale. The commercial introduction of new wave-division-multiplexing (WDM) systems and an increase in the demand for services at competitive prices is forcing telecommunications providers to consider increasing bandwidth in their networks.

In recent years several high capacity techniques for the upgrading of SSMF based links, have been investigated. Due to the advanced state of development of the Erbium-Doped Fibre Amplifier (EDFA) most of the effort has been concentrated in the 1.55 μm optical window where a number of record transmission experiments have been performed within the laboratory environment. Those performed with SSMF include, for example, 10×20 Gbit/s, WDM, over 1,000 km with the use of dispersion compensating fibre (DCF) [1] and 10 Gbit/s over 2,245 km [2], again with DCF. Nevertheless, the installation of DCF can be an expensive undertaking. At current prices DCF is roughly ten times the cost of SSMF.

Other techniques for increasing capacity with SSMF require a radical change in the signal format, for example dispersion supported transmission (DST) [3] duobinary transmission [4] or the use of fibre Bragg gratings. By contrast less effort has been focused on the minimum dispersion region of SSMF around 1.3 μm. No dispersion compensating techniques are required here.

In order to demonstrate that the present network may be upgraded the consortium chose to implement soliton transmission at 1.3 μm using polarisation-insensitive semiconductor optical amplifiers (SOA) as in-line optical repeaters. Soliton techniques at 1300 nm benefit considerably from the lower chromatic dispersion where, in principle, a transmission capacity, similar to soliton systems on dispersion shifted fibre (DSF) at 1.55 μm. The higher attenuation at 1.3 μm is not a hindrance to this approach. To be economically viable a repeater spacing of between 35 and 50 km is required. The project has already achieved a major milestone by demonstrating a 10 Gbit/s, SDH compatible, optically transparent link of 210 km between the German city of Kassel and CeBIT'97 exhibition centre in the city of Hannover.

2. Soliton Transmission on SSMF

The transmission of optical solitons in fibres with lumped amplification and their further application in ultra-long haul systems [5] has been well documented [6]. Link lengths of several thousand kilometres have been realised at data rates beyond 10 Gbit/s [7, 8]. These systems make sole use of dispersion shifted fibre (DSF) to achieve a low chromatic dispersion at 1.55 μm. Such systems are based on the well known concept of the guiding-centre

[9] or path-averaged [10] soliton where the soliton period must be larger than the repeater spacing in order to retain the pulse shape. Propagation is possible if the fibre loss is compensated by periodic amplification so that the dispersive wave can escape the soliton thus ensuring confinement via non-linear trapping. Furthermore, the pulse separation should be about five times as large as the pulse width to avoid non-linear pulse interaction.

In the case of SSMF at 1.55 μm soliton transmission corresponds to the quasi-resonant case for data rates above 5 Gbit/s [11]. In this case pulse distortions must be suppressed by techniques such as sliding filters [12], phase [11, 13] or amplitude [14] control. All the difficulties which appear in the quasi-resonant case for 1.55 μm soliton propagation on SSMF may be avoided by changing the optical window to the zero dispersion point of this fibre around 1.3 μm. Here the soliton period is much larger than the repeater spacing for a data rate of 10 Gbit/s. Furthermore, semiconductor optical amplifiers (SOA) with low polarisation sensitivity have been developed [15] for the 1.3 μm optical window.

It has been shown that soliton transmission up to bit rates of 20 Gbit/s are possible using a combination of SOAs and SSMF [16]. For higher data rates the centre guiding soliton approach is not valid and another approach should be considered. One such method is to use over-soliton pulses where non-linear compression during the initial stage of propagation will compensate the broadening at a later stage. A drawback is, however, the suppression of additional continuous radiation caused by the initial non-linear compression leading in turn to soliton interactions. To compensate for this effect soliton reshaping [17] is necessary whereby a minimisation of the contribution of dispersive wave components, and hence a reduction of soliton-soliton interactions, may be achieved through optimisation of the transmission system components.

3. Semiconductor Optical Amplifiers

In recent years the development of semiconductor optical amplifiers (SOA) has progressed such that they have now attained a sufficient performance [18, 19] for fibre based telecommunications in terms of optical gain, noise figure and fibre-coupled output power. The work reported here is based on the 1310 nm multiple quantum well (MQW) optical amplifier [15] which demonstrates good performance as, pre-amplifier [20], booster amplifier [20] and in-line amplifier [21].

Polarisation independent operation of a SOA is possible using either conventional bulk, almost-square, waveguides [22] or a ridge waveguide structure [23]. However, a higher gain and a significantly increased saturation output power are possible with a strained MQW structure. The amplifiers used in the following experiments are based on a MQW semi-insulating planar buried heterostructure [24]. In the booster amplifier, placed directly after a transmitter, four compressively strained quantum wells in an InGaAsP separate confinement layer are used as the gain medium. In the case of the polarisation insensitive in-line amplifier three tensile strained InGaAsP quantum wells are added where both types of quantum well have a bandgap around 1310 nm. The amplifier chips contain a 10° angled stripe with a length of 800 μm, window regions and AR coatings reduce the effective reflectivity to $\sim 3 \times 10^{-6}$. Both types of amplifier, in-line and booster, modules are housed in hermetically sealed 14-pin butterfly packages. The booster amplifier is connected to the transmitter via polarisation maintaining fibre (PMF).

Record performances have been obtained with the SOA. For the in-line amplifier a 33 dB fibre-to-fibre gain [25] has been demonstrated for a drive current of 400 mA. The resulting polarisation dependence at 1310 nm being 0.5 dB. The 3 dB bandwidth of the gain profile is ~40 nm in width. In the case of the single polarisation booster amplifier, a drive current of 500 mA is sufficient to produce a 3 dB saturation output power of 18 dBm [26]. The record noise figure of these pigtailed modules is 6 dB [27].

4. Optical Soliton Transmission in the Laboratory Environment

Experiments showed that RZ soliton transmission at 1310 nm on SSMF using periodically spaced SOAs was possible and that the technique exhibited sufficient stability for further investigation. First results showed that the transmission of 20 Gbit/s over 200 km of fibre with an amplifier spacing of 50 km was possible [16]. Analysis of a similar 250 km system at a bit rate of 6.5 Gbit/s showed that solitons were being generated in the transmission fibre [30, 31].

In Figure 1 the schematic of the experimental set-up for the 250 km transmission experiment is sketched. A 1316 nm, gain-switched, semiconductor diode laser was used to produce RZ optical pulses of 50 ps in width. After passing through an optical isolator, the pulses were compressed in a 2.75 km length of DSF with a dispersion of +19.5 ps/nm·km at 1316 nm. This compensated the chirp generated during the gain-switching process. The signal was boosted with amplifier A_1 to a level of 11.2 dBm, (peak power: 90 mW, pulse width: 20 ps). This signal was transmitted over a link with periodic amplification at 50 km intervals where the amplification (A_2–A_5) was exactly equal to the fibre loss. The average dispersion along the link was −0.8 ps/nm·km. Prior to detection the signal was again amplified (A_6) before being filtered by a bandpass filter of 2 nm full-width-half-maximum (FWHM). A low noise HEMT transimpedance front-end was used for signal

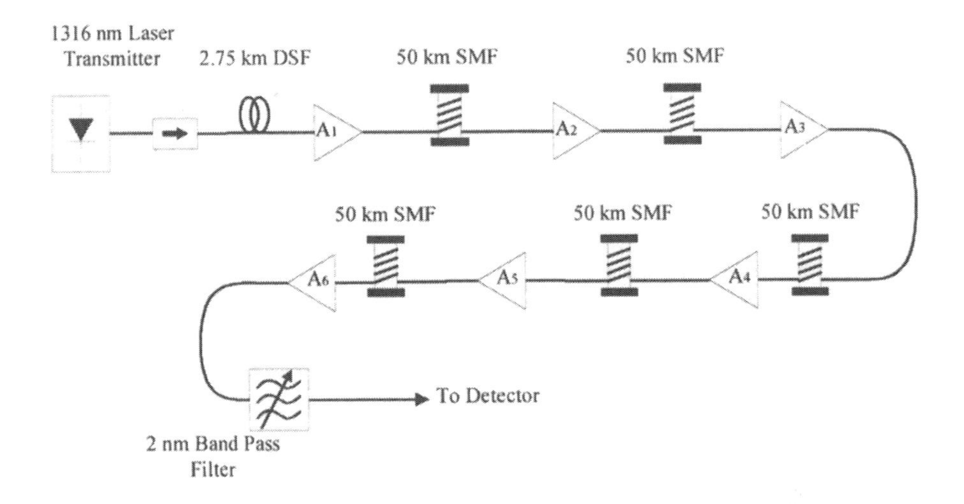

Figure 1. Schematic of 250 km transmission system with six concatenated MQW SOAs.

detection.

The temporal and spectral evolution of the pulses throughout the link was examined both experimentally and theoretically. The theoretical model was derived by solving the non-linear Schrödinger equation with the normal split-step Fourier method. As gain switching of a diode laser does not produce a Fourier-limited output, the input pulses for the numerical model were derived from the laser rate equations. Both theoretical and experimental results are shown in Figure 2(a)–(d). From these figures we concluded that we had observed the formation of optical solitons within the transmission fibre including the radiation of non-soliton components during propagation.

At this point it was assumed that the behaviour of soliton transmission over 1,000 km links could be predicted with above model including some rather optimistic values for the SOA characteristics. This was, however, not the case and the reason for this lies in the use of a novel transmitter constructed to overcome the effects of pattern dependent timing jitter associated with gain-switching if the laser is used as a data source. To compensate for this the concept of injection seeding, as illustrated in Figure 3, was developed [30].

One DFB laser, modulated with an NRZ data signal was injecting into a second DFB laser which was gain-switched at the clock frequency to generate an stream of RZ pulses. The injection process results in the transfer of the NRZ data to an RZ optical signal. Actually two RZ data signals are generated, one at 1312 nm and its compliment at 1316 nm. By plac-

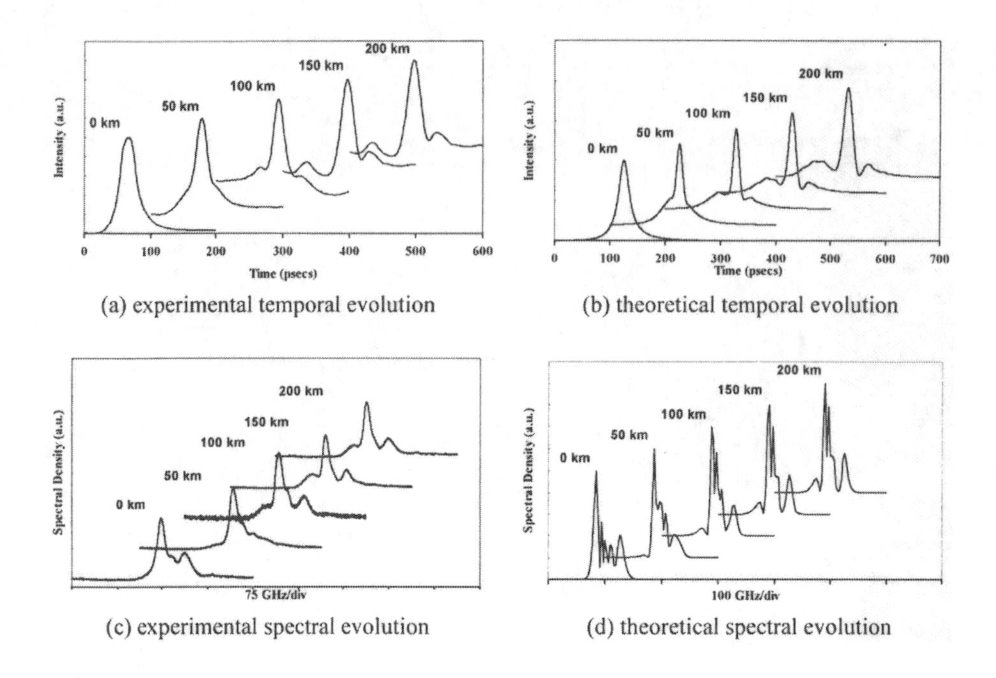

Figure 2. Experimental (a),(c) and theoretical (b),(d) evolution of the temporal and spectral shapes of the optical pulse at 50 km intervals along the transmission fibre.

ing the bandpass filter at the receiver rather than at the transmitter both data and compliment are transmitted through the line. Consequently the SOAs are subjected to a constant RZ pulse stream eliminating the effects of pattern-dependent gain-saturation. Unfortunately, this techniques while suitable for the laboratory does not lend itself to application at system level. Unwittingly, the use of this technique meant that an oversimplified theoretical treatment of the SOA saturation and response time still succeeded in producing results which closely matched the experimental observations as shown in Figure 2.

To date one of the best results obtained is the RZ transmission of 10 Gbit/s at 1309 nm over 420 km of SSMF using a chain of SOAs at 38 km intervals [31]. This surpassed previously reported efforts with SOAs operating at 1550 nm [32]. In this experiment, illustrated in Figure 4, a more suitable transmitter was utilised. An RZ optical clock signal was generated by directly driving a DFB laser module at 10 GHz to produce a pulse train of 40 ps pulses (FWHM). As indicated this source was connected to a lithium niobate modulator to produce a PRBS RZ data stream.

An optical booster amplifier was used to increase the transmitted signal to an average level of 0 to +2 dBm. The typical drive current of the SOA

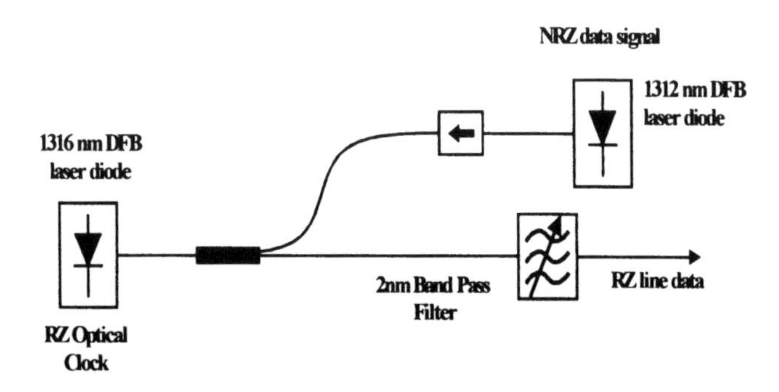

Figure 3. Injection seeding for the generation of data stream with a low timing jitter.

for compensating the fibre loss was between 120 mA and 160 mA. The SOAs used were designed for a maximum gain of >30 dB at a drive current of 400 mA. The SOA saturation power is current dependent, therefore a reduction in gain through a reduction in drive current leads to a much lower saturation output power. Correspondingly, a lower signal power must be used to avoid pattern effects in the data signal. Clearly, an amplifier with a higher saturation output power for a typical gain, as encountered in the network, is necessary. The simplest way to achieve this is through a reduction of the amplifier chip length leading to the required gain for a higher amplifier drive current.

It is important to consider the effects of amplified spontaneous emission (ASE), the primary noise source, on the system performance. In Figure 5 the accumulated penalty as a function of distance is shown for various settings of the amplifiers. The graph is compiled from a range of BER measurements, each point on the graph representing error-free transmission. The signal power at the receiver was measured within a 1 nm bandwidth. After 420 km there is a moderate penalty in the receiver sensitivity of 5 dB, showing that ultimately the transmission is limited by the accumulation of spontaneous emission within the signal bandwidth. This is the fundamental limitation of any optically amplified system.

It is unlikely that soliton transmission was demonstrated in this experiment as the signal wavelength was selected to be as close to the dispersion zero point of the fibre as possible, however this is a subject of ongoing investigation. The results of Figure 5 also show that attempts to increase the length of the line by concatenating further amplifiers would result in an unacceptable signal penalty. This being in contradiction to earlier theoretical predictions which indicated that 1,000 km should have presented no serious difficulties [29, 33]. The main difference being a much lower value for the

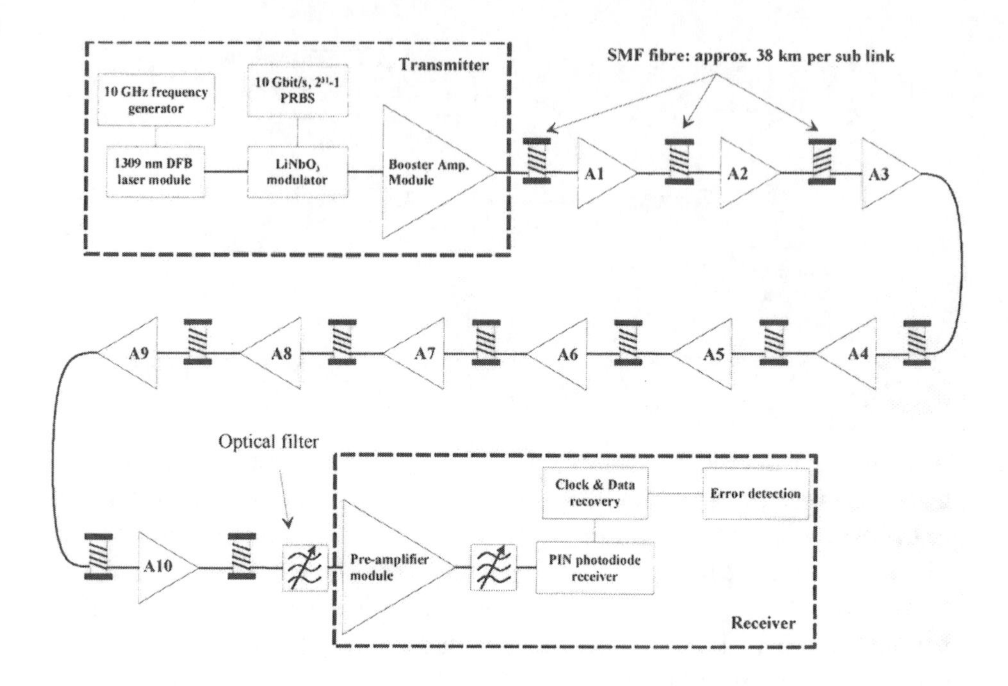

Figure 4. Set-up of the 420 km, 10 Gbit/s, RZ transmission experiment. The optical in-line amplifiers labelled A1 to A10 are all spaced at 38 km intervals.

amplifier saturation output power than the normal 12 dBm obtained with a drive current of 400 mW.

Improvement of the results through elimination of the ASE by the application of sliding filters has been a subject of theoretical investigation [34]. Normally sliding filters "slide" the soliton signal to longer wavelengths, however in the case of the SOA, amplifier saturation results in a continuous, intensity dependent, frequency blue shift of a guiding-centre soliton. Consequently direct application of sliding-filters would not appear to be beneficial to the system performance.

5. Design of an Optical Link and a Field Trial within the European Network

The main difficulty with the European network is that different countries have networked into it at different times resulting in a large variation of fibre parameters. Additionally the repeater stations are not equidistantly spaced. The result is that a high capacity optically transparent link based upon the exiting infrastructure has to be carefully optimised. Recently in

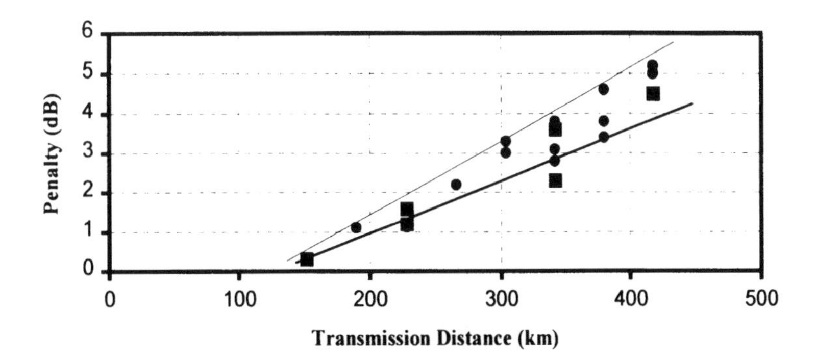

Figure 5. Penalties in receiver sensitivity as a function of transmission distance; accumulated from BER tests from 0–420 km with various amplifier settings and bandpass filter positions.

TABLE 1. Parameters for the fibre installed in the network data sample of 11,000 km.

	Average	Most	All
Amplifier Spacing	29 km	<36 km	<60 km
Attenuation			
1310 nm	0.37 dB/km		<0.45 dB/km
1550 nm	0.23 dB/km	<0.3 dB/km	<0.64 dB/km
Chromatic Dispersion	**Minimum**	**Average**	**Maximum**
Zero Disp. Wavelength	1303 nm	1311 nm	1318 nm
PMD(1321 nm & 1546 nm)	**Minimum**	**Most**	**Maximum**
New Fibres	0.04 ps/km$^{1/2}$		0.3 ps/km$^{1/2}$
Old Fibres	0.2 ps/km$^{1/2}$	<1.5 ps/km$^{1/2}$	6.4 ps/km$^{1/2}$

the German national network, members of the UPGRADE project team performed measurements on over 11,000 km of installed fibre [35]. The results of which are summarised in Table 1.

Table 1 also indicates that in a number of cases the fibre attenuation has clearly been higher in the 1.55 μm optical window compared with that of the 1.3 μm window. Polarisation mode dispersion (PMD) was found to be dependent upon the fibre lifetime.

A link composed of 210 km of the characterised fibre was installed between the German cities of Kassel and Hannover. This link was operated during the CeBIT'97 exhibition for the transmission of live video at a bitrate of 10 Gbit/s. The layout of the link is shown in Figure 6. The average fibre dispersion for the wavelength of operation, 1314 nm, was −0.28

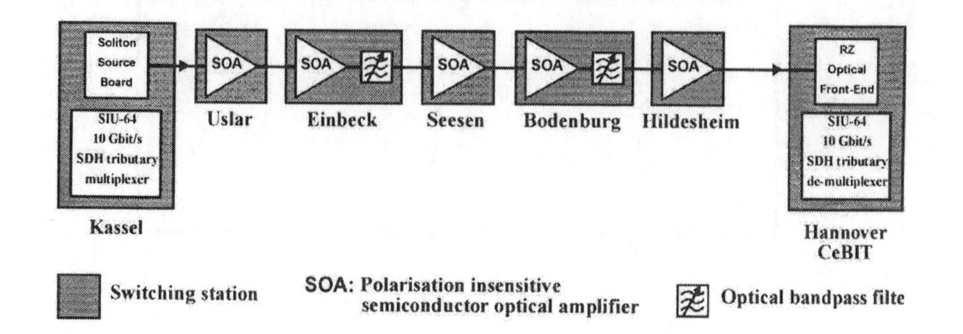

Figure 6. Layout of 210 km field trial between Kassel and Hannover. The distance between switching stations varies from 21 km to 53 km.

ps/nm·km. The attenuation in the fibre was of the order of 0.4 dB/km giving a total loss of 84 dB. The fibre PMD varied from a low of 0.96 ps (in a 22 km section) to 13.1 ps (in a 36 km section).

The optical repeater units (ORU), each designed to contain two SOAs for bi-directional operation, were specially designed to be self contained for installation at remote sites as each ORU includes control electronics and a supervision channel for remote management [36]. Based upon earlier experience [21] short length SOAs were selected for in-line amplification. In comparison to the previously used, longer, high gain devices (~30 dB at 400 mA) these SOAs produce a relatively higher saturation output power for a lower value of the gain (~20 dB at 200 mA). This gain was sufficient to compensate the losses in the link inclusive of the ~5 dB extra losses per ORU for the service channel and monitoring.

The SDH terminal multiplexers located at either end of the link were specially adapted for the transmission and detection of the RZ data format. The adapted RZ transmitter contained a DFB laser with integrated electro-absorption modulator (EAM) followed by a SOA booster to supply an output power of 6–10 dBm. RZ operation was achieved by driving the DFB with a 10 GHz clock signal while the modulator was driven with the data signal. The generated pulses were 42 ps FWHM with a minimum extinction ratio of 8.2 dB. In the receiver, conversion from the RZ to the NRZ data format was obtained by electrical post detection filtering. A minimum receiver sensitivity of −23 dBm was achieved. It was further established that the presence of PMD did not adversely effect system performance.

A comparison between what can be achieved in the laboratory [37] and what is in fact reality in the field is apparent in the evolution of the optical

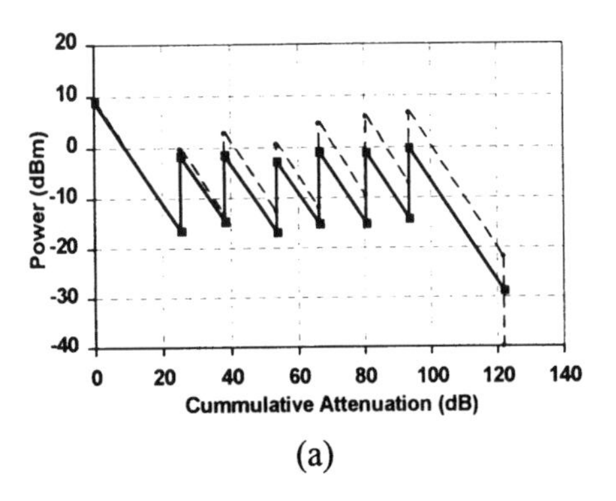

(a)

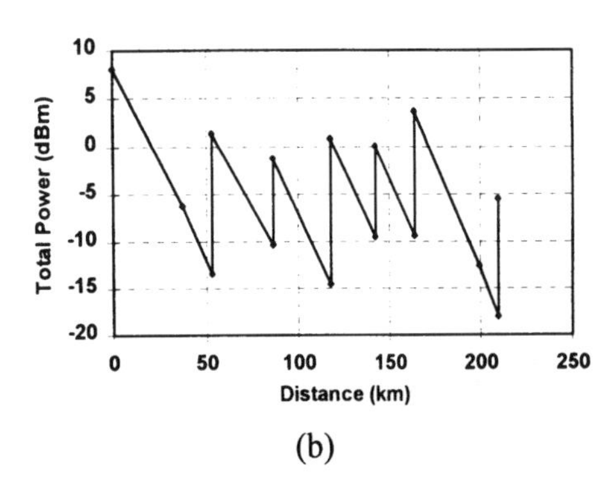

(b)

Figure 7. In (a) the power map obtained in the laboratory during initial component testing. Both the signal (solid line) and the total (dashed line) optical powers are indicated. In (b) the power map obtained in the field for the total optical power is indicated.

power throughout the link indicated in Figures 7. In Figure 7(a) the spacing between the amplifiers was a constant 38 km.

The main reason for the discrepancy in the two maps is that in the field there are many more sources of power loss. It is possible that owing to the physical layout of a switching station that several extra connectors are required within the switching station. In some cases the losses can be quite alarming. In the field error-free operation with a BER of 10^{-11} was obtained, while in the laboratory the system performance resulted in a BER of 10^{-12} with a PRBS test signal of $2^{31} - 1$. The operating range for the SOA in lies between 5 dB and -17 dB. If the optical power is too high then saturation results and if it is too low then the signal-to-noise ratio will be reduced to a level from which it cannot recover.

6. Conclusions

The successful CeBIT field trial was really only the initial preparation for a more ambitious field trial which will link the European capital cities Madrid and Lisbon for the duration of the world EXPO'98 scheduled to take place in Lisbon, a distance of 860 km. There is currently an ongoing discussion as to what role soliton transmission will play in this link, particularly in light of the results obtained to date. For this longer link an improved type of SOA has been developed which will help to increase the maximum possible transmission distance to one compatible with the original goal of covering 1000 km fibre spans. Clearly, we have developed a cost effective solution for the upgrading of the installed SSMF network without the necessity to install new cable.

Acknowledgements

The author acknowledges the valuable contributions made by his colleagues and fellow partners in the ACTS project UPGRADE. Part of this work was conducted within the framework of the European ACTS program sponsored by the European Union.

References

1. Le Guen, D., Favre, F., Moulinard, M. L., Henry, M., Michaud, G., Mace, L., Déveax, F., Charbonnier, B. and Georges T.: 200 Gbit/s 100 km-span soliton WDM transmission over 1000 km of standard fibre with dispersion compensation and pre-chirping, *OFC 97 Dallas*, paper **PD17**, (1997).
2. Kikuchi, N., Sasaki, S. and Sekine K. (1995) 10 Gbit/s dispersion-compensated transmission over 2245 km conventional fibres in a recirculating loop, *Electron. Lett.*, **31**, pp.375-377.
3. Wedding, B.: New method for optical transmission beyond dispersion limit, *Electron. Lett.*, **28**, (1992), pp.1298-1300.
4. Penninckx, D., Chbat M., Pierre L., Thiery J.-P.: The phase-shaped binary transmission (PBST): A new technique to transmit far beyond the chromatic dispersion limit, *IEEE Photon. Tech. Lett.*, **9**, (1997), pp.259-261.
5. Mollenauer, L. F., Lichtman, E., Neubelt, M. J. and Harvey, G. T.: Demonstration, using sliding frequency guiding filters, of error-free soliton transmission over more than 20 Mm at 10 Gbit/s single channel, and over more than 13 Mm at 20 Gbit/s in a two channel WDM, *Electron. Lett.*, **29**, (1993), pp.910-911.
6. Hasegawa, A. and Tappert, F. D.: Transmission of stationary nonlinear optical pulses in dispersive dielectric fibres. I: Anomalous dispersion, *Appl. Phys. Lett.*, **23**, (1973), pp.142-144.
7. Mollenauer, L. F., Gordon, J. P. and Islam, M. N.: Soliton propagation in long fibres with periodically compensated loss, *IEEE J. Quantum Electron.*, **22**, (1986), pp.157-173.
8. Mollenauer, L. F., Neubelt, M. J., Evangelides, S. G., Gordon, J. P. and Cohen, L. G.: Experimental study of soliton transmission over more than 10,000 km in dispersion shifted fibre, *Opt. Lett.*, **15**, (1990), pp.1203-1205.
9. Hasegawa, A. and Kodama, Y.: Guiding centre soliton in optical fibres, *Opt. Lett.*, **15**, (1990), pp.1443-1445.
10. Mollenauer, L. F., Evangelides, S. G. and Haus, H. A.: Long-distance soliton propagation using lumped amplifiers and dispersion shifted fibre, *J. Lightwave Tech.*, **9**, (1991), pp.194-197.
11. Mattheus, A. and Turitsyn, S. K.: Pulse interaction in nonlinear communication systems based on standard monomode fibres, *ECOC 93 Montreux*, paper **MoC2.3**, (1993).
12. Kodama, Y. and Wabnitz, S.: Reduction and suppression of soliton interactions by bandpass filters, *Opt. Lett.*, **18**, (1993), pp.1311-1313.
13. Haelterman, M. and Sheppard, A. P.: Cancelling soliton interaction in single mode optical fibres, *Electron. Lett.*, **29**, (1993), pp.1176-1177.
14. Nakazawa, M., Suzuki, K, Yamada, E., Kubota, H., Kimura, Y. and Takaya, M.: Experimental demonstration of soliton data transmission over unlimited distances with soliton control in time and frequency domains, *Electron. Lett.*, **29**, (1993), pp.729-730.
15. Tiemeijer, L. F., Thijs, P. J. A., van Dongen, T., Binsma, J. J. M. and Jansen, E. J.: Polarisation resolved, complete characterisation of 1310 nm fibre pigtailed multiple-quantum-well optical amplifiers, *J. Lightwave Tech.*, **14**, (1996), pp.1524-1533.

16. Reid, J. J. E., Liedenbaum, C. T. H. F., Tiemeijer, L. F. and Kuindersma, P. I.: Realisation of 20 Gbit/s long haul soliton transmission at 1300 nm on standard single mode optical fibre, *Proc. 20th Eur. Conf. on Opt. Comm ECOC 94 Firenze*, **4**, (1994), pp.61-64.
17. Matsumoto, M., Hasegawa, A. and Kodama, Y.: Adiabatic amplification of solitons by means of nonlinear amplifying loop mirrors, *Opt. Lett.*, **19**, (1994), pp.1019-1021.
18. van den Hoven, G. and Tiemeijer, L. F.: High performance semiconductor optical amplifiers, *Proc. Optical Amplifiers and their Applications*, Victoria, (1997), pp.108-111.
19. Doussiere, P.: Recent advances in conventional and gain clamped semiconductor optical amplifiers, *Proc. Optical Amplifiers and their Applications, Monterey*, **11**, (1996), pp.220-223.
20. Tiemeijer, L. F., Kuindersma, P. I., Cuijpers, G. P. J. M., Thijs, P. J. A., van Dongen, T., Binsma, J. J. M., Jansen, E. J., Walczyk, S., Reid, J. J. E. and Verboven, A. J. M.: 102 km 38.6 dB budget 10 Gbit/s NRZ repeaterless transmission at 1310 nm using semiconductor booster amplifier module and a semiconductor and a semiconductor pre-amplifier module, *Proc. 21st Eur. Conf. on Opt. Comm. ECOC 95 Brussels*, (1995), pp.275-278.
21. Kuindersma, P. I., Cuijpers, G. P. J. M., Jennen, J. G. L., Reid, J. J. E., Tiemeijer, L. F., de Waardt, H. and Boot, A. J.: 10 Gbit/s RZ transmission at 1309 nm over 420 km using a chain of multiple quantum well semiconductor optical amplifier modules at 38 km intervals, *Proc. 22nd Eur. Conf. on Opt. Comm. ECOC 96 Oslo*, paper **TuD2.1**, (1996).
22. Doussiere, P., Garabedian, P., Graver, C., Bonnevie, D., Fillion, T., Derouin, E., Monnot, M., Provost, J. G., Leclerc, D. and Klenk, M.: 1.55 μm polarisation independent semiconductor optical amplifier with 25 dB fibre to fibre gain, *IEEE Photon. Tech. Lett.*, **6**, (1994), pp.170-172.
23. Holtman, Ch., Besse, P. A. and Melchior, H.: Polarisation resolved complete noise characterisation of bulk ridge-waveguide semiconductor optical amplifiers, *Proc. Optical Amplifiers and their Applications Davos*, **18**, (1993), pp.115-118.
24. Thijs, P. J. A., Tiemeijer, L. F., Binsma, J. J. M. and van Dongen, T.: Progress in long-wavelength strained-layer InGaAs(P) quantum well semiconductor lasers and amplifiers, *J. Quantum Electron.*, **30**, (1994), pp.477-499.
25. Tiemeijer, L. F., Thijs, P. J. A., van Dongen, T., Binsma, J. J. M. and Walcyzk, S.: 33 dB fibre to fibre gain −13 dB fibre saturation power polarisation independent 1310 nm MQW laser amplifiers, *Proc. Optical Amplifiers and their Applications, Davos*, **18**, (1995), paper **PD1**.
26. Tiemeijer, L. F., Thijs, P. J. A., van Dongen, T., Binsma, J. J. M, Jansen, E. J., Kuindersma, P. I., Cuijpers, G. P. J. M. and Walczyk, S.: (1995) High-output-power (+15 dBm) unidirectional 1310-nm multiple-quantum-well booster amplifier module, *IEEE Photon. Tech. Lett.*, **7**, (1995), pp.1519-1521.
27. Tiemeijer, L. F., Thijs, P. J. A., van Dongen, T., Binsma, J. J. M. and Jansen, E. J.: Noise figure of saturated 1310-nm polarisation insensitive multiple-quantum-well optical amplifiers, *IEEE Photon. Tech. Lett.*, **8**, (1996), pp.873-875.
28. Liedenbaum, C. T. H. F., Reid, J. J. E., Tiemeijer, L. F., Boot, A. J., Kuindersma, P. I., Gabitov, I. R. and Mattheus, A.: Experimental long haul 1300 nm soliton transmission on standard single mode fibres using quantum well laser amplifiers, *Proc. 20th Eur. Conf. on Opt. Comm. ECOC 94*, Firenze, (1994), pp.233-236.
29. Mattheus, A., Gabitov, I. R., Boot, A. J., Liedenbaum, C. T. H. F., Reid, J. J. E. and Tiemeijer, L. F.: Analysis of periodically amplified soliton propagation on long-haul standard-monomode fibre systems at 1300 nm wavelength, *Proc. 20th Eur. Conf. on Opt. Comm ECOC 94 Firenze*, (1994), pp.491-494.
30. Liedenbaum, C. T. H. F. and Reid, J. J. E.: Unit for generating signal pulses, transmitter comprising such a unit, and multiplex transmission system comprising

such a transmitter, *US Patent 5,568,303*, (1996).
31. Kuindersma, P. I., Cuijpers, C. P. J. M., Jennen, J. G. L., Reid, J. J. E., Tiemeijer, L. F., de Waardt, H. and Boot, A. J.: 10 Gbit/s RZ transmission at 1309 nm over 420 km using a chain of multiple quantum well semiconductor optical amplifier modules at 38 km intervals, *Proc. 22nd Eur. Conf. on Opt. Comm. ECOC 96 Oslo*, paper **Tu.D.2.1**, (1996).
32. Yamamoto, S., Taga, H., Edagawa, N., Mochizuki, K. and Wakabayashi, H.: 516 km, 2.4 Gbit/s optical fibre transmission experiment using 10 semiconductor laser amplifiers and measurement of jitter accumulation, *7th Int. Conf. on Opt. And Fibre Comm. IOOC 89 Kobe*, paper **PDA-9**, (1989).
33. Settembre, M,. Matera, F., Hägele, V., Gabitov, I. R., Mattheus, A. and Turitsyn, S. K.: Cascaded optical communication systems with in-line semiconductor optical amplifiers, *J. Lightwave Tech.*, **15**, (1997), pp.134-139.
34. Wabnitz, S.: Soliton stabilisation in long distance fibre transmissions with semiconductor amplifiers, *Opt. Lett.*, **20**, (1995), pp.1979-1981.
35. Gruhl, H., Herchenr • er, G., Mattheus, A. and Vobian, J.: Characterisation of 11000 km of installed standard monomode fibre and statistical analysis in view of network design, *Proc. Eur. Conf. on Networks and Opt. Comm. NOC 97 Antwerp*, **1**, (1997), pp.59-64.
36. Reid, J. J. E.: (1997) High bitrate 1310 nm optical transmission in the field using cascaded semiconductor optical amplifiers, *Proc. Optical Amplifiers and their Applications*, Victoria, (1997), pp.82-85.
37. Kuindersma, P. I., Cuijpers, G. P. J. M., Reid, J. J. E., van den Hoven, G. N. and Walcyzk, S.: An experimental analysis of the system performance of cascades of 1.3 μm semiconductor optical amplifiers, *Proc. 23rd Eur. Conf. on Opt. Comm. ECOC 97 Edinburgh*, (1997), 1.79-1.82.

(Received November 17, 1997)

EUROPEAN UNION ACTS PROJECT MIDAS: OBJECTIVES AND PROGRESS TO DATE

D. J. RICHARDSON
MIDAS Consortium
Optoelectronics Research Center, Southampton University
Southampton, UK.

1. Introduction to the ACTS Program

Advanced Communications, Technology and Services, known simply as ACTS, is one of the specific programmes of the "Fourth Framework Programme of European Community activities in the field of research and technological development and demonstration (1994–1998)" [1]. It provides the main focus of the European Unions research effort to accelerate deployment of advanced communications infrastructures and services, and is complemented by extensive European research in the areas of information technology and telematics. The stated objectives of ACTS are to "develop advanced communication systems and services for economic development and social cohesion within Europe, taking account of the rapid evolution of technologies, the changing regulatory situation and opportunities for development of advanced transeuropean networks and services". Within ACTS, the emphasis of the work has shifted from the exploration of fundamental concepts and detailed system engineering, as it had been in earlier programs such as RACE (Research and development in Advanced Communication technologies for Europe), to issues relating to implementation of advanced systems and generic services, and applications which demonstrate the potential use of advanced communications in Europe. A key feature of the ACTS program is that the research be undertaken in the context of real-world trials. Work within the program is divided into six technical areas: Interactive digital multimedia services, photonic technologies, high speed networking, mobility and personal communication networks, intelligence in networks and services and quality, safety and security of communication systems and services. The total EU budget for the ACTS program is approximately 670 MECU, covering around 160 projects, with over 1,000 indi-

A. Hasegawa (ed.), New Trends in Optical Soliton Transmission Systems, 459–471.

vidual organisations participating within the program, thereby illustrating the scale of the activities.

MIDAS is one of five projects in the technical area of photonic technologies concerned with high speed transmission, the others being ESTHER, UPGRADE, HIGHWAY and SPEED, each concerned with various aspects or approaches to the development of 40 Gbit/s transmission systems within the European arena. A full list of project descriptions and objectives, as well as those of the ACTS program as a whole, are to be found in reference [1]. The MIDAS consortium consists of the following organisations: Chalmers University of Technology (Sweden), CSELT (Italy), Thomson LCR (France), United Monolithic Semiconductor (France), Telia (Sweden), Kings College London (UK), University of Athens (Greece) and ORC University of Southampton (UK). The project started in September 1995 and is currently scheduled to finish in September 1998.

2. MIDAS Project Objectives

MIDAS is primarily concerned with the field demonstration of transmission techniques suitable for the upgrade of existing fibre lines to 40 Gbit/s over length scales relevant to European requirements ($\sim$ 1,000 km). The ultimate project objectives are the realisation of two field trials over installed fibre lines, one a full 40 Gbit/s soliton field trial over dispersion shifted lines, and the other a 40 Gbit/s linear transmission experiment over high dispersion standard fibres using Mid Span Spectral Inversion (MSSI) to combat dispersion. Both field trials are fully supported by appropriate component development and system studies. In addition, the project has a small research element looking at advanced concepts and techniques which go beyond the demands of the current field trials and are targeted, mostly with a view to soliton transmission, at capacities between 40–100 Gbit/s.

3. Technical Scope and Status of the MIDAS Project

3.1. SOLITON FIELD TRIAL

The major field trial within the project is concerned with the upgrade of installed dispersion shifted fibres within the existing European fibre base to 40 Gbit/s through the use of soliton techniques. The soliton field trial is being undertaken principally by Chalmers University, and is to be performed over installed fibre lines belonging to the Swedish operator Telia. The trials are to use a fibre based approach within the transmitter and demultiplexer design. Considerable component development has been required to bring laboratory grade devices up to a standard suitable for field trial validation.

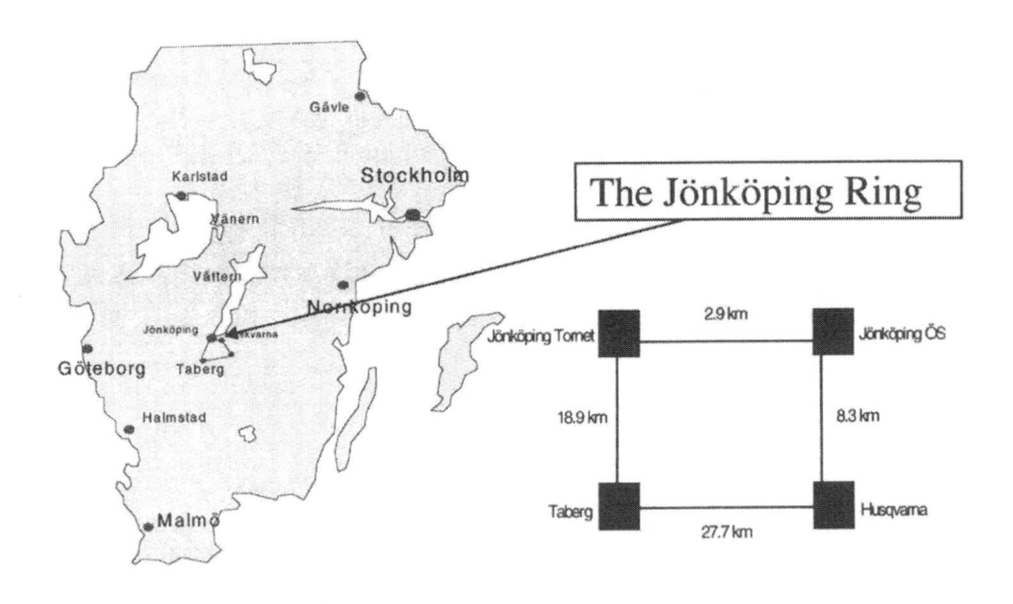

Figure 1. The geographic location of the installed Swedish dispersion-shifted fibre lines with inset of ring configuration.

The installed dispersion-shifted fibres to be used in the trial have been identified and are located near the town of Jönköping, Sweden (see Figure 1). The fibres, which were acquired by Telia from Sumitomo, Japan, consist of ribbons with four fibres in each ribbon, and were installed in 1991. The total length of installed fibre is 2,774 km.

The fibre is installed in a 57.8 km long ring structure (Figure 1) with four intermediate but (not equidistant) accessible hubs. Twenty four sections between each of the hubs, totalling in excess of 1,200 km, have been characterised with respect to attenuation, polarisation-mode dispersion (PAD), and zero-dispersion wavelength using standard techniques and commercial instruments. The average values with maximum variation from these were: loss = 0.27 dB/km +85 %/−26 %, PMD = 0.40 ps·km$^{-1/2}$+125 %/−92 %, and zero-dispersion wavelength = 1548 nm +/− 1 % respectively.

Critical to the success of the field trial is the simultaneous accommodation of the relatively large variations in zero dispersion wavelength and large values of PMD found within these fibres. The most critical of these parameters for 40 Gbit/s transmission is likely to be the PMD. Using numerical simulations to optimise our system configuration, we estimate that by appropriate fibre selection transmission distances in excess of 400 km

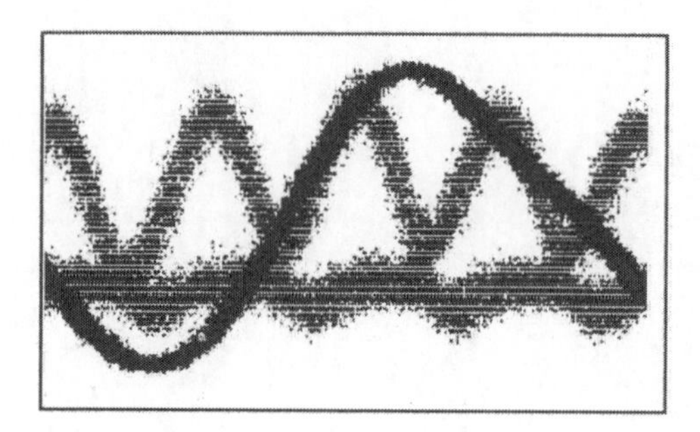

Figure 2. The soliton transmitter 40 Gbit/s eye pattern (measured with a 32 GHz detector) and 10 GHz clock extracted from the data.

are possible using orthogonally polarised solitons and a 57 km amplifier spacing. We have also briefly investigated other transmission line options and formats e.g. narrowband optical filtering, and dispersion management.

As previously mentioned the project is focused on the use of fibre based componentry. On the source side an environmentally stable, 10 GHz, polarisation maintaining, actively-modelocked erbium doped ring laser has been developed with regenerative feedback to accommodate cavity length drift due to thermal effects thereby maintaining optimal mode-locking conditions [2]. The cavity incorporates sections of dispersion shifted, polarisation maintaining fibre for the generation of soliton pulse shaping effects, thereby allowing the generation of pulses as short as 3.5 ps. An intracavity filter is also included within the cavity which, in conjunction with the nonlinear effects within the cavity, leads to passive pulse amplitude stabilisation and the suppression of supermode noise [2]. Suppression of all competing modes to < -80 dB has been achieved. Data is encoded onto the 10 GHz pulse stream using a $LiNbO_3$ modulator and the data sequence multiplexed up to 40 Gbit/s using a conventional passive fibre based multiplexer with appropriate delays to generate a true 40 Gbit/s psuedo random bit sequence. The multiplexer also facilitates the generation of a sub-multiple clock at 10 GHz.

The laser provides chirp-free sech^2 pulses with negligible timing jitter (< 1 ps) and exhibits excellent long term stability. Figure 2 shows an eye pattern at 40 Gbit/s. A slight intentional time-alteration between consecutive pulses is present and serves to generate a clock signal at 10 GHz. The extracted clock, after narrowband filtering, is also shown in Figure 3. The estimated timing jitter of the extracted clock is 0.9 ps.

Demultiplexing of the 40 Gbit/s at the receiver data will be achieved using electronic clock recovery and a nonlinear optical loop mirror based scheme, appropriately modified to achieve polarisation insensitive operation. A polarisation dependence of $< +/- 0.5$ dB over a wavelength range of 3.7 nm has been achieved. Error-free operation at 10 Gbit/s has been confirmed with no measurable penalty [3].

The project also has an element of work associated with the development of a direct 40 Gbit/s opto-electronic receiver preceded by an appropriate optical pre-amplifier. The design and initial characterisation of a low noise 980 nm pumped preamplifier of appropriate characteristics has been undertaken and the design completed for the electronics. The design comprises a side illuminated, multimode structure waveguide PIN photodetector with coplanar electrical access and integrated MMIC preamplifier. The expected performances for the photodetector are an overall sensitivity (including coupling) of at least 0.5 A/W with a 3 dB electrical bandwidth greater than 40 GHz. The amplifier gain is to be ~27 dB with 100 KHz to 35 GHz bandwidth.

A full 40 Gbit/s test bed incorporating a recirculating loop structure has been constructed at Chalmers University to allow a full simulation of the soliton field trial. The test bed has been designed to readily permit the consortium to perform system measurements on the advanced transmission concepts being developed within the project, thereby enhancing the potential of these aspects of the work.

3.2. 40 GBIT/S STANDARD FIBRE FIELD TRIAL

Most of the installed fibre lines within Europe contain standard step index fibre designs with high dispersion (~17 ps/nm·km at 1550 nm). The project incorporates a field trial concerned with upgrade of these lines to 40 Gbit/s. In order to get around the effects of dispersion we are employing the technique of Mid-Span Spectral Inversion (MSSI) in which the net effects of second order dispersion are cancelled by optical phase conjugation of the signal at a point close to the middle of the link [4].

The field trial is to take place in Spring 1998 in Italy, using installed lines in the vicinity of Turin. Southampton University and CSELT are the main contributors to this particular trial. Four standard telecommunication fibres have been obtained for the field trial from the ITINERA consortium, which manages this Italhost facility. The four fibres connect CSELT to the small towns of Ciriè and Casselle and are now available directly in the MIDAS laboratory. Two 70 km links are available, and the four ports will be connected to the transmitter (Tx), to the midpoint spectral inverter (MPSI), and to the receiver (Rx). The exact line layout is shown in Figure 3. The lines have been fully characterised in terms of loss, dispersion, PMD

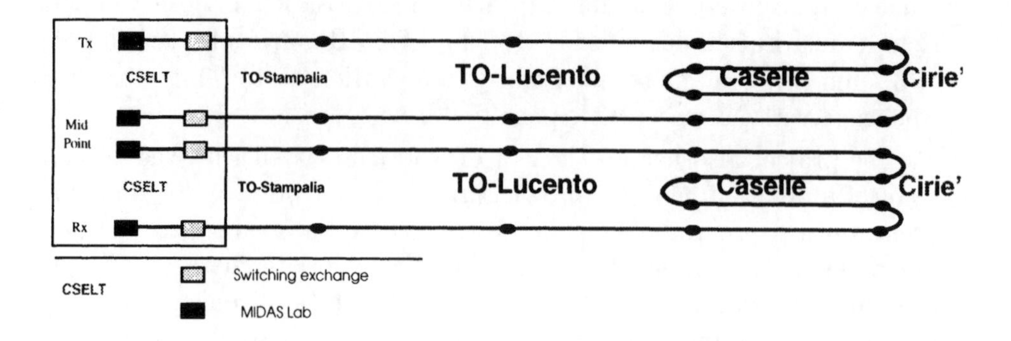

Figure 3. Layout of the CSELT-Ciriè-CSELT links (two links, each 70 km long, total transmission length 140 km).

and polarisation stability. The fibre selection for the field trial has already been made. The average fibre loss, which includes the loss of a considerable number of connectors, is 0.36 dB/km. The average PMD is 0.06 ps/km$^{1/2}$ and the average dispersion is 16.5 ps/nm·km at 1550 nm. Note that the PMD of the standard fibre lines is considerably less than that measured in the installed dispersion shifted lines to be used in the soliton trial. We will also conduct measurements of the spatial variation of the PMD along the trial lines using Polarimetric Optical Time Domain Reflectometry [5] within the project. Our initial experiments will focus on transmission over a distance of 140 km, although if time permits multiple circulations through the system will be attempted to extend the range.

As with the soliton field trial we follow a fibre based component approach to the system design. The transmitter is once again to be a 10 GHz, regeneratively mode locked fibre laser (in this instance a sigma-laser geometry [6]), modulated at 10 Gbit/s and passively multiplexed to 40 Gbit/s. A fibre demultiplexing scheme similar to that to be implemented in the soliton trial will be used at the receiver.

We have investigated both fibre and semiconductor options for the phase conjugation function comparing noise performance, signal distortion with signal power, and the application of fibre grating technology for optical noise reduction [7]. Due to the superior power performance, and the advantages in terms of noise reduction we have obtained from the design and fabrication of fibre Bragg grating based narrow-band optical filters, we will be using a fiber based conjugator within the field trial. Using such a conjugator we have now demonstrated dispersion cancellation over 1,000 second order dispersion lengths (100 km using 2.5 ps input pulses), the limit to pulse restoration quality being observed to be third order dispersion [8].

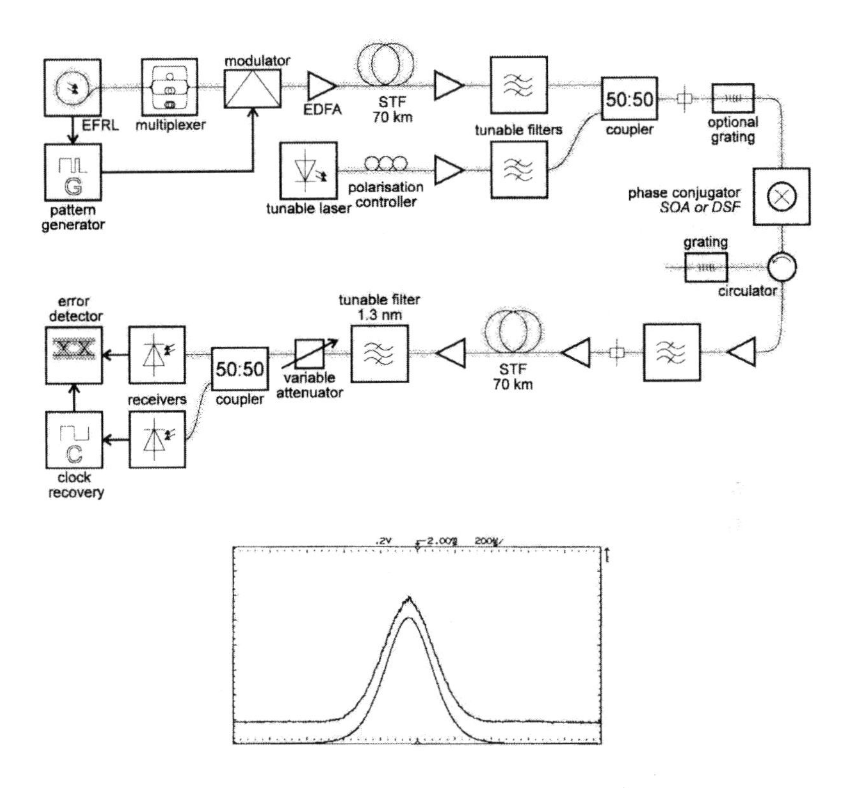

Figure 4. (a) Setup for the 10 Gbit/s MSSI field trial over 140 km of standard fibre. (b) Input (lower trace) and output (upper trace) autocorrelation function of 6 ps pulses before and after MSSI dispersion compensated transmission through 140 km of standard fibre.

To date most work within this activity has centred on component development and testing, line characterisation and both experimental and numerical simulations in support of the field trial. However, we have already performed a prefield trial over the selected lines at 10 Gbit/s using short ~6 ps pulses so as to experimentally confirm the viability of the full 40 Gbit/s trial. The experimental setup is shown in Figure 4(a) and is similar to that planned for the full trial other than that the base rate in this instance was only 2.5 Gbit/s and that there was no need for demultiplexing at the 10 Gbit/s receiver.

The experiments showed that excellent dispersion compensation could be achieved for 6 ps pulses (as will be used in the actual 40 Gbit/s trial) over the full 140 km transmission span as illustrated in Figure 4(b). This shows the autocorrelation function of the pulses both before and after transmission over the fibre line. Minimal pulse distortion is observed. Moreover, error free operation at a BER $< 10^{-11}$ over 140 km has been obtained with < 4 dB power penalty using a relatively unoptimised system in terms of ASE noise suppression and conjugator efficiency, giving us every confidence in the feasibility of the full 40 Gbit/s trial.

3.3. ADVANCED CONCEPTS FOR FUTURE SOLITON SYSTEMS

In addition to the field trial activities, the project also contains advanced concept activities exploring the key technologies of soliton control, distributed loss compensation and dispersion management for ≥ 40 Gbit/s soliton transmission. This element of the project has the potential to enhance the scope of the current 40 Gbit/s field trial through the incorporation of soliton control and specification of optimised dispersion maps, but is also critical for extending soliton techniques beyond the scope of the current field trial to far higher single/multi (WDM) channel data rates. The work contains elements of numerical simulation, novel component development and pulse transmission experiments. It is anticipated that further system measurements in this area will be made on the soliton field trial test bed in the near future.

3.3.1. *Soliton loss compensation*

Soliton communication systems based on lumped Erbium Doped Fibre Amplifiers (EDFAs) and uniform, dispersion shifted fibre offer great potential for high capacity soliton transmission. However, there are inherent limitations within such system implementations for both OTDM and WDM transmission. OTDM transmission is limited by the constraints of average soliton dynamics which require that the soliton period of the pulses (z_0) should be much greater than the amplifier spacing (L_a) [9]. At high data rates (~40 Gbit/s) the perturbation to the pulses due to the fibre loss becomes too severe resulting in a violation of 'average', or 'guiding centre', soliton dynamics and eventual decay of the pulses [10]. In WDM systems soliton pulses of different wavelengths suffer frequency shifts as they collide, which ultimately result in timing jitter and resultant loss of information unless the length scale over which they collide is kept greater than twice the amplifier spacing [11]. Moreover, and more significant from a practical perspective, is the fact that the collision process results in the uncontrollable growth of four wave mixing products which once again leads to the generation of timing jitter [12]. The only way to eliminate these deleterious effects is to eliminate the loss/gain cycle induced perturbations to the local balance between dispersion and nonlinearity by ensuring fundamental soliton propagation throughout the system. This can be achieved either by using a distributed amplifier to cancel the local fibre loss [13], or by the use of a Loss Compensating Dispersion Decreasing Fibre (LCDDF) [14]. In the later case the dispersion of the transmission fibre decays exponentially along its length so as to exactly follow the decrease in optical intensity and hence the strength of the nonlinear interaction. An exact balance between the two effects at all points along the fibre is thus ensured and as far as the soliton is concerned the fibre becomes lossless. Within MIDAS we have examined both the options of LCDDFs and distributed amplification through Raman scattering.

3.3.1.1. LCDDF fabrication, characterisation and transmission We have developed fabrication technology for long (40 km) LCDDF using control of the waveguide dispersion (and thereby total dispersion at a fixed wavelength) along the fibre length by control of the fibre's external diameter (core diameter), a technique first demonstrated over much shorter fibre lengths by workers at General Physics Institute, Moscow [15]. Critical to the fabrication of such long fibres is preform type and quality. We have examined the fabrication of such fibres from both step index and dispersion shifted preform types for both high and low average dispersions values and dispersion ranges [16]-[20]. Moreover, we have performed systematic distributed dispersion measurements, made using a backscattered four wave mixing technique [21], of the dispersion variation in such fibres to obtain estimates of the quality of the match to the fibre loss profiles and of the achieved dispersion accuracy [22]. In Figure 5 we show a plot of the measured dispersion variation in a 38 km LCDDF with dispersion designed to vary from 6 ps/nm·km at the input to 0.55 ps/nm·km and with an average loss of 0.265 dB/km. It can be seen that an excellent match to the fibre loss is obtained over the full fibre length.

The average deviation between the measured and the desired profile is less than 8 % for the whole length. This indicates that an excellent degree of control over dispersion is possible over significant fibre lengths. Of particular note is the control achieved at the low dispersion end of the fibre. In this regime an average error of $<$ 0.05 ps/nm·km is obtained over the 18 km of fibre comparable with the measurement uncertainty and is of the same magnitude as deviations from nominal uniform dispersion observed over 10 km length scales in commercial dispersion shifted fibre. Although this degree of dispersion error is small, it can still be significant for fibres with dispersion tapering close to the zero-dispersion wavelength as are likely to be required for most real world applications. Note that dispersion control of this form can be applied to generate more exotic dispersion profiles, as required for more complex dispersion managed systems e.g. quasi-soliton systems with a similar degree of absolute dispersion control.

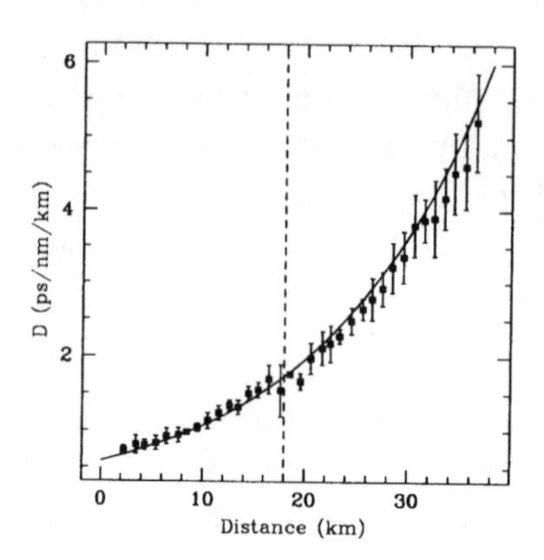

Figure 5. Experimentally determined (data points) and design dispersion profile (solid line) of a 38 km LCDDF fabricated by the diameter control technique.

The LCDDF fibres fabricated within the project have been used in single pass and recirculating loop experiments to provide demonstrations of high quality soliton loss compensation. In single pass experiments we have shown loss compensation over transmission distances in excess of 40 soliton periods (38 km) [16]. In loop experiments (in collaboration with A. D. Ellis, BT Laboratories) we have shown stable pulse transmission over total propagation distances in excess of 80 soliton periods and normalised amplifier spacings in excess of 5 soliton periods using 10 Gbit/s data on pulses of ~5 ps [18]. Furthermore we have experimentally demonstrated error free transmission of 10 Gbit/s, 4.8 ps pulses over 4,500 km at 10 Gbit/s in a 20 km, low dispersion LCDDF fibre with average dispersion 0.17 ps/nm·km [19]. The pulses could be transmitted error free over 1,000km at 40 Gbit/s. The limiting effects of Raman scattering, acousto-optic interactions and higher order dispersion to single channel, high bit-rate propagation in such fibres are currently under investigation.

3.3.1.2. Raman amplified soliton systems Raman amplifiers for soliton systems offer the attraction that the transmission fibre is its own amplifier, providing distributed amplification and elimination of the resonance affects associated with conventional lumped amplification. The first experiments with soliton transmission used Raman amplifiers and produced encouraging results [9], but the lack of inexpensive and reliable pump sources and the appearance of the more efficient erbium-doped fibre amplifiers put the Raman amplifier into the shade. However the development of highly efficient fibre lasers with output powers of up to 1 W could have a significant impact on the viability of Raman amplified systems and we have experimentally investigated this option.

We have experimentally demonstrated low distortion, subpicosecond pulse propagation over 33 km of dispersion shifted fibre using Raman gain for fibre loss compensation where the availability of efficient Er/Yb codoped fibres and fibre gratings makes it possible to design a simple and robust pump source operating at 1535 nm [23]. The fibre dispersion of 1.2 ps/nm·km at the signal wavelength of 1560 nm made the soliton propagation distance as long as 150 dispersion lengths – the longest distance for single span soliton transmission reported so far. A bidirectional scheme allowed us to obtain a more uniform distribution of the pump and to reduce the amount of shed non-soliton component, giving rise to the possibility of transmitting a 100 Gbit/s stream of solitons over 1,000's of km [24]. The Raman amplifier can therefore provide a real challenge to the EDFA in soliton transmission systems.

3.3.2. *Soliton Transmission and control using nonlinear gain*

Although dispersion management and the like can be used to ensure the stability of ultrashort soliton pulses propagating in an amplified transmission line, other interactions of the soliton can restrict the usefulness of such systems. During propagation along an optical fibre a soliton interacts with optical and acoustic fields generated by other solitons and optical amplifiers. The result of such interactions is a change of the soliton central frequency which translates into a change in group velocity and causes an uncertainty in the soliton arrival time that in turn results in error if the pulse fails to arrive in its assigned time slot.

Several techniques have been suggested to suppress or at least reduce the instability of a soliton data stream [25]-[27] by making the soliton transmission line more sophisticated i.e. consisting of not only transmission fibres and EDFAs but also spectral filters, modulators and (or) saturable absorbers.

Within MIDAS we are examining the development of nonlinear saturable absorbers (NSA) for use as soliton control elements in 40 Gbit/s systems. By using a NSA and appropriate spectral filtering one can provide additional loss for linear radiation e.g. ASE noise relative to the soliton. If the excess l inear gain at the fibre amplifier is high enough to compensate the soliton loss but not sufficient for the compensation of linear radiation excess loss, then in such a system one can expect stable soliton propagation with suppressed low-level non-soliton components and thereby reduced timing jitter [28]. To date we have made experimental proof of principle measurements using a fibre based (Nonlinear Amplifying Loop

Mirror) nonlinear saturable absorber [29] and are currently investigating the use of more practical options. In particular we are investigating both polymer and semiconductor MQW materials. Recovery time and saturation fluence are the key parameters for system performance. The current known limit for the operation speed of the polymer saturable absorbers for soliton control under development within the consortium is above 50 Gbit/s, and by suitable chemical modification of the molecular structure this may be extended to beyond 100 Gbit/s with a molecular charge transfer concept.

4. Conclusion

The MIDAS project is progressing well to its final objectives of field demonstrations of 40 Gbit/s data transmission techniques, both soliton and linear, and is making significant contributions to basic research studies in the area of advanced soliton systems. The undertaking of such trials illustrates the EU's strong commitment to the deployment of advanced communications infrastructures within Europe, it remains however to be seen if either of the approaches investigated within the project will be commercially developed.

Acknowledgements

I would like to acknowledge the contributions of my colleague within the MIDAS consortium to the development of the project: R. I. Laming, N. G. Broderick, P. Petropolous, A. B. Grudinin, H. Geiger, S. Y. Set, N. J. Traynor, S. Gray (Optoelectronics Research Centre, Southampton University), P. A. Andrekson, B. Bakhshi, J. S. Karlsson, P. O. Hedekvist, B. E. Ollson (Chalmers University), F. Cisterninio, P. Morra, M. Artiglia, A. Pagano (CSELT), W. Blau, A. Boyle, W. A. Pender (Trinity College, Dublin), T. Sphicoplous, D. Syvridis, D. Metin, I. Zaharopolous, I. Tomkos, C. Eleftherianos, Z. Tzelepis (University of Athens), I. P. Giles, V. Handerek, Y. R. Zhou (Kings College London), F. Deborgies, J. C. Renaud (Thomson LCR), G. Delaval (United Monolithic Semiconductors), J. Karlsson (Telia Network Services). The work is funded in Part by the European Union Advanced Communication Technologies (ACTS) program.

References

1. ACTS 96 Programme Guide, *EU document* Ref. No. AC962676-PG, (1996).
2. M. Nakazawa, E. Yoshida and Y. Kimura, *Electron. Lett.*, **30**, (1994), p.1603.
3. B. E. Olsson and P. A. Andrekson: *IEEE Phot. Tech. Lett.*, **9**, (1997), pp.764-766.
4. A. Yariv, D. Fekete and D. M. Pepper: *Opt. Letts.*, **4**, (1979), pp.52-54.
5. A. M. Vengsarkar and L. G. Cohen: *Electron. Lett.*, **29**, (1993), pp.848-850.
6. F. Cisternino, R. Girardi, A. Pagano, E. Riccardi and E. Castagna: *Proc. ECOC, Vol. 3*, Oslo, paper **WEP.19**, (1996), pp.3.285-3.288.

7. H. Geiger, S.Y. Set, R. I. Laming, M. J. Cole and L. Reekie: *Proc. Conference on Optical Fibre Communications*, paper **WH7**, (1997), pp.150-151.
8. A. Royset and R. I. Laming: *Proc. OFC 96*, San Jose, paper **THM3**, (1996).
9. L. F. Mollenauer, J. P. Gordon and M. N. Islam: *IEE J. Quant. El.*, **QE 22**, (1986), p.157.
10. A. Hasegawa and Y. Kodama, *Opt. Lett.*, **15**, (1990), p.1443.
11. L. F. Mollenauer, S. G. Evangelides and J. P. Gordon: *J. Lightwave Technol.*, **9**, (1991), p.362.
12. P. V. Mamyshev and L. F. Mollenauer: *Opt. Letts.*, **21**, (1996), pp.396-398.
13. K. Iwatsuki et al.: *IEEE Phot. Tech. Lett.*, **2**, (1990), pp.905-907.
14. K. Tajima: *Opt. Lett.*, **12**, (1987), pp.54-56.
15. V. A. Bogatyrev et al.: *J. Lightwave Technol.*, **9**, (1991), p.561.
16. D. J. Richardson et al.: *Electron Lett.*, **31**, (1995), pp.1681-1682.
17. A. J. Stentz et al.: *Opt. Lett.*, **20**, (1995), pp.1770-1772.
18. D. J. Richardson et al.: *Electron Lett.*, **32**, (1996), pp.373-374.
19. D. J. Richardson et al.: *Proc. International Symposium on Physics and Applications of Optical Solitons95*, Kyoto, (1995), pp.277-290.
20. D. J. Richardson, et al.: *Proc. OSA topical Meeting on Nonlinear Guided Wave Phenomena*, Cambridge, paper **FA3**(Sept. 1996) , pp.8-10.
21. L. F. Mollenauer and P. V. Mamyshev: *Optics Letters*, **21**, (1996), pp.1724-1726.
22. N. G. Broderick, D. J. Richardson and L. Dong: to appear *Phot. Tech. Lett.* (Nov. 1997).
23. A. B. Grudinin and S. Gray: *Proc. OFC96*, San Jose, paper **WA4**, (1996), p.94.
24. A. B. Grudinin, S. Gray and G. G. Vienne: *Electron. Letts.*, **32**, (1996), pp.573-575.
25. L. F. Mollenauer, J. P. Gordon and S. G. Evangelides: *Opt. Lett.*, **17**, (1992), p.1575.
26. H. Kubota and M. Nakazawa: *IEEE J. Quantum Electron*, **29**, (1993), p.2189.
27. M. Matsumoto, H. Ikeda and A. Haseqawa: *Opt. Lett.*, **19**, (1994), p.183.
28. A. B. Grudinin, I. A. Goncharenko, S. Gray and D. N. Payne: *Proc. ECOC 95*, Brussells, paper **Tu.L.I.3**, (1995), pp.295-298.
29. D. A. Atkinson, W. H. Loh, V. V. Afanasjev, A. B. Grudinin, A. J. Seeds and D. N. Payne: *Opt. Lett.*, **19**, (1994), pp.1514-1517.

(Received November 17, 1997)

INDEX

A

B

C

D

K

L

M

N

O

P

U

V

W

Z

List of Contributors (Speakers)

Abdullaev, Fatkhulla
Physical-Technical Institute
Uzbek Academy of Sciences
700084, Tashkent-84
G. Mavlyanov str. 2-b
Uzbekistan
Phone : +7 3712 35 43 38
Fax : +7 3712-35-42-91
E-mail : fatkh@pti.tashkent.su

Cotter, David
BT Laboratories
Martlesham Heath
Ipswich, Suffolk IP5 3RE
United Kingdom
Phone : +44 1473 645920
Fax : +44 1473 646885
E-mail : david.cotter@bt-sys.bt.co.uk

Doran, Nick J.
Division of Electronic Engineering
& Computer Science
Aston University
Aston Triangle
Birmingham B4 7ET
United Kingdom
Phone : +44 121 359 3611 Ext. 4961
Fax : +44 121 359 0156
E-mail : n.j.doran@aston.ac.uk

Elgin, John N.
Department of Mathematics
Imperial College
180 Queen's Gate
London SW7 2BZ
United Kingdom
Phone : +44 171 594 8508
Fax : +44 171 594 8517
E-mail : j.elgin@ic.ac.uk

Franco, Pierluigi
Pirelli Cavi S.p.A.
viale Sarca 222, 20146 Milano
Italy
Phone : +39-2-6442-2685
Fax : +39-2-6442-2205
E-mail : pierluigi.franco@pirelli.com

Gabitov Ildar
Permanent Address
L. D. Landau Institute for Theoretical Physics
2 Kosygin St., Moscow V-334
Russia
Phone : +7-95-742-0142 Ext. 3441
Fax : +7-95-938-2077
E-mail : ildar@cpd.landau.ac.ru or gabitov@itp.ac.ru
Temporary Address
Department of Mathematical Sciences
Rensselaer Polytechnic Institute
Troy, NY 12180-3590
United States of America
Phone : +1-518-899-7023
Fax : +1-518-276-4824
E-mail : gabiti@rpi.edu

Georges, Thierry
France Télécom
CNET Lannion DTD/RTO
2, av. P. Marzin
22307 Lannion Cedex
France
Phone : +33 2 96 05 31 12
Fax : +33 2 96 05 13 07
E-mail : thierry.georges@lannion.cnet.fr

Gupta, Gyaneshwar C.
Optical Network Technology Group
C&C Media Research Laboratories
NEC Corporation
4-1-1 Miyazaki, Miyazaki-ku
Kawasaki, Kanagawa 216
Japan
Phone : +81-44-856-2109
Fax : +81-44-856-2222
E-mail : G-Gupta@ccm.cl.nec.co.jp

Hamaide, Jean-Pierre
Alcatel Corporate Research Center
c/o Alcatel Alsthom Recherche
route de Nozay, F-91460 Marcoussis
France
Phone : +33 1 69 63 18 68
Fax : +33 1 69 63 18 65
E-mail : hamaide@aar.alcatel-alsthom.fr

Hasegawa, Akira
Present Office Address
ATR Bldg.
2-2 Hikaridai, Seikacho
Sorakugun, Kyoto-fu 619-0288
Japan
Phone : +81-774-98-3500
Fax : +81-774-98-3501
E-mail : haselab@mbox.kyoto-inet.or.jp

Hizanidis, Kyriakos
Department of Electrical and Computer Engineering
National Technical University of Athens
157 73 Athens
Greece
Phone : +30 1 772 3685
Fax : +30 1 772 3513
E-mail : kyriakos@central.ntua.gr

Ishikawa, George
Fujitsu Laboratories Ltd.
4-1-1 Kamikodanaka, Nakahara-ku
Kawasaki 211-88
Japan
Phone : +81-44-754-2641
Fax : +81-44-754-2640
E-mail : george@flab.fujitsu.co.jp

Jacob, John
Present Affiliation/Address
MCI Telecommunications Corporation
1167/107
2400 North Glenville Drive
Richardson, Texas 75082-4381
United States of America
Phone : +1-918-972-7118
Fax : +1-918-972-7820
E-mail : John.Jacob@mci.com

Karlsson, Magnus
Department of Optoelectronics and Electrical
Measurements
Chalmers University of Technology
SE-412 96, Gothenburg
Sweden
Phone : +46-31-772 1590
Fax : +46-31-772 1540
E-mail : magnus@elm.chalmers.se

Kodama, Yuji
Graduate School of Engineering
Osaka University
Suita, Osaka 565
Japan
Phone : +81-6-879-7692
Fax : +81-6-879-7688
E-mail : kodama@comm.eng.osaka-u.ac.jp

Kumar, Shiva
Institut für Theoretische Optik
University of Jena
Max-Wien-Platz 1
07743 Jena
Germany
Phone : +49-3641-94-7173
Fax : +49-3641-63-6682
E-mail : shiva@pinet.uni-jena.de

Kutz, J. Nathan
Hong Kong Polytechnic University
Hung Hom, Kowloon, Hong Kong
China
Phone : +852 2766-6248
Fax : +852 2362-8439
E-mail : kutz@amath.washington.edu

Midrio, Michele
Dipartimento di Elettronica e Informatica
Università degli Studi di Padova
via G. Gradenigo 6/A
35131 Padova
Italy
Phone : +39-49-827-7719
Fax : +39-49-827-7699
E-mail : michele@pcel40.dei.unipd.it

Morita, Itsurou
KDD R&D Laboratories
2-1-15 Ohara Kamifukuoka
Saitama 356
Japan
Phone : +81-492-78-7807
Fax : +81-492-78-7516
E-mail : morita@lab.kdd.co.jp

Murai, Hitoshi
Semiconductor Technology Laboratories
Oki Electric Industry Co, Ltd.
550-5 Higashi-asakawa
Hachioji, Tokyo 193
Japan
Phone : +81-426-62-6762
Fax : +81-426-67-6770
E-mail : murai@hlabs.oki.co.jp

Nakazawa, Masataka
NTT Optical Network Systems Laboratories
Tokai, Ibaraki-ken 319-11
Japan
Phone : +81-29-287-7333
Fax : +81-29-287-7874
E-mail : nakazawa@nttlsl.iecl.ntt.jp

Nishimura, Masayuki
Sumitomo Electric Industries, Ltd.
1 Taya-cho
Sakae-ku, Yokohama 244
Japan Phone : +81-45-853-7171
Fax : +81-45-851-1557
E-mail : nishi@yklab.sei.co.jp

Okamawari, Takao
Graduate School of Engineering
Osaka University
Suita, Osaka 565
Japan
Phone : +81-6-879-7692
Fax : +81-6-879-7688
E-mail : pan59785@pas.mei.co.jp

Rauschenbach, Kristin A.
MIT Lincoln Laboratory
C-237
244 Wood Street
Lexington, MA 02173
United States of America
Phone : +1-617-981-0229
Fax : +1-617-981-4129
E-mail : Rauschenbach@ll.mit.edu

Reid, John J. E.
Philips Optoelectronics Research
Prof. Holstlaan 4
5656 AA Eindhoven
The Netherlands
Phone : +31 40 27 43208
Fax : +31 40 27 44335
E-mail : reid@natlab.research.philips.com

Richardson, David
Optoelectronics Research Centre
Southampton University
Southampton
United Kingdom
Phone : +44 1703 594524
Fax : +44 1703 593142
E-mail : drj@orc.soton.ac.uk

Romagnoli, Marco
Fondazione Ugo Bordoni
via B. Castiglione,59
00142 Roma
Italy
Phone : +39 6 5480 2231
Fax : +39 6 5480 4402
E-mail : romag@fub.it

Shimoura, Kazuhiro
Technical Research Center
The Kansai Power Electric Co., Inc.
3-11-20 Nakoji
Amagasaki, Hyogo 661
Japan
Phone : +81-6-494-9748
Fax : +81-6-494-9728
E-mail : shimoura@rdd.kepco.co.jp

Sizmann, Andreas
Lehrstuhl für Optik
Physikalisches Institut der Universität–Erlangen-Nürnberg
Staudtstr. 7/B2
D-90158 Erlangen
Germany
Phone : +49-9131-858375
Fax : +49-9131-13508
E-mail : sizmann@physik.uni-erlangen.de

Tsuchiya, Masahiro
Department of Electronic Engineering
The University of Tokyo
Hongo 7-3-1, Tokyo 113-8656
Japan
Phone : +81-3-3812-2111 Ext. 7465
Fax : +81-3-5684-3274
E-mail : tsuchiya@ktl.t.u-tokyo.ac.jp

Turitsyn, Sergei
Institut für Theoretische Physik I
Heinrich-Heine-Universität-Düsseldorf
Universitätsstrasse 1
40225 Düsseldorf
Germany
Phone : +49 211 811 2473
Fax : +49 211 811 3117
E-mail : turitsyn@xerxes.thphy.uni-duesseldorf.de

Wabnitz, Stefan
Laboratoire de Physique
Université de Bourgogne
B. P. 400, 21011 Dijon
France
Phone : +33-3 80395932
Fax : +33-3 80395971
E-mail : Stefan.Wabnitz@u-bourgogne.fr

Wai, Ping-kong A.
Department of Electronic Engineering
The Hong Kong Polytechnic University
Hung Hom, Kowloon, Hong Kong
China
Phone : +852 2766 6231
Fax : +852 2362 8439
E-mail : enwai@polyu.edu.hk

SOLID-STATE SCIENCE AND TECHNOLOGY LIBRARY

1. A.J.P. Theuwissen: *Solid-State Imaging with Charge-Coupled Devices*. 1995
   ISBN 0-7923-3456-6
2. M.O. van Deventer: *Fundamentals of Bidirectional Transmission over a Single Optical Fibre*. 1996 ISBN 0-7923-3613-5
3. A. Hasegawa (ed.): *Physics and Applications of Optical Solitons in Fibres '95*. 1996 ISBN 0-7923-4155-4
4. M.A. Trishenkov: *Detection of Low-Level Optical Signals*. Photodetectors, Focal Plane Arrays and Systems. 1997 ISBN 0-7923-4691-2
5. A. Hasegawa (ed.): *New Trends in Optical Soliton Transmission Systems*. Proceedings of the Symposium held in Kyoto, Japan, 18–21 November 1997. 1998 ISBN 0-7923-5147-9

KLUWER ACADEMIC PUBLISHERS – DORDRECHT / BOSTON / LONDON

Physics and Chemistry of Materials with Low-Dimension Structures

9. L.J. de Jongh (ed.): *Magnetic Properties of Layered Transition Metal Compounds*. 1990 ISBN 0-7923-0238-9
10. E. Doni, R. Girlanda, G. Pastori Parravicini and A. Quattropani (eds.): *Progress in Electron Properties of Solids*. Festschrift in Honour of Franco Bassani. 1989 ISBN 0-7923-0337-7
11. C. Schlenker (ed.): *Low-Dimensional Electronic Properties of Molybdenum Bronzes and Oxides*. 1989 ISBN 0-7923-0085-8
12. Not published.
13. H. Aoki, M. Tsukada, M. Schlüter and F. Lévy (eds.): *New Horizons in Low-Dimensional Electron Systems*. A Festschrift in Honour of Professor H. Kamimura. 1992 ISBN 0-7923-1302-X
14. A. Aruchamy (ed.): *Photoelectrochemistry and Photovoltaics of Layered Semiconductors*. 1992 ISBN 0-7923-1556-1
15. T. Butz (ed.): *Nuclear Spectroscopy on Charge Density Wave Systems*. 1992 ISBN 0-7923-1779-3
16. G. Benedek (ed.): *Surface Properties of Layered Structures*. 1992 ISBN 0-7923-1961-3
17. W. Müller-Warmuth and R. Schöllhorn (eds.): *Progress in Intercalation Research*. 1994 ISBN 0-7923-2357-2
18. L.J. de Jongh (ed.): *Physics and Chemistry of Metal Cluster Compounds*. Model Systems for Small Metal Particles. 1994 ISBN 0-7923-2715-2
19. E.Y. Andrei (ed.): *Two-Dimensional Electron Systems*. On Helium and other Cryogenic Substrates. 1997 ISBN 0-7923-4738-2

KLUWER ACADEMIC PUBLISHERS – DORDRECHT / BOSTON / LONDON